Lecture Notes in Artificia　　　　　　　ﾉ

Edited by J. G. Carbonell and J. Siekmann

Subseries of Lecture Notes in Computer Science

Frithjof Dau   Marie-Laure Mugnier
Gerd Stumme (Eds.)

Conceptual Structures: Common Semantics for Sharing Knowledge

13th International Conference
on Conceptual Structures, ICCS 2005
Kassel, Germany, July 17-22, 2005
Proceedings

 Springer

Series Editors

Jaime G. Carbonell, Carnegie Mellon University, Pittsburgh, PA, USA
Jörg Siekmann, University of Saarland, Saarbrücken, Germany

Volume Editors

Frithjof Dau
Technische Universität Darmstadt, Fachbereich Mathematik, AG 1
Schloßgartenstr. 7, 64289 Darmstadt, Germany
E-mail: dau@mathematik.tu-darmstadt.de

Marie-Laure Mugnier
LIRMM (CNRS and University of Montpellier II)
161, rue ADA, 34392 Montpellier, France
E-mail: mugnier@lirmm.fr

Gerd Stumme
Universität Kassel, Fachbereich Mathematik und Informatik
Wilhelmshöher Allee 73, 34121 Kassel, Germany
E-mail: stumme@cs.uni-kassel.de

Library of Congress Control Number: 2005928703

CR Subject Classification (1998): I.2, G.2.2, F.4.1, F.2.1, H.4

ISSN   0302-9743
ISBN-10   3-540-27783-8 Springer Berlin Heidelberg New York
ISBN-13   978-3-540-27783-5 Springer Berlin Heidelberg New York

Springer is a part of Springer Science+Business Media

springeronline.com

© Springer-Verlag Berlin Heidelberg 2005
Printed in Germany

Typesetting: Camera-ready by author, data conversion by Boller Mediendesign
Printed on acid-free paper   SPIN: 11524564   06/3142   5 4 3 2 1 0

Preface

The 13th International Conference on Conceptual Structures (ICCS 2005) was held in Kassel, Germany, during July 17–22, 2005. Information about the conference can be found at `http://www.kde.cs.uni-kassel.de/conf/iccs05`.

The title of this year's conference, "Common Semantics for Sharing Knowledge", was chosen to emphasize on the one hand the overall aim of any knowledge representation formalism, to support the sharing of knowledge, and on the other hand the importance of a common semantics to avoid distortion of the meaning. We understand that both aspects are of equal importance for a successful future of the research area of conceptual structures. We are thus happy that the papers presented at ICCS 2005 addressed both applications and theoretical foundations.

"Sharing knowledge" can also be understood in a separate sense. Thanks to the German Research Foundation, DFG, we were able to invite nine internationally renowned researchers from adjacent research areas. We had stimulating presentations and lively discussions, with bidirectional knowledge sharing. Eventually the ground can be laid for establishing common semantics between the respective theories.

This year, 66 papers were submitted, from which 22 were selected to be included in this volume. In addition, the first nine papers present the invited talks. We wish to express our appreciation to all the authors of submitted papers, to the members of the Editorial Board and the Program Committee, and to the external reviewers for making ICCS 2005 a valuable contribution to the knowledge processing research field.

July 2005

Frithjof Dau
Marie-Laure Mugnier
Gerd Stumme

Organization

The International Conference on Conceptual Structures (ICCS) is the annual conference and principal research forum in the theory and practice of conceptual structures. Previous ICCS conferences were held at the Université Laval (Quebec City, 1993), at the University of Maryland (1994), at the University of California (Santa Cruz, 1995), in Sydney (1996), at the University of Washington (Seattle, 1997), at the University of Montpellier (1998), at Virginia Tech (Blacksburg, 1999), at Darmstadt University of Technology (2000), at Stanford University (2001), in Borovets, Bulgaria (2002), at Dresden University of Technology (2003), and at the University of Alabama (Huntsville, 2004).

General Chair

Gerd Stumme   University of Kassel, Germany

Program Chairs

Frithjof Dau   Darmstadt Technical University, Germany
Marie-Laure Mugnier   University of Montpellier, France

Editorial Board

Galia Angelova (Bulgaria)
Michel Chein (France)
Aldo de Moor (Belgium)
Harry Delugach (USA)
Peter Eklund (Australia)
Bernhard Ganter (Germany)
Mary Keeler (USA)
Sergei Kuznetsov (Russia)
Wilfried Lex (Germany)

Guy Mineau (Canada)
Bernard Moulin (Canada)
Peter Øhrstrøm (Denmark)
Heather Pfeiffer (USA)
Uta Priss (UK)
John Sowa (USA)
Rudolf Wille (Germany)
Karl Erich Wolff (Germany)

Program Committee

Anne Berry (France)
Tru Cao (Vietnam)
Dan Corbett (Australia)
Olivier Corby (France)
Pavlin Dobrev (Bulgaria)

David Genest (France)
Ollivier Haemmerlé (France)
Roger Hartley (USA)
Udo Hebisch (Germany)
Joachim Hereth Correia (Germany)

Richard Hill (UK)
Pascal Hitzler (Germany)
Kees Hoede (The Netherlands)
Julia Klinger (Germany)
Pavel Kocura (UK)
Robert Kremer (Canada)
Leonhard Kwuida (Germany)
M. Leclère (France)
Robert Levinson (USA)
Michel Liquière (France)
Carsten Lutz (Germany)
Philippe Martin (Australia)
Engelbert Mephu Nguifo (France)
Sergei Obiedkov (Russia)

Simon Polovina (UK)
Anne-Marie Rassinoux (Switzerland)
Gary Richmond (USA)
Olivier Ridoux (France)
Daniel Rochowiak (USA)
Sebastian Rudolph (Germany)
Eric Salvat (France)
Janos Sarbo (The Netherlands)
Henrik Schaerfe (Denmark)
Thanwadee T. Sunetnanta (Thailand)
William Tepfenhart (USA)
Petko Valtchev (Canada)
Sergei Yevtushenko (Germany)
G.Q. Zhang (USA)

External Reviewers

Sadok Ben Yahia (Tunisia)
Richard Cole (Australia)
Jon Ducrou (Australia)
Letha Etzkorn (USA)

Markus Krötzsch (Germany)
Boris Motik (Germany)
Anthony K. Seda (Ireland)

Table of Contents

Knowledge Engineering and Tools

Knowledge Acquisition and Ontologies

Patterns for the Pragmatic Web

Aldo de Moor

STARLab, Vrije Universiteit Brussel,
Pleinlaan 2, 1050 Brussels, Belgium
ademoor@vub.ac.be

Abstract. The Semantic Web is a significant improvement of the original World Wide Web. It models shared meanings with ontologies, and uses these to provide many different kinds of web services. However, shared meaning is not enough. If the Semantic Web is to have an impact in the real world, with its multiple, changing, and imperfect sources of meaning, adequately modeling context is essential. Context of use is the focus of the Pragmatic Web and is all-important to deal with issues like information overload and relevance of information. Still, great confusion remains about how to model context and which role it should play in the Pragmatic Web. We propose an approach to put ontologies in context by using pragmatic patterns in meaning negotiation processes, among other meaning evolution processes. It then becomes possible to better deal with partial, contradicting, and evolving ontologies. Such an approach can help address some of the complexities experienced in many current ontology engineering efforts.

1. Introduction

The World Wide Web has profoundly changed the way people collaborate. Whereas e-mail has lowered the threshold for interpersonal communication by providing a medium for fast, cheap, ubiquitous and global communication, the Web has become the metaphor and technology for doing the same with respect to linking and sharing knowledge resources. Even for the computing community, used to fast technological progress, the speed with which the Web has evolved from initial prototype to a foundation of daily life has been dazzling. It was only in 1991 that the following was announced by a then unknown employee from CERN:

"The WorldWideWeb application is now available as an alpha release in source and binary form from info.cern.ch. WorldWideWeb is a hypertext browser/editor which allows one to read information from local files and remote servers. It allows hypertext links to be made and traversed, and also remote indexes to be interrogated for lists of useful documents. Local files may be edited, and links made from areas of text to other files, remote files, remote indexes, remote index searches, internet news groups and articles … This project is experimental and of course comes without any warranty whatsoever. However, *it could start a revolution in information access* [my italics]"[1].

The rest, as they say, is history.

---

[1] Tim Berners-Lee, *comp.sys.next.announce* newsgroup, Aug.19, 1991.

F. Dau, M.-L. Mugnier, G. Stumme (Eds.): ICCS 2005, LNAI 3596, pp. 1-18, 2005.

The rise of the World Wide Web has led to many benefits to society. Documents, news, and results to queries can be obtained 24 hours a day from all over the world. The Web has given a huge boost to research, education, commerce and even politics. An interesting example of how deeply the Web has become embedded in the fabric of our globalizing society is the significant role web sites play in political reforms in less-than-democratic countries [17]. Still, not all is good. One serious consequence of the explosion of Web-accessible information resources is information overload. It is not uncommon to get hundreds, thousands, or even millions of hits when looking for a certain piece of information. Increasingly, the problem shifts from making information accessible, to delivering *relevant* information to the user.

The Semantic Web plays an important role in making the Web more relevant. Berners-Lee, et al. [1] present a cogent view of how the Semantic Web will structure meaningful content and add logic to the Web. In this web, data and rules for reasoning about data are systematically described, after which they can be shared and used by distributed agents. Granted, many of the basic theoretical ideas were already conceived by the AI community in the 1970s and 80s. The added value of the Semantic Web, however, is that this theory is finally being put into large scale-practice. The main components implementing this Web vision include techniques such as XML, for adding arbitrary structures to documents; RDF, to express meaning by simple statements about things having properties with values; and ontologies, to formally describe concepts and their relations. A typical ontology, in the sense of being an explicit specification of a conceptualization [10], consists of a taxonomy with a set of inference rules. Ontologies can be used to improve the accuracy of, for instance, Web search and service discovery processes. Ultimately, such an approach should lead to the evolution of human knowledge by scaling up collaboration from individual efforts to large, joint endeavors. Multiple ontologies then come into play. By selecting the right ontology for the right task, knowledge exchange, at least in theory, could become more effective and efficient.

In practice, however, the Semantic Web comes with its own set of problems. Voices are increasingly being heard that there is a need not only for explicitly taking into account the semantics, but also the pragmatics of the Web, e.g. [25,26,13,7,29,22]. Still, ideas and proposals are preliminary and sketchy and need further elaboration and integration. With this paper, we hope to contribute to the further maturation of thought on this important subject. We have two main objectives: finding out (1) what are fundamental conceptual elements of the Pragmatic Web and (2) how to use these elements in making meaning represented in semantic resources more relevant. In Sect. 2, we outline some contours of the Pragmatic Web that are becoming visible at the moment. This analysis results in a conceptual model of the Web in Sect. 3, outlining how the Semantic and the Pragmatic Web are interrelated. In Sect. 4, we focus on pragmatic patterns as a way to operationalize the pragmatics of the Web. In Sect. 5, we present a scenario of how a Pragmatic Web could look in practice. We end the paper with a discussion and conclusion.

2.   Contours of the Pragmatic Web

The Semantic Web, with all its (potential) benefits, still poses a number of difficult challenges, both with respect to the ontologies which contain the shared meanings and the services in which these are used.

Unlike data models, ontologies contain relatively generic knowledge that can be reused by different kinds of applications. Ontologies should therefore not be too tightly linked to a specific purpose or user group [30]. To select the right (parts of) ontologies, the communicative situation needs to be taken into account. To this purpose, a "mindshaking procedure" needs to be developed, in which a formal language for information exchange is determined (syntax), and a synchronisation of the meaning of concepts (semantics) takes place on the basis of a particular context, such as purpose, time, date, or profile [29]. An example of a (typically) manual version of such a procedure is described in [9]. There, a conceptual model supervisor regularly creates reports of existing classes. If concepts seem to be in conflict, and the conflicts are important enough, the model supervisor starts and controls a discussion among stakeholders, who can be either modelers or representatives from the involved departments. If the conflict remains unresolved, both concepts remain in the model marked with their own namespaces.

Ontologies are not an end in themselves. One of the major functions of the Semantic Web is to provide access to web services. These are often described and invoked through central registries. However, for describing, discovering, and composing web services, a semantic approach is not enough. Services cannot be described independently of how they are used, because communities of practice use services in novel, unexpected ways. Social mechanisms are therefore needed for evaluating and discovering trustworthy providers and consumers of services, taking into account contexts and interactions in the composition of service applications [25-26].

Clearly it is not sufficient to model semantics to resolve such issues related to the use of ontologies. Contextual elements like the community of use, its objectives and communicative interactions are important starting points for conceptualizing the pragmatic layer. These elements are combined in a conceptualist perspective. In such a view, meanings are elements of the internal cognitive structures of language users, while in communication, the conceptual structures of different views become attuned to each other [13]. We can therefore make a distinction among shared semantic resources, such as ontologies; individual pragmatic resources, i.e. the internal conceptual models of users applying the semantic resources to their own purposes; and common pragmatic resources, in which joint *relevant* meanings have been established through communication. In communication between users aiming at achieving joint objectives, concepts that are part of individual and common pragmatic resources are selected, defined, aligned, and used. Finding out how such a meaning negotiation process works is essential to understanding the pragmatics of the Web, and to developing (partially) automated support processes for meaning negotiation.

Developing sound and complete pragmatic perspectives, models, and methods can shed light on the confusing debates raging in the ontology and Semantic Web research communities. One fundamental question, for example, is whether the way to go is to develop large, detailed, standard ontologies such as Cyc[2] or myriad independent, domain-specific, micro-ontologies, one for each application. The answer is not either/or, but a mix of both approaches. A major reason why such a hybrid point of view cannot be easily adopted and defended, is that the real issues underlying these debates are not semantic, but pragmatic. The focus of many of these debates has thus been the wrong one, without the ontological engineering community making any significant progress on resolving the underlying issues.

Before further examining the Pragmatic Web, we first take a closer look at some of the finer details of pragmatics.

2.1   A Primer of Pragmatics

A traditional source of problems, often found in traditional conceptual modelling approaches, is to try and produce THE description of a joint reality. If members of a particular community disagree, the modellers, in the best case, keep negotiating explicit meanings until everybody agrees. If no agreement can be reached (or is not even sought) modellers often impose a meaning by choosing an ontology definition or system specification themselves.

A pragmatic approach, on the other hand, should allow for contradictions, different importance weights of information and subtle cultural differences [9]. Such differences, however, create problems of their own if not handled properly. Collaboration often fails, not because participants do not want to collaborate, but because pragmatic errors lead to the breakdown of the social and contextual components of a discourse [14]. To become successful, a pragmatic approach thus needs to acknowledge and adequately handle ambiguity and consequences of (differences in) semantics.

Facts only get their 'ultimate meaning' in their human context of use, and are always ambiguous. Such *ambiguities* are about shades of differences in meaning. The extent to and way in which ambiguities need to be resolved, depend on the context, including the points of view of the communicating agents, i.e. utterer and interlocutor, their common understanding of each other, and their (partially) shared goals [18].

But how to decide which ambiguities need to be resolved? A semantic approach, even when accepting different sources of meaning (i.e., ontologies), does not explicitly acknowledge the *consequences* of semantic choices. A pragmatic approach, on the other hand, assumes there are always conditions of difference, dependence, and novelty, and recognizes the need for an overall process for transforming existing knowledge to deal with negative consequences for community members [3]. We would argue that, in addition, the community should also examine the *positive* consequences, such as opportunities for action.

In a pragmatic approach, control over representation should shift from the information producer to the information consumer [22]. More precisely, we think control over how to *use* meaning representations should shift to the user, from which controlling representations follows.

---

[2] http://www.cyc.com/

The need to accept a necessary amount of ambiguity by communities of users assessing the consequences of semantic choices in a particular pragmatic context, implies that there needs to be some user-controlled *selection* process of semantic representations. In such a process, members of the community, using the knowledge for a particular purpose, are actively involved, and aim to reach agreement only on *relevant* knowledge issues. Pragmatically established changes in the implicit meaning of representations should in the end also lead to changes the *representation* of those meanings in ontologies. For instance, if users always ask for concepts that are not, or only insufficiently, described in an ontology, it may be worthwhile to add this concept to the ontology. Meaning selection and representation processes, however, do not occur in isolation, but are driven by a meaning *negotiation* process in a specific community of users. In such a process, stakeholders arrive at the requisite (as determined by their shared goals) amount of agreement on shared concepts.

3.   A Conceptual Model of the Web

Summarizing the previous discussion, we consider 'The Web' to consist of a Syntactic, a Semantic, and a Pragmatic web (Fig.1).

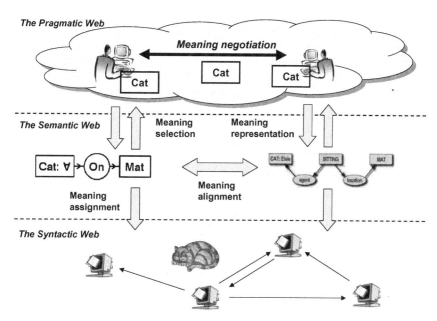

Fig. 1. A Conceptual Model of 'The Web'

The *Syntactic Web* consists of interrelated syntactic information resources, such as documents and web pages linked by HTML references. These resources describe many different domains.

The *Semantic Web* consists of a collection of semantic resources about the Syntactic Web, mainly in the form of ontologies. The ontologies contain semantic networks of concepts, relations, and rules that define the meaning of particular information resources.

The *Pragmatic Web* consists of a set of pragmatic contexts of semantic resources. We consider a *pragmatic context* to consist of a common context and a set of individual contexts. A *common context* is defined by the common concepts and conceptual definitions of interest to a community, the communicative interactions in which these concepts are defined and used, and a set of common context parameters (relevant properties of concepts, joint goals, communicative situation, and so on). Each community member also has an *individual context*, consisting of individual concepts and definitions of interest and individual context parameters. Common and individual context parameters are not discussed further in this paper, as we will focus on the meaning negotiation process in which these contexts play a role.

Meaning plays a central role in connecting the various Webs. *Meaning assignment* takes place when syntactic resources are semantically enriched, such as by XML-tags being added to HTML-pages. *Meaning alignment* has to do with interoperability between ontologies: to what extent do their semantic models agree? How can (parts of) ontologies be meaningfully linked? How to deal with definitions that partially overlap in meaning? Much recent work addresses these – very hard – issues, e.g. [24,4]. Such meaning alignment problems mostly focus on modeling representational and evolutionary aspects of ontologies. However, as we have seen what needs separate attention are issues of ontology *use*. In other words: how can the process of *meaning negotiation* be improved?   Meanings evolve not in the ontologies themselves, but in the pragmatic contexts where they are being used. Thus, a strong involvement of the community in ontology engineering processes is required, ensuring that individual and community changes in meaning are represented adequately in the ontologies.

3.1   The Complexity of Contexts

Our conceptual model allows us to examine a wide range of pragmatic contexts in the real world, and to identify commonalities and differences in problems with modelling, sharing and (re)using semantic resources such as ontologies.

Note that the sheer number of elements to analyze decreases as we move from the Syntactic to the Semantic Web, but strongly increases again when moving from the Semantic to the Pragmatic Web. There may be many thousands of (syntactic) information resources for a particular domain. In general, there will be many fewer ontologies defining the meanings of those resources. However, of pragmatic contexts there can be an infinite number. There are many dimensions of pragmatics to be taken into account, such as purposes, communicative situations, organizational norms, individual values, and so on. These contextual parameters lead to a great variety of contexts. The multiple pragmatic contexts are even harder to formalize and standardize than the semantics of the concepts they interpret. Individual context views may agree with each other, or differ. Community members may use different ontologies to define the meaning of a particular concept. Many concepts have rich tacit meanings for individuals that can, nor should, always be made explicit in collaborative situations [21]. To assess the consequences of meaning choices, fully-automated negotiation processes will therefore never be sufficient. *Augmentation*, not automation of human meaning negotiation processes is required, in the sense proposed by Doug Engelbart [27].

One strategy to deal with this pragmatic complexity is to only model those pragmatic constructs that are essential to reach joint objectives. The meaning negotiation process should be a consensus seeking process, balancing individual and common requirements. Different individual views on the meaning of common concepts should be allowed, as long as they do not endanger the quality of the communicative interaction. For example, in a business transaction, it is essential that both parties have the same view of crucial parts of the definition of their contract, such as legal obligations. Where and how to store copies of the contract internally does not need to be part of a common meaning, however, and can thus be left as a degree of freedom. If differences in meaning are inhibiting the accomplishment of common goals, however, meaning negotiation has to proceed until the necessary amount of consensus has been reached.

How to proceed? What is a scalable way to operationalize such a pragmatic approach? If pragmatic contexts are unique and very different, how to systematically support meaning negotiation and related processes like meaning selection and representation? What is a requisite amount of consensus? The approach we propose in this paper is to base meaning negotiation on a set of fundamental pragmatic patterns, which can made available in a *meta-ontology*. These formal patterns can be used to define pragmatic constraints on processes in which explicit meanings are being defined and applied in contexts of use. Such an approach can help to better understand the potential uses and limitations of particular ontology engineering efforts, by clarifying the 'meaning of those meanings' for particular contexts of use.

4. Pragmatic Patterns

In [6], we presented a method for collaboratory improvement. Collaboratories are evolving socio-technical systems of people and tools aimed at providing environments for effective and efficient collaboration. About collaboratories often only partial knowledge of different degrees of specificity is or can be represented. The method uses ontology-grounded *improvement patterns* to capture various levels of socio-technical context knowledge about information and communication processes in collaboratories, including knowledge about workflows, design processes and improvement processes. We view collaboratory improvement as a Peircean pragmatic inquiry process in which hypotheses about socio-technical improvements of the collaboratory are continuously constructed and tested in the community. This process, properly supported, should lead to more effective and efficient collaboratory evolution. Such an inquiry process could be a major driver of meaning selection in a community and hence form an important constituent of the Pragmatic Web [7].

A collaboratory improvement process is a good example of a community using patterns to evolve specification knowledge about its own socio-technical system. In the current paper, we want to develop a broader perspective. Instead of using patterns just to improve collaboratories, we intend to use patterns to 'improve semantics'. Given our conceptual model of the Web, what kind of patterns do we need? How do we represent them? How can we use them to deal with some of the problems inhibiting the progress of the Semantic Web?

4.1   Patterns

Humans use patterns to order the world and make sense of things in complex situations [15]. Patterns are often used in the construction of complex systems. An influential definition of patterns in architecture, also useful for information systems, was given by Christopher Alexander: "A pattern is a careful description of a perennial solution to a recurring problem within a building context, describing one of the configurations which brings life to a building (Alexander, et al., 1977, in [23]". A pattern thus contains elements of a solution to a problem, and applies within a particular context. Important is to focus on the words *recurring* problem and *perennial* solutions, indicating that the pattern definition of problems and solutions must be generic enough to cover a range of problem situations which in reality are always subtly different from the ideal, while being specific enough to offer useful solutions for the particular problem at hand.

Patterns are another view on domain models stored in ontologies. Developing ontologies for open environments like the Semantic Web is difficult, since more rules make ontologies less generic, while light-weight ontologies are not very useful [30]. This problem of finding the right degree of semantic specificity of ontologies to address problems in the domains they were created for, is not going to go away. The problem is not technical, but philosophical. If the types and number of applications of an ontology are infinite, and cannot be known beforehand, it will not work to try and produce the 'ultimate ontology' of semantic patterns. The usefulness of an ontology is always in the eye of the beholder, or more precisely, the eyes of many beholders: the many communities and individuals within communities using the ontology for their particular, changing, collaborative purposes.

Accepting this reality of eternal semantic partiality, conflict, and confusion, there is another, potentially more rewarding way to go. It consists of (1) making a strict conceptual separation between *modelling* and *using* ontologies, (2) identifying meta-patterns, i.e. *pragmatic patterns* that can (3) be used in *meaning evolution* processes in communities of users in order to make existing ontologies more useful and easier to change[3]. These processes include what we referred to in the previous section as meaning representation, assignment, selection, alignment, and negotiation. Only by tackling these pragmatic issues head-on can the vision of the Semantic Web assisting the evolution of human knowledge as a whole [1], be realized in practice.

---

[3] These processes concern the evolution of *explicated* meanings. Many meanings are implicit, in people's heads. Although they may, and probably should change as well, understanding these are more the focus of, for instance, psychological studies.

4.2   Core Pragmatic Patterns

To operationalize our vision of the Pragmatic Web, we need some core pragmatic patterns. We do not formalize the patterns in this article, but will outline some and describe their possible role in the scenario presented in the next section. Using conceptual graphs, it should be relatively easy to structure and reason about their (meta)semantics.

For a particular community, core pragmatic patterns include:

- *Pragmatic context:* a pattern that defines the speakers, hearers, type of communication, and identifiers of the individual and common contexts of a community.
- *Individual context:* a pattern that defines an individual community member, individual context parameters and an identifier of the individual context ontology.
- *Common context:* a pattern that defines the common context parameters and an identifier of the common context ontology of a community.
- *Individual pragmatic pattern*: a meaning pattern relevant to an individual community member. An individual context ontology consists of the total set of meaning patterns relevant to that individual.
- *Common pragmatic pattern*: a meaning pattern relevant to the community as a whole. The common context ontology consists of the total set of common meaning patterns relevant to the community.

Pragmatic patterns are template definitions that can be used as the basis of conceptual definitions used in meaning negotiation and other meaning evolution processes. These patterns can be refined and extended by communities if and when necessary.

Pragmatic patterns have a normative status, being either required, permitted, or forbidden. In the case of a pattern being required, this implies that the pattern must be satisfied in the process where it is used. If it is forbidden, it may not be matched in such a process. If permitted, it may be applicable, but not necessarily so. Such normative matching processes can provide powerful guidance of meaning evolution processes.

Earlier, we said that there is a much larger number and diversity of pragmatic contexts than of the ontologies which they use. Still, the number of pragmatic patterns, if chosen at the right level of specificity, can be relatively small. These patterns should not include the infinite number of details that make each pragmatic context unique, but only those that contribute to improving the effectiveness and efficiency of meaning evolution, with a focus on meaning negotiation. Of course, in this paper, we do not claim to solve the pragmatic puzzle. We will not provide the ultimate reference set of pragmatic patterns to be used in optimizing meaning evolution on the Semantic Web. Our aim is much more modest: showing proof of principle about what pragmatic patterns are and the role they could play in dealing with some of the meaning evolution issues mentioned. To this purpose, we introduce a hypothetical case very relevant to the conceptual graphs community: getting the famous cat its mat.

5. Using Pragmatic Patterns: How to Get a Mat for the Cat?

The mat producing company MatMakers wants to explore new markets. The grapevine has it that an interesting niche exists of cat lovers wanting nothing but the best for their furry friends. Its marketing officer Charles is commissioned to find new customers who will appreciate MatMaker's high-quality mats for their cats.

Charles decides to look for potential customers using the WYO=WTW (WhatYouOffer-is-WhatTheyWant) e-business broker. This broker is a web service that maximizes precision of advertising by using the latest Pragmatic Web-technologies. In particular, it mediates in meaning negotiation between sellers and prospective buyers by intelligent use of pragmatic pattern matching. The following type hierarchy is part of the WYO=WTW community context ontology (Fig.2):

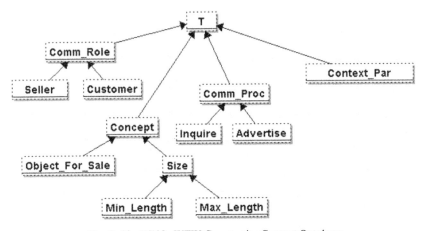

Fig. 2. The WYO=WTW Community Context Ontology

A most relevant concept in any advertising process is the *object for sale*. One important property of these objects, which is often discussed in the business negotiations of this particular community, is the *size* of the object being offered. Two important size indicators are the *minimum* and the *maximum length* of the object. Two communication roles in an e-business transaction are the *seller* and the *consumer*, referring to the parties who can play the speaker or hearer-roles. The community using the WYO-WTW service distinguishes two types of communication processes: *inquiring* about objects for sale, initiated by customers, and *advertising* objects, initiated by producers.

MatMakers has its own corporate ontology, from which Charles imports the Mat and Size-concepts (including their positions in the type hierarchy) into the individual context ontology of MatMakers for the WYO=WTW service. He also adds the Cat-concept, since that is what he wants to focus his particular potential customer search on. Since the maximum length of the mats produced by MatMakers is one meter, Charles adds to his individual context ontology the required pattern that to be of interest for an advertisement any cat for which the mat is bought should be at most one meter long (Fig.3):

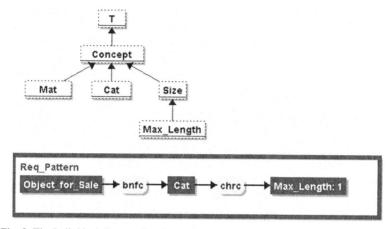

Fig. 3. The Individual Context Ontology of MatMakers for the WYO-WTW Service

The Cat-Lovers-Association-of-the-World (CLAW) is a worldwide virtual community of amateurs crazy about cats. They have interest groups studying not only small cats, like street cats and Siamese cats, but also large cats, like lions and tigers. The database of member addresses of such a highly motivated global community is of high potential value to corporations. In principle, CLAW is not adverse to their members being offered products for their pets. However, they are not interested in offers of products for large cats, since their members are amateurs only, not zoo owners. Therefore, they demand that any sales offer in an advertisement concerns small cats only (Fig.4):

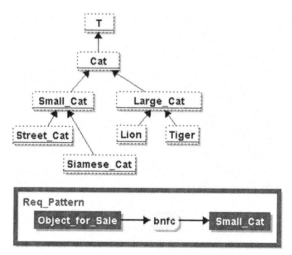

Fig. 4. The Individual Context Ontology of CLAW for the WYO-WTW Service

To find potential customers, Charles first sends a query identical to his required pattern to the web service (Fig.5):

Fig. 5. The Initial Query

WYO-WTW searches the individual context ontologies of all registered members of its services for a match with this required pattern, by projecting the required pattern on the individual context ontologies of the various members[4]. Nothing matches. Charles realizes that his query could have been too specific, not because no customers share his interest, but because their meanings have not yet been sufficiently specified in the ontologies they use with respect to Charles' purpose. He decides to relax the query by only looking for potential customers who are interested in products for cats, and try to find out about the length of their animals later. He therefore sends the following generalization of his required pattern (Fig.6):

Fig. 6. The Generalized Query

WYO-WTW projects this generalization again on the various individual context ontologies. It now matches with all (i.e. the only) required pattern of the CLAW context ontology, returning the following result (Fig.7):

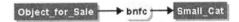

Fig. 7. The Result of the Generalized Query

CLAW's only (and all) pragmatic requirements on any seller have now been satisfied, and the association in principle is open to being sent the advertisement. However, MatMaker's own required pattern has not been satisfied yet. To see if it could, WYO-WTW goes on the Semantic Web, projecting MatMakers *core semantic pattern* (i.e. the essence of MatMakers' required pattern adapted to the semantic constraints of the potential customer party's required patterns) on the public interfaces of various ontologies that are in CLAW's list of *trusted semantic resources*[5]. It does this to see if these ontologies can be useful in enriching CLAW's ontology sufficiently for it to match with MatMaker's required pattern. The core semantic pattern in this case is the part that follows the bnfc-relation, since this indicates what MatMakers requires from its customers for them to be eligible candidates for advertisement. The Cat-concept is thereby specialized to Small_Cat, since that specialization is demanded by CLAW's required pattern. Furthermore, any instances are left out, since values may have to be calculated by inference rules, instead of being stored directly in ontologies. This would lead to queries failing, even though semantically they should match with an ontology. Thus, the WYO-WTW

---

[4] We do not go into the details of dealing with labels like Req_Pattern here, these can be dealt with by syntactically rewriting the queries.

[5] If a semantic resource like an ontology is trusted, a user is willing to accept the ontological commitments implied by the definitions of that resource.

service sends out the following core semantic pattern query to trusted ontologies on the Semantic Web (Fig.8):

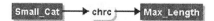

Fig. 8. The Core Semantic Pattern Query to the Semantic Web

Again nothing matches. WYO-WTW now automatically starts to look for similar concepts. It first tries to find synonyms for the *Small_Cat*-label by contacting the Cyc-URI (Uniform Resource Indicator) service. This Semantic Web-service finds *Felinae* as a synonym. It resends the query, but this time with *Small_Cat* replaced by *Felinae*. It turns out that this query matches with the ontology of Animal Diversity Web (ADW), a university zoological taxonomy server[6]. A part of this ontology is the following (Fig.9):

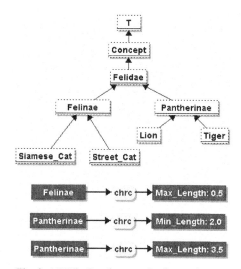

Fig. 9. ADW's Ontology on the Semantic Web

The fact in ADW's ontology that matches with the query (i.e. is a specialization) is the following (Fig.10):

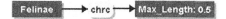

Fig. 10. The Result of the Revised Core Semantic Pattern Query

---

[6] In fact, the taxonomy server on which this (hypothetical) ontology is based really exists. It is hosted by the University of Michigan: http://animaldiversity.ummz.umich.edu/site/. Such relatively stable reference ontologies could play an important role in optimizing meaning negotiation processes on the future Pragmatic Web.

The following fact is automatically added to the WYO-WTW common context ontology (since *Felinae* is equivalent to the *Small_Cat*-label, and the latter is the terminology used by at least one of the community members) (Fig.11):

Fig. 11. The Common Pragmatic Pattern

This *common pragmatic pattern* forms the basis for starting the actual advertising process. It means that seller and customer share an interest in beginning an advertising process about objects for sale for small cats which have a maximum length of half a meter. The pattern is - necessarily - a specialization of the parts of the required patterns of both communicating parties that define the properties of the beneficiary of the object for sale.

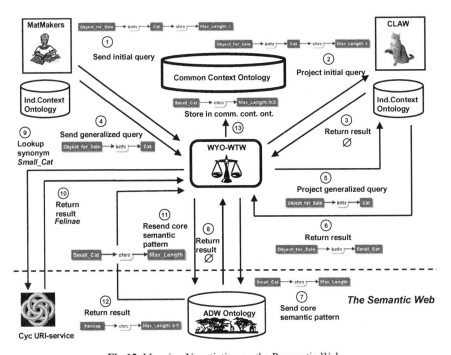

Fig 12. Meaning Negotiation on the Pragmatic Web

Fig. 12 summarizes the meaning negotiation process. This already complex scenario was just a simplified example of a realistic meaning negotiation process. Still, it should demonstrate the power of a combination of a conceptually clearly separated, yet interdependent Semantic and Pragmatic Web.

6. Discussion

The purpose of this paper was to expand current thinking on the Pragmatic Web by identifying some issues and presenting a sketch of one possible approach for its operationalization. There are many directions in which this work should be expanded, however. For example:

- The Pragmatic Web and the Semantic Web are strongly interdependent. Many open issues of the Semantic Web still need to be resolved, before a robust Pragmatic Web can be constructed. One example concerns useful and widely adopted URI (Uniform Resource Identifier) schemes. Still, it would already be very useful to systematically examine current and projected components of the Semantic Web through a pragmatic lens, in order to discover new applications. Vice versa, insights about the Pragmatic Web may help address some of the thorny issues currently blocking progress in the Semantic Web community. Many semantic approaches, for example, already have pragmatic components. However, semantics and pragmatics are often mixed up in confusing ways. Our approach could help disentangle some of these conceptual knots, allowing for optimizations with the right focus, e.g. with respect to either modelers' or users' needs in particular cases.
- We have mostly stuck to a rather practical and shallow interpretation of pragmatics. Philosophically, pragmatics is a very complex idea, however. Insights from philosophers with a strong focus on evolution of meaning, such as Peirce's pragmaticism [2] and Habermas' theory of communicative action [11] could be very useful in strengthening the theory of the Pragmatic Web.
- We only defined meaning negotiation and selection processes informally as sequences of graph projections. The (meta)-semantics of the various pragmatic patterns is still quite fuzzy. How to formalize individual and common contexts and pragmatic patterns? What role should they play in the various meaning evolution processes? When and how should recurring pragmatic patterns stored in meta-ontologies be included in domain ontologies on the Semantic Web? Conceptual graphs research can also make important contributions here, both in terms of advanced theoretical research like context modelling [20] and architectures for pragmatic graph application systems [28,5]. Also, the normative status of patterns is a complex issue. In the scenario, we only used required patterns. In realistic applications, these may conflict with prohibited and permitted patterns. Deontic logic is one theoretical field that help clarify some of these issues [19].
- Human communication is crucial in meaning negotiation on the Pragmatic Web. Conceptual approaches such as proposed in this paper can only augment, not automate human meaning interpretation and negotiation processes. A theoretical foundation for modelling more complex and realistic communicative interactions is the Language/Action Perspective, which stresses the coordinating role of language. This perspective has led to various proposals for human/agent communication-based collaborative models and systems e.g. [16,28,31,12].

Another rich source of ideas for designing pragmatic systems supporting human communication is (business) negotiation theory (e.g. [8]).

- Ontologies play a crucial role, at both the Semantic and Pragmatic Web levels. The ontologies presented in the scenario were exceedingly simple, since the focus was on proof of concept, not on the finer semantic details. Much ontology research focuses on these representation and reasoning issues. Although valuable and necessary, ontology research on the Pragmatic Web level should also focus much more on ontology *methodology* issues. These include the (partially) human processes of modeling, selecting, using and changing meanings for collaborative purposes. The DOGMA-methodology being developed at STARLab consists of a set of methods, techniques, and tools to arrive at scalable ontologies that can actually be made useful in practice [29]. One of our projects in which a strong focus will be on exploring the relations between ontologies and pragmatics is the CODRIVE project[7]. Our aim is to develop a methodology for negotiating a common competence ontology by key stakeholders in the European labor market. These parties, representing the educational sector, public employment agencies, and industry have a need for a common competency ontology that can be used for collaborative applications such as doing job matches and developing individual training pathways. Given the widely varying interests and definitions of competence concepts, this should be a very interesting test case to further develop theory and practice of the Pragmatic Web.

7. Conclusion

The Pragmatic Web is the next phase in the evolution of the Web. Most research attention currently focuses on the Semantic Web. However, for the Semantic Web to truly realize its potential, much more work needs to be done on its pragmatics aspects. This entails that the context of use of explicated meanings that are stored in ontologies need to be much better understood. The driver of the Pragmatic Web are meaning negotiation processes. These processes are connected to the Semantic Web by meaning selection and representation processes.

In this paper, we have explored the contours and some fundamental concepts of the Pragmatic Web. By means of a scenario we have explored what the Pragmatic Web in a few years time might look in practice. The aim of this paper was not to solve existing problems, but to help open up an exciting new territory for intellectual and practical exploration. Moving the research focus from semantics to pragmatics, from representing to *using* meaning, is the next step on the way to network applications that help communities of people realize their full collaborative potential.

---

[7] EU Leonardo da Vinci project BE/04/B/F/PP-144.339

References

1. Berners-Lee, T., Hendler, J. and Lassila, O. (2001), The Semantic Web, *Scientific American*, May 2001: 35-43.
2. Buchler, J. (1955), *Philosophical Writings of Peirce*. Dover Publ., New York.
3. Carlile, P. R. (2002), A Pragmatic View of Knowledge and Boundaries: Boundary Objects in New Product Development, *Organization Science*, 13(4): 442-455.
4. Corbett, D. (2004), Interoperability of Ontologies Using Conceptual Graph Theory. In *Proc. of the 12th Intl. Conference on Conceptual Structures (ICCS 2004), Huntsville, AL, USA, July 2004*, LNAI 3127. Springer, Berlin, pp. 375-387.
5. Delugach, H. S. (2003), Towards Building Active Knowledge Systems With Conceptual Graphs, in *Proc. of the 11th Intl. Conf. on Conceptual Structures (ICCS 2003), Dresden, Germany, July 2003*, LNAI 2746. Springer, Berlin, pp. 296-308
6. de Moor, A. (2004), Improving the Testbed Development Process in Collaboratories. In *Proc. of the 12th Intl. Conference on Conceptual Structures (ICCS 2004), Huntsville, AL, USA, July 2004*, LNAI 3127. Springer, Berlin, pp. 261-274.
7. de Moor, A., Keeler, M. and Richmond, G. (2002), Towards a Pragmatic Web. In *Proc. of the 10th Intl. Conference On Conceptual Structures (ICCS 2002), Borovets, Bulgaria, July 2002*, LNAI 2393. Springer, Berlin, pp. 235-249.
8. de Moor, A. and Weigand, H. (2004), Business Negotiation Support: Theory and Practice, *International Negotiation*, 9(1):31-57.
9. Fillies, C., Wood-Albrecht, G. and Weichhardt, F. (2003), Pragmatic Applications of the Semantic Web Using SemTalk, *Computer Networks*, 42: 599-615.
10. Gruber, T. (1994), Towards Principles for the Design of Ontologies Used for Knowledge Sharing. In N. Guarino and R. Poli (eds.) *Formal Ontology in Conceptual Analysis and Knowledge Representation*. Kluwer.
11. Habermas, J. (1981) *Theorie des kommunikativen Handelns* (2 vols.). Suhrkamp, Frankfurt.
12. Harper, L.W., and Delugach, H.S. (2004), Using Conceptual Graphs to Represent Agent Semantic Constituents. In *Proc. Of the 12th Intl. Conference on Conceptual Structures (ICCS 2004), Huntsville, AL, USA, July 2004*, LNAI 3127. Springer, Berlin pp. 325-338.
13. Kim, H. and Dong, A. (2002), Pragmatics of the Semantic Web. In *Semantic Web Workshop 2002, Hawaii*.
14. Kreuz, R. J. and Roberts, R. M. (1993), When Collaboration Fails: Consequences of Pragmatic Errors in Conversation, *Journal of Pragmatics*, 19: 239-252.
15. Kurtz, C. F. and Snowden, D. J. (2003), The New Dynamics of Strategy: Sense-Making in a Complex and Complicated World, *IBM Systems Journal*, 42(3): 462-483.
16. McCarthy, J. (1996), Elephant 2000: A Programming Language Based on Speech Acts, Technical Report, Stanford University.
17. McLaughlin, W. S. (2003), The Use of the Internet for Political Action by Non-State Dissident Actors in the Middle East, *First Monday*, 8(11).
18. Mey, J. L. (2003), Context and (Dis)ambiguity: a Pragmatic View, *Journal of Pragmatics*, 35: 331-347.
19. Meyer, J. J.-C. and Wieringa, R., eds. (1993), *Deontic Logic in Computer Science: Normative System Specification*. John Wiley & Sons, Chichester.
20. Mineau, G.W. and Gerbe, O. (1997). Contexts: A Formal Definition of Worlds of Assertions. In *Proc. of the 5th Intl. Conference on Conceptual Structures (ICCS 1997), Seattle, Washington, USA, August 1997*, LNCS 1257. Springer, Berlin, pp.80-94.
21. Nonaka, I., Reinmoeller, P. and Senoo, D. (1998), The 'ART' of Knowledge: Systems to Capitalize on Market Knowledge, *European Management Journal*, 16(6): 673-684.

22. Repenning, A. and Sullivan, J. (2003), The Pragmatic Web: Agent-Based Multimodal Web Interaction with no Browser in Sight. In *Human-Computer Interaction - INTERACT'03.* IOS Press, IFIP, pp. 212-219.
23. Schuler, D. (2002), A Pattern Language for Living Communication. In *Participatory Design Conference (PDC'02), Malmo, Sweden, June 2002.*
24. Shanks, G., Tansley, E. and Weber, R. (2003), Using Ontology to Validate Conceptual Models, *Communications of the ACM,* 46(10): 85-89.
25. Singh, M. P. (2002a), The Pragmatic Web, *IEEE Internet Computing,* May/June: 4-5.
26. Singh, M. P. (2002b), The Pragmatic Web: Preliminary Thoughts. In *Proc. of the NSF-EU Workshop on Database and Information Systems Research for Semantic Web and Enterprises, April 3-5, Amicalolo Falls and State Park, Georgia.*
27. Skagestad, P. (1993), Thinking with Machines: Intelligence Automation, Evolutionary Epistemology, and Semiotic, *Journal of Social and Evolutionary Systems,* 16(2): 157-180.
28. Sowa, J. (2002), Architectures for Intelligent Systems, *IBM Syst. Journal,* 41(3):331-349.
29. Spyns, P. and Meersman, R. A. (2003), From Knowledge to Interaction: from the Semantic to the Pragmatic Web. Technical Report STAR-2003-05, STARLab , Brussels.
30. Spyns, P., Meersman, R. A. and Jarrar, M. (2002), Data Modelling versus Ontology Engineering, *ACM SIGMOD Record,* 31(4): 12-17.
31. Weigand, H. and de Moor, A. (2003), Workflow Analysis with Communication Norms, *Data & Knowledge Engineering,* 47(3):349-369.

Conceptual Graphs for Semantic Web Applications

Rose Dieng-Kuntz, Olivier Corby

INRIA Sophia-Antipolis, ACACIA Project
2004 route des Lucioles, BP 93, 06902 Sophia-Antipolis Cedex, France
{Rose.Dieng, Olivier.Corby}@sophia.inria.fr

Abstract. In this paper, we aim at showing the advantages of Conceptual Graph formalism for the Semantic Web through several real-world applications in the framework of Corporate Semantic Webs. We describe the RDF(S)-dedicated semantic search engine, CORESE, based on a correspondence between RDF(S) and Conceptual Graphs, and we illustrate the interest of Conceptual Graphs through the analysis of several real-world applications based on CORESE.

1 Introduction

"The Semantic Web is an extension of the current web in which information is given well-defined meaning, better enabling computers and people to work in cooperation." [5]. The W3C works for *"defining standards and technologies that allow data on the Web to be defined and linked in a way that it can be used for more effective discovery, automation, integration, and reuse across applications. The Web will reach its full potential when it becomes an environment where data can be shared and processed by automated tools as well as by people."* [6].

When Tim Berners-Lee presented his vision of the Semantic Web [4], several research communities studied thoroughly how their research field results could contribute to reach this ambitious goal. In particular, researchers working in Knowledge Representation (KR) recognised the potential important role that their KR formalisms could play for representing the ontologies needed in Semantic Web. Object-oriented (OO) representation formalisms, Description Logics (DL) and Conceptual Graphs (CG) were the main candidates to achieve this purpose. The DL community was strongly involved in the definition of the Ontology Web Language (OWL)[1] [14] [33], that W3C recommended in 2004 for representing ontologies. However some researchers of the CG community had also brought their contributions very early, with various strategies. Some researchers adopted CG directly as formalism for representing ontologies and annotations in Semantic Web context: e.g. WebKB [32]. Others preferred to rely on a correspondence between CG and RDF(S) [2] – the language recommended by W3C for describing Web resources [29]: the ACACIA team thus proposed and implemented a translation of RDF(S) into CG and built a semantic search engine, CORESE, based on this correspondence [11] while in [24], the authors sug-

---

[1] http://www.w3.org/sw/WebOnt.
[2] Resource Description Framework, http://www.w3.org/RDF

F. Dau, M.-L. Mugnier, G. Stumme (Eds.): ICCS 2005, LNAI 3596, pp. 19-50, 2005.
© Springer-Verlag Berlin Heidelberg 2005

gested to use CG as the Ontolingua for allowing the automatic translation of knowledge structures between different KR formalisms, and they described RDF(S) meta-model in CG.

In this paper, we summarise the ACACIA team approach for the Semantic Web and emphasise the role of CG in this approach. Our objective is to show that in the framework of Semantic Web, Conceptual Graphs have enough expressivity for KR and enough reasoning capabilities for real-world applications.

2 Corporate Semantic Web Approach

2.1 ACACIA Project Evolution

The ACACIA project[3] is a multidisciplinary team that aims at offering methodological and software support for knowledge management (KM), and in particular for building, managing, evaluating and evolving corporate memories.

Historically, in 1992, we were focusing on corporate memories materialised in documents and knowledge bases (KBs). Our main topics were: (1) multi-expertise, (2) "intelligent" information retrieval and (3) management of links between documents and KBs. Therefore we chose CG as privileged KR formalism since it offered reasoning capabilities interesting for intelligent information retrieval, and it seemed a good basis for tackling multi-expertise or for representing a KB associated to texts. The main results of our previous research in CG formalism were the following:

- The CGKAT system integrated a CG-represented ontology extending WordNet and enabled to associate a base of CGs to a structured document. The user could ask queries about either the base of CGs or the document contents and CGKAT could retrieve relevant document elements through a projection of the user's query on the CG base associated to the document [31].
- The MULTIKAT system offered an algorithm for comparing KBs modelling knowledge of two experts in a given domain. MULTIKAT enabled comparison and merging of two CG ontologies [19] as well as comparison and integration of two CG bases, the integration being guided by different integration strategies [18];
- The C-VISTA and CG-VISTA models enabled to represent viewpoints in CG formalism (both in a CG support and in a CG base) [39].

In 1998, Tim Berners-Lee proposed his vision of the Semantic Web [4]. The main approach suggested was to use an ontology for making explicit semantic annotations on Web resources. According to us, CG was clearly a relevant formalism for representing such ontology and annotations: there was an analogy between on the one hand, a Web document annotated by semantic annotations w.r.t. an ontology and on the other hand, a structured document associated to a base w.r.t. the CGKAT ontology. But W3C had already started to define RDF as the future language for describing Web resources. So, as there was also an analogy between RDF and CG, and as a W3C-recommended language was more likely to be widely adopted by the different

---

[3] http://www-sop.inria.fr/acacia/

research communities or industrial companies than CG, our strategy was to rely on this RDF – CG correspondence, so as to take advantage of both the standard feature of RDF and our competence in CG. Therefore, instead of building an RDF-dedicated tool from scratch, we preferred to rely on this RDF – CG correspondence for developing CORESE, an RDF-dedicated search engine based on CG: thus, the first version of CORESE was quickly implemented and tested using the API of Notio [44]. Moreover, our KM approach evolved towards our so-called "Corporate Semantic Web" approach. From research viewpoint, using CORESE as the kernel of our research and applications enables us to check the validity of our hypothesis that CG is a good internal KR formalism for "Corporate Semantic Web" applications.

2.2 Corporate Semantic Webs

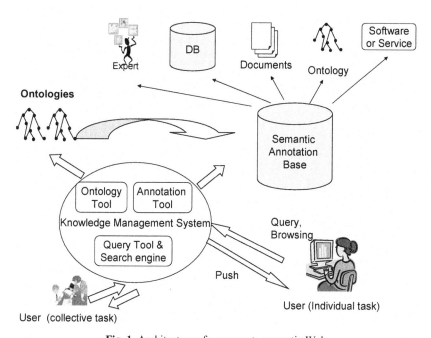

Fig. 1. Architecture of a corporate semantic Web

In 1999, we proposed an approach called « Corporate Semantic Web » approach, that relies on a natural analogy between Web resources and corporate memory resources, and on a generalisation of CGKAT approach. We thus proposed to materialise a corporate memory through:

– *resources*: they can be documents (in various formats such as XML, HTML, or even classic formats), but they can also correspond to people, services, software or programs,

–. *ontologies*, describing the conceptual vocabulary shared by one or several communities in the company,

– *semantic annotations on these resources* (i.e. contents of documents, skills of persons or characteristics of services / software / programs), based on these ontologies.

with diffusion on the Intranet or the corporate Web.

A Corporate Semantic Web has some specificities w.r.t. the Semantic Web: the fact that an organisation is bounded should allow an easier agreement on a corporate policy, an easier creation of ontologies and annotations, an easier verification of validity and reliability of information sources, a description of more precise user profiles, a smaller scale for the document corpora and for the ontologies.

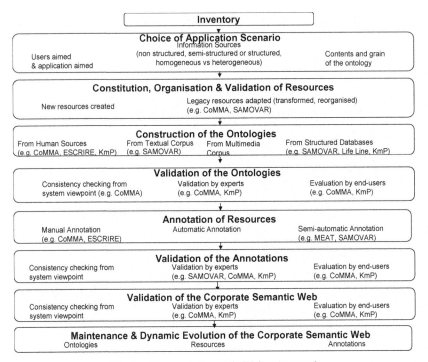

Fig. 2. « Corporate Semantic Web » Approach

Figure 1 shows the architecture of a corporate semantic Web and figure 2 summarises our method for building it, method stemming from our synthesis and abstraction after analysis of all our semantic web applications. In [17], we study thoroughly the components of a corporate semantic Web (resources, ontologies, annotations).

In this paper, we will illustrate this "Corporate Semantic Web" approach by several ACACIA team research results: the CORESE semantic search engine, project memory (SAMOVAR system), distributed memory (CoMMA system), information retrieval from Medline (ESCRIRE system), support to interpretation or validation of DNA micro-array experiments (MEAT project), support to collaborative reasoning in a healthcare network (Life Line project), support to skills cartography (KmP project). Table 1 summarises their contributions to the ACACIA team research program.

Table 1. Contribution of the different projects to ACACIA research program

Research questions	Contributing projects
How to build or enrich an ontology from textual sources?	SAMOVAR
How to build or enrich an ontology from a structured database?	SAMOVAR, Life Line, KmP
What existing KR formalisms are the most relevant for the semantic Web?	ESCRIRE
How to create semantic annotations manually through an editor?	CoMMA
How to create semantic annotations (concepts or relations) semi-automatically from texts?	SAMOVAR, MEAT
How to offer ontology-guided information retrieval?	CORESE, ESCRIRE, SAMOVAR, CoMMA
How to offer approximate reasoning?	CORESE, KmP
How to distribute annotations in a memory? How to distribute query processing among agents? How to use agents and ontologies for information retrieval from a distributed memory?	CoMMA
How to offer scenario-guided, user-centred evaluation of a corporate memory?	CoMMA, KmP
How to offer friendly interfaces (for ontology browsing, querying or result presentation)?	KmP

Table 2. Scenarios studied in the different applications

Application Scenario	Contributing project
Project memory	SAMOVAR, Aprobatiom
Integration of a new employee	CoMMA
Support to technological watch	CoMMA
Experiment memory for a research community	MEAT
Support to cooperative work	Life Line
Skills cartography	KmP

3 CORESE Semantic Search Engine

As detailed in [11][12][13], CORESE[4] is a semantic search engine, i.e. an ontology-based search engine for the semantic web: it enables to retrieve web resources anno-

---

[4] http://www-sop.inria.fr/acacia/soft/corese

tated in RDF. CORESE ontology representation language is built upon RDFS, that enables to represent an ontology by a class hierarchy and a property hierarchy. CORESE thus takes into account subsumption links between concepts and between relations when it needs to match a query with an annotation. CORESE ontology representation language also enables to represent domain axioms on which CORESE relies when matching a query with an annotation.

Annotations are represented in RDF and related to the RDF Schema representing the ontology they are built upon. The CORESE query language is also based on RDF; for each query, an RDF graph is generated, related to the same RDF Schema as the one of the annotations to which the query will be matched.

The CORESE engine works on CG internally. When matching a query with an annotation according to their common ontology, both the RDF graphs and their schema are translated into the CG model [45] [10]. Figure 3 summarises the principle of CORESE. Through this translation, CORESE takes advantage of previous work of the KR community on reasoning with CG [41] [42] [37] [3].

3.1 RDF(S) and Conceptual Graphs

As stressed in [11][12][13], the RDF(S) and CG models share many common features and a mapping can easily be established between RDFS and a large subset of the CG model. An in-depth comparison of both models was the starting point of the development of CORESE. Both models distinguish ontological knowledge and assertional knowledge. In both models, the assertional knowledge is positive, conjunctive and existential and it is represented by directed labelled graphs. In CORESE, an RDF graph representing an annotation or a query is thus translated into a CG.

Concerning the ontological knowledge, the class (resp. property) hierarchy in an RDF Schema is translated into a concept (resp. relation) type hierarchy in a CG support. RDF properties are declared as first class entities as RDFS classes, just as relation types are declared independently of concept types in a CG support. This common handling of properties makes the mapping of RDFS and CG models relevant, contrarily to OO formalisms where properties must be defined as attributes inside classes.

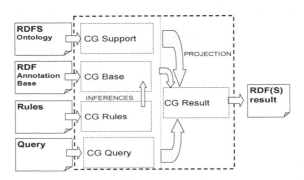

Fig. 3. Principle of CORESE

There are a few differences between the RDF(S) and CG models in their handling of classes and properties but such differences can be easily dealt with:

– In RDF(S), a resource can be instance of several classes while in CG, an individual marker has a unique concept type corresponding to the lowest concept type associated to the instance referred by this individual marker (i.e. a concept is exact instance of only one concept type). The declaration of a resource as an instance of several classes in RDF is translated in CG model by generating the concept type corresponding to the highest common subtype of the concept types translating these classes.
– Similarly, an RDF property can have several domains (resp. ranges), while in CG, a relation type signature is unique. Moreover in RDF, properties are binary while in CG, relations are n-ary. The multiple domains (resp. ranges) constraint of an RDF property is translated into a single domain (resp. range) constraint in CG by generating the concept type corresponding to the highest common subtype of the concept types constraining the domain (resp. range).

The projection operation is the basis of reasoning in the CG model. A query is thus processed in the CORESE engine by projecting the corresponding CG into the CGs obtained by translation of the RDF annotations. The retrieved web resources are those for which there exists a projection of the query graph into their annotation graph.

For example, for the ontology shown in figure 4, the query graph G in figure 5 can be projected on the two annotation graphs G_1 and G_2. The web resources annotated by these graphs will be found as answers of the G query and will be retrieved by CORESE when processing this query.

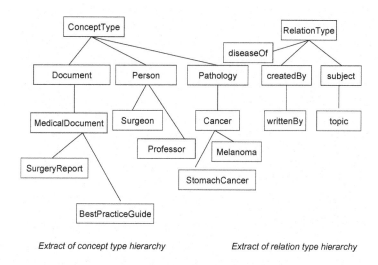

Extract of concept type hierarchy   *Extract of relation type hierarchy*

Fig. 4. Example of CG support

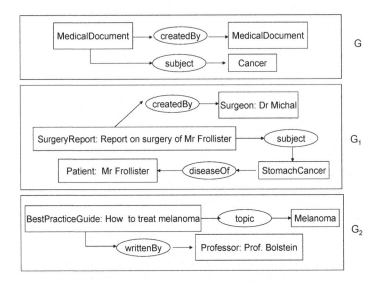

Fig. 5. Examples of CG graphs based on the previous CG support

3.2 Domain Axioms

In addition to a concept type hierarchy and a relation type hierarchy, a more expressive ontology can contain domain axioms enabling to deduce new knowledge from existing one. However RDF Schema does not offer such a feature. Therefore an RDF Rule extension to RDF was proposed in [13] and CORESE integrates an inference engine based on forward chaining. The rules are applied once the annotations are loaded in CORESE and before the query processing. Hence, the annotation graphs are augmented by rule conclusions before the query graph is projected on them.

CORESE production rules implement CG rules [41] [42] [37] [3]: a rule $G_1 \Rightarrow G_2$ (also noted IF G_1 THEN G_2) is a pair of lambda abstractions ($\lambda x_1, ..., \lambda x_n G_1, \lambda x_1, ..., \lambda x_n G_2$) where the x_i are co-reference links between generic concepts of G_1 and corresponding generic concepts of G_2 that play the role of rule variables.

For instance, the following CG rule states that if a person ?p suffers from an allergy to a molecule ?m and if a drug ?d contains this molecule, then this drug must not be prescribed to this patient:

```
IF [Person: ?p] - (allergic_to) - [Molecule:?m] - (con-
tained_in) - [Drug: ?d]
THEN [Drug: ?d] - (forbidden_to) - [Person:?p]
```

A rule $G_1 \Rightarrow G_2$ applies to a graph G if there exists a projection π from G_1 to G, i.e. G contains a specialisation of G_1. The resulting graph is built by joining G and G_2 while merging each $\pi(x_i)$ in G with the corresponding x_i in G_2. Joining the graphs may lead to specialise the types of some concepts, to create relations between concepts and to create new individual concepts (i.e. concepts without variable).

Remark: In the first implementation of CORESE rule language and rule engine, in order to avoid possible loops during the execution of the forward-chaining engine, we restricted the rule language by the following constraint: no new generic concept could be created in a rule conclusion. However, later applications required the need to use RDF anonymous resources in rule conclusions. So we decided to suppress this restriction, so as to allow the creation of generic nodes for expressing such graphs as:

```
IF  [Patient:  ?p]  -  (taken-care-in)  -  [Hospital:  ?h]
THEN  [Patient:  ?p]  -  (attended-by)  -  [Doctor:  ?d]  -
(working-in) - [Hospital: ?h].
```

But the rule engine keeps track of the variable values that led to the triggering of such a rule, so as to avoid to apply twice the same rule with the same variable values.

3.3 CORESE Query Language

CORESE had several successive query languages, that evolved through the various applications of CORESE. The last version of the CORESE query language enables to express queries in the form of RDF triples or of a Boolean combination of RDF triples. It is an SQL–like query language for RDF, compatible with the W3C SPARQL RDF query language [38].

The query processor is the CG projection. It relies on the RDF Schema by using subsumption links (*rdf:subClassOf* and *rdfs:subPropertyOf*) and it processes datatype values. The query language can also query the RDF Schema itself (i.e. the CG support). Last, it is able to return the best approximate answers by relaxing types.

For example, the following query enables to retrieve any doctor author of a medical document; the name of this doctor as well as the title of the medical document must be returned in the answer:

```
select ?n ?t where
?p rdf:type lv:Doctor
?p lv:name ?n
?p lv:author ?doc
?doc rdf:type lv:MedicalDocument
?doc lv:Title ?t
```

(lv is the namespace where the classes and properties are defined; as a notation, the variables are characterised by a name starting by an interrogation point.)

This query is translated into the following CG:

```
[Doctor: ?p] - {
   - (name) - [Literal: ?n]
   - (author) - [MedicalDocument: ?doc] - (Title) -
[Literal: ?t]}
```

3.4 Approximate Reasoning

In [12], CORESE approximate reasoning capabilities are detailed. The principle is to calculate the semantic distance of concept types or of relation types in the ontology hierarchies, so that two brothers are closer at the ontology deepest levels. When the user asks an approximate search about a query, the query projection on the annotation base is performed independently of the subsumption link between concept types (resp. relation types). For each retrieved resource, the distance to the query is computed and only the resources having a semantic distance to the query lower than a given threshold are presented to the user; these results are sorted by increasing distance.

3.5 Conclusions

CORESE works with domain ontologies represented through RDF(S) extended with domain axioms (which corresponds to simple CG extended by graph rules). It can process queries expressed in a query language close to SPARQL, and it offers approximate reasoning in case of need. CORESE has been tested with several existing RDF Schemas such as the Gene ontology, IEEE LOM, W3C CC/PP, etc.

Let us now present several approaches illustrating corporate semantic web applications developed in the ACACIA team and based on CORESE.

4 Memory of a Vehicle Project: SAMOVAR Project

The first application of CORESE was the SAMOVAR system. The objective of the SAMOVAR project was to capitalise knowledge on the problems encountered during a vehicle design project at Renault. The capitalisation was already initiated at Renault through a database (DB) called "Problem Management System" and describing the problems detected during the validation phases in a vehicle project: in the textual fields of this DB, the participants discussed about these problems, about the possible solutions for solving them, with their constraints and their advantages, as well as the solution finally chosen. But due to the volume of this DB, these textual fields of the Problem Management System constituted an unused information mine.

4.1 Methodological Approach

The SAMOVAR ontology was made up by using both (1) a corpus of texts constituted by the textual fields of the Problem Management System and (2) a structured database describing the nomenclature of all parts that can be used in a vehicle project at Renault. The SAMOVAR approach [25] consisted of the following steps:

- Make an inventory, through discussions with the experts, for analysing the organisation structure, a vehicle project development, and the corporate data available.
- Apply a *natural language processing tool* (more precisely, the *term extractor* Nomino) on the textual fields of the Problem Management System.
- Analyse the *structure of the obtained candidate terms* for determining their *linguistic regularities* to be used in *heuristic rules* allowing to enrich the ontology.

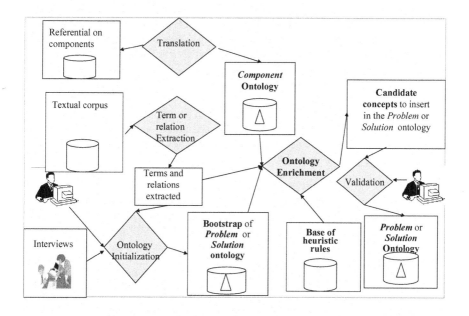

Fig. 6. Method of construction of the SAMOVAR ontology [25]

- Build an *ontology with several components*: *Problem, Part, Project, Service*. The *Part* sub-ontology was built automatically from a part nomenclature, available in the company. The *Problem* sub-ontology, manually initialised from the analysis of the discussions with the experts and of the Nomino-obtained candidate terms, was then enriched semi-automatically using the previous heuristic rules (cf. figure 6).
- *Annotate semi-automatically descriptions of problems with this ontology.*
- Use the *semantic search engine* CORESE for information retrieval (once the ontology represented in RDFS and the annotations in RDF): in particular, CORESE enables to retrieve problem descriptions satisfying given features.

4.2 Example of Query

The user searches all the problems of Geometry encountered on the *Driver_cockpit _crossmember*:

```
[GeometryProblem: ?pb] - (concerning) - [Part:
Driver_cockpit _crossmember]
```

which corresponds in CORESE query language to:

```
?pb rdf:type GeometryProblem
?pb concerning Driver_cockpit _crossmember
Driver_cockpit _crossmember rdf:type Part
```

Using the projection, the SAMOVAR system is able to find not only descriptions of problems that are annotated by « *GeometryProblem* concerning *Driver_cockpit _crossmember* » but also those annotated by « *CenteringProblem* (resp. *InterferenceProblem* or *ClearanceProblem*) concerning *Driver_cockpit_crossmember* », since *CenteringProblem*, *InterferenceProblem* and *ClearanceProblem* are subtypes of *GeometryProblem*.

The SAMOVAR system is an illustration of a corporate semantic Web, implementing a project memory through resources constituted by problem descriptions annotated semantically w.r.t. the ontology. This ontology represented in RDFS then allows CORESE to guide information retrieval in this memory.

Table 3. Summary of SAMOVAR project

System	SAMOVAR: vehicle project memory system
Context or scenarios	Memory of problems encountered during a vehicle project
Domain	Automotive sector
Company	Renault car manufacturer
Scope of Semantic Web (SW) Approach	Corporate semantic web on an internal web
Resources	Database of problem descriptions
Information sources	Human experts + Structured DB on part nomenclature describing all the parts used for vehicle manufacturing + Textual corpus constituted by textual comments of the database of problem descriptions
Ontology	792 concepts and 4 relations in the SAMOVAR ontology.
Annota- tions	4483 problem descriptions annotated
Expert validation	Validation by some experts at Renault
Typical user's query	"Find all problems of a given (resp. any) type that occurred on a given (resp. any) part of the vehicle in a given (resp. any) project"
Used reasoning	Classic projection + Browsing of the concept type hierarchy
CORESE functions used	CORESE past query language (consisting of RDF + variables and operators)
End-user evaluation	Evaluation by some experts at Renault
Research progress	Use of linguistic tools for semi-automatic construction / enrichment of an ontology and of semantic annotations

4.3 Conclusions

SAMOVAR was the first application of CORESE in a real world concrete problem. It proved the interest of ontology-guided information retrieval. It was based on CORESE first query language (that was similar to RDF with variables). SAMOVAR illustrated how legacy systems can be used as sources of the memory. It offered an example of knowledge acquisition from heterogeneous information sources: texts, databases, experts. It enabled us to study thoroughly the problem of creation of an ontology and of semantic annotations from textual sources, such ontology and annotations being then represented in RDF(S) in order to serve as inputs of CORESE.

The corpus-based approach for building the ontology and the annotations is generic. It relies on the two following stages: (1) Analyse the structures of the candidate terms obtained from a term extractor, in order to determine their regularities that will be used to create heuristic rules; (2) Use these rules in order to enrich the ontology.

The SAMOVAR method was generalised in a method for construction of a memory of problems encountered during a project of design of a complex system [25].

After SAMOVAR, several applications enabled to improve CORESE query language, CORESE Graphical User Interface (both for browsing the ontology, for querying and for presenting the answers).

5 Distributed Memory: COMMA Project

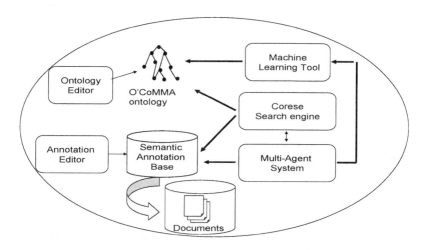

Fig. 7. Architecture of the CoMMA system

The management of a distributed memory was studied thoroughly in the CoMMA IST project that enabled to develop a multi-agent system in order to manage a distributed memory [23]. This memory was constituted by a corporate semantic Web made of:

- O'CoMMA corporate ontology and user models,
- resources constituted by the corporate members or by documents useful for a new employee or handled for technological watch,
- semantic annotations on these resources.

The CoMMA system [23] includes five main components: (1) the O'CoMMA ontology, (2) the multi-agent knowledge management system, (3) the CORESE semantic search engine, (4) machine learning algorithms enabling to improve relevance of retrieved documents according to the users' interest centres, (5) graphical interfaces.

5.1 The O'CoMMA Ontology

The method of construction of the O'CoMMA ontology [23] relied on the following phases: knowledge elicitation from human sources, manual terminological analysis phase, ontology structuration, validation by experts, ontology formalisation in RDFS. The O'CoMMA ontology is structured in three levels [23]:

- A high level comprising abstract types of concepts, very reusable but not very usable by end-users in their daily work, and having to be hidden during the ontology browsing by the end-user: e.g. *Entity* and *Situation* types of concepts.
- An intermediate level comprising concept types useful for the processed scenarios of corporate memory and for the considered domains and reusable for other scenarios and similar domains: e.g . concepts related to the aspects *Document, Organisation, Person, Telecommunications* and *Building*, useful for support to integration of new employees and technological watch in telecommunications (cf T-NOVA and CSELT) and in building industry (CSTB) but generic enough to be reusable for other similar scenarios and domains.
- A specific level comprising concepts typical to the considered enterprise and therefore very useful for the end-users but not very reusable outside this enterprise: e.g. concepts also related to the aspects *Document, Organisation, Person, Telecommunications* and *Building*, but typical of the company considered (for example, *guide of the route of the new employee, technological watch file* or *thematic referent*).

The ontology construction method and the ontology structure are reusable. The reusability of the O'CoMMA ontology was proven through the Aprobatiom project where the O'CoMMA ontology was extended by concepts useful for building a project memory in building sector for CSTB.

Let us note that this O'CoMMA ontology was not only aimed at annotating documents and people, but also aimed at being browsed by human users and used by software agents (helping in the search of information in a distributed memory).

5.2 Examples

Here is an example of query in CoMMA for integration of a new employee:
"Find the titles and authors of all documents aimed at newcomers".
In the new CORESE query language:

```
select ?t ?auth
?doc rdf:type comma:Document
?doc comma:title ?t
?auth comma:author ?doc
?doc comma:target ?targ
?targ rdf:type comma:Newcomer
```

An example of query in CoMMA for technological watch is:
"Find all the persons for which the documents found on fire detection systems are interesting".

```
select ?pers
?doc rdf:type comma:Document
?doc comma:Subject comma:FireDetectionSystem
?pers rdf:type comma:Person
?pers comma:HasForWorkInterest
comma:FireDetectionSystem
```

An example of use of rule is:
"If a team includes a person professionally interested in a subject, the team can be considered as professionally interested in the subject and the documents on the subject are relevant for being sent to the team."

```
IF [Team:?team] - (Includes) - [Person:?pers] - (Has-
ForWorkInterest) - [Topic:?theme]
    [Document:?doc] - (Subject) - [Topic:?theme]
THEN [Team:?team] - (HasForWorkInterest) -
[Topic:?theme]
    [Document:?doc] - (RelevantFor) - [Team:?team]
```

In CORESE new rule language, it is expressed as:

```
IF ?team rdf:type comma:Team
    ?team comma:Includes ?pers
    ?pers rdf:type comma:Person
    ?pers comma:HasForWorkInterest ?theme
    ?theme rdf:type comma:Topic
    ?doc rdf:type comma:Document
    ?doc comma:Subject ?theme
THEN ?team comma:HasForWorkInterest ?theme
    ?doc comma:RelevantFor ?team
```

Table 4. Summary of CoMMA project

System	CoMMA, a multi-agent system for corporate memory management
Context or scenario	Support to integration of a new employee and to technological watch
Domain	Telecommunications and Building sector
Companies	T-NOVA (Deutsche Telecom), CSELT (Italian Telecom) and CSTB (French Centre for Science and Technique of Building)
Scope of Semantic Web Approach	Distributed corporate semantic Web, both on internal and external web, with agents used for search and for push towards the end-user
Resources	Documents describing the organisation, the corporate members and the domain
Information sources	Human experts + Document manual analysis
Ontology	The O'CoMMA ontology comprises 472 concepts, 80 relations and 13 levels of depth
Annotations	Annotations on documents or on people in an organisation
Expert validation	Validation by T-NOVA, CSELT, CSTB experts
Typical User query	"Find the document of this type speaking about this subject or interesting for this kind of user" "Find the persons of this type that have these characteristics"
Used reasoning	Classic projection + Use of global objects + Graph rules + Ontology browsing
CORESE functions used	CORESE past query language + Use of rules
End-user evaluation	Evaluation by end-users at T-NOVA, CSELT and CSTB
Research progress	O'CoMMA ontology represented in RDFS + DRDF(S) a CG-based extension of RDF(S) for expressing contextual knowledge + Concept learning from RDF annotations + Annotation and query distribution + Multi-agent system for distributed information retrieval

5.3 CORESE Extensions Designed for CoMMA.

In the CoMMA project, CORESE was extended with:

- a global graph representing global knowledge true for all annotations, so that this global graph would be joined to each annotation during query processing,
- the expression and processing of reflexive, symmetric or transitive properties,

- extensions of the query language with type operators enabling to specify more precisely the type of the requested resources,
- rule graphs and a forward-chaining rule engine.

5.4 DRDF(S): Extensions of RDFS for Expressing Contextual Knowledge

One lesson of the CoMMA project was that RDF(S) was insufficient to express contextual knowledge such as definitions or axioms: therefore we took inspiration of CG for proposing DRDFS (Defined Resource Description Framework), an extension of RDF(S) with class, property and axiom definitions [16]. DRDFS more generally enables to express contextual knowledge on the Web. The RDF philosophy consists of letting anybody free to declare anything about any resource. Therefore the knowledge of by whom and in which context an annotation is stated is crucial. DRDF(S) enables to assign a context to any cluster of annotations. The representation of axioms or class and property definitions is just a particular use of DRDFS contexts. DRDFS is a refinement of the core RDFS which remains totally compliant with the RDF triple model. More precisely, DRDFS is an RDF Schema extending RDFS with new primitives. This extension of RDFS is inspired of CG features. DRDFS is built upon the CORESE mapping established between RDF(S) and the Simple CG model. A DRDFS context corresponds to a CG; a DRDFS class (resp. property) definition corresponds to a concept (relation) type definition; a DRDFS axiom corresponds to a graph rule.

5.5 Learning Concepts from RDF Annotations

We offered a method for classifying documents and capturing knowledge by learning concepts from the RDF annotations of documents [15]. Our approach consists of extracting descriptions of the documents from the RDF graph gathering all the document annotations and then forming concepts by generating the most specific generalisation of these descriptions for each possible set of documents. In order to deal with the intrinsic exponential complexity of such a task, the concept hierarchy is built incrementally by increasing at each step the maximum size of the RDF document descriptions extracted from the whole RDF graph gathering all the annotations.

5.6 Annotation Distribution and Query Distribution

In [22], algorithms enabling the CoMMA multi-agent system to allocate and retrieve semantic annotations in a distributed corporate semantic web were proposed. The agents used the underlying graph model when they needed to allocate a new annotation and when solving distributed queries. As the agents dedicated to ontology or to the annotations used the CORESE API, they used CG reasoning.

5.7 Conclusions

CoMMA project was a very important step for improving CORESE. This application also showed the interest of RDF(S) but also its insufficiencies. From CG viewpoint, it enabled to propose multiple improvements of RDF(S) inspired of CG: graph rules, expression of contexts, concept learning.

6 Comparison of KR Formalisms: The ESCRIRE Project

Table 5. Summary of ESCRIRE project

System	EsCorServer, system for querying annotated abstracts of Medline database
Context or scenarios	ESCRIRE project, cooperative project among INRIA teams
Domain	Biology
Company	None
Scope of SW Approach	External, open Web
Resources	Abstracts of Medline database on genetics
Information sources	A researcher (computer scientist) that had taken part in projects representing knowledge on biology and genetics
Ontology	ESCRIRE ontology has 28 classes (enabling to describe 163 genes as instances of these classes) and 17 relations
Expert validation	No expert available for validation
Typical User query	"Find the articles describing interactions where a given gene acts as target and where the instigator is either this other gene or that other gene"
Used reasoning	Classic projection + Processing of "OR queries" + Use of global objects (for representing global knowledge)
CORESE functions used	Another CORESE query language a la *select from where* + Use of "OR queries" + Use of rules for handling reflection, symmetry and transitivity properties and inverse relations of relations
End-user evaluation	No evaluation by end-users
Research progress	Techniques of translation between a pivot language and conceptual graphs via RDFS + Virtual document composition

The ESCRIRE project [1] [34] was a cooperative project launched in 1999 - before the existence of OWL - among three INRIA teams, in order to compare DL, CG and OO languages for representing the contents of documents and for querying about them. It relied on the annotation of abstracts of Medline DB on genetics. It enabled to answer "OR queries" such as: *Find the articles describing interactions where the **Ubx** gene acts as target and where the instigator is either **en** gene or **dpp** gene?"*

For comparing the three languages, chosen test articles were annotated in order to constitute a test base to be queried. An XML-based pivot language was defined, for describing the ESCRIRE ontology, expressing queries and describing the answers. So each team had to translate this pivot language into the target language (DL, CG, OO KR). ACACIA developed the translator from ESCRIRE language to CG.

Example of ESCRIRE rule:

```
IF  [Interaction: ?int1] - {
    - (promoter) - [Gene: ?x]
    - (target) - [Gene:?y]
    - (effect) - [Inhibition]}
    [Interaction: ?int2] - {
    - (promoter) - [Gene: ?y]
    - (target) - [Gene:?x]
    - (effect) - [Inhibition]}
THEN [Gene: ?x]-(mutually-repressive-with)-[Gene: ?y]
```

As a conclusion, the ESCRIRE project illustrates how using a translation from different KR formalisms to this pivot language enables then to work with CGs.

7 Experiment Memory in Biology: The MEAT Project

Table 6. Summary of MEAT project

System	MEAT, a memory of experiments of biologists on DNA micro-arrays
Context	Memory of DNA-micro-array experiments
Domain	Biology
Company	IPMC
Scope of SW Approach	External, open web
Re-sources	Scientific articles useful for interpretation or validation of results of DNA micro-array experiments
Information sources	Human experts
Ontology	UMLS semantic network that contains 134 concept types and 54 relations, and is linked to millions of terms via UMLS metathesaurus.
Expert validation	Validation of extracted relations and of generated annotations, by biologists of IPMC

Typical user query	" Find all the articles asserting a given (resp. any) relation between a given biological entity (gene, protein…) and another biological entity" " Find all the articles asserting any relation between a given gene and any disease"
Used reasoning	Classic projection
CORESE functions used	CORESE new query language + Use of rules + Use of approximate reasoning
End-user evaluation	Evaluation by biologists of IPMC
Research progress	Automatic extraction of relations from texts + Automatic generation of RDF annotations

The MEAT project aims at building a memory of experiments in DNA micro-array, and at supporting biologists in their interpretation and validation of the results of their experiments, through analysis of semantically annotated Medline scientific articles. We consider the UMLS semantic network as a general ontology for the biomedical domain: the UMLS hierarchy of semantic types can be regarded as a hierarchy of concept types and the terms of the metathesaurus can be considered as instances of these concept types.

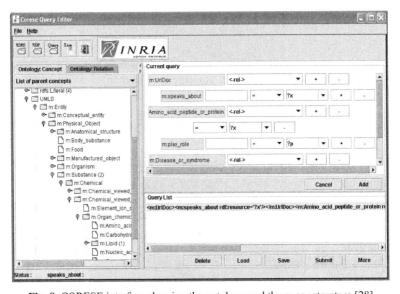

Fig. 8. CORESE interface showing the ontology and the query structure [28]

The MEAT-Annot system [28] relies on analysis of scientific articles through linguistic tools in order to generate automatically RDF annotations (not only concepts as in SAMOVAR but also relations among concepts). In the text, MEAT-Annot recognises terms corresponding to UMLS concepts and then it uses a relation extraction grammar for extracting automatically relations linking terms denoting UMLS concepts. It generates an RDF annotation that is stored, once validated by the biologist.

The annotation base is then used by CORESE for retrieving the articles possibly relevant for answering the biologist's query and supporting him/her in the interpretation of a DNA micro-array experiment.

Figure 8 shows the CORESE interface for asking a query *"Find the URL of documents speaking about an amino-acid-peptide-or-protein playing a role in a disease"* and figure 9 shows the answers to this query.

As a conclusion, the MEAT project illustrates the reuse of an existing ontology and the use of linguistic tools for generating RDF annotations.

Fig. 9. CORESE answers for the previous query [28]

8 Support to Cooperative Work in a Healthcare Network: The Life Line Project

Table 7. Summary of Life Line project

System	A Virtual staff in the framework of the "Life Line" project
Context or scenario	Support to cooperative reasoning of members of a health care network
Domain	Medicine
Company	Nautilus (specialised in marketing medical software)
SW Scope	Medical semantic Web among distributed health partners
Resources	Medical documents: patient records, guide of best practices
Info. sources	A medical database translated automatically into RDF(S)

Ontology	Nautilus ontology with 26432 concepts and 13 relations
Expert validation	Validation by our industrial partner Nautilus
Typical User query	"Find the past sessions of virtual staff where a given therapy was chosen for the patient and indicate the arguments that were given in favour of his solution" "Find a past session of virtual staff where the patient suffered from a given symptom and indicate the disease diagnosed and the therapy protocol decided"
Used reasoning	Classic projection
CORESE functions used	CORESE past query language
End-user evaluation	Evaluation by our industrial partner
Research progress	Translation of a database into an RDF(S) ontology + Integration of an ontology with SOAP and QOC graphs

The Life Line project [20] [40] aims at developing a knowledge management tool for a care network.

8.1 Nautilus Ontology

We developed a translator of the Nautilus medical DB from its internal format towards RDF(S), by using an approach of "reverse engineering" relying on the analysis of the principle of coding of this DB. The obtained Nautilus ontology was then extended by a classification of medical professions and by classes relevant for a healthcare network, and represented in RDFS. It was used in a tool called "Virtual Staff", that allows the members of a healthcare network to visualise their reasoning when formulating diagnosis assumptions or when making decisions of therapeutic procedures [20]. This application corresponds to an organisational semantic Web dedicated to a medical community cooperating in the context of a health care network.

In the Virtual Staff, the dependencies between the various diagnostic and therapeutic hypotheses are represented through a graph using the concepts defined in the Nautilus ontology. The doctor reasons by linking the health problems to the symptoms, the clinical signs and the observations in order to propose care procedures.

The Virtual Staff relies on the SOAP model (Subjective, Objective, Assessment, Plan) used by the medical community [46]. In this model, the S nodes describe current symptoms and clinical signs of the patient, the O nodes describe analyses or observations of the physician, the A nodes correspond to the diseases or health problems of the patient, and the P nodes correspond to the procedures or action plans set up in order to solve the health problems.

Sometimes, the doctor may need to visualise all the possible solutions and the arguments in their favour or against them. The QOC model (Question Options Criteria) [30]) is useful for support to decision-making. In this model, a question Q corresponds to a problem to solve. To solve the question Q, several Options are thought out, with, for each option, the criteria in its favour and the criteria against it. Two types of questions are possible for the Virtual Staff: (1) *Diagnosis of a pathology:*

Which pathology explains the clinical signs of the patient? (2) *Search of a prescription:* Which action plan will enable to treat the diagnosed pathology?

In the Virtual Staff, the SOAP graph enables to visualise the patient's record and in phase of decision, QOC graphs enable to choose between pathologies or between action plans. Using the Nautilus ontology, the system can propose a list of possible concept types to help the users to build SOAP and QOC graphs [20]. Table 8 indicates the concept types among the subtypes of which each category of node must be chosen.

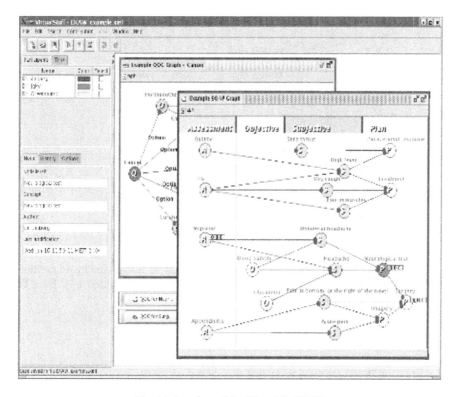

Fig. 10. Interface of the Virtual Staff [40]

Table 8. Nodes of Virtual Staff graphs and Nautilus ontology concept types

Node Category	Possible concept types
S node in a SOAP graph	Symptom
O node in a SOAP graph	LaboratoryTest, PathologicalAgent or ForeignBody
A node in a SOAP graph	Malformation, Pathology or PsychologicalProblem
P node in a SOAP graph	Treatment or DiagnosticGesture
Option in a QOC graph	Pathology or Treatment
Criterion in a QOC graph	Symptom, LaboratoryTest, Pathology or Treatment

The arcs between the nodes correspond to relations among concepts:

```
[Symptom]   - (has_for_cause)   - [Pathology]
[Pathology] - (has_for_consequence) - [Symptom]
[Pathology] - (confirmed_by)    - [LaboratoryTest]
[Pathology] - (treated_by)      - [Treatment]
```

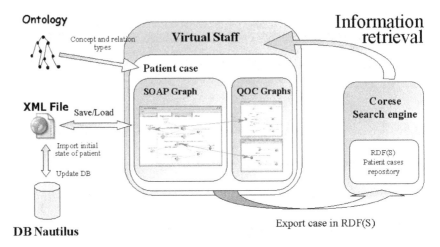

Fig. 11. Architecture of the Virtual Staff [40]

The arcs between the nodes of a QOC graph can be interpreted by « Question has-solution Option » or by « Option has-positive-criterion Criterion » or by « Option has-negative-criterion Criterion ».

8.2 Examples of Queries

"Find the past sessions of virtual staff where a given therapy was chosen for the patient and indicate the arguments that were given in favour of this solution"

```
[VirtualStaff Session: ?session] - (has-QOC) -
[QOCGraph: ?graph] - (chosen-therapy) - [Therapy:
?therap] - (positivecriteria) -[ConceptNautilus: ?cri-
terion]
```

"Find a past session of virtual staff where the patient suffered from a given symptom and indicate the disease diagnosed and the therapy protocol decided"

```
[VirtualStaff Session: ?session] - (has-SOAP) - [SOAP-
Graph: ?graph] - (has_symptom) - [Symptom: ?symp] -
(has_for_cause) - [Pathology: ?patho] - (treated_by) -
[Therapy: ?therap]
```

8.3 Conclusions

The Virtual Staff illustrates an application where CORESE helps to retrieve the relevant subclasses for editing the SOAP and QOC graphs and enables to retrieve information on past virtual staff sessions stored in RDF(S).

9 KMP (Knowledge Management Platform) Project

The KmP project[5] is an RNRT project for skills cartography of Sophia Antipolis firms in telecommunications. It is so far the largest application using CORESE and its new features such as new query language, new rule language, approximate reasoning. CORESE is the kernel of the semantic web server developed for KmP. Design of the KmP system was characterised by a user-centred, participative approach with a special care for GUI interfaces developed in SVG (see figure 12).

Table 8. Summary of KmP project

System	KmP system for cartography of skills in telecommunications for Sophia Antipolis firms
Context or scenarios	1) Increase visibility of Telecom Valley 2) Cartography of skills in order to enhance inter-firms cooperation 3) Support to cooperation between industrial companies and research laboratories
Domain	Telecommunications
Companies	Companies of Telecom Valley: Amadeus, ATOS Origin, Coframi, Elan IT, Eurecom, France Télécom R&D, HP, IBM, INRIA, Philips Semiconductors, Qwan System, Transiciel, UNSA-CNRS
Semantic Web Approach	Semantic Web server
Resources	Documents describing collective competencies of firms or of research laboratories
Information sources	Human experts
Ontology	The KMP ontology comprises 1136 concepts and 83 relations and 15 levels of depth
Expert validation	User-centred participative design and user-centred validation
User query	"Find one company having this type of skills" "Show me the poles of complementary competencies in this field" "Show me the clusters of similar competencies in this field"

---

[5] http://www-sop.inria.fr/acacia/soft/kmp

Used reasoning	Classic projection + Approximate reasoning
CORESE functions used	CORESE past and new query language + Use of rules + Use of approximate reasoning
End-user evaluation	User-centred evaluation
Research Progress	Competence ontology represented in RDF(S) + New ontological distance for approximate reasoning + SVG graphical user interfaces

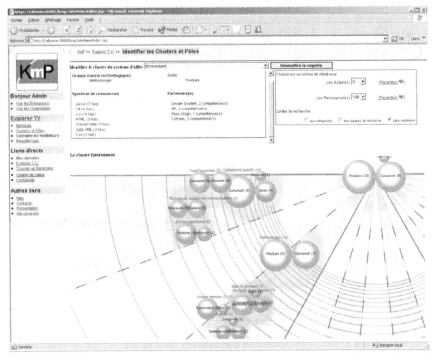

Fig. 12. Interface of KmP showing clusters of complementary competences (credit: Gronnier)

10 Conclusions

10.1 Evolution of CORESE

Let us summarise the evolution of CORESE through all these applications:
- In SAMOVAR, we used CORESE first query language and we used a linguistic tool for generating both the ontology concept types and the annotations from texts.
- In CoMMA, CORESE was enhanced by representation of graph rules enabling to express property relations such as reflexivity, symmetry, transitivity, inverse rela-

tion and to express domain rules; a forward-chaining engine was developed for using such rules in order to enlarge the annotation base.
- In the ESCRIRE project, a new *select ... from ... where* query language was introduced, and the "or queries" were processed, as well as reflexive, symmetric, transitive and inverse relations.
- In Life Line and MEAT projects, the new query language was used.
- In KmP, a semantic web server was developed, as well as new interfaces with SVG and the new CORESE query language and approximate reasoning were used.

CORESE can be compared to query languages or tools dedicated to RDF such as RQL [27], Triple [43], SquishQL [35], Sesame [7], or tools such as [21]. But CORESE is the only RDF(S)-dedicated engine that offers both inference rules and approximate search and the only RDF(S)-dedicated engine relying on CG. A few researchers use CG in semantic Web context: a link between CG and RDF(S) meta-model is proposed in [24]; CG is also the KR formalism used by Ontoseek (a content-based information retrieval system for online yellow page and product catalogs) [26] and by WebKB (an ontology server and Web robot) [32]. With respect to the synthesis on CG applications [9], CORESE is a prototype that has been used in ten applications and by several dozens of users. It proves a real usefulness of CG reasoning for corporate semantic webs. Moreover, the applications MEAT, Life Line and KmP are not restricted to a single organisation and are relevant for the (open) semantic Web.

10.2 Evaluation Issues

In the previous applications, evaluation was carried out from three viewpoints [17]:
- *Checking from system viewpoint:* we can check whether the system actually realises the functions it was intended for, and whether it satisfies classic criteria of software quality (performance, robustness, etc). We performed such a verification partially in most of our applications.
- *Validation from expert viewpoint:* the experts must validate the correctness, quality and relevance of the knowledge included in the system (mainly the ontology and the semantic annotations): we performed such validation for all our applications except ESCRIRE project where no expert was available.
- *Evaluation from end-user viewpoint:* the end-users evaluate whether the system is useful and usable according to them. We performed a very detailed, scenario-guided, end-user evaluation in two projects: CoMMA and KmP.

In our past applications such as SAMOVAR or CoMMA, we did not organise experiments for calculating systematically information retrieval recall and precision. However, during SAMOVAR evaluation, the Renault experts appreciated the ontology-guided retrieval offered by CORESE: for example, CORESE enabled them to discover similarity between different problems described in the description problem base, similarity that could not have been found automatically using classic SQL queries since it depended on the semantic contents of textual fields and not on structured fields of the DB. As the SAMOVAR ontology had been created from these textual fields, it enabled to capture this similarity between different problems (in the same project or in different projects). Likewise, during the CoMMA project, T-NOVA considered CORESE as retrieving far more relevant documents than its previous search engine and in a more efficient way. More generally, in most of our past applications, the evaluation was ergonomics-based, qualitative and based on human assessment. However, we performed a quantitative evaluation in some projects not detailed here (e.g. ontology alignment [2]). we are now preparing quantitative evaluation experiments in the MEAT project and in a technological watch application.

10.3 Discussion

The main advantages of CG in the applications previously described are:

- *The expressiveness at the level of KR:* Sowa's CGs "a la Pierce" [45] enable to represent first-order logic; in CORESE, we translated RDF(S) into simple CG [10]; SG family [3] is more expressive than RDF(S): for example, it enables to represent n-ary relations. Even though some expressions of OWL cannot be represented in SG family, however, in most of our applications, RDF(S) extended with rules was sufficient for expressing what was needed, which means that simple CGs with graph rules – as studied in the SG family– were sufficient.
- *The power at the level of reasoning:* CORESE takes advantage of the classic CG projection (that we optimised for dealing with large ontologies), and offers forward-chaining on graph rules. The approximate search based on the calculation of an ontological distance requires the implementation of an "approximate projection" [12]. It would be interesting to study its complexity as for the SG family [3].
- *The power at the level of queries:* CORESE query language is close to SPARQL specification [38] – that should become a recommendation of W3C.

As a further work, we intend to study systematically (1) what can be expressed with CGs (KR and queries) that could not be expressed with the other languages and (2) what cannot be expressed in CGs while it could be expressed with other languages.

One frequent argument in favour of CG formalism is its greater readability thanks to its graphical notation. However in CORESE, it is RDF(S) that is handled and visible externally and not CGs, since RDF(S) is the W3C-recommended language, that plays the role of exchange language between CORESE and external world (e.g. the user or other applications). However, adequate graphical user interfaces can represent RDF(S) through graphs (as CGs): in the MEAT project, our validation interface for the biologists presents the generated RDF annotations through user-friendly graphs.

We distinguish the internal language handled by the semantic web tools (i.e. RDF(S) and OWL), the language handled by the developer (in our case, RDF(S) and CG), and last, what can be seen by the end-user through user interfaces.

Our choice of RDFS is mainly historical since the first implementations of CORESE preceded the emergence of OWL. But as, in 2004, W3C recommended OWL for representing more complex ontologies, our strategy can be either to continue to focus on RDF(S) – i.e. privilege applications needing only simple ontologies since most existing Web-available ontologies are still in RDF(S) – or to evolve towards OWL. The ACACIA team is studying CORESE extensions to handle annotations represented in OWL Lite. Handling simple ontologies should be an advantage in the framework of the open semantic Web, where heterogeneity and scalability issues are more crucial than for corporate semantic Webs.

10.4 Towards the Semantic Web

In [17], we emphasise the most important research topics needed to be performed on construction, management and evolution of the different elements of a corporate semantic Web: automation, heterogeneity, evolution, evaluation and scalability.

In addition to the topics described in this paper, the ACACIA team also studies thoroughly RDF(S) ontology alignment ontology [2], multi-viewpoint ontologies [39] support to technological watch with ontology-guided search on the external Web [8] and support to e-Learning (a new KM scenario) using semantic Web technologies. As the results of this research will be used to extend CORESE search engine, this research on RDF(S) indirectly works with CGs. We hope to have shown through this paper that more than 10 years of research in CG are useful for contributing to reach Tim Berners-Lee's vision of the Semantic Web.

Acknowledgements.
We deeply thank all our colleagues of the ACACIA team for their enthusiasm in our collective work on corporate memories and on corporate semantic Webs. We are specially grateful to Khaled Khelif, Nicolas Gronnier, and Marek Ruzicka for the hard-copies of MEAT-Annot, KmP and Virtual Staff integrated in this paper. CORESE was mainly designed and developed by Olivier Corby, with a support of Olivier Savoie for implementation. Rules were added to CORESE by Alexandre Delteil and Catherine Faron-Zucker, that also proposed the DRDF(S) language and the algorithm for concept learning from RDF annotations. Joanna Golebiowska developed the SAMOVAR system, Fabien Gandon the O'CoMMA ontology and the annotation and query distribution algorithms in the CoMMA multi-agent system, Carolina Medina-Ramirez developed the EsCorServer server of biological knowledge, Khaled Khelif the MEAT-Annot system for generation of annotations in the framework of the MEAT project initiated by Leila Alem on biochip experiment memory, Frédéric Corby and David Minier reconstituted the Nautilus ontology from the medical database, David Minier and Marek Ruzicka implemented the Virtual Staff, Nicolas Gronnier and Cécile Guigard developed the KmP system specified by Alain Giboin, Fabien Gandon and Olivier Corby. Bach Thanh-Le developed the Asco algorithm for RDF(S) ontology alignment, Cao Tuan-Dung an algorithm for supporting technological watch guided by an ontology, Laurent Alamarguy an algorithm for relation extraction from texts, Sylvain Dehors semantic Web techniques for eLearning, and Luong Hiep-Phuc studies the evolution of a corporate semantic Web. Many thanks also to the previous members of the ACACIA team, since their past research was an important step towards corporate semantic Webs: Philippe Martin developed the CGKAT system linking structured documents, WordNet ontology and CG, Sofiane Labidi, Krystel Amergé, Laurence Alpay contributed to knowledge acquisition from multiple experts, Stéphane Lapalut to reasoning on CGs, Christophe Cointe, Nada Matta, Norbert Glaser and Roberto Sacile to research on CommonKADS, Kalina Yacef to research on eLearning, Stefan Hug developed the MultiKat system for

CG ontology comparison and Myriam Ribière proposed models for integrating viewpoints in a CG ontology.

References

1. Al-Hulou, R., Corby, O., Dieng-Kuntz, R., Euzenat, J., Medina Ramirez, C., Napoli, A. and Troncy, R. Three knowledge representation formalisms for content-based manipulation of documents, KR'2002 Workshop on "Formal Ontology, Knowledge Representation and Intelligent Systems for the World Wide Web". Toulouse, France, April 2002.
2. Bach, T.-L., Dieng-Kuntz, R., Gandon, F. On Ontology Matching Problems (for building a corporate Semantic Web in a multi-communities organization). ICEIS 2004. Porto, 2004.
3. Baget, J.-F., Mugnier, M.-L. Extensions of Simple Conceptual Graphs: The Complexity of Rules and Constraints, JAIR, 16:425-465, 2002
4. Berners-Lee, T. Semantic Web Road Map. http://www.w3.org/DesignIssues/Semantic.html, September 1998.
5. Berners-Lee, T., Hendler, J., Lassila, O. The Semantic Web. Scientific American, May 2002.
6. Berners-Lee T. , Miller E. The Semantic Web lifts off. ERCIM News No. 51, Oct. 2002.
7. Broekstra, J., Kampman, A., van Harmelen, F. Sesame: A Generic Architecture for Storing and Querying RDF and RDF Schema. Proc. of ISWC'2002, pp. 54-68, Sardinia, Italy, 2002.
8. Cao, T.-D., Dieng-Kuntz, R., Fiès, B. An Ontology-Guided Annotation System For Technology Monitoring, IADIS Int. Conf. WWW/Internet 2004, Madrid, October 2004.
9. Chein, M., Genest, D. CGs Applications: Where are we 7 years after the first ICCS. In ICCS'2000, Darmstadt, Germany, August 14-18, Springer LNAI 1867, p. 127-139.
10. Chein, M., Mugnier, M.-L. Conceptual graphs: fundamental notions. RIA, 6(4): 365-406. 1992.
11. Corby, O., Dieng, R., Hébert, C. A Conceptual Graph Model for W3C Resource Description Framework, In Proc. of ICCS'2000, Darmstadt, 2000, LNAI 1867, p. 468-482.
12. Corby, O., Dieng-Kuntz, R., Faron-Zucker, C. Querying the Semantic Web with the CORESE Search Engine. ECAI'2004, Valencia, August 2004, IOS Press, p. 705-709.
13. Corby, O., Faron, C., CORESE: A Corporate Semantic Web Engine. WWW'2002 Workshop on Real World RDF and Semantic Web Appl., Hawaii, USA, May 2002. http://paul.rutgers.edu/~kashyap/workshop.html
14. Dean, M., Schreiber, G. (eds). OWL Web Ontology Language Reference. W3C Recommendation, 10 February 2004, http://www.w3.org/TR/owl-ref/
15. Delteil A., Faron C., Dieng R. Learning Ontologies from RDF Annotations. In Proc. of IJCAI'01 Workshop on Ontology Learning, Seattle, USA, August 2001.
16. Delteil A., Faron C., Dieng R. Extensions of RDFS Based on the Conceptual Graph Model. ICCS'2001, LNAI 2120, Springer-Verlag, pp. 275-289, Stanford, CA, USA, 2001.
17. Dieng-Kuntz, R. Corporate Semantic Webs. To appear in D. Schwartz ed, Encyclopaedia of Knowledge Management, Idea Publishing, July 2005.
18. Dieng, R., Hug, S. MULTIKAT, a Tool for Comparing Knowledge from Multiple Experts. Proc. of ICCS'98, Montpellier, 1998, Springer-Verlag, LNAI 1453
19. Dieng, R., Hug, S. Comparison of "personal ontologies" represented through conceptual graphs. Proc. of ECAI'98, Wiley & Sons, p. 341-345, Brighton, UK, 1998.
20. Dieng-Kuntz, R., Minier, D., Corby, F., Ruzicka, M., Corby, O., Alamarguy, L. & Luong, P.-H. Medical Ontology and Virtual Staff for a Health Network, Proc. of EKAW'2004, Whittlebury Hall, UK, October 2004, p. 187-202.
21. Eberhart, A. Automatic Generation of Java/SQL Based Inference Engines from RDF Schema and RuleML. Proc. of ISWC'2002, pp. 102-116, Sardinia, Italy, 2002.

22. Gandon, F., Berthelot, L., Dieng-Kuntz, R., A Multi-Agent Platform for a Corporate Semantic Web, AAMAS'2002, p. 1025-1032, July 15-19, 2002, Bologna, Italy.

23. Gandon, F., Dieng-Kuntz, R., Corby, O., Giboin, A. Semantic Web and Multi-Agents Approach to Corporate Memory Management, Proc. of the 17th IFIP World Computer Congress IIP Track, p. 103-115, August 25-30, 2002, Montréal, Canada.

24. Gerbé O., Mineau G. W. The CG Formalism as an Ontolingua for Web-Oriented Representation Languages. ICCS'2002, Borovetz, July 2002, Springer, p. 205-219.

25. Golebiowska, J., Dieng, R., Corby, O., Mousseau, D. Building and Exploiting Ontologies for an Automobile Project Memory, K-CAP, Victoria, Oct. 2001, p. 52-59.

26. Guarino, N., Masolo, C., Vetere, G. Ontoseek: Content-based access to the Web. In IEEE Intelligent Systems, vol. 14(3), pp. 70-80, 1999.

27. Karvounarakis, G., Alexaki, S., Christophides, V., Plexousakis, D., Scholl M. RQL: a declarative query language for RDF. In Proc. of WWW'2002, Honolulu, pp. 592-603.

28. Khelif, K., Dieng-Kuntz, R. Ontology-Based Semantic Annotations for Biochip Domain, EKAW'2004 Workshop on Application of Language and Semantic Technologies to support Knowledge Management Processes, UK, October 2004, http://CEUR-WS.org/Vol-121/

29. Lassila, O., Swick, R. R. Resource Description Framework (RDF) Model and Syntax Specification. W3C Recomm., February 22, http://www.w3.org/tr/rec-rdf-syntax/ , 1999.

30. Maclean, A., Young, R., Bellotti, V., Moran T.: Questions, Options, and Criteria: Elements of a Design Rationale for User Interfaces. IJHCI, 6(3/4):201-250. 1991.

31. Martin, P. CGKAT: A Knowledge Acquisition and Retrieval Tool Using Structured Documents and Ontologies. ICCS'97, 1997, Springer, LNAI 1257, p. 581-584.

32. Martin, P., Eklund, P. Knowledge Retrieval and the World Wide Web. IEEE Intelligent Systems, 15(3):18-25, 2000.

33. McGuinness, D. L., van Harmelen, F. (eds) OWL Web Ontology Language 57. Overview, W3C Recommendation, February 10, 2004. http://www.w3.org/TR/owl-features/

34. Medina Ramirez, R. C., Dieng-Kuntz, R., Corby, O. Querying a heterogeneous corporate semantic web: a translation approach, Proc. of the EKAW'2002 Workshop on KM through Corporate Semantic Webs. Sigüenza, Spain, October 2002.

35. Miller, L., Seaborne, A., Reggiori, A. Three Implementations of SquishQL, a Simple RDF Query Language. In Proc. of ISWC'2002, 2002, pp. 423-435, Sardinia, Italy.

36. Mineau, G. W. A First Step toward the Knowledge Web: Interoperability Issues among Conceptual Graph Based Software Agents, Part I. Proc. of ICCS'2002, p. 250-260.

37. Mugnier, M.-L. Knowledge Representation and Reasonings Based on Graph Homomorphism. ICCS'2000, Darmstadt, August 2000, Springer 1867, p. 172-192.

38. Prud'hommeaux E., Seaborne A., SPARQL Query Language for RDF, W3C Working Draft, 17 February 2005, http://www.w3.org/TR/2005/WD-rdf-sparql-query-20050217/

39. Ribière, M., Dieng-Kuntz, R. A Viewpoint Model for Cooperative Building of an Ontology, ICCS'2002, Borovets, July 2002, Springer LNAI 2393, p. 220-234.

40. Ruzicka, M., Dieng-Kuntz, R., Minier, D. Virtual Staff - Software Tool for Cooperative Work in a Health Care Network INRIA Research Report RR-5390, November 2004.

41. Salvat E. Theorem Proving Using Graph Operations in the Conceptual Graph Formalism, In Proc. of ECAI'98, pp. 356-360, Brighton, UK, 1998.

42. Salvat É., Mugnier M.-L. Sound and Complete Forward and Backward Chaining of Graph Rules. ICCS '96, Sydney, 1996, Springer, LNAI 1115, p. 248-262.

43. Sintek, M., Decker, S. Triple: A Query, Inference and Transformation Language for the Semantic Web. Proc. of ISWC'2002, pp. 364-378, Sardinia, 2002.

44. Southey, F. and Linders, J. G., Notio - A Java API for developing CG tools, ICCS'99, 1999.

45. Sowa, J. Conceptual Graphs: Information Processing in Mind and Machine. Reading, Addison Wesley, 1984.

46. Weed, L. D. The Problem Oriented Record as a Basic Tool in Medical Education, Patient Care and Clinical Research. Ann Clin Res 3(3):131-134. 1971.

47. Wielemaker J., Schreiber G., Wielinga B. Prolog-Based Infrastructure for RDF: Scalability and Performance. ISWC'2003, pp. 644-658, 2003.
48. Zhong J., Zhu, H., Li, J., Yu, Y. Conceptual Graph Matching for Semantic Search, ICCS'2002, pp. 92-106, Borovets, July 2002.

Knowledge Representation and Reasoning in (Controlled) Natural Language

Norbert E. Fuchs

Department of Informatics
University of Zurich, Switzerland
fuchs@ifi.unizh.ch
http://www.ifi.unizh.ch/attempto/

Abstract. Attempto Controlled English (ACE) is a controlled natural language, i.e. a precisely defined, tractable subset of full English that can be automatically and unambiguously translated into first-order logic. ACE seems completely natural, but is actually a formal language, concretely it is a first-order logic language with the syntax of a subset of English. Thus ACE is human and machine understandable. While the meaning of a sentence in unrestricted natural language can vary depending on its — possibly only vaguely defined — context, the meaning of an ACE sentence is completely and uniquely defined. As a formal language ACE has to be learned, which — as experience shows — takes about two days.

ACE was originally developed to specify software programs, but has since been used as a general knowledge representation language. For instance, we specified in ACE an automated teller machine, Kemmerer's library data base, Schubert's Steamroller, data base integrity constraints, and Kowalski's subway regulations. ACE served as natural language interface for the model generator EP Tableaux, for a FLUX agent, and recently for MIT's Process Handbook. We partially investigated applying ACE to knowledge assimilation, medical reports, planning, and as input language for a synthesiser of constraint logic programs. Other people suggested to express in ACE ontologies, legal texts, or standards. There were first attempts to use ACE to teach logic. Our current focus of application is the semantic web within the EU Network of Excellence REWERSE.

To support automatic reasoning in ACE we have developed the Attempto Reasoner (RACE). RACE proves that one ACE text is the logical consequence of another one, and gives a justification for the proof in ACE. If there is more than one proof then RACE will find all of them. Variations of the basic proof procedure permit query answering and consistency checking. Reasoning in RACE is supported by auxiliary first-order axioms and by evaluable functions. The current implementation of RACE is based on the model generator Satchmo. As a consequence, RACE cannot only be used for theorem proving but also for model generation.

ACE and RACE are powerful and general tools that do not require a priori world knowledge or a domain ontology — though both can be expressed in ACE — and they are neutral with regard to particular applications or methods.

More information on ACE and RACE, demo versions, and publications can be found at: http://www.ifi.unizh.ch/attempto.

F. Dau, M.-L. Mugnier, G. Stumme (Eds.): ICCS 2005, LNAI 3596, pp. 51–51, 2005.

What Is a Concept?

Joseph Goguen

University of California at San Diego, Dept. Computer Science & Engineering
9500 Gilman Drive, La Jolla CA 92093–0114 USA

Abstract. The lattice of theories of Sowa and the formal concept analysis of Wille each address certain formal aspects of concepts, though for different purposes and with different technical apparatus. Each is successful in part because it abstracts away from many difficulties of living human concepts. Among these difficulties are vagueness, ambiguity, flexibility, context dependence, and evolution. The purpose of this paper is first, to explore the nature of these difficulties, by drawing on ideas from contemporary cognitive science, sociology, computer science, and logic. Secondly, the paper suggests approaches for dealing with these difficulties, again drawing on diverse literatures, particularly ideas of Peirce and Latour. The main technical contribution is a unification of several formal theories of concepts, including the geometrical conceptual spaces of Gärdenfors, the symbolic conceptual spaces of Fauconnier, the information flow of Barwise and Seligman, the formal concept analysis of Wille, the lattice of theories of Sowa, and the conceptual integration of Fauconnier and Turner; this unification works over any formal logic at all, or even multiple logics. A number of examples are given illustrating the main new ideas. A final section draws implications for future research. One motivation is that better ways for computers to integrate and process concepts under various forms of heterogeneity, would help with many important applications, including database systems, search engines, ontologies, and making the web more semantic.

1 Introduction

This paper develops a theory of concepts, called the Unified Concept Theory (**UCT**), that integrates several approaches, including John Sowa's lattice of theories [61] (abbreviated **LOT**), the formal concept analysis (**FCA**) of Rudolf Wille [15], the information flow (**IF**) of Jon Barwise and Jerry Seligman [3], Gilles Fauconnier's logic-based mental spaces [13], Peter Gärdenfors geometry-based conceptual spaces [16], the conceptual integration (**CI**, also called blending) of Fauconnier and Turner, and some database and ontology approaches, in a way that respects their cognitive and social aspects, as well as their formal aspects. UCT does not claim to be an empirical theory of concepts, but rather a mathematical formalism that has applications to formalizing, generalizing, and unifying theories in cognitive linguistics, psychology, and other areas, as well as to various kinds of engineering and art; it has implementations and is sound engineering as well as sound mathematics.

F. Dau, M.-L. Mugnier, G. Stumme (Eds.): ICCS 2005, LNAI 3596, pp. 52–77, 2005.

Although the question in the title does not get a final answer, I hope that readers will enjoy the trip through some perhaps exotic seeming countries that lie on the borders between the sciences and the humanities, and return to their home disciplines with useful insights, such as a sense of the limitations of disciplinary boundaries, as well as with some new formal tools.

Section 2 illustrates several of the difficulties with concepts mentioned above, through a case study of the concept "category" in the sense of contemporary mathematics; Section 2.1 outlines some sociology of this concept, and Section 2.2 discusses some methodological issues, drawing particularly on theories of Bruno Latour. Section 3 reviews research on concepts from several areas of cognitive semantics, especially cognitive psychology and cognitive linguistics. This section provides a technical reconciliation of two different notions of "conceptual space" in cognitive semantics, Fauconnier's symbolic notion and Gärdenfors' geometric notion; it also gives an extended illustration of the use of this new theory, and an example using fuzzy convex sets to represent concepts. In addition, it suggests an approach to the symbol grounding problem, which asks how abstract symbols can refer to real entities. Section 4 reviews research from sociology, concerning how concepts are used in conversation and other forms of interaction; values are also discussed. Section 5 explains the UCT approach to semantic integration, using tools from category theory to unify LOT, FCA, IF and CI, and generalize them to arbitrary logics, based on the theory of institutions; an ontology alignment example is given using these methods. Finally, Section 6 considers ways to reconcile the cognitive, social and technical, and draws some conclusions for research on formal aspects of concepts and their computer implementations. Unfortunately, each section is rather condensed, but examples illustrate several main ideas, and a substantial bibliography is provided.

Acknowledgements I thank John Sowa, Mary Keeler and Rod Burstall for constructive suggestions, references, and remarks, though of course any remaining errors are my own. This paper is based on work supported by the National Science Foundation under Grant No. 9901002, SEEK, Science Environment for Ecological Knowledge.

2 A Case Study: "Category"

Category theory has become important in many areas of modern mathematics and computer science. This section presents a case study, exploring the mathematicians' concept of category and its applications, in order to help us understand how concepts work in practice, and in particular, how they can evolve, and how they can interact with other concepts within their broader contexts. This socio-historical situation is made more complex and interesting by the fact that the mathematical notion of category formalizes a certain kind of mathematical concept; Section 5 makes considerable use of categories in this sense.

The word "category" has a complex history, in which it manifested as many different concepts, several of which are of interest for this paper. Aristotle introduced the term into philosophy, for the highest level kinds of being, based on basic grammatical categories of classical Greek. The idea was later taken up and modified by others, including the medieval scholastics, and perhaps most famously, by Immanual Kant. Later Friedrich Hegel and Charles Sanders Peirce criticized Kant, and put forth their own well known triads, the three elements of which are commonly called "categories."

In the early 1940s, Samuel Eilenberg and Saunders Mac Lane developed their theory of categories [12] to solve certain then urgent problems in algebraic topology[1]. They borrowed the word from Kant, but their concept is very different from anything in Kant's philosophy. They gave a semantic formalization of the notion of a mathematical structure: the category for a structure contains all mathematical objects having that structure; but the real innovation of category theory is that the category also contains all structure preserving "morphisms" between its objects, along with an operation of composition for these morphisms. This concept, which is made explicit in Definition 1 below, allows an amazing amount of mathematics to be done in a very abstract way, e.g., see [51, 57]. For example, one can define notions of "isomorphism," "product," "quotient," and "subobject," at the abstract categorical level, and then see what form they take (if they exist) in particular categories; many general theorems can also be proved about such notions, which then apply to an enormous range of concrete situations. Even more can be done once functors are introduced, and that is not the end of the story by any means. Thus "category" is not an isolated concept, but part of a large network of inter-related concepts, called "category theory."

The phrase "all mathematical objects having that structure" raises foundational issues, because Zermelo-Fraenkel (**ZF**) set theory, the most commonly used foundation for mathematics, does not allow such huge collections. However, Gödel-Bernays set theory does allow them under its notion of "classes," which it distinguishes from the small "sets." Further foundational issues arise when one considers the category *Cat* of all categories. This has also been solved, by so-called universes of sets, along lines developed by either Alexander Grothendieck or Solomon Feferman. Some of these issues are discussed in [51]. This illustrates how a concept can generate difficulties when pushed to its limits, and then stimulate further research to resolve those difficulties[2].

Such seemingly esoteric issues can have significant social consequences. For example, in the 1970s, I had several perfectly good papers rejected, due to referee reports claiming that the mathematics used, namely category theory, lacked adequate foundations! Such incidents no longer occur, but there is lingering

---

[1] There were several different homology theories, some of which seemed equivalent to others, but it was very unclear how to define an appropriate notion of equivalence. Eilenberg and Mac Lane characterized homology theories as functors from the category of topological spaces to the category of groups, and defined two such theories to be equivalent if there was a natural equivalence between their functors.

[2] An example that followed a somewhat different trajectory is Gottlob Frege's formalization of Cantor's informal set theory, which was shown inconsistent by Russell's paradox and then supplanted by theories with a more restricted notion of set; see [39] for details.

disquiet due to the fact that foundations more sophisticated than standard ZF set theory are required.

Definition 1. *A* **category** $\mathbb{C}$ *consists of: a class, denoted* $|\mathbb{C}|$, *of* **objects***; for each pair* A, B *of objects a set* $\mathbb{C}(A, B)$ *of* **morphisms** *from* A *to* B*; for each object* A*, a morphism* 1_A *in* $\mathbb{C}(A, A)$*, called the* **identity** *at* A*; and for each three objects* A, B, C*, an operation called* **composition***,* $\mathbb{C}(A, B) \times \mathbb{C}(B, C) \to \mathbb{C}(A, C)$*, denoted ";" such that* $f; (g; h) = (f; g); h$ *and* $f; 1_A = f$ *and* $1_A; g = g$ *whenever these compositions are defined. Write* $f : A \to B$ *when* $f \in \mathbb{C}(A, B)$ *and call* A *the* **source** *and* B *the* **target** *of* f.

Example 2. Perhaps the easiest example to explain is the category $\mathfrak{Set}$ of sets. Its objects are sets, its morphisms $f : A \to B$ are functions, 1_A is the identity function on the set A, and composition is as usual for functions. Another familiar example is the category of Euclidean spaces, with $\mathbb{R}^n$ for each natural number n as objects, with morphisms $f : \mathbb{R}^m \to \mathbb{R}^n$ the $n \times m$ real matrices, with identity morphisms the $n \times n$ diagonal matrices with all 1s on the diagonal, and with matrix multiplication as composition.

There are also other ways of defining the category concept. For example, one of these involves only morphisms, in such a way that objects can be recovered from the identity morphisms. The precise senses in which the various definitions are equivalent can be a bit subtle. Moreover, a number of weakenings of the category concept are sometimes useful, such as semi-categories and sesqui-categories.

A powerful approach to understanding the category concept is to look at how it is used in practice. It turns out that different communities use it in different ways. Following the pioneering work of Grothendieck, category theory has become the language of modern algebraic geometry, and it is also much used in modern algebraic topology and differential geometry, as well as other areas, including abstract algebra. Professional category theorists generally do rather esoteric research within category theory itself, not all of which is likely to be useful elsewhere in the short term. Many mathematicians look down on category theory, seeing it more as a language or a tool for doing mathematics, rather than an established area of mathematics. Within theoretical computer science, category theory has been used to construct and study models of the lambda calculus and of type theories, to unify the study of various abstract machines, and in theories of concurrency, among many other places. It is also often used heuristically, to help find good definitions, theorems, and research directions, and to test the adequacy of existing definitions, theorems and directions; such uses have an aesthetic character. I tried to formulate general principles for using category theory this way in "A Categorical Manifesto" [21].

Let us consider two results of this approach in my own experience: initial algebra semantics [34] and the theory of institutions [28]. As discussed in [21], initial algebras, and more generally initial models, formalize, generalize and simplify the classical Herbrand Universe construction, and special cases include abstract syntax, induction, abstract data types, domain equations, and model theoretic semantics for functional, logic and constraint programming, as well as all of their combinations. The first application was an algebraic theory of abstract data types; it spawned a large literature, including several textbooks. But its applications to logic programming and other areas attracted much less attention, probably because these areas already had well established semantic frameworks, whereas abstract data types did not; moreover, resistance to high levels of abstraction is rather common in computer science.

Institutions [28, 53] are an abstraction of the notion of "a logic," originally developed to support powerful modularization facilities over any logic; see Section 5 for more detail. The modularization ideas were developed for the Clear specification language [7] before institutions, but institutions allow an elegant and very general semantics to be given for them [28, 35]; these facilities include parameterized modules that can be instantiated with other modules. This work influenced the module systems of the programming languages Ada, C++, ML, LOTOS, and a large number of lesser known specification languages[3]. Among the programming languages, the Standard ML (or SML) variant of ML comes closest to implementing the full power of these modularization facilities, while the specification languages mentioned in footnote 3 all fully implement it. ML is a functional programming language with powerful imperative features (including assignment and procedures)[4].

2.1   Sociology of Concepts

The evolution of concepts called "category" from Aristotle's ontological interpretation of syntax, through a multitude of later philosophies, into mathematics and then computer science, has been long and complex, with many surprising twists, and it is clear that the mathematical concept differs from Aristotle's original, as well as from the intermediate philosophical concepts. This evolution has been shaped by the particular goals and values of the communities involved, and views of what might be important gaps in then current philosophical, mathematical, or technical systems, have been especially important; despite the differences, there is a common conceptual theme of capturing essential similarities at a very abstract level.

The heuristic use of the category theoretic concepts for guiding research in computer science suggested in [21] is not so different from that of Aristotle and later ordinary language philosophers, who sought to improve how we think by clarifying how we use language, and in particular by preventing "category errors," in which an item is used in an inappropriate context. However, the approach of [21] is much more pragmatic than the essentially normative approach of Aristotle and the ordinary language philosophers. Many theoretical computer scientists have reported finding [21] helpful, and it has a fairly high citation index.

---

[3] These include CafeOBJ [11], Maude [8], OBJ3 [36], BOBJ [31], and CASL [54].

[4] Rod Burstall reports that much of the design work for the module system of the SML version of ML was done in discussions he had with David MacQueen. Perhaps the most potentially useful ideas still not in SML are the renamings of Clear [28] and the default views of OBJ3 [36].

On the other hand, beyond its initial success with abstract data types, the initial model approach, despite its elegance and generality, has not been taken up to any great extent. In particular, the programming languages community continues to use more concrete approaches, such as fixed points and abstract operational semantics, presumably because they are closer to implementation issues.

The extent to which a concept fits the current paradigm (in the general sense of [44]) of a community is crucial to its success. For example, Rod Burstall and I expected to see applications of institutions and parameterized theories to knowledge representation (**KR**), but that did not happen. Perhaps the KR community was not prepared to deal with the abstractness of category theory, due to the difficulty of learning the concepts and how to apply them effectively. Although we still believe this is a promising area, we were much more successful with the specification of software systems, and the most notable impact of our work has been on programming languages.

Let us now summarize some of what we have seen. Concepts can evolve over very long periods of time, but can also change rather quickly, and the results can be surprising, e.g., the "categories" of Aristotle, Kant, Hegel, and Peirce. Multiple inconsistent versions of a concept can flourish at the same time, and controversies can rage about which is correct. Concepts can become problematic when pushed further than originally intended, requiring the invention of further concepts to maintain their life, as when category theory required new foundations for mathematics. Otherwise, the extended concept may be modified or even abandoned, as with Frege's logic. New concepts can also fail to be taken up, e.g., initial model semantics and the security model mentioned in footnote 5.

The same concept can be used very differently in different communities: the uses of categories by working mathematicians, category theorists, and computer scientists are very different, so much so that, despite the mathematical definitions being identical, one might question whether the concepts should be regarded as identical. These differences include being embedded in different networks of other concepts, other values, and other patterns of use; to a large extent, the values of the communities determine the rest. Moreover, these differences can lead to serious mutual misunderstandings between communities. For example, mathematicians value "deep results," which in practice means hard proofs of theorems that fit well within established areas, whereas computer scientists must be concerned with practical issues such as efficiency, cost, and user satisfaction. These differences make it difficult for category theorists and computer scientists to communicate, as I have often found in my own experience.

In retrospect, it should not be so surprising that our research using category theory had its biggest impact in programming languages rather than formal logic or mathematics, because of the great practical demand for powerful modularization in contemporary programming practice. To a great extent, science today is "industrial science" or "entrepreneurial science," driven by the goals of large organizations and the pressures of economics, rather than a desire for simple general abstract concepts that unify diverse areas.

It seems likely that the above observations about the "category" concept also hold for many other concepts used in scientific and technical research, but further case studies and comparative work would be needed to establish the scope and limitations of these observations. There is also much more in the category theory literature that could be considered, such as the theory of topoi [4].

2.2   Methodology

Because much of the intended audience for this paper is unlikely to be very interested in social science methodology, this subsection is deliberately quite brief. Science Studies is an eclectic new field that studies science and technology in its social context, drawing on history, philosophy and anthropology, as well as sociology; its emphasis is qualitative understanding rather than quantitative reduction, although quantitative methods are not excluded. The "strong program" of David Bloor and others [5] calls for a "symmetric" approach, which disallows explaining "true" facts rationally and "false" facts socially; it is constructionist but not anti-realist. Donald McKenzie has a fascinating study of the sociology of mechanized proof [52][5]. The actor network theory of Bruno Latour [49, 48] and others, emphasizing the interconnections of human and non-human "actants" required to produce scientific knowledge and technology, grew out of the strong program. Symbolic interactionism (e.g., [6]) is concerned with how meaning arises out of interaction through interpretation; it was strongly influenced by the pragmatism of Peirce[6]. Ethnomethodology [17], a more radical outgrowth of symbolic interactionism, provides useful guidelines when the analyst is a member of the group being studied; Eric Livingston has done an important study of mathematics in this tradition [50]. Among more conventional approaches are the well known historical theories of Thomas Kuhn [44] and the grounded theory of Anselm Strauss and others [19]. Some related issues are discussed in Section 4.

3   Cognitive Science

This section is a short survey of work in cognitive semantics, focused on the notion of "concept." In a series of papers that are a foundation for contemporary cognitive semantics, Eleanor Rosch designed, performed, and carefully analyzed innovative experiments, resulting in a theory of human concepts that differs greatly from the Aristotelian tradition of giving necessary and sufficient conditions, based on properties. Rosch showed that concepts exhibit *prototype effects*, e.g., degrees of membership that correlate with similarity to a central member.

---

[5] While reading this book, I was startled to encounter a sociology of the concept "security" which (along with many other things) explained that the early 1980s Goguen-Meseguer security model was largely ignored at the time, due to pressure from the National Security Agency, which was promoting a different (and inferior) model. This illustrates how social issues can interact with formal concepts.

[6] Interpretation in this tradition is understood in essentially the sense of Peirce's thirdness [56].

Moreover, she found that there are what she called *basic level concepts*, which tend to occur in the middle of concept hierarchies, to be perceived as gestalts, to have the most associated knowledge, the shortest names, and to be the easiest to learn. Expositions in [46, 47, 37] give a concise summary of research of Rosch and others on conceptual categories. This work served as a foundation for later work on metaphor by George Lakoff and others. One significant finding is that many metaphors come in families, called *basic image schemas*, that share a common sensory-motor pattern. For example, MORE IS UP is grounded in our everyday experience that higher piles contain more dirt, or more books, etc. Metaphors based on this image schema are very common, e.g., "That raised his prestige." or "This is a high stakes game."

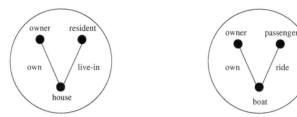

Fig. 1. Two Simple Conceptual Spaces

Fauconnier's *mental spaces* [13] (called *conceptual spaces* in [14]) do not attempt to formalize concepts, but instead formalize the important idea that concepts are used in clusters of related concepts. This formalization uses a very simple special case of logic, consisting of individual constants, and assertions that certain relations (mostly binary) hold among certain of those individuals; it is remarkable how much natural language semantics can be encoded with this framework (see [13, 14]). Figure 1 shows two simple conceptual spaces, the first for "house" and the second for "boat." These do not give all possible information about these concepts, but only the minimal amount needed for a particular application, which is discussed in Section 5. The "dots" represent the individual constants, and the lines represent true instances of relations among those individuals. Thus, the leftmost line asserts `own(owner, house)`, which means that the relation `own` holds between these two constants.

Goguen [23] proposed *algebraic theories* to handle additional features that are important for user interface design. In particular, many signs are complex, i.e., they have parts, and these parts can only be put together in certain ways, e.g., consider the words in a sentence, or the visual constituents of the diagram in Figure 1. Algebraic theories have *constructor functions* build complex signs from simpler signs; for example, a window constructor could have arguments for a scrollbar, label, and content. Then one can write $W1 = window(SB1, L1, C1)$; there could also be additional arguments for color, position, and other details of how these parts are put together to constitute a particular window. This approach conveys information about the relations between parts and wholes in a much more explicit and useful way than just saying `has-a(window, scrollbar)`, and it also seems to avoid many problems that plague the `has-a` relation and its axiomatizations in formal mereology[7] (see [58] for some discussion of these problems).

Algebraic theories also have *sorts* (often called "types"), which serve to restrict the structure of signs: each individual has a sort, and each relation and function has restrictions on the sorts that its arguments may take. For example, `owner` and `resident` might have sort `Person` while `house` has sort `Object` and `own` takes arguments only of sorts `Person, Object`. Allowing sorts to have *subsorts* provides a more effective way to support inheritance than the traditional `is-a` relation. For example, `Person` might have a subsort `Adult`. *Order sorted algebra* [32] provides a mathematical foundation for this approach, integrating inheritance with whole/part structure (using constructor functions instead of the `has-a` relation) in an elegant and computationally tractable algebraic formalism that also captures some subtle relations between the two[8].

Algebraic theories may have conditional equations as axioms (though arbitrary first order sentences could be used if needed) to further constrain the space of possible signs; for example, certain houses might restrict their residents to be adults. Fauconnier's mental spaces are the special case of order sorted algebraic theories with no functions, no sorts or subsorts, and with only atomic relation instances as axioms. There is extensive experience applying algebraic theories to the specification and verification of computer-based systems (e.g., [36, 31, 11]). Hidden algebra [31] provides additional features that handle dynamic systems with states, which are central to computer-based systems. *Semiotic spaces* extend algebraic theories by adding priority relations on sorts and constructors, which express semiotic information that is vital for applications to user interface design [23, 24]. Semiotic spaces are also called *sign systems*, because they define systems of signs, not just single signs, e.g., all possible displays on a particular digital clock, or a particular cell phone.

Gärdenfors [16] proposes a notion of "conceptual space" that is very different from that of Fauconnier, since it is based on geometry rather than logic. One of the most intriguing ideas in [16] is that all conceptual spaces are *convex*[9]; there is also a nice application of Voronoi tessellation to a family of concepts defined by multiple prototypes[10]. Although [16] aims to reconcile its geometric conceptual spaces with symbolic representations like those of Fauconnier, it does not provide a unified framework. However, such a unification can be done in two relatively

---

[7] Mereology is the study of whole/part relations.

[8] E.g., the monotonicity condition on overloaded operations with respect to subsorts of argument sorts [32].

[9] A subset of Euclidean space is *convex* if the straight line between any two points inside the subset also lies inside the subset; this generalizes to non-Euclidean manifolds by using geodesics instead of straight lines, but it is unclear what convexity means for arbitrary topological spaces.

[10] Given a set of n "prototypical points" in a convex space, the Voronoi tessellation divides the space into n convex regions, each consisting of all those points that are closest to one of the prototypes.

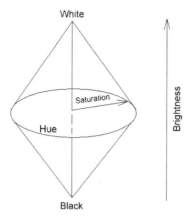

Fig. 2. Human Color Manifold

straightforward steps. The first step is to introduce *models* in addition to logical theories, where a model provides a set of instances for each sort, a function for each function symbol, and a relation for each relation symbol; since we are interested in the models that satisfy the axioms in the theory, an explicit notion of satisfaction is also needed[11]. The second step is to fix the interpretations in models of certain sorts to be particular geometrical spaces (the term "standard model" is often used in logic for such a fixed interpretation). Sorts with fixed interpretation give a partial solution to the *symbol grounding problem*[12] [38], while those without fixed interpretation are open to arbitrary interpretations. This way of combining logic with concrete models can be applied to nearly any logic[13].

For example, a sort `color` might be interpreted as the set of points in a fixed 3D manifold representing human color space, coordinatized by hue, saturation and brightness values, as in Figure 2, which is shaped like a "spindle," i.e., two cones with a common base, one upside down. This provides a precise framework within which one can reason about properties that involve colors, as in the following:

Example 3. A suggestive example in [16] concerns the (English) terms used to describe human skin tones (red, black, white, yellow, etc.), which have a very

---

[11] These items are the usual ingredients of a logic, but Section 5 argues that the extra ingredients provided by institutions, e.g., variable context and context morphisms, are also needed.

[12] This is the problem in classic logic-based AI, of how abstract computational symbols can be made to refer to entities in the real world. The approach in this paragraph is partial because it only shows how abstract symbols can refer to geometrical models of real world entities.

[13] More precisely, to any *concrete institution*, which is a specialization of the notion of institution described in Section 5, in which symbols have sorts and models are many-sorted sets; see e.g. [53] for details.

different meaning in that context than e.g., in a context of describing fabrics. Gärdenfors explains this shift of meaning by embedding the space of human skin tones within the larger color manifold, and showing that in this space, the standard regions for the given color names are the closest fits to the corresponding skin tone names. Technically, it is actually better to view the spaces as related by a canonical projection from the spindle to the subspace, because Gärdenfors' assumption that all such spaces are convex guarantees that such a canonical projection exists. Gärdenfors does not give a formal treatment of the color terms themselves, but we can view them as belonging to a mental space in the sense of Fauconnier, and view their relationship as a classification between the space of skin color terms and the space of all colors; note that many colors will not have any corresponding skin tone name.

Unified Concept Theory (see Section 5) uses the term **frame** for a combination of a context, a symbolic space, a geometrical space (or spaces), and a relation between them for the given context. Thus, there are two frames in the above example, and the embedding and projection of color spaces mentioned above can be seen as *frame morphisms*.

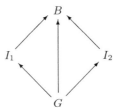

Fig. 3. Information Integration over a Shared Subobject

The most important recent development of ideas in the tradition of Rosch, Lakoff, and Fauconnier is *conceptual blending*, claimed in [14] to be a (previously unrecognized) fundamental cognitive operation, which combines different conceptual spaces into a unified whole. The simplest case is illustrated in Figure 3, where for example I_1, I_2 might be mental spaces for "house" and "boat" (as in Figure 1), with G containing so-called "generic" elements such that the maps $G \rightarrow I_i$ indicate which individuals should be identified. Some "optimality principles" are given in Chapter 16 of [14] for judging the quality of blends, and hence determining which blends are most suitable, although these distillations of common sense are far from being formal. In contrast to the categorical notion of colimit[14], blends are not determined uniquely up to isomorphism; for example, B could be "houseboat," or "boathouse," or some other combination of the two input spaces I_i (see [30] for a detailed discussion of this example).

---

[14] Colimits abstractly capture the notion of "putting together" objects to form larger objects, in a way that takes account of shared substructures; see [21] for an intuitive discussion.

Blending theory as in [14] also refines the metaphor theory of Lakoff, by proposing that a metaphorical mapping from I_1 to I_2 is really a kind of "side effect" of a blend B of I_2 and I_2, since a metaphor really constructs a new space in which only certain parts of I_1 and I_2 appear, and in which some new structure found in neither I_1 nor I_2 may also appear; the usual formulation of metaphor as a "cross space mapping" $m\colon I_1 \to I_2$ is the reflection of the identifications that are made in B, i.e., if i_1, i_2 are constants in I_1, I_2 respectively, that map to the same constant in B, then we set $m(i_1) = i_2$.

Example 4. In the metaphor "the sun is a king," the constants "sun" and "king" from the input spaces are identified in the blend, so "sun" maps to "king," but the hydrogen in the sun is (probably) not mapped up or across, nor is that fact that kings may collect taxes. But if we add the clause "the corona is his crown," then another element is added to the cross space map.

Example 5. One of the most striking examples in [14], called "the Buddhist monk," is not a metaphor. It can be set up as follows: A Buddhist monk makes a pilgrimage to a sacred mountain, leaving at dawn, reaching the summit at dusk, spending the night there in meditation, then departing at dawn the next day, and arriving at the base at dusk. A question is then posed: is there a time such that the ascending monk and the descending monk are at the same place at that time? This question calls forth a blend in which the two days are merged into one, but the one monk is split into two! The reasoning needed to answer the question cannot be done in a logic-based blend space, because some geometrical structures are needed to model the path of the monk(s), in addition to the individuals and relations that are given logically. The table below shows the semiotic spaces for the first and second day in its first and second columns, respectively; notice the explicitly given types, which are needed to constrain possible interpretations of the declared elements.

$Time = [6, 18]$	$Time = [6, 18]$
$Loc = [0, 10]$	$Loc = [0, 10]$
$m\colon Time \to Loc$	$m\colon Time \to Loc$
$m(6) = 0$	$m(6) = 10$
$m(18) = 10$	$m(18) = 0$
$(\forall\, t, t'\colon Time)\ t > t' \Rightarrow$	$(\forall\, t, t'\colon Time)\ t > t' \Rightarrow$
$m(t) > m(t')$	$m(t) < m(t')$

The first two lines of each theory are type definitions. A model for the first day will interpret $Time$ as the fixed interval [6,18] (for dusk and dawn, in hours); it will also interpret Loc as another fixed interval, [0,10] (for the base and summit locations, in miles). Then m is interpreted as some continuous function $[6, 18] \to [0, 10]$, giving the monk's distance along the path as a function of time. The key axiom is the last one, a monotonicity condition, which asserts that the monk always makes progress along the path, though without saying how quickly or slowly. Each such function m corresponds to a different model of the theory. The theory for the second day is the same except for the last three axioms, which assert that the monk starts at the top and always descends until reaching the bottom. Notice that the types $Time$ and Loc must be given exactly the same interpretations on the two days, but the possible paths are necessarily different. The blended theory is shown in the array below, in which m indicates the monk's locations on the first day and m' on the second day.

$$
\begin{array}{|l|}
\hline
Time = [6, 18] \qquad Loc = [0, 18] \\
m, m', m^* : Time \to Loc \\
t^* : Time \\
m(6) = 0 \qquad\qquad m(18) = 10 \\
m'(6) = 10 \qquad\quad\, m'(18) = 0 \\
(\forall\, t, t' : Time)\ t > t' \Rightarrow\ m(t) > m(t') \\
(\forall\, t, t' : Time)\ t > t' \Rightarrow\ m'(t) < m'(t') \\
m^*(t) = M'(t) - M(t) \\
m * (t^*) = 0 \\
\hline
\end{array}
$$

To answer the question, we have to solve the equation $m(t) = m'(t)$. If we let t^* denote a solution and let $m^* = m' - m$, then the key "emergent" structure added to the blend space is $m^*(t^*) = 0$, since this allows us to apply (the strict monotone version of) the Intermediate Value Theorem, which says that a strict monotone continuous function which takes values a and b with $a \neq b$ necessarily takes every value between a and b exactly once. In this case, m^* is strict monotone decreasing, $m^*(6) = 10$ and $m^*(18) = -10$, so there is a unique time t^* such that $m^*(t^*) = 0$.

It is interesting to notice that if we weaken the monotonicity axioms to become non-strict, so that the monk may stop and enjoy the view for a time, as formally expressed for the first day by the axiom

$(\forall\, t, t' : Time)\ t > t' \Rightarrow m(t) \geq m(t')$

then (by another version of the Intermediate Value Theorem) the monk can meet himself on the path for any fixed closed proper subinterval $[a, b]$ of $[6,18]$ (i.e., with $6 \leq a \leq b \leq 18$ with either $a \neq 6$ or $b \neq 18$). Moreover, if we drop the monotonicity assumption completely but still assume continuity, then (by the most familiar version of the Intermediate Value Theorem) there must still exist values t^* such that $m(t^*) = m'(t^*)$, but these t^* are no longer confined to a single interval, and can even consist of countably many isolated intervals. It seems safe to say that such observations would be very difficult to make without a precise mathematical analysis like that given above[15].

Finally, it is important to notice that the structures used in this example, consisting of a many-sorted logical theory and a class of models that satisfy that theory, such that some sorts have fixed interpretations in the models, can be seen as a classification in the sense of Barwise and Seligman [3]; however, it is better to consider these structures as frames in the sense of the theory of institutions (in Section 5), because in this example the geometrical structure of the models is important, and the different theories involve different vocabularies.

---

[15] Of course, some results of this analysis are unrealistic, due to assuming that the monk can move arbitrarily quickly; a velocity restriction be added, but at some cost in complexity.

Notice that it is not just the theories that are blended, but the frames, including the vocabularies over which they are defined.

Concepts with prototype effects have a fuzzy logic, in that similarity to a prototype determines a grade of membership. A formal development of first order many sorted fuzzy logic is given in [20], where membership can be measured by values in the unit interval, or in more general ordered structures; it is not difficult to generalize semiotic spaces to allow fuzzy predicates, and it is also possible to develop a comprehensive theory of convex fuzzy sets[16]. The example below demonstrates how these ideas are useful in extending Gärdenfors' approach.

Example 6. The "pet fish" problem from cognitive semantics is, in brief, how can a guppy be a bad exemplar for "pet" and for "fish," but a very good exemplar for "pet fish"? Let A be a space for animals, coordinatized by k measurable (real valued) quantities, such as average weight at maturity, average length at maturity, number of limbs, etc. For each point a in A that represents an animal, let $p(a)$ denote the extent to which a is judged to be a "pet" (in some set of experiments), and similarly $f(a)$ for "fish"; for convenience, we can interpolate other values for p and f between those with given experimental values, so that they are continuous functions on A, which can be assumed to be a rectilinear subset of $\mathbb{R}^k$. In fuzzy logic, "pet fish" is the intersection of p and f, which we an write as $pf(a) = p(a) \wedge f(a)$, where $\wedge$ gives the minimum of two real values; see Figure 4. Then "guppy" can be have a maximal value for pf even though it is far from maximal for either p or f. (It makes sense that the maximum value of pf is smaller than those for p or f, because "pet fish" is a somewhat rare concept, but if desired, pf can be renormalized to have maximum 1.) Moreover, if Gärdenfors is right, then the level sets $p_\ell = \{a \in A \mid p(a) \geq \ell\}$ are convex for each $0 \leq \ell \leq 1$, and similarly for f_ℓ. It follows from this that pf is also convex, because $pf_\ell = \{a \in A \mid p(a) \geq \ell \text{ and } f(a) \geq \ell\} = p_\ell \cap f_\ell$ and the intersection of convex subsets of $\mathbb{R}^k$ is convex.

Fig. 4. A fuzzy intersection

As a final remark, our answer to the symbol grounding problem (in the sense of footnote 12) follows Peirce, who very clearly said that signs must be interpreted in order to refer, and interpretation only occurs in some pragmatic context of signs being actually used; this means that the symbol grounding problem is artificial, created by a desire for something that is not possible for purely

---

[16] The author has done so in an unpublished manuscript from 1967. A fuzzy set $f \colon X \to L$ is *convex* iff for each $\ell \in L$, the level set $f^\ell = \{x \mid f(x) \geq \ell\}$ is convex, where L is any partially ordered set (usually at least a complete lattice).

symbolic systems, as in classic logic-based AI, but which is easy for interpreted systems[17].

4   Social Science

There is relatively little work in the social sciences addressing concepts in the sense of this paper. However, there are some bright spots. *Ethnomethodology* emphasizes the situatedness of social data, and closely examines how competent members of a group actually organize their interactions. A basic principle of *accountability* says that members may be required to account for certain actions by their social groups, and that exactly those actions are socially significant to those groups. From this follows a principle of *orderliness*, which says that social interaction can be understood, because the participants understand it due to accountability; therefore analysts can also understand it, once they discover how members make sense of their interactions.

To understand interaction, ethnomethodology looks at the *categories* (i.e., concepts) and the *methods* that members use to render their actions intelligible to each other; this contrasts with presupposing that the categories and methods of the analyst are necessarily superior to those of members. Harvey Sacks' *category systems* [59] are collections of concepts that members distinguish and treat as naturally co-occurring. Some rules that govern the use of such systems are given in [59], which also demonstrates how category systems provide a rich resource for interpreting ordinary conversation. A similar approach is used in a study reported in [22] which aimed to understand the workflow and value systems of a corporate recruitment agency. Ethnomethodology has many commonalities with Peirce's pragmatism and semiotics that would be well worth further exploration.

The *activity theory* of Lev Vygotsky [64, 65] and others is an approach to psychology emphasizing human activity, its material mediation (taken to include language), and the cultural and historical aspects of communication and learning; Michael Cole's *cultural psychology* [9] builds on this, and is particular concerned with social aspects of learning. The *distributed cognition* of Edwin Hutchins [40] and others claims that cognition should be understood as distributed rather than purely individual, and studies "the propagation of representational states across media." Leigh Star's boundary objects [62], in the tradition of science studies, provide interesting examples in which different subgroups use the same material artifact in quite different ways. It might be objected that this discussion concerns how concepts are used in social groups, rather than what they "are." But the authors cited in this section would argue that concepts cannot be separated from the social groups that use them.

An important ingredient that seems to me under-represented in all these theories is the role of *values*, i.e., the motivations that people have for what

---

[17] From an engineering perspective, one can say that sensors, effectors, and a world model are needed to ground the constants in conceptual spaces; this is of course what robots have, and our previously discussed partial solution can provide intermediate non-symbolic (geometric) representations for systems with such capabilities.

they do. It is argued in [22] that these can be recovered using the principle of accountability, as well as through discourse analysis, in the socio-linguistic tradition of William Labov [45] and others; some examples are also given in [22].

5 Logical and Semantic Heterogeneity and Integration

This section explains Unified Concept Theory; although mainly mathematical, it is motivated by non-mathematical ideas from previous sections, as the final section makes more explicit.

We have already shown how frames unify the symbolic concept spaces of Fauconnier [13] with the geometrical conceptual spaces of Gärdenfors [16] (see Example 3). This section shows how LOT, FCA, IF and CI can be included, and also shows how these theories can make sense over any logic, although they are usually done over first order, or some closely related logic (such as conceptual graphs [61]). For example, Section 3 argued that order sorted algebra is an interesting alternative for applications to ontologies, databases and so on. One might think a lot of work is required to transport these theories to such a different logic. But the theory of institutions [28, 53] actually makes this quite easy [26].

Institutions formalize the notion of "a logic", based on a triadic satisfaction relation, which like Peirce's signs, includes a context for interpretation. Intuitively, institutions capture the two main ingredients of a logic, its sentences and its models, as well as the important relation of satisfaction between them, with the novel ingredient of parameterization by the vocabulary used; this is important because different applications of the same logic in general use different symbols. The formalization is very general, allowing vocabularies to be objects in a category; these objects are called "signatures," and they serve as "contexts," or "interpretants" in the sense of Peirce. The possible sentences (which serve as "descriptions," or "representamen" in Peirce-speak) are given by a functor from signatures. The possible models ("tokens," or "objects" for Peirce) are given by a (contravariant) functor from the signature category. Finally, given a signature Σ, satisfaction is a binary relation between sentences and models, denoted $\models_\Sigma$, required to satisfy a condition asserting invariance of satisfaction (or "truth") under context morphisms. The following makes this precise:

Definition 7. *An* **institution** *consists of an abstract category* Sign, *the objects of which are signatures, a functor* $Sen\colon \mathsf{Sign} \to \mathsf{Set}$, *and a functor* $Mod\colon \mathsf{Sign}^{op} \to \mathsf{Set}$ *(technically, we might uses classes instead of sets here).* **Satisfaction** *is then a parameterized relation* $\models_\Sigma$ *between* $Mod(\Sigma)$ *and* $Sen(\Sigma)$, *such that the following* **Satisfaction Condition** *holds, for any signature morphism* $\varphi\colon \Sigma \to \Sigma'$, *any* Σ'-*model* M', *and any* Σ-*sentence* e,

$$M' \models_{\Sigma'} \varphi(e) \quad \text{iff} \quad \varphi(M') \models_\Sigma e$$

where $\varphi(e)$ *abbreviates* $Sen(\varphi)(e)$ *and* $\varphi(M')$ *abbreviates* $Mod(\varphi)(M')$.

Much usual notation of model theory generalizes, e.g., we say $M \models_\Sigma T$ where T is a set of Σ-sentences, if $M \models_\Sigma \varphi$ for all $\varphi \in T$, and we say $T \models_\Sigma \varphi$ if for all Σ-models M, $M \models_\Sigma T$ implies $M \models_\Sigma \varphi$. Moreover, if V is a class of models, let

$V \models_\Sigma T$ if $M \models_\Sigma T$ for all models M in V. We call the structure that results when a signature is a fixed frame[18]. It is now very interesting to notice that the Buddhist monk example is really a blending of two frames, not just of two conceptual spaces.

Example 8. First order logic is a typical example: its signatures are sets of relation names with their arities; its signature morphisms are arity preserving functions that change names; its sentences are the usual first order formulae; its models are first order structures; and satisfaction is the usual Tarskian relation. See [28] for details. It is straightforward to add function symbols and sorts.

Other logics follow a similar pattern, including modal logics, temporal logics, many sorted logics, equational logics, order sorted logics, description logics, higher order logics, etc. Database systems of various kinds are also institutions, where database states are contexts, queries are sentences, and answers are models [25]. The theory of institutions allows one to explore formal consequences of Peirce's triadic semiotic relation, which are elided in the dyadic formalisms of IF and FCA; for example, one insight with significant consequences is that Peirce's "objects" (i.e., an institution's models) are contravariant with respect to "interpretants" (i.e., signatures).

In the special case where the signature category has just one object and one morphism (the identity on that object), institutions degenerate to the classifications of [3], which are the same as the formal contexts of [15] (as well as the frames of institutions). The translation is as follows (where we use the prefix "I-" to indicate institutions): I-models, IF-tokens, and FCA-objects correspond; I-sentences, IF-types, and FCA-attributes correspond; and I-satisfaction, IF-classification, and FCA formal concept correspond. Chu spaces (see the appendix in [2]), in the usual case where $Z = \{0, 1\}$, are also a notational variant of one signature institutions, and general Chu spaces (and their categories) are the one signature case of the "generalized institutions" of [27], where satisfaction takes values in an arbitrary category V (but [63] gives a better exposition).

Institution morphisms can be defined in various ways [33], each yielding a category, in which many important relations among institutions can be expressed, such as inclusion, quotient, product, and sum. This gives a rich language for comparing logics and for doing constructions on logics. In the special case of institutions with just one signature, institution morphisms (actually "comorphisms") degenerate to the "infomorphisms" of [3], and the formal context morphisms of [15]. Categories of institutions support *logical heterogeneity*, i.e., working in several logics at the same time, as developed in some detail in [11, 26] and many other papers.

Given an institution $\mathbb{I}$, a *theory* is a pair (Σ, T), where T is a set of Σ-sentences. The collection of all Σ-theories can be given a lattice structure, under

---

[18] In [27], the word "room" is used, and the more precise mathematical concept is called a "twisted relation" in [28]); in fact, institutions can be seen as functors from signatures to frames.

inclusion[19]: $(\Sigma, T) \leq (\Sigma', T')$ iff $\Sigma \subseteq \Sigma'$ and $T \subseteq T'$; this is the lattice of theories (LOT) of Sowa [61]. However, a richer structure is also available for this collection of theories, and it has some advantages: Let $Th(\mathbb{I})$ be the category with theories as objects, and with morphisms $(\Sigma, T) \rightarrow (\Sigma', T')$ all those signature morphisms $f \colon \Sigma \rightarrow \Sigma'$ such that $T' \models_{\Sigma'} f(\varphi)$, for all $\varphi \in T$.

The category $Th(\mathbb{I})$ was invented to support flexible modularity for knowledge representation, software specification, etc., as part of the semantics of the Clear specification language [28]. The most interesting constructions are for parameterized theories and their instantiation. An example from numerical software is CPX[X :: RING], which constructs the complex integers, CPX[INT], the ordinary complexes, CPX[REAL], etc., where RING is the theory of rings, which serves as an *interface theory*, in effect a "type" for modules, declaring that any theory with a ring structure can serve as an argument to CPX. For another example, if TRIV is the trivial one sort theory, an interface that allows any sort of any theory as an argument, then LIST[X :: TRIV] denotes a parameterized theory of lists, which can be instantiated with a theory of natural numbers, LIST[NAT], of Booleans, LIST[BOOL], etc. Instantiation is given by the categorical pushout construction (a special case of colimits) in $Th(\mathbb{I})$, where $\mathbb{I}$ is an institution suitable for specifying theories, such as order sorted algebra.

Other operations on theories include renaming (e.g., renaming sorts, relations, functions) and sum; thus, compound "module expressions," such as NAT + LIST[LIST[CPX[INT]]] are also supported. This module system inspired those of C++, Ada, ML, and numerous specification languages (e.g., those in footnote 3). Colimits are suggested in [28] for evaluating module expressions such as $E = \text{REAL} + \text{LIST}[\text{CPX}[\text{INT}]]$. For example, we may instantiate Figure 3 with B the module expression E above, $I_1 = \text{REAL}$, $I_2 = \text{LIST}[\text{CPX}[\text{INT}]]$, and $G = \text{INT}$ as a common subobject, noting that INT is a subtheory of REAL.

For any signature Σ of an institution $\mathbb{I}$, there is a *Galois connection* between its Σ-theories and its sets of Σ-models: $(\Sigma, T)^{\bullet} = \{M \mid M \models_{\Sigma} T\}$, and if $\mathcal{M}$ is a collection of Σ-models, let $\mathcal{M}^{\bullet} = \{\varphi \mid \mathcal{M} \models_{\Sigma} \varphi\}$. A theory T is *closed* if $T^{\bullet\bullet} = T$, and it is natural to think of the closed theories as the "*concepts*" of $\mathbb{I}$, or following FCA [15], to define *formal concepts* to be pairs $(T, \mathcal{M})$ such that $T^{\bullet} = \mathcal{M}$ and $\mathcal{M}^{\bullet} = T$. Much of FCA carries over to arbitrary institutions. For example, the closed theories form a complete lattice under inclusion which is (anti-) isomorphic to the lattice of closed model sets; this lattice is called the **formal concept lattice** in [15]. The Galois connection also gives many identities connecting the various set theoretic operations with closure, such as $(T \cup T')^{\bullet} = T^{\bullet} \cap T'^{\bullet}$. Example 9 below applies these ideas to an ontology problem, and a more elegant, categorical version of these ideas is given just after that example.

Ontologies are a promising application area for ideas in this section. First, numerous logics are being proposed and used for expressing ontologies, among which description logics, such as OWL, are the most prominent, but by no means

---

[19] The converse ordering is used by Sowa, and seems to have more intuitive appeal, because a larger theory has a smaller collection of models.

the only, examples. Second, a given semantic domain will often have a number of different ontologies. Thus both logical and semantic heterogeneity are quite common, and integration at both levels is an important challenge. Fortunately, category theory provides appropriate tools for this. The "IFF" approach of Robert Kent has pioneered work in this area, integrating FCA and IF in [42], and more recently, using institutions to unify the IEEE Standard Upper Ontology [43], which defines a set of very high level concepts for use in defining and structuring specific domain ontologies.

Example 9. A simple but suggestive example, originally from [61], but elaborated in [60] and further elaborated here, illustrates applications to the problems of "ontology alignment" and "ontology merging," where the former refers to how concepts relate, and the latter refers to creating a joint ontology. The informal semantics that underlies this example is explained in the following quote from [61]:

> In English, size is the feature that distinguishes "river" from "stream"; in French, a "fleuve" is a river that flows into the sea, and a "riviére" is a river or a stream that runs into another river.

We can now set up the problem as follows: there are two (linguistic) contexts, French and English, each of which has two concepts; let us also suppose that each context has three instances, as summarized in the two classification relations shown in the following tables.

English	river	stream
Mississippi	1	0
Ohio	1	0
Captina	0	1

French	fleuve	riviére
Rhône	1	0
Saône	0	1
Roubion	0	1

Although these tables are insufficient to recover the informal relations between concepts given in the quotation above, if we first align the instances, a surprising amount of insight can be obtained. So let us further suppose it is known that the corresponding rows of the two tables have the same conceptual properties, e.g., that the Mississippi is a fleuve but not a riviére, that the Saône is a river but not a stream, etc.

One might expect that this correspondence of rows should induce some kind of a blend, and indeed, we will formalize the example in such a way that a blend is obtained in our formal sense. We first describe the very simple logic involved. Its signatures are sets of unary predicates. Its sentences over a signature Σ are the nullary predicate *false*, and the unary predicates in Σ. Signature morphisms are maps of the unary predicate names. A model for Σ is a one point set, with some subset of predicates in Σ designated as being satisfied; the satisfaction relation for this model is then given by this designation. Note that for this logic, $Th(\Sigma)$ has as its objects the set of all sets of Σ-sentences.

The English signature Σ^E in this example is just $\{river, stream\}$ and the French signature is $\Sigma^F = \{fleuve, riviére\}$, while the blend signature Σ^B is their union. The two tables above can be seen as defining 6 models, and also as defining the two satisfaction relations, or better, two frames.

It is possible to recover some interesting relationships among the concepts represented by the predicates if we extend the satisfaction relation to sets of sentences and sets of models as indicated just after Definition 7. Since the entities corresponding to the three rows of the two tables have the same properties, we can merge them when we blend the signatures. If we denote these merged entities by MR, OS and CR, then the merged models and their satisfaction relation are described by the following:

	river	stream	fleuve	riviére
MR	1	0	1	0
OS	1	0	0	1
CR	0	1	0	1

Figure 5 shows the formal concept lattice of this merged context over the merged signature Σ^B, with its nodes labeled by the model sets and theories to which they correspond. It is interesting to notice that any minimal set of generators for this lattice gives a canonical formal vocabulary for classifying models. Although in this case, there are $2^3 = 8$ possible sets of models, only 7 appear in the concept lattice and a subset of 3 are sufficient to generate the lattice. (The reason there are only 7 elements is that there is no model that both is small and flows into the sea.)

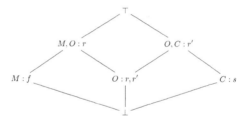

Fig. 5. A formal concept lattice

A more sophisticated institutional view is that we have blended the English and French contexts into the single context of the final table, by a pushout construction (as in Figure 3) in the category of frames, where the contravariant nature of the model part of the institution means that the frame pushout does a pullback on models, giving the identifications (or "alignment") that we made, whereas the covariant nature of the signature and sentence components gives a union. The Buddhist monk example can be understood in the same way.

Given an institution I, we define another institution I^G, the **Galoisification** of I, as follows: its signature category is the same as that of I, its sentences are the closed theories of I, its models are the closed model classes of I, and its satisfaction relation is the natural extension of that of I. We may call a frame of I^G a **Galois frame**, or **G-frame** for short. Then the satisfaction relation of each G-frame is a bijective function, the corresponding pairs of which are the formal concepts of [15] in the special case of their logic, though this construction applies to any institution.

The category $Th(\mathbb{I})$ is also an example of a very general construction, called *Grothendieck flattening*; see [26] for the definition and many examples, as well as many more details about information flow and channel algebra over an arbitrary institution. Grothendieck flattening supports *semantic heterogeneity*, i.e., working in several different contexts at the same time, by defining new morphisms, called *heteromorphisms*, between objects from different categories; this gives rise to the **Grothendieck category**, of which $Th(\mathbb{I})$ is an example. A number of useful general theorems are known about Grothendieck categories, for example, conditions for limits and colimits to exist.

Colimits are not an adequate formalization of blending in the sense of cognitive linguistics, because more than one blend is possible for two input spaces (e.g., "houseboat" and "boathouse"). This raises the mathematical challenge of weakening colimits so that non-multiple isomorphic solutions are allowed. Another challenge is to discover precise "optimality principles" that measure the quality of the blend space. These problems are addressed in *algebraic semiotics* [23, 24], which is a general theory of representation, based on the semiotic spaces discussed in Section 3) and *semiotic morphisms*, which (partially) preserve the structure of semiotic spaces, and which model representations, such as an index to a book, a graph of a dataset, or a GUI for UNIX. A basic principle in [23] is that the quality of a representation depends on how well its semiotic morphism preserves structure, and [23] proposes a number of quality measures on that basis; [23] also introduces 3/2-category theory, including 3/2-colimits as a model of blending, under the hypothesis that optimal blends are achieved by using semiotic morphisms that score well on the quality measures.

This theory is tested against some simple examples in [23], and has also been implemented in a *blending algorithm* called "Alloy" [30, 29], which generates novel metaphors on the fly as part of a poetry generation system (called "Griot") developed by Fox Harrell. The optimality principles used are purely formal (since they could not otherwise be implemented); they measure the degree to which the injection morphisms $I_i \to B$ in Figure 3 preserve structure, including constants, relation instances, and types[20].

6   Reconciliations and Conclusions

This paper considers cognitive, social, pragmatic, and mathematical perspectives on concepts. Despite this diversity, there has been a single main goal, which is to facilitate the design and implementation of systems to support information

---

[20] An interesting sidelight is that recent poetry (e.g., Neruda and Rilke) often requires what we call "disoptimality principles" to explain certain especially striking metaphors [29]. For example, type coercions are needed when items of very different types are blended.

integration. Good mathematical foundations are of course a great help, in providing precise descriptions of what should be implemented. But without some understanding of broader issues, it is impossible to understand the potential limitations of such systems. In addition, models from cognitive psychology and sociology can inspire new approaches to information integration. So the huge gaps that currently exist between disciplines are really counter-productive. This section is devoted to sketching various forms of reconciliation among them, along with some general conclusions.

It can be argued that some concepts (scientific concepts) have a hard physical reality, manifesting as perceivable regularities of behavior, or in a more sophisticated language, as invariants over perceptions, e.g., in James Gibson's "ecological" approach to perception [18]. It can also be argued that other concepts (mathematical) are formal transcendental, existing independently of humans and even physical reality. But as realized long ago by the Indian philosopher Nagarjuna [55], phenomenological human concepts are not like that: they are elastic, situated, evolving, relative, pragmatic, fuzzy, and strongly interconnected in domains with other concepts; the thoughts we actually have cannot be pinned down as scientific or mathematical concepts. Contemporary cognitive science has explored many reasons for this, and Section 4 suggests that there are also many social reasons.

There are several views on how phenomenological, social and scientific approaches to concepts can be reconciled; in general, they argue that the phenomenological and the social are interdependent, in that concepts necessarily exist at both levels, and that scientific and mathematical concepts are not essentially different from other concepts, though it is convenient to use these more precise languages to construct models, without needing to achieve perfect fidelity. For example, Gärdenfors [16] argues that formal entities, such as equations, "are not elements of the internal cognitive process, but function as *external devices* that are, literally, manipulated"; this is similar to the material anchors of Hutchins [40], which represent shared concepts in concrete form to facilitate cooperative work. Terrence Deacon [10] argues that concepts evolve in social environments in much the same way that organisms evolve in natural environments, as part of an ambitious program that (among other things) seeks a biological reconstruction of Richard Dawkins' implausible reification of concepts as "memes".

In a brilliant critique of modernism, Bruno Latour [48] proposes to reconcile society, science, and the myriad hybrid "quasi-objects" that have aspects of both (among which he apparently includes natural language) through a "symmetry principle" that refuses to recognize the "modern" distinctions among these areas, and that grounds explanations in the quasi-objects, which inherently mix cognitive, social, and formal aspects. Latour rightly criticizes the Saussurian dyadic semiotics that dominated French cultural thought for a time, but seems unaware of Peirce's [56] triadic semiotics, which I believe offers a better solution, because its signs are already defined to be hybrid "quasi-objects"; note that one of Peirce's goals for his triadicity was to reconcile nominalism and realism [41], an enterprise with a flavor similar to that of Latour's. A Peircean perspective is also the basis of Deacon's theories [10] concerning how concepts evolve within social contexts.

Ideas in this paper have implications for the study of formal aspects of concepts, including knowledge representation, analysis and integration. Section 3 unified two notions of conceptual space, due to Fauconnier and to Gärdenfors using frames. The Unified Concept Theory of Section 5 generalized this, as well as LOT, FCA, IF and CI, to arbitrary logics by using institutions. Institutions can be considered to formalize Peirce's triadic semiotics, as well as Latour's quasi-objects. It could be interesting to apply UCT to description logics [1], which are commonly used for ontologies, or alternatively, as argued in Section 3, to an order sorted algebra approach to ontologies. I believe there are also fruitful applications to database systems, e.g., the problem of integrating information from multiple databases; see [25]. These examples illustrate how the exploration of ways to reconcile cognitive, social, pragmatic, and formal approaches to concepts can be useful in suggesting new research directions.

Perhaps the most important conclusion is that research on concepts should be thoroughly interdisciplinary, and in particular, should transcend the boundaries between sciences and humanities. Unfortunately, such efforts, including those of this paper, are likely to attract criticism for blurring distinctions between established disciplines, which indeed often operate under incompatible assumptions, using incomparable methods. It is my hope that the reconciliations and unifications sketched above may contribute to the demise of such obstructions, as well as to a better understanding of concepts and their applications.

References

1. Franz Baader, Diego Calvanese, Deborah McGuinness, Daniele Nardi, and Peter Patel-Schneider, editors. *Description Logic Handbook*. Cambridge, 2003.
2. Michael Barr. ⋆-autonomous categories and linear logic. *Mathematical Structures in Computer Science*, 1:159–178, 1991.
3. Jon Barwise and Jerry Seligman. *Information Flow: Logic of Distributed Systems*. Cambridge, 1997. Tracts in Theoretical Computer Science, vol. 44.
4. John Lane Bell. *Toposes and Local Set Theories: An Introduction*. Oxford, 1988. Oxford Logic Guides 14.
5. David Bloor. *Knowledge and Social Imagery*. Chicago, 1991. Second edition.
6. Herbert Blumer. *Symbolic Interactionism: Perspective and Method*. California, 1986.
7. Rod Burstall and Joseph Goguen. Putting theories together to make specifications. In Raj Reddy, editor, *Proceedings, Fifth International Joint Conference on Artificial Intelligence*, pages 1045–1058. Department of Computer Science, Carnegie-Mellon University, 1977.
8. Manuel Clavel, Steven Eker, Patrick Lincoln, and José Meseguer. Principles of Maude. In José Meseguer, editor, *Proceedings, First International Workshop on Rewriting Logic and its Applications*. Elsevier Science, 1996. Volume 4, *Electronic Notes in Theoretical Computer Science*.
9. Michael Cole. *Cultural Psychology: A once and future discipline*. Harvard, 1996.

10. Terrence Deacon. Memes as signs in the dynamic logic of semiosis: Beyond molecular science and computation theory. In Karl Wolff, Heather Pfeiffer, and Harry Delugach, editors, *Conceptual Structures at Work: 12th International Conference on Conceptual Structures*, pages 17–30. Springer, 2004. Lecture Notes in Computer Science, vol. 3127.

11. Răzvan Diaconescu and Kokichi Futatsugi. *CafeOBJ Report: The Language, Proof Techniques, and Methodologies for Object-Oriented Algebraic Specification*. World Scientific, 1998. AMAST Series in Computing, Volume 6.

12. Samuel Eilenberg and Saunders Mac Lane. General theory of natural equivalences. *Transactions of the American Mathematical Society*, 58:231–294, 1945.

13. Gilles Fauconnier. *Mental Spaces: Aspects of Meaning Construction in Natural Language*. Bradford: MIT, 1985.

14. Gilles Fauconnier and Mark Turner. *The Way We Think*. Basic, 2002.

15. Bernhard Ganter and Rudolf Wille. *Formal Concept Analysis: Mathematical Foundations*. Springer, 1997.

16. Peter Gärdenfors. *Conceptual Spaces: The Geometry of Thought*. Bradford, 2000.

17. Harold Garfinkel. *Studies in Ethnomethodology*. Prentice-Hall, 1967.

18. James Gibson. *An Ecological Approach to Visual Perception*. Houghton Mifflin, 1979.

19. Barney Glaser and Ansolm Strauss. *The Discovery of Grounded Theory: Strategies for qualitative research*. Aldine de Gruyter, 1999.

20. Joseph Goguen. The logic of inexact concepts. *Synthese*, 19:325–373, 1969.

21. Joseph Goguen. A categorical manifesto. *Mathematical Structures in Computer Science*, 1(1):49–67, March 1991.

22. Joseph Goguen. Towards a social, ethical theory of information. In Geoffrey Bowker, Leigh Star, William Turner, and Les Gasser, editors, *Social Science, Technical Systems and Cooperative Work: Beyond the Great Divide*, pages 27–56. Erlbaum, 1997.

23. Joseph Goguen. An introduction to algebraic semiotics, with applications to user interface design. In Chrystopher Nehaniv, editor, *Computation for Metaphors, Analogy and Agents*, pages 242–291. Springer, 1999. Lecture Notes in Artificial Intelligence, Volume 1562.

24. Joseph Goguen. Semiotic morphisms, representations, and blending for interface design. In *Proceedings, AMAST Workshop on Algebraic Methods in Language Processing*, pages 1–15. AMAST Press, 2003.

25. Joseph Goguen. Data, schema and ontology integration. In *Proceedings, Workshop on Combinination of Logics*, pages 21–31. Center for Logic and Computation, Instituto Superior Tecnico, Lisbon, Portugal, 2004.

26. Joseph Goguen. Information integration in instutions. In Lawrence Moss, editor, *Memorial volume for Jon Barwise*. Indiana, to appear.

27. Joseph Goguen and Rod Burstall. A study in the foundations of programming methodology: Specifications, institutions, charters and parchments. In David Pitt, Samson Abramsky, Axel Poigné, and David Rydeheard, editors, *Proceedings, Conference on Category Theory and Computer Programming*, pages 313–333. Springer, 1986. Lecture Notes in Computer Science, Volume 240; also, Report CSLI-86-54, Center for the Study of Language and Information, Stanford University, June 1986.

28. Joseph Goguen and Rod Burstall. Institutions: Abstract model theory for specification and programming. *Journal of the Association for Computing Machinery*, 39(1):95–146, January 1992.

29. Joseph Goguen and Fox Harrell. Style as a choice of blending principles. In Shlomo Argamon, Shlomo Dubnov, and Julie Jupp, editors, *Style and Meaning in Language, Art Music and Design*, pages 49–56. AAAI Press, 2004.

30. Joseph Goguen and Fox Harrell. Foundations for active multimedia narrative: Semiotic spaces and structural blending, 2005. To appear in *Interaction Studies: Social Behaviour and Communication in Biological and Artificial Systems*.

31. Joseph Goguen and Kai Lin. Behavioral verification of distributed concurrent systems with BOBJ. In Hans-Dieter Ehrich and T.H. Tse, editors, *Proceedings, Conference on Quality Software*, pages 216–235. IEEE Press, 2003.

32. Joseph Goguen and José Meseguer. Order-sorted algebra I: Equational deduction for multiple inheritance, overloading, exceptions and partial operations. *Theoretical Computer Science*, 105(2):217–273, 1992. Drafts exist from as early as 1985.

33. Joseph Goguen and Grigore Roşu. Institution morphisms. *Formal Aspects of Computing*, 13:274–307, 2002.

34. Joseph Goguen, James Thatcher, Eric Wagner, and Jesse Wright. Initial algebra semantics and continuous algebras. *Journal of the Association for Computing Machinery*, 24(1):68–95, January 1977.

35. Joseph Goguen and William Tracz. An implementation-oriented semantics for module composition. In Gary Leavens and Murali Sitaraman, editors, *Foundations of Component-based Systems*, pages 231–263. Cambridge, 2000.

36. Joseph Goguen, Timothy Winkler, José Meseguer, Kokichi Futatsugi, and Jean-Pierre Jouannaud. Introducing OBJ. In Joseph Goguen and Grant Malcolm, editors, *Software Engineering with OBJ: Algebraic Specification in Action*, pages 3–167. Kluwer, 2000.

37. Rebecca Green. Internally-structured conceptual models in cognitive semantics. In Rebecca Green, Carol Bean, and Sung Hyon Myaeng, editors, *The Semantics of Relationships*, pages 73–90. Kluwer, 2002.

38. Stevan Harnad. The symbol grounding problem. *Physica D*, 42:335–346, 1990.

39. William S. Hatcher. *Foundations of Mathematics*. W.B. Saunders, 1968.

40. Edwin Hutchins. *Cognition in the Wild*. MIT, 1995.

41. Mary Keeler. Hegel in a strange costume. In Aldo de Moor, Wilfried Lex, and Bernhard Ganter, editors, *Conceptual Structures for Knowledge Creation and Communication*, pages 37–53. Springer, 2003. Lecture Notes in Computer Science, vol. 2746.

42. Robert Kent. Distributed conceptual structures. In Harre de Swart, editor, *Sixth International Workshop on Relational Methods in Computer Science*, pages 104–123. Springer, 2002. Lecture Notes in Computer Science, volume 2561.

43. Robert Kent. Formal or axiomatic semantics in the IFF, 2003. Available at suo.ieee.org/IFF/work-in-progress/.

44. Thomas Kuhn. *The Structure of Scientific Revolutions*. Chicago, 1962.

45. William Labov. *Language in the Inner City*. University of Pennsylvania, 1972.

46. George Lakoff. *Women, Fire and Other Dangerous Things: What categories reveal about the mind*. Chicago, 1987.

47. George Lakoff and Mark Johnson. *Philosophy in the Flesh: The Embodied Mind and its Challenge to Western Thought*. Basic, 1999.

48. Bruno Latour. *We Have Never Been Modern*. Harvard, 1993. Translated by Catherine Porter.

49. Bruno Latour and Steve Woolgar. *Laboratory Life*. Sage, 1979.

50. Eric Livingston. *The Ethnomethodology of Mathematics*. Routledge & Kegan Paul, 1987.

51. Saunders Mac Lane. *Categories for the Working Mathematician.* Springer, 1998. Second Edition.
52. Donald MacKenzie. *Mechanizing Proof.* MIT, 2001.
53. Till Mossakowski, Joseph Goguen, Razvan Diaconescu, and Andrzej Tarlecki. What is a logic? In Jean-Yves Beziau, editor, *Logica Universalis.* Birkhauser, 2005. *Proceedings*, First World Conference on Universal Logic.
54. Peter Mosses, editor. *CASL Reference Manual.* Springer, 2004. Lecture Notes in Computer Science, Volume 2960.
55. Nagarjuna. *Mulamadhyamika Karaka.* Oxford, 1995. Translated by Jay Garfield.
56. Charles Saunders Peirce. *Collected Papers.* Harvard, 1965. In 6 volumes; see especially Volume 2: Elements of Logic.
57. Benjamin C. Pierce. *Basic Category Theory for Computer Scientists.* MIT, 1991.
58. Simone Pribbenow. Merenymic relationships: From classical mereology to complex part-whole relations. In Rebecca Green, Carol Bean, and Sung Hyon Myaeng, editors, *The Semantics of Relationships*, pages 35–50. Kluwer, 2002.
59. Harvey Sacks. On the analyzability of stories by children. In John Gumpertz and Del Hymes, editors, *Directions in Sociolinguistics*, pages 325–345. Holt, Rinehart and Winston, 1972.
60. Marco Schlorlemmer and Yannis Kalfoglou. A channel-theoretic foundation for ontology coordination. In *Proceedings, 18th European Workshop on Multi-Agent Systems.* 2004.
61. John Sowa. *Knowledge Representation: Logical, Philosophical and Computational Foundations.* Brooks/Coles, 2000.
62. Susan Leigh Star. The structure of ill-structured solutions: Boundary objects and heterogeneous problem-solving. In Les Gasser and Michael Huhns, editors, *Distributed Artificial Intelligence*, volume 2, pages 37–54. Pitman, 1989.
63. Andrzej Tarlecki, Rod Burstall, and Joseph Goguen. Some fundamental algebraic tools for the semantics of computation, part 3: Indexed categories. *Theoretical Computer Science*, 91:239–264, 1991. Also, Monograph PRG–77, August 1989, Programming Research Group, Oxford University.
64. Lev Vygotsky. *Thought and Language.* MIT, 1962.
65. Lev Vygotsky. *Mind in Society.* Harvard, 1985.

Applications of Description Logics: State of the Art and Research Challenges

Ian Horrocks

School of Computer Science, University of Manchester
Oxford Road, Manchester M13 9PL, UK
horrocks@cs.man.ac.uk

Abstract. Description Logics (DLs) are a family of class based knowledge representation formalisms characterised by the use of various constructors to build complex classes from simpler ones, and by an emphasis on the provision of sound, complete and (empirically) tractable reasoning services. They have a range of applications, but are mostly widely known as the basis for ontology languages such as OWL. The increasing use of DL based ontologies in areas such as e-Science and the Semantic Web is, however, already stretching the capabilities of existing DL systems, and brings with it a range of challenges for future research.

1 Introduction

Description Logics (DLs) are a family of class (concept) based knowledge representation formalisms. They are characterised by the use of various constructors to build complex concepts from simpler ones, an emphasis on the decidability of key reasoning tasks, and by the provision of sound, complete and (empirically) tractable reasoning services.

Description logics have been used in a range of applications, e.g., configuration [1], and reasoning with database schemas and queries [2,3,4]. They are, however, best known as the basis for ontology languages such as OIL, DAML+OIL and OWL [5]. As well as DLs providing the formal underpinnings for these languages (i.e., a declarative semantics), DL systems are widely used to provide computational services for a rapidly expanding range of ontology tools and applications [6,7,8,9,10,11].

Ontologies, and ontology based vocabularies, are used to provide a common vocabulary together with computer-accessible *descriptions* of the *meaning* of relevant terms and relationships between these terms. Ontologies play a major role in the Semantic Web [12,13], and are widely used in, e.g., knowledge management systems, e-Science, and bio-informatics and medical terminologies [14,15,16,17]. They are also of increasing importance in the Grid, where they may be used, e.g., to support the discovery, execution and monitoring of Grid services [18,19,20].

The success of the current generation of DLs and DL reasoning brings with it, however, requirements for reasoning support which may be beyond the capability of existing systems. These requirements include greater expressive power,

F. Dau, M.-L. Mugnier, G. Stumme (Eds.): ICCS 2005, LNAI 3596, pp. 78–90, 2005.

improved scalability and extended reasoning services. Satisfying these requirements presents a major research challenge, not only to the DL community, but
to the logic based Knowledge Representation community as a whole.

2   Ontology Languages and Description Logics

The OWL recommendation actually consists of three languages of increasing
expressive power: OWL Lite, OWL DL and OWL Full. Like OWL's predecessor
DAML+OIL, OWL Lite and OWL DL are basically very expressive description
logics with an RDF syntax. OWL Full provides a more complete integration
with RDF, but its formal properties are less well understood, and key inference
problems would certainly be *much* harder to compute.[1] For these reasons, OWL
Full will not be considered in this paper.

More precisely, OWL DL is closely related to the well known $\mathcal{SHIQ}$ DL [21];
it restricts the form of $\mathcal{SHIQ}$ number restrictions to be unqualified (see [22]),
and extends $\mathcal{SHIQ}$ with *nominals* [23] (i.e., concepts having exactly one instance) and *datatypes* (often called concrete domains in DLs [24]). Following the
usual DL naming conventions, the resulting logic is called $\mathcal{SHOIN}(\mathbf{D})$ (where
$\mathcal{O}$ stands for nominals, $\mathcal{N}$ stands for unqualified number restrictions and $(\mathbf{D})$
stands for datatypes). OWL Lite is equivalent to the slightly simpler $\mathcal{SHIF}(\mathbf{D})$
DL. These equivalences allow OWL to exploit the considerable existing body of
description logic research, e.g.:

- to define the semantics of the language and to understand its formal properties, in particular the decidability and complexity of key inference problems [25];
- as a source of sound and complete algorithms and optimised implementation
  techniques for deciding key inference problems [21,26];
- to use implemented DL systems in order to provide (partial) reasoning support [27,28,29].

2.1 $\mathcal{SHOIN}$ Syntax and Semantics

The syntax and semantics of $\mathcal{SHOIN}$ are briefly introduced here (we will ignore
datatypes, as adding a datatype component would complicate the presentation
and has little affect on reasoning [30]).

Definition 1. *Let* $\mathbf{R}$ *be a set of* role names *with both transitive and normal role
names* $\mathbf{R}_+ \cup \mathbf{R}_\mathsf{P} = \mathbf{R}$, *where* $\mathbf{R}_\mathsf{P} \cap \mathbf{R}_+ = \emptyset$. *The set of* $\mathcal{SHOIN}$-*roles (or* roles
for short) is $\mathbf{R} \cup \{R^- \mid R \in \mathbf{R}\}$. *A* role inclusion axiom *is of the form* $R \sqsubseteq S$,
for two roles R *and* S. *A* role hierarchy *is a finite set of role inclusion axioms.*

---

[1] Inference in OWL Full is clearly undecidable as OWL Full does not include restrictions on the use of transitive properties which are required in order to maintain
   decidability [21].

An interpretation $\mathcal{I} = (\Delta^{\mathcal{I}}, \cdot^{\mathcal{I}})$ *consists of a non-empty set* $\Delta^{\mathcal{I}}$, *called the domain of* $\mathcal{I}$, *and a function* $\cdot^{\mathcal{I}}$ *which maps every role to a subset of* $\Delta^{\mathcal{I}} \times \Delta^{\mathcal{I}}$ *such that, for* $P \in \mathbf{R}$ *and* $R \in \mathbf{R}_+$,

$$\langle x, y \rangle \in P^{\mathcal{I}} \text{ iff } \langle y, x \rangle \in P^{-\mathcal{I}},$$
$$\text{and if } \langle x, y \rangle \in R^{\mathcal{I}} \text{ and } \langle y, z \rangle \in R^{\mathcal{I}}, \text{ then } \langle x, z \rangle \in R^{\mathcal{I}}.$$

An interpretation $\mathcal{I}$ *satisfies a role hierarchy* $\mathcal{R}$ *iff* $R^{\mathcal{I}} \subseteq S^{\mathcal{I}}$ *for each* $R \sqsubseteq S \in \mathcal{R}$; *such an interpretation is called a* model *of* $\mathcal{R}$.

Definition 2. *Let* N_C *be a set of* concept names *with a subset* $N_I \subseteq N_C$ *of* nominals. *The set of* $\mathcal{SHOIN}$-*concepts (or concepts for short) is the smallest set such that*

1. *every concept name* $C \in N_C$ *is a concept,*
2. *if* C *and* D *are concepts and* R *is a role, then* $(C \sqcap D)$, $(C \sqcup D)$, $(\neg C)$, $(\forall R.C)$, *and* $(\exists R.C)$ *are also concepts (the last two are called universal and existential restrictions, resp.), and*
3. *if* R *is a simple role[2] and* $n \in \mathbb{N}$, *then* $\leqslant nR$ *and* $\geqslant nR$ *are also concepts (called atmost and atleast number restrictions).*

The interpretation function $\cdot^{\mathcal{I}}$ *of an interpretation* $\mathcal{I} = (\Delta^{\mathcal{I}}, \cdot^{\mathcal{I}})$ *maps, additionally, every concept to a subset of* $\Delta^{\mathcal{I}}$ *such that*

$$(C \sqcap D)^{\mathcal{I}} = C^{\mathcal{I}} \cap D^{\mathcal{I}}, \qquad (C \sqcup D)^{\mathcal{I}} = C^{\mathcal{I}} \cup D^{\mathcal{I}}, \qquad \neg C^{\mathcal{I}} = \Delta^{\mathcal{I}} \setminus C^{\mathcal{I}},$$
$$\sharp o^{\mathcal{I}} = 1 \qquad \text{for all } o \in N_I,$$
$$(\exists R.C)^{\mathcal{I}} = \{x \in \Delta^{\mathcal{I}} \mid \text{There is a } y \in \Delta^{\mathcal{I}} \text{ with } \langle x, y \rangle \in R^{\mathcal{I}} \text{ and } y \in C^{\mathcal{I}}\},$$
$$(\forall R.C)^{\mathcal{I}} = \{x \in \Delta^{\mathcal{I}} \mid \text{For all } y \in \Delta^{\mathcal{I}}, \text{ if } \langle x, y \rangle \in R^{\mathcal{I}}, \text{ then } y \in C^{\mathcal{I}}\},$$
$$\leqslant nR^{\mathcal{I}} = \{x \in \Delta^{\mathcal{I}} \mid \sharp\{y \mid \langle x, y \rangle \in R^{\mathcal{I}}\} \leqslant n\},$$
$$\geqslant nR^{\mathcal{I}} = \{x \in \Delta^{\mathcal{I}} \mid \sharp\{y \mid \langle x, y \rangle \in R^{\mathcal{I}}\} \geqslant n\},$$

where, for a set M, *we denote the cardinality of* M *by* $\sharp M$.

For C *and* D *(possibly complex) concepts,* $C \sqsubseteq D$ *is called a* general concept inclusion *(GCI), and a finite set of GCIs is called a* TBox.

An interpretation $\mathcal{I}$ *satisfies a GCI* $C \sqsubseteq D$ *if* $C^{\mathcal{I}} \subseteq D^{\mathcal{I}}$, *and* $\mathcal{I}$ *satisfies a TBox* $\mathcal{T}$ *if* $\mathcal{I}$ *satisfies each GCI in* $\mathcal{T}$; *such an interpretation is called a* model *of* $\mathcal{T}$.

A concept C *is called* satisfiable *with respect to a role hierarchy* $\mathcal{R}$ *and a TBox* $\mathcal{T}$ *if there is a model* $\mathcal{I}$ *of* $\mathcal{R}$ *and* $\mathcal{T}$ *with* $C^{\mathcal{I}} \neq \emptyset$. *Such an interpretation is called a* model *of* C *w.r.t.* $\mathcal{R}$ *and* $\mathcal{T}$. *A concept* D *subsumes a concept* C *w.r.t.* $\mathcal{R}$ *and* $\mathcal{T}$ *(written* $C \sqsubseteq_{\mathcal{R},\mathcal{T}} D$) *if* $C^{\mathcal{I}} \subseteq D^{\mathcal{I}}$ *holds in every model* $\mathcal{I}$ *of* $\mathcal{R}$ *and* $\mathcal{T}$. *Two concepts* C, D *are* equivalent *w.r.t.* $\mathcal{R}$ *and* $\mathcal{T}$ *(written* $C \equiv_{\mathcal{R},\mathcal{T}} D$) *iff they are mutually subsuming w.r.t.* $\mathcal{R}$ *and* $\mathcal{T}$. *(When* $\mathcal{R}$ *and* $\mathcal{T}$ *are obvious from the context, we will often write* $C \sqsubseteq D$ *and* $C \equiv D$.) *For an interpretation* $\mathcal{I}$, *an individual* $x \in \Delta^{\mathcal{I}}$ *is called an* instance *of a concept* C *iff* $x \in C^{\mathcal{I}}$.

---

[2] A role is simple if it is neither transitive nor has any transitive subroles. Restricting number restrictions to simple roles is required in order to yield a decidable logic [21].

Note that, as usual, subsumption and satisfiability can be reduced to each other, and reasoning w.r.t. *general TBoxes* and role hierarchies can be reduced to reasoning w.r.t. role hierarchies only [21,26].

2.2 Practical Reasoning Services

Most modern DL systems use *tableaux* algorithms to test concept satisfiability. These algorithms work by trying to construct (a tree representation of) a model of the concept, starting from an individual instance. Tableaux expansion rules decompose concept expressions, add new individuals (e.g., as required by $\exists R.C$ terms),[3] and merge existing individuals (e.g., as required by $\leqslant nR.C$ terms). Nondeterminism (e.g., resulting from the expansion of disjunctions) is dealt with by searching the various possible models. For an unsatisfiable concept, all possible expansions will lead to the discovery of an obvious contradiction known as a *clash* (e.g., an individual that must be an instance of both A and $\neg A$ for some concept A); for a satisfiable concept, a complete and clash-free model will be constructed [31].

Tableaux algorithms have many advantages. It is relatively easy to design provably sound, complete and terminating algorithms, and the basic technique can be extended to deal with a wide range of class and role constructors. Moreover, although many algorithms have a higher worst case complexity than that of the underlying problem, they are usually quite efficient at solving the relatively easy problems that are typical of realistic applications.

Even in realistic applications, however, problems can occur that are much too hard to be solved by naive implementations of theoretical algorithms. Modern DL systems, therefore, include a wide range of optimisation techniques, the use of which has been shown to improve typical case performance by several orders of magnitude [32,33,34,29,35,36]. Key techniques include lazy unfolding, absorption and dependency directed backtracking.

Lazy Unfolding In an ontology, or DL Tbox, large and complex concepts are seldom described monolithically, but are built up from a hierarchy of named concepts whose descriptions are less complex. The tableaux algorithm can take advantage of this structure by trying to find contradictions between concept names before adding expressions derived from Tbox axioms. This strategy is known as *lazy unfolding* [32,34].

The benefits of lazy unfolding can be maximised by lexically *normalising* and *naming* all concept expressions and, recursively, their sub-expressions. An expression C is normalised by rewriting it in a standard form (e.g., disjunctions are rewritten as negated conjunctions); it is named by substituting it with a new concept name A, and adding an axiom $A \equiv C$ to the Tbox. The normalisation step allows lexically equivalent expressions to be recognised and identically named, and can even detect syntactically "obvious" satisfiability and unsatisfiability.

---

[3] Cycle detection techniques known as *blocking* may be required in order to guarantee termination.

Absorption Not all axioms are amenable to lazy unfolding. In particular, so called *general concept inclusions* (GCIs), axioms of the form $C \sqsubseteq D$ where C is non-atomic, must be dealt with by explicitly making every individual in the model an instance of $D \sqcup \neg C$. Large numbers of GCIs result in a very high degree of non-determinism and catastrophic performance degradation [34].

Absorption is another rewriting technique that tries to reduce the number of GCIs in the Tbox by absorbing them into axioms of the form $A \sqsubseteq C$, where A is a concept name. The basic idea is that an axiom of the form $A \sqcap D \sqsubseteq D'$ can be rewritten as $A \sqsubseteq D' \sqcup \neg D$ and absorbed into an existing $A \sqsubseteq C$ axiom to give $A \sqsubseteq C \sqcap (D' \sqcup \neg D)$ [37]. Although the disjunction is still present, lazy unfolding ensures that it is only applied to individuals that are already known to be instances of A.

Dependency Directed Backtracking Inherent unsatisfiability concealed in sub-expressions can lead to large amounts of unproductive backtracking search known as thrashing. For example, expanding the expression $(C_1 \sqcup D_1) \sqcap \ldots \sqcap (C_n \sqcup D_n) \sqcap \exists R.(A \sqcap B) \sqcap \forall R.\neg A$ could lead to the fruitless exploration of 2^n possible expansions of $(C_1 \sqcup D_1) \sqcap \ldots \sqcap (C_n \sqcup D_n)$ before the inherent unsatisfiability of $\exists R.(A \sqcap B) \sqcap \forall R.\neg A$ is discovered. This problem is addressed by adapting a form of dependency directed backtracking called *backjumping*, which has been used in solving constraint satisfiability problems [38].

Backjumping works by labelling concepts with a dependency set indicating the non-deterministic expansion choices on which they depend. When a clash is discovered, the dependency sets of the clashing concepts can be used to identify the most recent non-deterministic expansion where an alternative choice might alleviate the cause of the clash. The algorithm can then jump back over intervening non-deterministic expansions *without* exploring any alternative choices. Similar techniques have been used in first order theorem provers, e.g., the "proof condensation" technique employed in the HARP theorem prover [39].

3   Research Challenges for Ontology Reasoning

Ontology based applications will critically depend on the provision of efficient reasoning support: on the one hand, such support is required by applications in order to exploit the semantics captured in ontologies; on the other hand, such support is required by ontology engineers to design and maintain sound, well-balanced ontologies. Experience with a wide range of applications and the development of user-oriented environments has highlighted a number of key requirements that will need to be met by the next generation of DL reasoners if they are to provide the basis for this support:

Greater expressive power For example, in ontologies describing complex physically structured domains such as biology [40] and medicine [41], it is often important to describe aggregation relationships between structures and their component parts, and to assert that certain properties of the component parts transfer to the structure as a whole (a femur with a fractured shaft is a fractured femur) [42]. The importance of this kind of knowledge can be gauged from the fact that various "work-arounds" have been described for use with ontology languages that cannot express it directly [43].

Similarly, in grid and web services applications, it may be necessary to describe composite processes in terms of their component parts, and to express relationships between the properties of the various components and those of the composite process. For example, in a sequential composition of processes it may be useful to express a relationship between the inputs and outputs of the composite and those of the first and last component respectively, as well as relationships between the outputs and inputs of successive components [13].

Improved scalability Practical ontologies may be very large—tens or even hundreds of thousands of classes. Dealing with large-scale ontologies already presents a challenge to the current generation of DL reasoners, in spite of the fact that many existing large-scale ontologies are relatively simple. In the 40,000 concept Gene Ontology (GO), for example, much of the semantics is currently encoded in class names such as "heparin-metabolism"; enriching GO with more complex definitions, e.g., by explicitly modelling the fact that heparin-metabolism is a kind of "metabolism" that "acts-on" the carbohydrate "heparin", would make the semantics more accessible, and would greatly increase the value of GO by enabling new kinds of query such as "what biological processes act on glycosaminoglycan" (heparin is a kind of glycosaminoglycan) [40]. However, adding more complex class definitions can cause the performance of existing reasoners to degrade to the point where it is no longer acceptable to users. Similar problems have been encountered with large medical terminology ontologies, such as the GALEN ontology [41]. As well as using a conceptual model of the domain, many applications will also need to deal with very large volumes of instance data—the GO, for example, is used to annotate millions of individuals, and practitioners want to answer queries that refer both to the ontology and to the relationships between these individuals, e.g., "what DNA binding products interact with insulin receptors". Answering this query requires a reasoner not only to identify individuals that are (perhaps only implicitly) instances of DNA binding products and of insulin receptors, but also to identify which pairs of individuals are (perhaps only implicitly) instances of the interactsWith role. For existing ontology languages it is possible to use DL reasoning to answer such queries, but dealing with the large volume of GO annotated gene product data is far beyond the capabilities of existing DL systems [44]. A requirement to store and query over large numbers of individuals is common in many application areas, e.g., in the Semantic Web, where ontologies are to be used in the annotation of web resources, and where users may want to answer queries such as "which bioinformatics researchers work in a university department where there is also a DL researcher?".

Extended reasoning services The ability to explain unexpected inferences is also crucial in ontology development and would be useful in query answering: when the DL reasoner returns an unexpected answer, such as an unintended sub-class relationship, users often find it difficult to understand (and if necessary fix) the causes of the unexpected inference; expert users may even doubt the validity of such inferences, and lose confidence in the inference system. Moreover, existing DL reasoners provide only a limited range of reasoning services, such as class subsumption and instance retrieval. In practice, users often want to ask questions for which it may only be possible to provide an approximate answer. For example, the user may want to be told "everything" that can be inferred about a class or individual, what it is "reasonable" to assert about a class or individual, or what the difference is between two classes or two individuals.

The following sections highlight interesting work in progress that addresses some of the above problems; they do *not* constitute an exhaustive survey.

3.1   Expressive Power

Some of the above mentioned requirements may be met by tableau algorithms that have recently been developed for DLs that are more expressive than those currently implemented in state-of-the-art reasoners. These algorithms are able to deal with, e.g., nominals (singleton classes) [26], complex role inclusion axioms [45], the use of datavalues as keys [46], the representation of temporal constraints [47], and the integration of reasoning over datatypes and built-in predicates [48]. The increased expressive power of these DLs would satisfy (at least partially) key application requirements, e.g., supporting the description of complex structures and the transfer of properties to and from structures and their component parts.

Some expressive requirements will, however, call for very expressive ontology languages based on (larger fragments of) FOL, where key reasoning problems are no longer decidable in general, e.g., the recently proposed SWRL language [49]. How to provide practical reasoning support for such languages is still an open problem, but encouraging results have already been obtained using a state-of-the art first order theorem prover with special optimisation and tuning designed to help them cope with the large number of axioms found in realistic ontologies [50].

3.2   Scalability

Even for $\mathcal{SHIQ}$, class consistency/subsumption reasoning is ExpTime-complete, and for $\mathcal{SHOIN}$ this jumps to NExpTime-complete [26]. There is encouraging evidence of empirical tractability and scalability for implemented DL systems [34,51], but this is mostly w.r.t. logics that do not include inverse properties (e.g., $\mathcal{SHF}$ [4]). Adding inverse properties makes practical implementations more problematical as several important optimisation techniques become much less

---

[4] $\mathcal{SHF}$ is equivalent to $\mathcal{SHIQ}$ without inverse properties and with only functional properties instead of qualified number restrictions [21].

effective. Work is required in order to develop more highly optimised implementations supporting inverse properties, and to demonstrate that they can scale as well as $\mathcal{SHF}$ implementations. It is also unclear if existing techniques will be able to cope with large numbers of class/property instances [52].

Coping with the large volumes of instance data that will be required by many applications (i.e., millions of individuals) will be extremely challenging, given that existing DL implementations cannot deal with more than (in the order of) a few thousand individuals, even when the relational structure is relatively simple [44]. It seems doubtful that, in the case of instance data, the necessary improvement in performance can be achieved by optimising tableaux based algorithms, which are inherently limited by the need to build and maintain a model of the whole ontology (including all of the instance data).

Several alternative approaches are currently under investigation. One of these involves the use of a hybrid DL-DB architecture in which instance data is stored in a database, and query answering exploits the relatively simple relational structure encountered in typical data sets in order minimise the use of DL reasoning and maximise the use of database operations. A successful prototype of this architecture, the so-called *instance store*, has already been developed [44]. This prototype is, however, only able to deal with data that has *no* relational structure (i.e., in which the instance data does not include any role assertions), and so cannot answer queries involving relationships between individuals. Work is underway to extend the prototype to deal with arbitrary instance data, but it is too early to say if this will be successful.

Another technique that is under investigation is to use reasoning techniques based on the encoding of $\mathcal{SHIQ}$ ontologies in Datalog [53]. On the one hand, theoretical investigations of this technique have revealed that data complexity (i.e., the complexity of answering queries against a fixed ontology and set of instance data) is significantly lower than the complexity of class consistency reasoning (i.e., NP-complete for $\mathcal{SHIQ}$, and even polynomial-time for a slight restriction of $\mathcal{SHIQ}$) [54]; on the other hand, the technique would allow relatively efficient Datalog engines to be used to store and reason with large volumes of instance data. Again, it is still too early to determine if this technique will be useful in practice.

3.3   Extended Reasoning Services

In addition to solving problems of class consistency/subsumption and instance checking, explaining how such inferences are derived may be important, e.g., to help an ontology designer to rectify problems identified by reasoning support, or to explain to a user why an application behaved in an unexpected manner.

Work on developing practical explanation systems is at a relatively early stage, with different approaches still being developed and evaluated. One such technique involves exploiting standard reasoning services to identify a small set of axioms that still support the inference in question, the hope being that presenting a much smaller (than the complete ontology) set of axioms to the user will help them to understand the "cause" of the inference [55]. Another (possibly complementary) technique involves explaining the steps by which the inference was derived, e.g., using a sequence of simple natural deduction style inferences [56,57].

As well as explanation, so-called "non-standard inferences" could also be important in supporting ontology design; these include matching, approximation, and difference computations. Non-standard inferences are the subject of ongoing research [58,59,60,61]; it is still not clear if they can be extended to deal with logics as expressive as those that underpin modern ontology languages, or if they will scale to large applications ontologies.

4   Summary

Description Logics are a family of class based knowledge representation formalisms characterised by the use of various constructors to build complex classes from simpler ones, and by an emphasis on the provision of sound, complete and (empirically) tractable reasoning services. They have been used in a wide range of applications, but perhaps most notably (at least in recent times) in providing a formal basis and reasoning services for (web) ontology languages such as OWL.

The increasing use of DL based ontologies in areas such as e-Science and the Semantic Web is, however, already stretching the capabilities of existing DL systems, and brings with it a range of challenges for future research. The extended ontology languages needed in some applications may demand the use of more expressive DLs, and even for existing languages, providing efficient reasoning services is extremely challenging.

Some applications may even call for ontology languages based on larger fragments of FOL. The development of such languages, and reasoning services to support them, extends these challenges to the whole logic based Knowledge Representation community.

Acknowledgements

I would like to acknowledge the contribution of the many collaborators with whom I have been privileged to work. These included Franz Baader, Sean Bechhofer, Dieter Fensel, Carole Goble, Frank van Harmelen, Carsten Lutz, Alan Rector, Ulrike Sattler, Peter F. Patel-Schneider, Stephan Tobies and Andrei Voronkov.

References

1. McGuinness, D.L., Wright, J.R.: An industrial strength description logic-based configuration platform. IEEE Intelligent Systems (1998) 69–77
2. Calvanese, D., De Giacomo, G., Lenzerini, M., Nardi, D., Rosati, R.: Description logic framework for information integration. In: Proc. of the 6th Int. Conf. on Principles of Knowledge Representation and Reasoning (KR'98). (1998) 2–13

3. Calvanese, D., De Giacomo, G., Lenzerini, M.: On the decidability of query containment under constraints. In: Proc. of the 17th ACM SIGACT SIGMOD SIGART Symp. on Principles of Database Systems (PODS'98). (1998) 149–158
4. Horrocks, I., Tessaris, S., Sattler, U., Tobies, S.: How to decide query containment under constraints using a description logic. In: Proc. of the 7th Int. Workshop on Knowledge Representation meets Databases (KRDB 2000), CEUR (http://ceur-ws.org/) (2000)
5. Horrocks, I., Patel-Schneider, P.F., van Harmelen, F.: From $\mathcal{SHIQ}$ and RDF to OWL: The making of a web ontology language. J. of Web Semantics **1** (2003) 7–26
6. Knublauch, H., Fergerson, R., Noy, N., Musen, M.: The protégé OWL plugin: An open development environment for semantic web applications. In McIlraith, S.A., Plexousakis, D., van Harmelen, F., eds.: Proc. of the 2004 International Semantic Web Conference (ISWC 2004). Number 3298 in Lecture Notes in Computer Science, Springer (2004) 229–243
7. Liebig, T., Noppens, O.: Ontotrack: Combining browsing and editing with reasoning and explaining for OWL Lite ontologies. In McIlraith, S.A., Plexousakis, D., van Harmelen, F., eds.: Proc. of the 2004 International Semantic Web Conference (ISWC 2004). Number 3298 in Lecture Notes in Computer Science, Springer (2004) 244–258
8. Rector, A.L., Nowlan, W.A., Glowinski, A.: Goals for concept representation in the GALEN project. In: Proc. of the 17th Annual Symposium on Computer Applications in Medical Care (SCAMC'93), Washington DC, USA (1993) 414–418
9. Visser, U., Stuckenschmidt, H., Schuster, G., Vögele, T.: Ontologies for geographic information processing. Computers in Geosciences (to appear)
10. Oberle, D., Sabou, M., Richards, D.: An ontology for semantic middleware: extending daml-s beyond web-services. In: Proceedings of ODBASE 2003. (2003)
11. Wroe, C., Goble, C.A., Roberts, A., Greenwood, M.: A suite of DAML+OIL ontologies to describe bioinformatics web services and data. Int. J. of Cooperative Information Systems (2003) Special Issue on Bioinformatics.
12. Berners-Lee, T., Hendler, J., Lassila, O.: The semantic Web. Scientific American **284** (2001) 34–43
13. The DAML Services Coalition: DAML-S: Web service description for the semantic web. In: Proc. of the 2003 International Semantic Web Conference (ISWC 2003). Number 2870 in Lecture Notes in Computer Science, Springer (2003)
14. Uschold, M., King, M., Moralee, S., Zorgios, Y.: The enterprise ontology. Knowledge Engineering Review **13** (1998)
15. Stevens, R., Goble, C., Horrocks, I., Bechhofer, S.: Building a bioinformatics ontology using OIL. IEEE Transactions on Information Technology in Biomedicine **6** (2002) 135–141
16. Rector, A., Horrocks, I.: Experience building a large, re-usable medical ontology using a description logic with transitivity and concept inclusions. In: Proceedings of the Workshop on Ontological Engineering, AAAI Spring Symposium (AAAI'97), AAAI Press, Menlo Park, California (1997)
17. Spackman, K.: Managing clinical terminology hierarchies using algorithmic calculation of subsumption: Experience with SNOMED-RT. J. of the Amer. Med. Informatics Ass. (2000) Fall Symposium Special Issue.
18. Emmen, A.: The grid needs ontologies—onto-what? (2002) http://www.hoise.com/primeur/03/articles/monthly/AE-PR-02-03-7.h% tml.
19. Tuecke, S., Czajkowski, K., Foster, I., Frey, J., Graham, S., Kesselman, C., Vanderbilt, P.: Grid service specification (draft). GWD-I draft , GGF Open Grid Services Infrastructure Working Group (2002) http://www.globalgridforum.org/.

20. Foster, I., Kesselman, C., Nick, J., Tuecke, S.: The physiology of the grid: An open grid services architecture for distributed systems integration (2002) http://www.globus.org/research/papers/ogsa.pdf.
21. Horrocks, I., Sattler, U., Tobies, S.: Practical reasoning for expressive description logics. In Ganzinger, H., McAllester, D., Voronkov, A., eds.: Proc. of the 6th Int. Conf. on Logic for Programming and Automated Reasoning (LPAR'99). Number 1705 in Lecture Notes in Artificial Intelligence, Springer (1999) 161–180
22. Baader, F., Calvanese, D., McGuinness, D., Nardi, D., Patel-Schneider, P.F., eds.: The Description Logic Handbook: Theory, Implementation and Applications. Cambridge University Press (2003)
23. Blackburn, P., Seligman, J.: Hybrid languages. J. of Logic, Language and Information **4** (1995) 251–272
24. Baader, F., Hanschke, P.: A schema for integrating concrete domains into concept languages. In: Proc. of the 12th Int. Joint Conf. on Artificial Intelligence (IJCAI'91). (1991) 452–457
25. Donini, F.M., Lenzerini, M., Nardi, D., Nutt, W.: The complexity of concept languages. Information and Computation **134** (1997) 1–58
26. Horrocks, I., Sattler, U.: Ontology reasoning in the $\mathcal{SHOQ}$(D) description logic. In: Proc. of the 17th Int. Joint Conf. on Artificial Intelligence (IJCAI 2001). (2001) 199–204
27. Horrocks, I.: The FaCT system. In de Swart, H., ed.: Proc. of the 2nd Int. Conf. on Analytic Tableaux and Related Methods (TABLEAUX'98). Volume 1397 of Lecture Notes in Artificial Intelligence., Springer (1998) 307–312
28. Patel-Schneider, P.F.: DLP system description. In: Proc. of the 1998 Description Logic Workshop (DL'98), CEUR Electronic Workshop Proceedings, http://ceur-ws.org/Vol-11/ (1998) 87–89
29. Haarslev, V., Möller, R.: RACER system description. In: Proc. of the Int. Joint Conf. on Automated Reasoning (IJCAR 2001). Volume 2083 of Lecture Notes in Artificial Intelligence., Springer (2001) 701–705
30. Pan, J.Z.: Description Logics: Reasoning Support for the Semantic Web. PhD thesis, University of Manchester (2004)
31. Horrocks, I., Sattler, U., Tobies, S.: Practical reasoning for very expressive description logics. J. of the Interest Group in Pure and Applied Logic **8** (2000) 239–264
32. Baader, F., Franconi, E., Hollunder, B., Nebel, B., Profitlich, H.J.: An empirical analysis of optimization techniques for terminological representation systems or: Making KRIS get a move on. Applied Artificial Intelligence. Special Issue on Knowledge Base Management **4** (1994) 109–132
33. Bresciani, P., Franconi, E., Tessaris, S.: Implementing and testing expressive description logics: Preliminary report. In: Proc. of the 1995 Description Logic Workshop (DL'95). (1995) 131–139
34. Horrocks, I.: Using an expressive description logic: FaCT or fiction? In: Proc. of the 6th Int. Conf. on Principles of Knowledge Representation and Reasoning (KR'98). (1998) 636–647
35. Patel-Schneider, P.F.: DLP. In: Proc. of the 1999 Description Logic Workshop (DL'99), CEUR Electronic Workshop Proceedings, http://ceur-ws.org/Vol-22/ (1999) 9–13
36. Horrocks, I., Patel-Schneider, P.F.: Optimizing description logic subsumption. J. of Logic and Computation **9** (1999) 267–293

37. Horrocks, I., Tobies, S.: Reasoning with axioms: Theory and practice. In: Proc. of the 7th Int. Conf. on Principles of Knowledge Representation and Reasoning (KR 2000). (2000) 285–296
38. Baker, A.B.: Intelligent Backtracking on Constraint Satisfaction Problems: Experimental and Theoretical Results. PhD thesis, University of Oregon (1995)
39. Oppacher, F., Suen, E.: HARP: A tableau-based theorem prover. J. of Automated Reasoning **4** (1988) 69–100
40. Wroe, C., Stevens, R., Goble, C.A., Ashburner, M.: A methodology to migrate the Gene Ontology to a description logic environment using DAML+OIL. In: Proc. of the 8th Pacific Symposium on Biocomputing (PSB). (2003)
41. Rogers, J.E., Roberts, A., Solomon, W.D., van der Haring, E., Wroe, C.J., Zanstra, P.E., Rector, A.L.: GALEN ten years on: Tasks and supporting tools. In: Proc. of MEDINFO2001. (2001) 256–260
42. Rector, A.: Analysis of propagation along transitive roles: Formalisation of the galen experience with medical ontologies. In: Proc. of DL 2002, CEUR (http://ceur-ws.org/) (2002)
43. Schulz, S., Hahn, U.: Parts, locations, and holes - formal reasoning about anatomical structures. In: Proc. of AIME 2001. Volume 2101 of Lecture Notes in Artificial Intelligence., Springer (2001)
44. Horrocks, I., Li, L., Turi, D., Bechhofer, S.: The instance store: DL reasoning with large numbers of individuals. In: Proc. of the 2004 Description Logic Workshop (DL 2004). (2004) 31–40
45. Horrocks, I., Sattler, U.: Decidability of $\mathcal{SHIQ}$ with complex role inclusion axioms. Artificial Intelligence **160** (2004) 79–104
46. Lutz, C., Areces, C., Horrocks, I., Sattler, U.: Keys, nominals, and concrete domains. J. of Artificial Intelligence Research (2004) To Appear.
47. Wolter, F., Zakharyaschev, M.: Temporalizing description logics. In Gabbay, D., de Rijke, M., eds.: Frontiers of Combining Systems II. Studies Press/Wiley (2000) 379–401
48. Pan, J.Z., Horrocks, I.: Extending Datatype Support in Web Ontology Reasoning. In: Proc. of the 2002 Int. Conference on Ontologies, Databases and Applications of SEmantics (ODBASE 2002). Number 2519 in Lecture Notes in Computer Science, Springer (2002) 1067–1081
49. Horrocks, I., Patel-Schneider, P.F., Boley, H., Tabet, S., Grosof, B., Dean, M.: SWRL: A semantic web rule language combining owl and ruleml. W3C Member Submission (2004) Available at http://www.w3.org/Submission/SWRL/.
50. Tsarkov, D., Riazanov, A., Bechhofer, S., Horrocks, I.: Using Vampire to reason with OWL. In McIlraith, S.A., Plexousakis, D., van Harmelen, F., eds.: Proc. of the 2004 International Semantic Web Conference (ISWC 2004). Number 3298 in Lecture Notes in Computer Science, Springer (2004) 471–485
51. Haarslev, V., Möller, R.: High performance reasoning with very large knowledge bases: A practical case study. In: Proc. of the 17th Int. Joint Conf. on Artificial Intelligence (IJCAI 2001). (2001) 161–168
52. Horrocks, I., Sattler, U., Tobies, S.: Reasoning with individuals for the description logic $\mathcal{SHIQ}$. In McAllester, D., ed.: Proc. of the 17th Int. Conf. on Automated Deduction (CADE 2000). Volume 1831 of Lecture Notes in Computer Science., Springer (2000) 482–496
53. Hustadt, U., Motik, B., Sattler, U.: Reducing SHIQ-description logic to disjunctive datalog programs. In: Proc. of the 9th Int. Conf. on Principles of Knowledge Representation and Reasoning (KR 2004). (2004) 152–162

54. Motik, B., Sattler, U., Studer, R.: Query answering for OWL-DL with rules. In: Proc. of the 2004 International Semantic Web Conference (ISWC 2004). (2004) 549–563
55. Schlobach, S., Cornet, R.: Explanation of terminological reason-ing: A preliminary report. In: Proc. of the 2003 Description Logic Workshop (DL 2003). (2003)
56. McGuinness, D.L.: Explaining Reasoning in Description Logics. PhD thesis, Rutgers, The State University of New Jersey (1996)
57. Borgida, A., Franconi, E., Horrocks, I.: Explaining $\mathcal{ALC}$ subsumption. In: Proc. of the 14th Eur. Conf. on Artificial Intelligence (ECAI 2000). (2000)
58. Baader, F., Küsters, R., Borgida, A., McGuinness, D.L.: Matching in description logics. J. of Logic and Computation **9** (1999) 411–447
59. Brandt, S., Turhan, A.Y.: Using non-standard inferences in description logics — what does it buy me? In: Proc. of KI-2001 Workshop on Applications of Description Logics (KIDLWS'01). Volume 44 of CEUR (http://ceur-ws.org/). (2001)
60. Küsters, R.: Non-Standard Inferences in Description Logics. Volume 2100 of Lecture Notes in Artificial Intelligence. Springer Verlag (2001)
61. Brandt, S., Küsters, R., Turhan, A.Y.: Approximation and difference in description logics. In: Proc. of the 8th Int. Conf. on Principles of Knowledge Representation and Reasoning (KR 2002). (2002) 203–214

Methodologies for the Reliable Construction of Ontological Knowledge

Eduard Hovy

Information Sciences Institute, University of Southern California
4676 Admiralty Way, Marina del Rey, CA 90292-6695, USA
hovy@isi.edu

Abstract. This paper addresses the methodology of ontology construction. It identifies five styles of approach to ontologizing (deriving from philosophy, cognitive science, linguistics, AI/computational linguistics, and domain reasoning) and argues that they do not provide the same results. It then provides a more detailed example of one of the approaches.

1 Introduction

After nearly a decade in which statistical techniques made "ontology" a bad word in various computational communities, there are encouraging signs that the pendulum is swinging back. But ontologies will be most readily accepted by their traditional critics only if at least two conditions are met: good methodologies for building and evaluating them are developed, and ontologies prove their utility in real applications. This paper addresses the question of the/a methodology for ontology building, a topic that has received relatively little attention.

Since ontologies have many aspects, Section 2 outlines the aspect of interest for this paper, namely ontology content. Section 3 describes five alternative content construction methodologies that have been adopted in the past. Section 4 provides a generic ontology construction procedure, and Section 5 puts the previous two sections together by illustrating the language-oriented methodology with a more detailed account of how ontology builders might proceed, and what they might produce. Finally, Section 6 expresses the hope that further work on developing this and other ontology construction methodologies can be of use.

2 Ontology Content: Shallow Semantics

The construction and use of ontologies for computational purposes has a long and varied history. In natural language processing (NLP), for example, ontologies were initially seen as the ultimate answer to many problems, but later rejected by almost everyone when it became clear that building adequate ones at the time was impossible. Recently, influential NLP figures are beginning to recognize that a certain type

F. Dau, M.-L. Mugnier, G. Stumme (Eds.): ICCS 2005, LNAI 3596, pp. 91-106, 2005.
© Springer-Verlag Berlin Heidelberg 2005

of semantics—very *shallow* semantics—is probably necessary to help statistical NLP systems overcome the quality performance ceilings many of them seem to have reached. Since statistical systems learn their operations from suitably prepared training material, the argument is generally that the nature and quality of these systems' performance is limited by the nature of the information in the material. Just as you cannot make gold from stone, you cannot obtain semantically adequate machine translation, text summarization, information retrieval, question answering, dialogue management, etc., without giving the system some access to semantics in its training.

A less extreme variant of this history was experienced in several other computational areas. Eventually, they all face the same core problem: semantics is important, but *which semantics*? Turning to the Knowledge Representation (KR) community, whose origin lies in the traditions of mathematics, logic, and philosophy, does not help. It has not yet been able to build large, widely used, general-purpose semantic theories or semantic resources required for practical use at the scale of NLP and similar applications; such semantic data and theories as do exist are almost always limited to small-scale (or toy) applications. KR work has been excellent in developing formalisms for representation, and for investigating the properties and requirements of various classes of deductive systems. But for practical applications such formalisms need *content*; the deductive systems need to work *on* something. A semantic theory of the kind needed to support NLP and other applications requires at least a collection of (unambiguous) semantic symbols, each carrying a clear denotation; a set of rules for composing these symbols, using a set of relations, into non-atomic representations of more-complex meanings; and some method of validating the results of composition, deduction, and other semantic operations.

Because of the complexity involved, content building of this sophistication has mostly occurred at a smaller scale. So despite the fact that the world is certainly complex enough that it is reasonable to expect more than 20,000 individual 'atoms' of meaning to be used as building blocks, very few term collections of this size are more than flat enumerations (for example, Standard Industrial Classification (SIC) codes, lists of geographical entities and locations, lists of chemicals, or pumps, or plants, all with their properties, etc.). Sets of symbols, taxonomized to enable inheritance of information and to support inference, are often called ontologies. But except for CYC [29], large-scale term sets (over 20,000 nodes) tend to contain little reasoning knowledge, providing mostly lexically anchored networks of words; the best-known examples are WordNet [10,35] and its immediate derivates such as EuroWordNet [46] and other languages' WordNets (http://www.globalwordnet.org/), though a few more-distant derivatives such as SENSUS [26] are also available. For composing atomic meaning symbols into more-complex structures, most theories provide relations, and typical sets of relations range from a dozen or so, such as Fillmore's early case roles [11], to maybe fifty by Sowa [43]. But no accepted standard set of relations exists either.

Probably the most troublesome aspect of conceptual 'content' semantics, however, is the near-complete absence of methodological discussion and emerging 'methodology theory' that would provide to the general enterprise of ontology building and relation creation the necessary rigor, systematicity, and eventually, methods for verification that would turn this work from an art to a science. (A notable exception is the work on DOLCE, which makes a good beginning; see www.loa-cnr.it/DOLCE.html and [12].) Without at least some ideas about how to validate semantic resources, both the semantics builder and the eventual semantics user are in trouble. The builder does not know how to ensure that what is built today is consistent, in a deep sense, with what was built yesterday (and indeed, problems of inconsistency have plagued all larger ontology-building efforts since their inception; for example, CYC at one point discarded more than half its content and started anew). The user does not know how to choose between various alternative semantic theories and resources, and is forced to rely on unverifiable claims, the builders' reputation and/or erudition, or subjective preferences.

What to do?

3   Five Methodologies for Ontology Construction

The rest of this paper focuses on the problem of creating and organizing into an ontology a set of terms, primitive or not, with which to define meaning in some domain. I use the word "ontology" quite informally here to denote any set of terms organized hierarchically according to the general property inheritance relation following subclass, but without additional requirements that logical entailment or other inferences be defined or that the terms obey such entailment. That is, "ontology" here includes terminology taxonomies such as WordNet. While it would be nice to adopt such requirements, the reality is that most people build ontologies to support their knowledge representation needs in some practical application, and that they are more concerned about the computational effectiveness and correctness of their application than about the formal completeness, correctness, or consistency of the ontology per se. (Should the system work well even if the ontology is somehow formally deficient, the practical projects I know of would be quite satisfied.) Naturally, completeness, consistency, etc., are ways to ensure that the ontology will not lead to unwelcome surprises in system behavior. But unfortunately, the strictures introduced by these requirements are usually so onerous that they make adopting the requirements tantamount to placing a severe limitation on the eventual scope of the ontology and hence of the whole practical enterprise. That is, especially for NLP applications, many people build their ontologies as relatively simple term taxonomies with some inheritance inference, but do not enforce stricter logical requirements.

When constructing an ontology (or domain model; the terms are used here interchangeably) one can either build everything de novo or one can start with the ontologies of others and combine, prune, and massage them together as needed. In recent work, several ontology building projects have created interfaces to assist with the manual merging of ontologies. Typically, these tools extend ontology building interfaces such as those of Ontolingua [97], Intraspect ([22] http://polaris-md.pepperdine.edu/Overview.html) and the Stanford CML Editor [23] by incorporating one or more variants of name matching and other heuristics plus validation routines that check for consistency of edited results [26,20,38,34].

With regard to the creation or extension of new domain models, some work in manual knowledge acquisition is developing interfaces that assist the knowledge entry worker by continually verifying what is entered, actively eliciting information to complete partial specifications, etc., using strategies modeled after human tutoring procedures [24,5].

However, what is still lacking in ontology construction is a systematic and theoretically motivated methodology that guides the builder and facilitates consistency and accuracy, at all levels. The reason for this lack is evident: today, we still do not have an adequate theory on which to base such a methodology. It is not even clear how one would begin to approach the problem of designing theoretically motivated procedures from a suitably general point of view. Consequently, although many people build ontologies today—see for example the OntoSelect website at http://views.dfki.de/Ontologies/ for over 750 ontologies in various domains—not one of the builders would be able to provide a set of operationalizable tests that could be applied to every concept and relation to inform in which cases his or her choices were wrong.

How then would one approach such a methodology? Well, we can consider what ontology builders actually do—the core operation(s)—and study how they justify their actions. As discussed in [1,45,21], three situations can arise when aligning terms from two ontologies: either the two terms are exactly equivalent, or one term is more general than the other, or the terms are incompatible[1]. As soon as inconsistencies are found, one has to make a choice[2]. It helps if one understands *why* the creators of the source ontologies did what they did. Based on lessons learned from practical experience in merging parts of several ontologies [20] and discussions with numerous individuals, the author has identified five types of motivation, which can be identified with five different research approaches: the *philosophers* [16,15,43]; the *cognitive scientists* [35,10,27,2]; the *linguists* [31,39]; the *Artificial Intelligence reasoners* [30,13], which includes the *computational linguists* [3,36]; and the *domain specialists* (too numerous to list). Each of these types of individuals operates in a distinct way, resolving questions with arguments that appeal to different authorities and patterns of reasoning, and (not unexpectedly) lead to very different results. (It is important to point out right away that none of these are *correct*, compared to the others; the notion of correctness is itself a point of methodological discussion. But within each mode of thought, it is of course possible to be correct or wrong, to be more or less elegant, and to develop a more or less satisfying solution.)

In generating each new (candidate) ontology item, the ontologizer performs an act of creation. Stated as simply as possible, the ontologizer has to decide *whether* to create a term, and if so, *how to place it* with regard to the other existing terms (which constitutes some portion of the act of defining the term, of course). Then follows *additional specification and definition*. This decision process plays out as follows for the five 'personality types' of ontologizer:

---

[1] This is not quite true; concepts can share parts of their meanings; see discussion later, and in [28] and [8]. However, following general practice, in this paper we limit the discussion to 'discrete' ontologies, in which concepts do not overlap.

[2] A reviewer of the paper points out that one does not *have* to make a choice; one can include alternatives, separating them in different namespaces, as long as one is not tied to a specific application. The point here is that when one is tied to an application, then one does have to make this choice, and since one builds ontologies for applications, the problem is prevalent.

Type 1: Abstract feature recombination (the philosophers). The procedure of concept creation by additive feature specification—systematically adding new differentiae—is the historical method of ontologization; interesting examples can be found all the way back to Aristotle. A modern version is provided in [43], who defines several highly abstract features (*Concrete–Abstract*; *Positive–Negative*; etc.) and then more or less mechanically forms combinations of them as separate concepts, using these features as differentiae. Sowa illustrates this procedure by generating the topmost few dozen concepts and arranging them in their combinatory lattice structure, under which he proposes a more traditionally-derived concept taxonomy be arranged. With the DOLCE ontology (www.loa-cnr.it/DOLCE.html, [12] employ so-called identity criteria to determine whether two concepts are the same or not, and how they may differ; these differences help establish the appropriate differentiae. The rigor adopted in this work, especially the use of identity criteria as a driving methodology, is a model for others. In general, the ontologizers adopting this approach are of course the philosophers; once they believe they have found the essential set of semantic primitives, the rest follows by logic. The approach is elegant, but unfortunately doesn't work beyond the very most abstract levels, and is hence not very useful for practical domain ontologies. Defining a list of the most abstract notions underlying our conceptualizations is a complex enough task; but it is truly scary to consider creating a list of all the differentiae one would have to specify (and arrange in some order, so as to avoid the full combinatory complexity) in order to define such notions as Love, Democracy, and (even) Table.

Type 2: Intuitive ontological distinctions (the cognitive scientists). The oldest and most natural reason for creating a new concept is simply the intuitive feeling that it is not the same as anything else already defined, which means that one has to split it off from its near-siblings and begin a new variation or subspecies. Unfortunately people are not consistent in doing so, and 'split off' new 'concepts' quite actively as the occasion demands, creating ad hoc subgroupings differentiated by whatever feature(s) are relevant to their purposes. This playful freedom is useful for communication and no doubt for thought in general; its results are sometimes recorded (by having words that name the concepts), and sometimes not. The result is a hodge-podge of word-sense families whose meanings partly overlap but differ on arbitrary dimensions, and for which no regular correspondences are found across languages in general (see for example [8] for a very nice paper on plesionyms[3] in various languages). Determining this kind of concept formation is the specialty of the cognitive scientist (especially the one interested in language), whose methodology (and proof) turns to devising clever experiments to measure how people make distinctions between close concepts. But the fluidity of the distinction process, being dependent on the person's interests, knowledge, task, and other circumstances, make this approach to ontology building fraught with inconsistency to the point of hopelessness.

Type 3: Cross-linguistic phenomena (the linguists). For some people, concepts can be motivated simply because words or expressions for them appear in many languages. When many cultures independently name a thought, is that not evidence for the existence of that thought as a separate concept? Whether one believes Vygotsky

---

[3] Near-synonyms, like *jungle, forest,* and *woods.*

[47], Sapir [42] and Whorf [49], or Piaget [41] as to which of language and thought is primary (if either), the very close intertwining of them in the mind is generally accepted. As shown for example in EuroWordNet [46] or the plesionym study [8], analysis of cross-language differences uncovers complex and fascinating interrelationships among meanings and meaning facets. Naturally, this approach fits the linguistic tradition, and for some NLP applications, especially cross-lingual ones, paying attention to many languages for ontology construction and lexicon development can be rewarding (see for example [7], about which more below). But since there are many concepts for which no words exist, and since it is easy to demonstrate shadings of meaning from one concept to another that suggest continua of unimagined dimensionality, no-one will accept the argument that "because it's so in language(s), it has to be exactly so in thought" as a final arbiter. Nonetheless, words, as the richest component in our arsenal of tools to attack meaning, remain central. Therefore language-inspired ontology work usually produces word networks such as WordNet that are strongest in the (typically large) middle region, corresponding approximately to language lexicons; at both the abstract (upper) and domain-specific and particular (lower) region, they tend to lose expressive utility.

Type 4: Inference-based concept generalizations (the computational reasoners). For computational systems, one creates in effect a data model. The ontology (and domain model) is so arranged that those items in the domain that should be treated similarly are grouped together and typed, so that they can in fact be recognized and treated similarly. Such groupings tend to emphasize domain-specific concepts and produce more abstract concepts (i.e., upper ontologies) only insofar as they are required for grouping. For this methodology the validation method is simple and direct: Do I need separate treatment of some thing(s) by the system? If I create an appropriate new type, does the system work as required? The domain terms thus tend to mirror the metadata and the system variables. They must be defined in enough detail to support a reasonably powerful set of operators or rules, but not be so differentiated as to require too many rules. This methodology is relatively clean, depending on the elegance of the computational solution to the problem. Unfortunately, however, the decision justifications offered by systems builders today are seldom interesting to philosophers, psychologists, and linguists. •Given how many ways there are to achieve a computational result, system builders are seldom able to express their analysis of the problem and its solution in terms of necessary information transformation operations, i.e., in compelling general information-theoretic terms that would convince the other disciplines' specialists of the logical necessity of organizing one's knowledge one way or another. This fact lies at the heart of the inevitable communication breakdown and decoupling that occurs when systems builders and any of the other disciplines' researchers set out jointly to build an ontology for some application, a breakdown that requires significant effort to overcome.

Type 5: Inherited domain distinctions (the domain specialists). In many ontology building enterprises, the reason for creating and arranging concepts stems neither from abstract theoretical analysis nor experimentation, but from existing domain theory and practice. Biologists, neuroscientists, aircraft builders, pump manufacturers, legal scholars, and anyone else in knowledge-intensive enterprises find it perfectly natural to construct ontologies that reflect the way their fields view their worlds; this class of ontologies is thus one of the most common in practice; see for example the over 700 ontologies listed on the OntoSelect website (http://views.dfki.de/ontologies/), covering space, travel, music, wine, sports, science, etc. [4]. Often the exercise of actually building an ontology is prompted by the desire to work in a computational setting, and frequently the organizational discipline imposed by ontology software causes an experience of some enjoyment and even some reorganization of the builder's own understanding. Connecting a domain ontology to a generic computational system (such as a sentence generator or parser) sometimes requires realignment and/or reconceptualization of the domain terms into the categories interpretable by the computational engine; a typical solution is to embed such domain model(s) under an Upper Model that supports the computation, as illustrated for sentence generation in [3].

Addendum: Type 6: Taxonomic clarity. There is another motivation for introducing concepts, one that almost all ontology builders employ. Sometimes it is simply useful for an ontology builder to insert some mid-level concepts in order to create organizational clarity, without explicitly formulating the criteria that justify their existence (esthetics and/or clarity of display are reasons not generally deemed sufficient to measure up to serious insights derived from psychological experiment, philosophical argument, computational necessity, or cross-linguistic comparison).

It is tempting to consider these approaches as complementary; one could for example ask the philosophers to build the uppermost, most abstract, regions, the cognitive scientists to provide some overall ontology framework that the computationalists and domain specialists can then flesh out and refine, etc. But there is no guarantee that the distinctions natural to ontology builders of one type will in fact correspond to or be useful for others' purposes. In practice, such admixture tends to require that all parties learn a little about every approach, and that one of them becomes the ultimate arbiter, usually on irrelevant grounds such as personality or loudness of argument.

4   Ontology Construction Procedure

Mismatches between ontologies are a source of never-ending discussion and wonder. But they are not surprising; when concept creation decisions can be justified on such different grounds as listed above, mismatches are to be expected and are not really very interesting on an individual basis. The ontologies simply differ in content and 'focus'. What is interesting is when one discipline delivers no insight and another must come to its aid. To prevent disaster, a methodology of ontology creation should recognize this fact and assign relative priorities to the various concept creation methods and justification criteria *a priori*, before any actual ontology building is done.

The above considerations apply for all ontology building efforts (although upper ontologies, given their abstraction from domain particulars, are a somewhat special case). To create a domain model, the methodology generally adopted (see for example [14]), which can be called *continual graduated refinement*, is:

1. Determine the general characteristics of the ontology to be built. A list of such characteristics (and additional ones) is provided in [19], and includes the domain

of interest, the purpose of the ontology, the target level of granularity, the conceptual and theoretical antecedents, etc. Central to these decisions is selecting the principal criteria of ontologization (concept creation and justification methods) and specifying their order. In this step the task/enterprise is determinate: is this domain model to be used in a computational system? Or is it a conceptual product for domain analysis and description? Who are the intended users/readers of the ontology, and what is their purpose with it? What justification criteria will they find most understandable, and do these criteria match the purpose or task in mind?

2. Gather all additional knowledge resources, including starter ontologies, upper structures or microtheories (of, say, time and space), glossaries of domain terms, supporting descriptive and definitional material, algorithms and tools, existing theoretical descriptions, etc.

3. Delimit the major phenomena for consideration: identify the core concepts, types of features allowed, principal differentiae, etc. To lay out the general area, starting with an existing upper ontology, even one with just some dozen nodes, can be helpful.

4. List all readily apparent terms/concepts important for the task or enterprise. These terms may be derived from a (meta-)data model, from the algorithm of the system (to be) built, from experts' reports on the major components and processes in the domain, etc.

5. For each concept, explicitly record the principle(s) and factors that justify its creation. The definition may still be incomplete and informal, but should contain the principal differentiae and features of interest. Also identify interrelationships between the concept and related concepts (including subclass hierarchicalization, part-wholes, equivalence/synonymy, etc.), and specify/define them.

6. Inspect the nascent domain model for (ir)regularity, (im)balance, etc. Then for each major region (types of entity, types of action, types of state, etc.) repeat steps 3 to 5, refining existing concepts as needed. During this iterative refinement, record all problematic issues; they may require extensions to the upper ontology or even to the basic criteria of ontologization.

7. When done, characterize the ontology or domain model by recording its essential parameters, as spelled out in [19].

Working out the details of this methodology takes time and effort. Not all aspects apply in all cases, and not to all domains or ontologization styles [44]. Careful study of how domain ontologizers actually instantiate this procedure will help flesh out a systematic methodology of ontologizing.

5   Example Language-Based Methodology: Annotator-Driven Concept Granularity Using Wordsenses

The core ontologization decisions outlined in the preceding section can be viewed as a question of concept granularity: given some semantic notion circumscribed by one or more near-synonymous words, how many concepts should one define for them, how should one organize the concepts, and how should one validate these choices? This section outlines, as example, a language-based methodology that starts by creating and validating wordsenses and then uses them to suggest concepts. It contains two parts: experiences with wordsense annotation and converting wordsenses into concepts.

5.1 Experiences in Wordsense Annotation

The OntoBank project (Weischedel et al., in prep.) is an ongoing attempt by researchers at BBN (Weischedel, Ramshaw, et al.), the University of Pennsylvania (Marcus, Palmer, et al.) and ISI (Hovy et al.) to construct a shallow semantic version of a collection of texts. Should continuation funding be achieved, the goal is to build by hand a set of one million sentences with their associated shallow semantic frames. In this project, shallow semantics includes disambiguated concept/wordsense symbols for each noun, verb, and adjective; basic verb frames with relations to constituents; resolved anaphoric links; canonicalized representations for straightforward dates, numerical expressions, proper named entities, and a few other phenomena. Central to OntoBank is the PropBank wordsense differentiation and annotation procedure [25].

The IAMTC project [7], which ended early for lack of continuation funding after one year, had as goal to uncover the representational machinery required to support an interlingua notation from an analysis of differences across seven languages. Project members annotated some 150 translated texts, each one in both English and its source language (one of Hindi, Arabic, Korean, Japanese, French, and Spanish). Similar to OntoBank, annotation included selection of a disambiguated concept/wordsense for each noun, verb, and adjective; the determination of an appropriate verb frame (in this case, LCS theta role frames [6]) and its connections to sentence constituents; and the design of a series of incrementally deepening representations en route toward the interlingua.

The author participated in both OntoBank and IAMTC, in more or less the same role, as ontology maintainer and developer. In both cases, ISI's Omega ontology at http://omega.isi.edu [40,18] was used as repository of all semantic symbols. In both projects, all nouns, verbs, and adjectives were annotated by multiple people, who selected the appropriate concept(s) to express the words' meaning(s) in context. Both projects paid considerable attention to the annotation interface, annotator training, post-annotation reconciliation discussions, and annotator agreement measures.

Of primary interest for this paper is the ontological considerations that arise when such annotation efforts are conducted. It is relatively straightforward, though not always easy, to build ontologies of specific well-circumscribed domains for computational purposes. But the picture changes somewhat when the focus is annotation of wide-coverage newspaper text in the interest of creating shallow semantic representations. In particular, the ontology maintainer is confronted with a stream of seemingly unrelated decisions about concept granularity and ontology placement, more or less one for every verb, noun, and adjective encountered. The OntoBank methodology is illustrative. Following the PropBank annotation procedure [25], the most frequent N words of a given type (say, verbs) are selected for annotation. For each verb, 100 sentences containing it are extracted from the corpus. Two or more annotators each see the same hundred sentences plus a list of candidate concept (sense) choices extracted from the ontology. Their task is to select just those concepts that express the meaning(s) of the verb in a given context.

It is apparent that the nature of the concept alternatives and the quality of their definition are of central importance. Omega, which for a large part is derived from the lexical network WordNet, usually contains too many close alternatives, confusing the annotators (annotator selection agreement when given WordNet senses as options is only around 70% for nouns). For example, the verb "drive" has 22 senses in WordNet, including separate senses for driving a car as chauffeur and driving a car as one's work. In an employment domain, the difference between chauffeur, taxi driver, and other kind of employed driver may be important, but in general texts this distinction is often not made, or it is so implicit that many people don't make it, leading to different annotator choices.

In contrast, the MIKROKOSMOS ontology [33,32], another source of concepts for Omega, almost always offers too little granularity; it has only one symbol for all vehicles, including cars, buses, airplanes, etc. Ideally, one would like something in between: just enough concepts to express the semantic differences that most people easily agree on in the context of the text.

5.2 From Words to Concepts: Hierarchical Graduated Refinement

Palmer and colleagues use sense creation/compression to build PropBank; they have developed a well-tested procedure. A slightly more elaborated and formalized procedure was developed at ISI to mirror appropriate wordsenses in the Omega ontology, which for this work was extended to accommodate separately wordsenses and concepts [18]. This sense creation procedure is interesting to perform, and perfectly illustrates the problem of the ontology builder, and the need for a strict methodology. For example, how many senses (concepts) should one define for the word "drive"? Different members of OntoBank produced quite different results, which differed from the decisions of PropBank's expert sense creator. The continual graduated refinement procedure outlined above is as follows: starting with the word and a number of example sentences representing all/most of its meanings, identify and split out the most semantically different sense (cluster) and create a branch point in the evolving tree; then for each branch, repeat the process downward. At each split write down the criteria, or at least a description of the difference; these give a hint at the thought process for later discussion and may eventually allow one to define more-formal differentiae. For example, consider the following sentences for "drive":

1. *Drive the demons out of her and teach her to stay away from my husband!!*

2. *Shortly before nine I drove my jalopy to the street facing the Lake and parked the car in shadows.*

3. *He drove carefully in the direction of the brief tour they had taken earlier.*

4. *Her scream split up the silence of the car, accompanied by the rattling of the freight, and then Cappy came off the floor, his legs driving him hard.*

5. With an untrained local labor pool, many experts believe, that policy could drive businesses from the city.

6. Treasury Undersecretary David Mulford defended the Treasury's efforts this fall to drive down the value of the dollar.

7. Even today range riders will come upon mummified bodies of men who attempted nothing more difficult than a twenty-mile hike and slowly lost direction, were tortured by the heat, driven mad by the constant and unfulfilled promise of the landscape, and who finally died.

8. Cows were kept in backyard barns, and boys were hired to drive them to and from the pasture on the edge of town.

How many concepts/senses should one create? In WordNet, "drive" has 22 senses. Employing our procedure of hierarchical graduated refinement on these (and additional) sentences, the author and a student separately found 7 major senses, in the order of the wordsense hierarchy below (hints for differentiate are indicated in angle brackets, as well as sentence numbers with focal words):

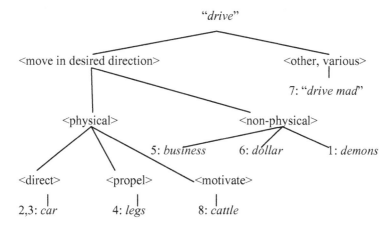

Given more sentences, additional subsenses can be found, including driving a tool ("*drive the hammer*") under <propel>, driving non-cars ("*drive a bulldozer*") under <direct>, employment ("*drive a taxi for work*") or phrasal expressions ("*drive a hard bargain*") under <other>, etc.

Adopting the above hierarchical graduated refinement procedure[4] is useful for several reasons. It supports human analysis and agreement by systematically removing the most glaringly different cases. It helps suggest differentiae. It allows one to vary the granularity at will by simply ending the differentiation process sooner or later, and by grouping together as undifferentiated all senses lower than the chosen level.

---

[4] Sense differentiation in PropBank is not performed hierarchically as shown here; the expert produces a flat list. Nonetheless, trained staff can easily estimate the level of granularity that will ensure an inter-annotator agreement of over 90% (the target value in OntoBank; this figure, on average, required a 50% reduction of WordNet senses, for example).

This capability is extremely useful when one separates word senses from ontology concepts. As argued in for example [37] and discussed in [17], not all wordsenses in the sense hierarchy need (or should) be converted into ontological categories (concepts). The sense hierarchicalization for "drive" above, for example, requires at least three distinct concepts, namely "drive mad" (i.e., something like Cause-Mental-Instability), the nonphysical sense group (rooted in something like Cause-State-Change-toward-Desired-Value), and the physical group (rooted in Cause-Movement-in-Desired-Direction) respectively. Being so different, the first of these will be inserted into the ontology at a point quite remote from the other two. Further, sentence 1, "drive the demons out of her", may be treated at two levels: the surface-metaphorical (in which case "demons" metaphorically stands for "illness" and the driving is nonphysical), or the 'true' semantic (in which case the meaning is something like Heal-of-Mental-Disorder, and there is no driving, even metaphorically). In the latter case, sentence 1 would also be classified in a region of the ontology quite remote from the other concepts.

One can continue ontologizing the sense hierarchy to the extent one wishes to (or can) formalize the differentiae and one believes further distinction provides valuable additional explanatory or computational utility. Each step of ontologization requires placing the newly formed concept into the growing ontology where appropriate. This procedure makes apparent that there is no direct homomorphism of the sense hierarchy into the ontology, though as senses are 'closer' lower down in the hierarchy, one would expect their ontological equivalents also to be closer in the ontology.

It is a good reason to stop ontologizing the sense hierarchy when no obviously most-different sense or sense group can be identified, that is, when it is possible to split the remaining group in several equally good ways according to different criteria. At this point, one has most probably reached a level of semantic homogeneity at which several features combine in equal measure to form the concept's unique identity.

Cross-language studies are useful in helping to identify a division between 'sense space' and 'ontology space'. In the IAMTC project, when the languages in question (Hindi, Arabic, Korean, French, Spanish, and Japanese) provided different words (and hence usually also senses) after translation (by various professional-quality translators) into English, then the granularity of the concept in question had to be such as to represent the senses common across the various translations, while their individual, language-idiosyncratic, facets of difference remained in the sense hierarchy. One can thus think of 'ontology space' as the *interlingual* representation symbols (symbols capturing common, or common enough, meaning aspects); of 'sense space' as the *multi-lingual* representation symbols (symbols for senses that may or may not co-occur across languages, but that are mapped to meanings no more specific than they denote themselves), and of 'lexical space' as the *monolingual* representation symbols (namely, the words of each language). There is a complex many-to-many mapping across both gaps.

6 Conclusion

The two-stage methodology—creating wordsenses and annotating them to determine granularity, followed by conversion of (part of) the sense hierarchy into ontology concepts has several desirable properties. It is 'empirical', in the sense that the semantic distinctness of a set of wordsense (concept) candidates can be validated through annotator agreement. This gives an upper bound on the granularity of concepts, and interestingly blends linguistic and cognitive motivations in a computational setting. The hierarchical graduated refinement procedure is easily extensible to finer detail, and provides suggested differentiae for formalization. The overall methodology accommodates comparison across and incorporation of different languages by separating language-dependent wordsenses from language-independent concepts.

It is not claimed that the methodology described here is a proven, or even adequately explored, methodology for ontology creation. But given the increasingly pressing need for more attention to methodology in current-day ontology creation, it is to be hoped that these thoughts will inspire further exploration. In addition, hopefully the new interest in ontologies will usher in more work on methodology as well. It is a fascinating and rewarding direction of research.

Acknowledgements

I am indebted to Graeme Hirst and Patrick Pantel for their comments, both in print and in person.

References

1. Aguado, G., A. Balon, J. Bateman, S. Bernardos, M. Fernández, A. Gómez-Pérez, E. Nieto, A. Olalla, R. Plaza, and A. Sánchez. 1998. ONTOGENERATION: Reusing Domain and Linguistic Ontologies for Spanish Text Generation. *Proceedings of the ECAI Workshop on Applications of Ontologies and Problem Solving Methods*, 1–10. ECAI Conference. Brighton, England.
2. Alberdi, E., D.H. Sleeman, and M. Korpi. 2000. Accommodating Surprise in Taxonomic tasks: The Role of Expertise. *Cognitive Science* 24(1), 53–92.
3. Bateman, J.A., Kasper, R.T., Moore, J.D., and Whitney, R.A. 1989. A General Organization of Knowledge for Natural Language Processing: The Penman Upper Model. Unpublished research report, USC/Information Sciences Institute, Marina del Rey, CA. A version of this paper appears in 1990 as: Upper Modeling: A Level of Semantics for Natural Language Processing. *Proceedings of the 5th International Workshop on Language Generation*. Pittsburgh, PA.
4. Buitelaar, P., T. Eigner, and T. Declerck. 2004. OntoSelect: A Dynamic Ontology Library with Support for Ontology Selection. *Proceedings of the Demo Session at the International Semantic Web Conference.* Hiroshima, Japan.
5. Chklovski, T., V. Ratnakar, and Y. Gil. 2005. User Interfaces with Semi-Formal Representations: A Study in Designing Augmentation Systems. *Proceedings of the Conference on Intelligent User Interfaces (IUI05)*. San Diego, CA.

6. Dorr, B.J., M. Olsen, N. Habash, and S. Thomas. 2001. The LCS Verb Database. Technical Report Online Software Database, University of Maryland, College Park. http://www.umiacs.umd.edu/bonnie/LCS_Database_Docmentation.html.

7. Dorr, B., D. Farwell, R. Green, N. Habash, S. Helmreich, E.H. Hovy, L. Levin, K. Miller, T. Mitamura, O. Rambow, F. Reeder, A. Siddharthan. (submitted). Interlingual Annotation of Parallel Corpora. Submitted to *Journal of Natural Language Engineering*.

8. Edmonds, P. and G. Hirst. 2002. Near-synonymy and Lexical Choice. *Computational Linguistics* 28(2), 105–144.

9. Farquhar, A., R. Fikes, and J. Rice. 1997. The Ontolingua Server: A Tool for Collaborative Ontology Construction. *International Journal of Human-Computer Studies* 46, 707–727.

10. Fellbaum, C. 1998. (ed.) *WordNet: An On-Line Lexical Database and Some of its Applications*. Cambridge: MIT Press.

11. Fillmore, C. 1976. The Case for Case. In E. Bach and R. Harms (eds) *Universals in Linguistic Theory*, 1–88. Holt, Rinehart, and Winston.

12. Gangemi, A., N. Guarino, C. Masolo, A. Oltramari, and L. Schneider. 2002. Sweetening Ontologies with DOLCE. In A. Gomez-Perez and V.R. Benjamins (eds), *Knowledge Engineering and Knowledge Management*. Proceedings of Ontologies and the Semantic Web at the 13th EKAW Conference. Siguenza, Spain. Springer Verlag, 166–181.

13. Gil, Y., R. MacGregor, K. Myers, S. Smith, and W.R. Swartout. 1999. CommonP: A Common Plan Representation for Air Campaign Plans. USC/ISI Technical Report available at http://www.isi.edu/isd/HPKB/planet/alignment.

14. Gruber, T.R. 1993. Toward principles for the design of ontologies used for knowledge sharing. In N. Guarino and R. Poli (eds.), *International Workshop on Formal Ontology*, Padova, Italy. Revised August 1993. Published in *International Journal of Human-Computer Studies*, special issue on Formal Ontology in Conceptual Analysis and Knowledge Representation (guest editors: N. Guarino and R. Poli) (to appear). Available as technical report KSL-93-04, Knowledge Systems Laboratory, Stanford University.

15. Guarino, N. 1998. Some Ontological Principles for Designing Upper Level Lexical Resources. *Proceedings of the First International Conference on Lexical Resources and Evaluation* (LREC), 527–534. Granada, Spain.

16. Guarino, N. and C. Welty. 2004. An Overview of OntoClean. In S. Staab and R. Studer (eds), *Handbook on Ontologies*. Springer Verlag, 151–159.

17. Hirst, G. 2004. Ontology and the Lexicon. In S. Staab and R. Studer (eds), *Handbook on Ontologies*. Springer Verlag, 209–229.

18. Hovy, E.H., A. Philpot, P. Pantel, M.B. Fleischman. (in prep.) The Omega Ontology.

19. Hovy, E.H. 2002. Comparing Sets of Semantic Relations in Ontologies. In R. Green, C.A. Bean, and S.H. Myaeng (eds), *The Semantics of Relationships: An Interdisciplinary Perspective*, 91–110. Dordrecht: Kluwer.

20. Hovy, E.H. 1998. Combining and Standardizing Large-Scale, Practical Ontologies for Machine Translation and Other Uses. *Proceedings of the 1st International Conference on Language Resources and Evaluation* (LREC), 535–542. Granada, Spain.

21. Hovy, E.H. and S. Nirenburg. 1992. Approximating an Interlingua in a Principled Way. *Proceedings of the DARPA Speech and Natural Language Workshop*. Arden House, NY.

22. Intraspect Corp. 1999. *http://www.intraspect.com/product_info_solution.htm*.

23. Iwasaki, Y., A. Fraquhar, R. Fikes, and J. Rice. 1997. *A Web-Based Compositional Modeling System for Sharing of Physical Knowledge*. Nagoya: Morgan Kaufmann.

24. Kim, J., Y. Gil, and M. Spraragen. 2004. A Knowlesdge-Based Approach to Interactive Workflow Composition. *Proceedings of the Workshop on Planning and Scheduling for Web and Grid Services* at the International Conference on Automatic Planning and Scheduling (ICAPS 04). Whistler, Canada.

25. Kingsbury P. and M. Palmer. 2002. From Treebank to PropBank. *Proceedings of the 3rd International Conference on Language Resources and Evaluation (LREC-2002)*. Las Palmas, Spain.
26. Knight, K. and S.K. Luk. 1994. Building a Large-Scale Knowledge Base for Machine Translation. *Proceedings of the AAAI Conference*, 773–778.
27. Korpi, M. 1988. *Making Conceptual Connections: An Investigation of Cognitive Strategies and Heuristics for Inductive Categorization with Natural Concepts*. Ph.D. dissertation, Stanford University.
28. Lakoff, G. 1987. *Women, Fire, and Dangerous Things: What Categories Reveal about the Mind.* Chicago: University of Chicago Press.
29. Lenat, D.B. and R.V. Guha. 1990. Building Large Knowledge-Based Systems. Reading: Addison-Wesley.
30. Lenat, D.B. 1995. CYC: A Large-Scale Investment in Knowledge Infrastructure. *Communications of the ACM* 38(11), 32–38.
31. Levin, B. 1993. *English Verb Classes and Alternations*. Chicago: University of Chicago Press.
32. Mahesh, K. and S. Nirenburg. 1995. A Situated Ontology for Practical NLP. *Proceedings of the Workshop on Basic Ontological Issues in Knowledge Sharing, International Joint Conference on Artificial Intelligence (IJCAI-95).* Montreal, Canada.
33. Mahesh, K. 1996. Ontology Development for Machine Translation: Ideology and Methodology (CRL report MCCS-96-292). Las Cruces: New Mexico State University.
34. McGuinness, D.L., R. Fikes, J. Rice, and S. Wilder. 2000. An Environment for Merging and Testing Large Ontologies. *Proceedings of the Seventh International Conference on Principles of Knowledge Representation and Reasoning (KR2000)*.
35. Miller, G.A. 1990. WordNet: An Online Lexical database. International Journal of Lexicography 3(4) (special issue).
36. Nirenburg, S., V. Raskin and B. Onyshkevych. 1995. Apologiae Ontologia. *Proceedings of the International Conference on Theoretical and Methodological Issues* (TMI). Leuven, Belgium.
37. Nirenburg, S. and Y. Wilks. 2001. What's in a Symbol: Ontology, Representation and Language. *Journal of Experimental and Theoretical Artificial Intelligence* 13(1), 923.
38. Noy, N.F. and M.A. Musen. 1999. An Algorithm for Merging and Aligning Ontologies: Automation and Tool Support. *Proceedings of the Workshop on Ontology Management at the Sixteenth National Conference on Artificial Intelligence (AAAI-99)*. Orlando, FL. Also available a <http://www-smi.stanford.edu/pubs/SMI_Reports/SMI-1999-0799.pdf>.
39. Palmer, M., J. Rosenzweig, and W. Schuler. 1998. Capturing Motion Verb Generalizations with Synchronous TAGs. In P. St. Dizier (ed.) *Predicative Forms in NLP*. Boston, MA: Kluwer Academic Press.
40. Philpot, A., M. Fleischman, E.H. Hovy. 2003. Semi-Automatic Construction of a General Purpose Ontology. *Proceedings of the International Lisp Conference*. New York, NY. October 2003. Invited.
41. Piaget, J. 1959 (English translation). *Language and Thought of the Child*. London: Routledge Classics.
42. Sapir, E. 1929/1964. The Status of Linguistics as a Science. *Language* 5, 207–214. Reprinted in D.G. Mandelbaum (ed) *Culture, Language, and Personality: Selected Essays of Edward Sapir*. Berkeley, CA: University of California Press.
43. Sowa, J. 1999. *Knowledge Representation*.
44. Uschold, M., P. Clark, M. Healy, K. Williamson, and S. Woods. 1998. Ontology Reuse and Application. *Proceedings of the International Workshop on Formal Ontology in Information Systems*, 179–192. Held in association with the KR-98 Conference. Trento, Italy.

45. Visser, P., D. Jones, T. Bench-Capon, and M. Shave. 1998. Assessing Heterogeneity by Classifying Ontology Mismatches. *Proceedings of the International Workshop on Formal Ontology in Information Systems*, 148–162. Held in association with the KR-98 Conference. Trento, Italy.
46. Vossen, P. (ed) 1998. *EuroWordNet: A Multilingual Database with Lexical Semantic Networks*. Dordrecht: Kluwer Academic Publishers.
47. Vygotsky, L.S. 1978 (English edition). *Mind in Society: The Development of Higher Psychological Processes*. Cambridge, MA: Harvard University Press.
48. Weischedel, R., E.H. Hovy, M. Marcus, M. Palmer, L. Ramshaw. (in prep.) The Onto-Bank Project.
49. Whorf, B.L. 1940/1972. Science and Linguistics. *Technology Review* 42(6), 227–231, 247–248. Reprinted in J.B. Carroll (ed) *Language, Thought, and Reality: Selected Writings of Benjamin Lee Whorf*. Cambridge, MA: MIT Press.

Using Formal Concept Analysis and Information Flow for Modelling and Sharing Common Semantics: Lessons Learnt and Emergent Issues

Yannis Kalfoglou[1] and Marco Schorlemmer[2]

[1] Advanced Knowledge Technologies (AKT), School of Electronics and Computer Science,
University of Southampton, UK
y.kalfogou@ecs.soton.ac.uk
[2] Institut d'Investigació en Intel·ligència Artificial,
Consell Superior d'Investigacions Científiques, Spain
marco@iiia.csic.es

Abstract. We have been witnessing an explosion of user involvement in knowledge creation, publication and access both from within and between organisations. This is partly due to the widespread adoption of Web technology. But, it also introduces new challenges for knowledge engineers, who have to find suitable ways for sharing and integrating all this knowledge in meaningful chunks. In this paper we are exposing our experiences in using two technologies for capturing, representing and modelling semantic integration that are relatively unknown to the integration practitioners: Information Flow and Formal Concept Analysis.

1 Introduction

Since the early nineties there exists a continuing effort to produce machine-processable common sense knowledge models that aim at capturing real-world conceptualisations for the benefit of reusing and sharing knowledge-base and information-system components. Well into the second decade of this research and development endeavour we have already profited from outcomes such as ontologies (as understood within the information systems and artificial intelligence communities, see [11]), and we have also seen significant advances in knowledge engineering technology. In recent years, we have also witnessed a renewed interest in globally accessed conceptual structures by using WWW technology, in particular, by using the Web's ambitious extension, Semantic Web (SW).

The advances in knowledge engineering technology and the SW are evidently intertwined in that they use and depend on each other. Knowledge engineering technologies nowadays use the SW to support knowledge management for dispersed and distinct systems, whereas the SW depends on these technologies as the means to reason over and deliver semantic information about a topic of interest.

Another sociological change that is emerging is the role that users and their communities of practice can play in knowledge sharing and reuse. In an open and distributed environment, like the SW, anyone can publish, retrieve, and access semantic information about a topic of interest. The challenge for engineers is then to ensure that communities are provided with the right means for: (a) capturing and attaching semantic

F. Dau, M.-L. Mugnier, G. Stumme (Eds.): ICCS 2005, LNAI 3596, pp. 107–118, 2005.
© Springer-Verlag Berlin Heidelberg 2005

information about their topics of interest, (b) publishing and accessing semantic information in a distributed environment that facilitates sharing and reuse (SW), and (c) reasoning over this information.

This is indeed the holy grail for knowledge engineering: to capture, represent, model and share semantics among diverse communities in a open and distributed environment like the SW. In this paper we elaborate on our own experiences in dealing with technologies that could be harnessed to help us achieve some of these goals: Information Flow (IF) as the means to capture, represent and model semantic integration (section 3), Formal Concept Analysis (FCA) as the means for modelling and analyzing semantic information (section 4). We also speculate on emergent issues with respect to the adoption of these relatively unknown technologies to the larger SW and knowledge engineering community (section 5).

Initially though, we elaborate in the next section on the role of common semantics as they are exposed by communities in an environment like the SW, and on the need for semantically integrating these communities for the benefits of knowledge sharing and reuse.

2   From Common Semantics to Semantically Integrated Communities

Communities of practice (or communities of interest) have always been indispensable for knowledge management systems. Their use in various parts of a knowledge artifact's life-cycle, from creation to expiration, is vital, because communities represent the knowledge of a group and incorporate individuals' expertise. Most of knowledge management has been using computational means for assisting these communities in closed or controlled environments, like organisational intranets. This made it possible for knowledge engineers to design, develop and deploy conceptual structures upon which knowledge sharing processes are based. Ontologies are the most used example of these structures as they are suitable for capturing and representing common semantics.

The situation becomes somewhat more complicated though, when we face the emergent SW and operate in an open and distributed environment like the Web. There, we no longer have the luxury of a centrally controlled repository of semantics. As users are encouraged to participate in knowledge management processes (either as members of a community or individually) the danger of flooding the Web (or the SW) with semantically-rich information is becoming a reality.

One has to find ways of extracting meaningful chunks of knowledge from user information as these are disseminated in all forms using all possible mediums. For example, one of the least anticipated trends for disseminating knowledge, and one that is witnessing unprecedented success, is the use of *blogs* with more than 50 million blogs available online. Most of this information would probably not be of interest for a specific system; however, this is something that we can only tell once we capture the semantics of this information and represent it in a way that allows us to reason with it.

In order to move from an environment where semantics are exposed and published *en masse*, to an environment where common semantics are identified—and hence semantic integration is possible—the focus has to be on technologies that can capture, model, and share semantics. We also need methodologies that go beyond the use of traditional constructs found in most conceptual structures (such as classes and attributes). For example, in his work for sharing ontologies in distributed environments, Kent advocates for the use of instances as the main piece of information that is passed around [18].

In the next two sections, we will elaborate on how we used IF and FCA to capture, model, and represent common semantics for the sake of semantic integration.

3   Information Flow

In this paper we refer to IF in accordance with Barwise and Seligman's theory of information flow, as put forth in [4]. Their work has put within the context of the general endeavour to develop a *mathematics of information*. The first mathematical theory addressing the idea of information in a rigorously formal way was Shannon's communication theory [24]; but his was a *quantitative*, syntactic theory of amounts of information and channel capacity that did not focus on the semantic content of communicated messages.

Building upon Shannon's probabilistic theory, and using his insight of seeing information as an objective commodity that can be studied independently of the means of transmission, Dretske developed a *qualitative*, semantic theory of information, in which he was able to formulate a definition of information content of concrete messages [9]. However, the probabilistic approach did not captured satisfactorily the semantic link between the information generated at the source and that arriving at the receiver.

What was needed was a theory that accounted for the mechanism by which signals encode information. Such mechanism was addressed by Barwise and Perry in situation semantics [3], by abandoning Shannon's and Dretske's probabilistic approach. Devlin further developed situation semantics in order to shift the emphasis that classical logic was putting on the mathematical concepts of *truth* and *proof* (which proved ill-suited for tackling problems which lay outside the scope of the mathematical realm, such as common-sense reasoning, natural language processing, or planning) to address the issues of information and information flow [8].

3.1   The Logic of Distributed Systems

The latest comprehensive theory within the effort towards a mathematics of information is *channel theory* [4], which constitutes an abstract account of Dretske's theory of information flow in which Barwise and Seligman assume some of Dretske's most fundamental observations and principles, but also abandon the problematic probabilistic approach. Barwise and Seligman see flow of information as a result of the regularities in a distributed system of components, and they use techniques borrowed from category theory and algebraic logic to formalise these regularities in their theory. Barwise and Seligman's is a mathematical model that succeeds in describing partial flow of information between components. Like Dretske's theory, but unlike Shannon's communication theory, it was not originally developed as a tool for engineers facing real world needs; rather it is a descriptive theory of information flow in distributed systems. Later in this paper though, we report on how we have been using the Barwise-Seligman theory of information flow to address realistic scenarios of semantic heterogeneity in large-scale distributed environments such as the Web.

In channel theory, each component of a distributed system is represented by an *IF classification* $\mathbf{A} = \langle tok(\mathbf{A}), typ(\mathbf{A}), \models_{\mathbf{A}} \rangle$, consisting of a set of *tokens*, $tok(\mathbf{A})$, a set of *types*, $typ(\mathbf{A})$, and a *classification relation*, $\models_{\mathbf{A}} \subseteq tok(\mathbf{A}) \times typ(\mathbf{A})$, that classifies tokens to types.[3]

The flow of information between components in a distributed system is modelled in channel theory by the way the various IF classifications that represent the vocabulary and context of each component are connected with each other through *infomorphisms*. An infomorphism $f = \langle f^\wedge, f^\vee \rangle : \mathbf{A} \rightleftarrows \mathbf{B}$ from IF classifications $\mathbf{A}$ to $\mathbf{B}$ is a contravariant pair of functions $f^\wedge : typ(\mathbf{A}) \rightarrow typ(\mathbf{B})$ and $f^\vee : tok(\mathbf{B}) \rightarrow tok(\mathbf{A})$ satisfying, for each type $\alpha \in typ(\mathbf{A})$ and token $b \in tok(\mathbf{B})$, the fundamental property that $f^\vee(b) \models_{\mathbf{A}} \alpha$ iff $b \models_{\mathbf{B}} f^\wedge(\alpha)$:

$$
\begin{array}{ccc}
\alpha & \xrightarrow{\ f^\wedge\ } & f^\wedge(\alpha) \\
\models_{\mathbf{A}} \Big| & & \Big| \models_{\mathbf{B}} \\
f^\vee(b) & \xleftarrow{\ f^\vee\ } & b
\end{array}
$$

The basic construct of channel theory is that of an *IF channel*—two IF classifications $\mathbf{A}_1$ and $\mathbf{A}_2$ connected through a core IF classification $\mathbf{C}$ via two infomorphisms f_1 and f_2:

$$
\begin{array}{ccccc}
 & & typ(\mathbf{C}) & & \\
 & \overset{f^\wedge_1}{\nearrow} & | & \overset{f^\wedge_2}{\nwarrow} & \\
typ(\mathbf{A}_1) & & |\models_{\mathbf{C}} & & typ(\mathbf{A}_2) \\
| & & | & & | \\
\models_{\mathbf{A}_1}| & & tok(\mathbf{C}) & & |\models_{\mathbf{A}_2} \\
| & & & & | \\
tok(\mathbf{A}_1) & \underset{f^\vee_1}{\searrow} & & \underset{f^\vee_2}{\swarrow} & tok(\mathbf{A}_2)
\end{array}
$$

According to Barwise and Seligman, this basic construct captures the information flow between components $\mathbf{A}_1$ and $\mathbf{A}_2$. In Barwise and Seligman's model it is by virtue of the existing connection between particular tokens (captured by projections $f^\vee_1$ and $f^\vee_2$) that components carry information of other components in a distributed system: information flow crucially involves both types and tokens.

3.2 Duality in Knowledge Sharing

Our own interest in the Barwise-Seligman theory of information flow arose from the observation by Corrêa da Silva and his colleagues [6] that, although ontologies were proposed as a silver bullet for knowledge sharing, for some knowledge-sharing scenarios the integration of ontologies by aligning, merging, or unifying concepts and relations as specified by their respective theories alone turned out to be insufficient. A closer analysis of these scenarios through the lenses of Barwise and Seligman's approach to information flow revealed that successful and reliable knowledge sharing between two

---

[3] We are using the prefix 'IF' in front of some channel-theoretic constructions to distinguish them from their usual meaning.

systems went closely together with an agreed understanding of an existing duality between the merging of local ontologies into a global one, and the identification of particular situations in which the sharing of knowledge was going to take place. Actually, such duality is a recurrent theme in logic and mathematics, which has been thoroughly studied within category theory by means of Chu spaces [12,2,20]. Chu spaces lie at the foundations of both FCA and IF, and the relationship between FCA and IF resulting from this common foundation has also been explored by Wolff [26] and Kent [19].

Consequently, Schorlemmer proposed in [22] categorical diagrams in the Chu category as a formalisation of knowledge-sharing scenarios. This approach made the duality between merged terminology and shared situations explicit, which accounted for the insufficiencies put forth in Corrêa da Silva's work, and provided a deeper understanding and more precise justification of sufficient conditions for reliable flow of information in a scenario for sharing knowledge between a probabilistic logic program and Bayesian belief networks, proposed by Corrêa da Silva in [7].

3.3   Information-Flow Theory for Semantic Interoperability

Our research has been driven by the observation that the insights and techniques gained from an information-theoretic analysis of the knowledge sharing problem could also help us in tackling the increasing challenge of semantic heterogeneity between ontologies in large-scale distributed environments such as the Web. A thorough survey on existing ontology mapping techniques in this domain revealed a surprising scarcity of formal, theoretically-sound approaches to the problem [16]. Consequently we set out to explore information-flow theoretic methods in various ways and scenarios:

IF-Map: In [15] we describe a novel ontology mapping method and a system that implements it, IF-Map, which aims to (semi-)automatically map ontologies by representing them as IF classifications and automatically generate infomorphisms between them. We demonstrated this approach by using the IF-Map system to map ontologies in the domain of computer science departments from five UK universities. The underlying philosophy of IF-Map follows the assumption that the way communities classify their instances with respect to local types reveals the semantics which could be used to guide the mapping process. The method is also complemented by harvesting mechanisms for acquiring ontologies, translators for processing different ontology representation formalisms, and APIs for Web-enabled access of the generated mappings (all in the form of infomorphisms).

Theory of Semantic Interoperability: Beyond the application of information-flow theory to guide the automatic mapping of ontologies, we have also explored the suitability of the theory to define a framework that captures semantic interoperability without committing to any particular semantic perspective (model-theoretic, property-theoretic, proof-theoretic, etc.), but which accommodates different understandings of semantics [17]. We articulated this framework around four steps that, starting from a characterisation of an interoperability scenario in terms of IF classifications of tokens to types, defines an information channel that faithfully captures the scenario's semantic interoperability. We used this framework in an e-Government alignment scenario, where we used our four-step methodology to align UK and US governmental departments using their ministerial units as types and their respective set of responsibilities as tokens which were classified against those types.

Ontology Coordination: Our most recent work in this front applies information-flow theory to address the issues arising during ontology coordination [23]. Since *a priori* aligned common domain ontologies need to be as complete and as stable as possible, they are mostly useful in clearly delimited and stable domains, but they are untenable and even undesirable in highly distributed and dynamic environments such as the Web. In such an environment, it is more realistic to progressively achieve certain levels of semantic interoperability by coordinating and negotiating the meaning attached to syntactic constructs on-the-fly. We have been modelling ontology coordination with the concept of a *coordinated information channel*, which is an IF channel that states how ontologies are progressively coordinated, and which represents the semantic integration achieved through interaction between two agents. It is a mathematical model of ontology coordination that captures the *degree of participation* of an agent at any stage of the coordination process, and is determined both, at the type and at the token level. Although not yet a fully-fledged theory of ontology coordination, nor an ontology coordination methodology or procedure, we have illustrated our ideas in a scenario taken from [25] where one needs to coordinate different conceptualisations in the English and French language for the concepts of *river* and *stream* on one side, and *fleuve* and *reivière* on the other side.

4   Formal Concept Analysis

Formal Concept Analysis (FCA)[10] provides a fertile ground for exploitation with its generic structure of lattice building algorithms to visualise the consequences of partial order that the underlying mathematical theory builds on. However, there is little support for the modeller to help in identifying appropriate conceptual structures to capture common, domain, semantics.

FCA has been applied at various stages of a system's life cycle: for example, in the early stages, when analyzing a domain for the purpose of building and using a knowledge-rich representation of that domain—like the work of Bain in [1] where FCA was used to assist building an ontology from scratch—or at later stages, in order to enhance an existing system for the purpose of providing a specific service—like the *CEM* email management system described in [5].

The core modelling ingredient underpinning FCA is a formal context: *objects* and *attributes*[4] related by an *incidence relation*. This stems from predicative interpretations of set theory (notice the common underlying mathematical foundation of FCA contexts and IF classifications as already pointed out in section 3). Thus, for a given object, one

---

[4] Priss points out in [21] these can be *elements, individuals, tokens, instances, specimens* and *features, characteristics, characters, defining elements*, respectively.

performs a "closure" operation to form a set of objects which is the intersection of the extension of the attributes that the object is characterised by. These are defined as the concepts in any particular formal context, with the order ideal (or down set) $\downarrow m$ of any attribute m.

In the AKT project [5], we experimented with scenarios taken from the scientific knowledge management realm, in which we were confronted with loosely defined objects and attributes. We describe the scenarios in detail in [14] but here we recapitulate on our experiences using FCA. Our aim was to use FCA to help us performing certain knowledge management tasks, such as:

Analyzing Programme Committee Memberships: One could assume that programme committee membership for a conference or similar events requires that those on the programme committee (PC) are the current and prominent figures in the field at question. Using this as a working hypothesis and the year in which they served at a specific PC as temporal marker of recognised prominence, we then applied FCA techniques like concept lattice exploration to visualise the distribution of PC members over a number of years. This could, arguably, give us an idea of how the specific event evolved over a period of time by virtue of the changes (or otherwise) in their PCs.

In our experiments the objects were PC members and attributes were EKAW conferences in which these members served. A visual inspection of this sort of lattice can reveal trends of how the event has evolved over the years. For example, we can identify people who where in PCs of early EKAWs but not in more recent EKAWs, whereas others have a continuous presence in PCs throughout the whole period of 1994 to 2002. If we correlate this information with information regarding the research interests of the PC' members, we could end up with a strong indication of the evolution of research themes for EKAW conferences.

Analyzing the Evolution of Research Themes: This analysis can be supported by another lattice which depicts the evolution of research themes in EKAW conferences, based on the designated conference session topics. We show this lattice in figure 1. From the lattice drawing point of view, and in contrast with conventions followed when drawing these sort of lattices, we deliberately changed the position of the nodes in the line diagrams produced. We did that to enhance its readability and ease its illustration when depicted on paper as we wanted to include full textual descriptions for all labels for objects and attributes. That compromised the grid projection property of the diagram without, however, affecting the representation of partial order between nodes.

Again, a close inspection shows some trends which are evident in today's research agendas in many organisations: *knowledge modelling frameworks* and *generic components* were popular in the early 90s whereas nowadays the research focus is on *semantic web* and *knowledge management*. The inherited taxonomic reasoning of concept lattices can also reveal interesting relationships between research topics, as for instance, the subsumption of *ontologies* from *knowledge management*, *knowledge acquisition* and *semantic web* topics.

---

[5] http://www.aktors.org

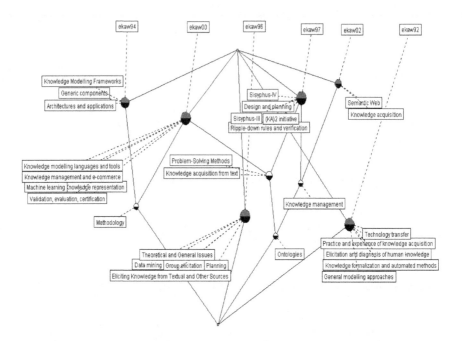

Fig. 1. Concept lattice depicting session topics of the EKAW conferences from 1994 to 2002.

Analyzing Research Areas Attributed to Published Papers: We also applied FCA techniques, in particular context reduction algorithms like those described in the FCA textbook (p.27 in [10]), to analyze the formal context of online academic journals. Our aim was to expose relationships between research areas used to classify published papers. The premise of our analysis is the very algorithm that Ganter and Wille describe in [10] for clarifying and reducing formal contexts: "[...] we merge objects with the same intents and attributes with the same extents. Then we delete all objects, the intent of which can be represented as the intersection of other object intents, and correspondingly all attributes, the extent of which is the intersection of other attributes extents.". This process, if captured in a step-wise fashion, will expose the objects and attributes that are about to be merged with others, hence allowing us to infer that they are related.

For our data sets, we used a small number of articles from the *ACM Digital Library* portal[6] focusing on the *ACM Intelligence* journal[7]. The formal context consists of 20 objects (articles) and 58 attributes (research areas). The research areas originate from a standard classification system, the *ACM Computing Classification System*[8]. We also used a second data set, the *Data and Knowledge Engineering (DKE)* journal from

---

[6] http://portal.acm.org

[7] http://www.acm.org/sigart/int/

[8] http://www.acm.org/class/1998/

Elsevier[9]. In this context we had the same articles (objects) as in the ACM context, but this time we classified them against the *DKE*'s own classification system, Elsevier's classification of *DKE* fields[10], which uses 27 research areas (attributes) for classifying the aforementioned articles.

For both data sets we chose as objects for their context, papers that appeared in the journals. For instance, for the *ACM Intelligence* journal we chose papers that appeared over a period of three years, from 1999 to 2001 and were accessible from the *ACM Digital Library* portal. As these were already classified according to the *ACM Computing Classification System*, we used their classification categories as attributes. We then applied typical context reduction techniques in a step-wise fashion. While we were getting a reduced context, we captured the concepts that are deemed to be interrelated by virtue of having their extents (objects that represent articles in the journal) intersected. For instance, the *ACM classification category H.3.5* on *Web-based services* is the intersection of *H.5* on *Information Interfaces and Presentation* and *H.3* on *Information Storage and Retrieval* by virtue of classifying the same articles. This sort of analysis supports identification of related research areas by using as supporting evidence the classification of articles against standardised categories (originating from the *ACM Computing Classification System*), and the inherited taxonomic reasoning which FCA concept lattices provide.

5   Emergent Issues

The main focus of the SW research community is, at the moment, on infrastructure with concrete deliverables such as the OWL family of ontology web languages [11], and metadata description formats like RDF [12], which are backed by standardisation bodies like the W3C (www.w3c.org). Despite these deliverables and progress made in the infrastructure front, it was argued in [13] that in order to realise the SW vision we need to tackle four dark areas which remain relatively untouched: (a) agency coordination, (b) mechanise trust, (c) robust reasoning, and (d) semantic interoperability. These areas are concerned with designing, developing and most importantly, operationalising services on the SW which could potentially change the way we use the Web.

These challenges, however, cover a broad area of scientific research and it is not realistic to expect them to be fully resolved before the SW will be available and commercially exploitable. It will take time to come up with sound scientific and practical solutions to the problems of robust reasoning, agency coordination, and semantic interoperability, to name a few. In the meantime, the SW will continue to grow and attract attention based on short to medium term solutions. We see this maturity phase of the SW as an opportunity for technologies like IF and FCA. As most of the SW machinery is Description Logic (DL) based, there are calls for something that goes beyond or at least complements DLs. For example with respect to IF, in [13] the authors point out that we need:

---

[9] http://www.elsevier.com/locate/issn/0169023X/

[10] http://www.elsevier.com/homepage/sac/datak/dke-classification-2002.pdf

[11] http://www.w3c.org/2004/OWL/

[12] http://www.w3c.org/RDF/

"[...] alternative approaches for a logic based on precise mathematical models of information as a necessary requirement for designing and operating information-processing systems have been advocated [...] We have recently explored how mathematical theories of information may provide a different angle from which to approach the distributive nature of semantics on the Semantic Web. As it seems, an information-theoretic approach such as that of Barwise and Seligman's channel theory may be suitable to accommodate various understandings of semantics like those occurring in the Web [...]"

On the other hand, FCA provides a set of tools that allows for formalizing a set of informal descriptions of a domain, thus providing the basis for ontology building. As the availability of (semi-) formal descriptions of vast amounts of data on the Web will become the key to any successful SW endeavour, FCA could play a vital role in helping the knowledge engineer to automate the task of processing these data. That could lead to automation of ontology building methods which in turn will make ontologies—the cornerstone of semantically-rich services on the SW—readily available.

Once these ontologies are built and made available on the SW, then the need for semantically integrating them will naturally arise. At this stage a number of technologies to assist achieve this ambitious goal are available (see, for example the survey in [16]), but IF-based approaches occupy a promising part of this landscape. Therefore, both FCA and IF based tools could be valuable components of an engineer's toolkit in order to tackle SW challenges.

A key issue that emerges seems to be the slow adoption and low profile that these technologies have in the larger SW community. This is not surprising, as IF is still at a premature phase for being technologically exploited, and FCA is mostly known and used in other fields. However, this shouldn't stop us using them to tackle SW challenges as it is the best way for raising their awareness among the SW researchers. As with the adoption of DLs[13], it is not the community (or research field) that will change to accommodate a new technology, but the technology itself has to be adopted in order to be appealing for a fields' practitioners. In the context of the SW that will mean incorporating SW standards, like OWL and RDF, into the mechanisms that FCA and IF are based on.

For example, it should be possible to adopt a popular technique in FCA, conceptual scaling, to accommodate the various representations of class information that are possible with the OWL family of languages. As there are different degrees of detail that each OWL version allows you to express, this granularity could be captured in a many-valued context (G,M,W,I) that FCA conceptual scaling provides. It has been discussed already in [14] that FCA contexts (G,M,I) could be used to represent OWL class information, then similarly, the many-valued context could be used to represent extra information that an OWL class has when encoded in more expressive versions of the language (OWL Full).

Similarly, as we discussed in section 3, IF could be used to assist mapping OWL ontologies where the mapped constructs are modelled as infomorphisms and represented

---

[13] DLs evolved from a purely theoretical AI-based exercise in the early nineties, to a mainstream tool for the SW researchers nowadays.

with OWL's `sameAs` construct. This construct is not the only one to express equivalence between OWL constructs. Others like `equivalentClass` could express more detailed equivalent conditions. A possible use of IF's infomorphisms would be to represent different semantic understandings of the intuitive notion of equality. For example, in OWL Lite and OWL DL, `sameAs` declares two individuals identical [14] whereas in OWL Full `sameAs` could be used to equate anything (a class to an individual, a property to a class, etc.).

6   Conclusions

Both FCA and IF have been developed and used by communities which are not closely related to the SW (or its predecessor). They have also been used in closed, controlled environments where the assurances of consistency, and possibly completeness made it possible to explore them in knowledge representation. However, they are also suitable for tackling some of the most prevalent problems for the ambitious SW endeavour: that of semantic integration. In this paper we exposed some of our experiences in using them to tackle this problem. Their adoption by the wider community depends on their ability to evolve and incorporate emerging standards.

Acknowledgments. This work is supported under the Advanced Knowledge Technologies (AKT) Interdisciplinary Research Collaboration (IRC), which is sponsored by the UK Engineering and Physical Sciences Research Council under grant number GR/N15764/01 and which comprises the Universities of Aberdeen, Edinburgh, Sheffield, Southampton and the Open University; and it is also supported under the UPIC project, sponsored by Spain's Ministry of Science and Technology under grant number TIN2004-07461-C02-02. M. Schorlemmer is also supported by a *Ramón y Cajal* Fellowship from Spain's Ministry of Science and Technology.

References

1. M. Bain. Inductive construction of ontologies from formal concept analysis. In *Proceedings of the 11th International Conference on Conceptual Structures (ICCS'03), Springer LNAI 2746, Dresden, Germany*, July 2003.
2. M. Barr. The Chu construction. *Theory and Applications of Categories*, 2(2):17–35, 1996.
3. J. Barwise and J. Perry. *Situations and Attitudes*. MIT Press, 1983.
4. J. Barwise and J. Seligman. *Information Flow: the Logic of Distributed Systems*. Cambridge Tracts in Theoretical Computer Science 44. Cambridge University Press, 1997. ISBN: 0-521-58386-1.
5. R. Cole and G. Stumme. CEM - a conceptual email manager. In *Proceedings of the 8th International Conference on Conceptual Structures (ICCS'00), Darmstadt, Germany*, Aug. 2000.
6. F. Corrêa da Silva, W. Vasconcelos, D. Robertson, V. Brilhante, A. de Melo, M. Finger, and J. Agusti. On the insufficiency of ontologies: Problems in knowledge sharing and alternative solutions. *Knowledge Based Systems*, 15(3):147–167, 2002.

---

[14] http://www.w3.org/TR/2004/REC-owl-guide-20040210/

7. F. Corrêa da Silva. Knowledge Sharing Between a Probabilistic Logic and Bayesian Belief Networks. In *in Proceedings of the International Conference on Processing and Management of Uncertainty*, 2000.

8. K. Devlin. *Logic and Information*. Cambridge University Press, 1991.

9. F. Dretske. *Knowledge and the Flow of Information*. MIT Press, 1981.

10. B. Ganter and R. Wille. *Formal Concept Analysis: mathematical foundations*. Springer, 1999. ISBN: 3-540-62771-5.

11. T. Gruber. A Translation Approach for Portable Ontology Specifications. *Knowledge Engineering*, 5(2):199–220, 1993.

12. V. Gupta. *Chu Spaces: A Model of Concurrency*. PhD thesis, Stanford University, 1994.

13. Y. Kalfoglou, H. Alani, M. Schorlemmer, and C. Walton. On the emergent semantic web and overlooked issues. In *Proceedings of the 3rd International Semantic Web Confernece (ISWC'04), LNCS 3298, Hiroshima, Japan*, page 576, 2004.

14. Y. Kalfoglou, S. Dasmahaptra, and J. Chen-Burger. FCA in Kowledge Technologies: Experiences and Opportunities. In *Proceedings of the 2nd International Confernece on Formal Concept Analysis (ICFCA'04), Sydney, Australia*, Feb. 2004.

15. Y. Kalfoglou and M. Schorlemmer. IF-Map: an ontology-mapping method based on information-flow theory. *Journal on Data Semantics*, 1:98–127, Oct. 2003. LNCS2800, Springer, ISBN: 3-540-20407-5.

16. Y. Kalfoglou and M. Schorlemmer. Ontology mapping: the state of the art. *The Knowledge Engineering Review*, 18(1):1–31, 2003.

17. Y. Kalfoglou and M. Schorlemmer. Formal Support for Representing and Automatic Semantic Interoperability. In *Proceedings of 1st European Semantic Web Symposium (ESWS'04), Crete, Greece*, May 2004.

18. R. Kent. The Information Flow Foundation for Conceptual Knowledge Organization. In *Proceedings of the 6th International Conference of the International Society for Knowledge Organization (ISKO), Toronto, Canada*, Aug. 2000.

19. R. Kent. Distributed Conceptual Structures. In *Proceedings of RelMiCS 2001*, LNCS 2561, pp. 104–123. Springer, 2002.

20. V. Pratt. The Stone gamut: A coordinatization of mathematics. *Logic in Computer Science*, pages 444–454, 1995.

21. U. Priss. Formalizing botanical taxonomies. In *Proceedings of the 11th International Conference on Conceptual Structures (ICCS'03), Springer LNAI 2746, Dresden, Germany*, July 2003.

22. M. Schorlemmer. Duality in Knowledge Sharing. In *Proceedings of the 7th International Symposium on Artificial Intelligence and Mathematics, Fort Lauderdale, Florida, USA*, Jan. 2002.

23. M. Schorlemmer and Y. Kalfoglou. A channel theoretic foundation for ontology coordination. In *Proceedings of the Meaning, Negotiation and Coordination workshop (MCN'04) at the ISWC04, Hiroshima, Japan*, Nov. 2004.

24. C. Shannon. A Mathematical Theory of Communication. *Bell Systems Technical Journal*, 27:379–243, 1948.

25. J. Sowa. *Knowledge Representations: Logical, Philosophical, and Computational Foundations*. Brooks/Cole, 2000.

26. K. E. Wolff. Information Channels and Conceptual Scaling. In *Working with Conceptual Structures. Contributions to ICCS 2000*, Shaker Verlag, 2000.

On the Need to Bootstrap Ontology Learning with Extraction Grammar Learning

Georgios Paliouras

Institute of Informatics and Telecommunications, NCSR "Demokritos", Greece

Abstract. The main claim of this paper is that machine learning can help integrate the construction of ontologies and extraction grammars and lead us closer to the Semantic Web vision. The proposed approach is a bootstrapping process that combines ontology and grammar learning, in order to semi-automate the knowledge acquisition process. After providing a survey of the most relevant work towards this goal, recent research of the Software and Knowledge Engineering Laboratory (SKEL) of NCSR "Demokritos" in the areas of Web information integration, information extraction, grammar induction and ontology enrichment is presented. The paper concludes with a number of interesting issues that need to be addressed in order to realize the advocated bootstrapping process.

1 Introduction

The task of information extraction from text has been the subject of significant research in the past two decades. Primarily, this work has focussed on the extraction of very specific pieces of information from documents that belong in a very narrow thematic domain. The typical example is the extraction of information about mergers and acquisitions from business news articles, e.g. the information:

{Buying-company: "MacroHard Corp",
Company-bought: "Africa Off-Line Ltd",
Amount: "3 billion rupees"}

could be extracted from the text:

"MacroHard Corp bought Africa Off-Line Ltd for 3 billion rupees."

or from the text

"Africa Off-Line was sold to MacroHard. ... The acquisition has costed three Bil. Rup."

Based solely on this limited example, one can understand the difficulty of the information extraction task, which is arguably as hard as full text understanding. However, when limiting the domain and the information to be extracted there are various ways to avoid full understanding and produce good results with shallow parsing techniques. These techniques usually involve lexico-syntactic patterns,

F. Dau, M.-L. Mugnier, G. Stumme (Eds.): ICCS 2005, LNAI 3596, pp. 119–135, 2005.
© Springer-Verlag Berlin Heidelberg 2005

coupled with a conceptual description of the domain and domain-specific lexicons. The manual construction and maintenance of these resources is a time-consuming process that can be partially automated with the use of learning techniques. For that reason, a significant part of the research in information extraction has refocussed on learning methods for the automatic acquisition of grammatical patterns, lexicons and even conceptual descriptions.

The rapid growth of the Web has brought significant pressure for practical information extraction solutions that promise to ease the problem of the user's overload with information. This has led to a new research direction, which aimed to take advantage of the uniform presentation style followed typically within particular Web sites. Information extraction systems designed for specific Web sites and based mainly on the HTML formatting information of Web pages have been termed wrappers. Despite the apparent ease of constructing wrappers, as opposed to free-text information extraction, the knowledge acquisition bottleneck remained, due to the frequent change in the presentation style of a specific Web site and most importantly due to the large number of different wrappers needed for any practical information integration system. This has led again to linguistically richer information extraction solutions and the use of learning methods.

More recently, a new goal was set for the information society: to move from the Web to the Semantic Web, which will contain many more resources than the Web and will attach machine-readable semantic information to all of these resources. The first steps towards that goal addressed the issue of knowledge representation for all this semantic information, which translated to the development of ontologies. Realizing the difficulty of designing the grand ontology for the world, research on the Semantic Web has focussed on the development of domain or task-specific ontologies which have started making their appearance in fairly large numbers. Having provided an ontology for a specific domain, the next step is to annotate semantically all related Web resources. If done manually, this process is very time-consuming and error-prone. Information extraction is the most promising solution for automating the annotation process. However, it comes along with the aforementioned knowledge acquisition bottleneck and the need for learning. At the same time, constructing and maintaining ontologies for various domains is also a hard knowledge acquisition task. In order to automate this process, the new research activity of ontology learning and population has emerged, which combines information extraction and learning methods.

Thus, information extraction makes use of various resources, among which a conceptual description of the domain, while at the same time ontology construction and maintenance is partly based on information extraction. The study of this interaction between the two processes and the role of learning in acquiring the required knowledge is the subject of this paper. The paper aims to initiate the interdisciplinary discussion and research that will lead to the uniform treatment of the problem of knowledge acquisition for information extraction and ontology maintenance. This effort is driven by the Semantic Web vision and the proposed vehicle is machine learning methods.

The rest of the paper is organized as follows: Section 2 highlights the state of the art in the related fields, focussing on key contributions that could facilitate the interdisciplinary effort. Section 3 presents related current research effort at the Software and Knowledge Engineering Laboratory (SKEL) of NCSR "Demokritos", where the author belongs[1]. Section 4 discusses key issues that should be addressed, in order to move this discussion forward. Finally section 5 summarizes the presented ideas.

2   State of the Art

This section presents recent approaches in the areas of information extraction and ontology learning. Rather than providing an extensive survey of these areas, the focus is on those efforts that seem most promising in terms of the desired convergence of the different approaches.

2.1   Information Extraction

Practical information extraction systems focus on a particular domain and a narrow extraction task. Within this domain, they require a number of resources, in order to achieve the required analysis of the text and identify the pieces of information to be extracted. These resources mainly consist of grammars, lexicons and semantic models. The number and complexity of the resources that are used varies according to the approach. Early approaches focussed on linguistically rich resources, with the hope that they can capture a wide variety of linguistic phenomena (e.g. [22], [18]). This approach did not prove effective in practice, as the construction and the use of the resources was very expensive. As a result, a turn towards "lighter" task-specific approaches took place (e.g. [1], [2]). These approaches combined simple grammars, e.g. regular expressions, with existing generic dictionaries, e.g. the Wordnet [17], task-specific list names, known as gazetteers, and rather simple semantic models, template schemata. As a solution to the narrow scope of these approaches, the use of machine learning methods was proposed, which allowed for the quick customization of the resources to new extraction tasks (e.g. [53], [36], [44]). This approach was taken to the extreme with the introduction of Web site wrappers and the automatic learning of these (e.g. [25], [29]).

More recently, a move towards deeper analysis of the text has started, in order to improve the performance of the extraction systems, which seemed to have exhausted their capabilities, reaching what is known as the 60% performance barrier. These new efforts include the use of more complex grammars, e.g. HPSG, for deeper structural analysis, and the use of semantic models, e.g. domain-specific ontologies, for semantic disambiguation. These developments were made possible by the improvement in deep analysis methods and the increase in available computational power.

---

[1] SKEL: http://www.iit.demokritos.gr/skel

DFKI[2] research on information extraction [31] provides an interesting example of the progress towards deeper analysis. Starting with the use of finite-state transducers [32] for shallow analysis and moving onto the incorporation of domain-specific ontologies and Head-driven Phrase Structure Grammars (HPSG) for deep structural and semantic analysis [11]. The main aim of this work is to combine the best of shallow and deep analysis, i.e., speed and accuracy. In order to achieve that, various integration strategies have been studied, focussing the use of deep analysis to those cases that are most likely to help improve the accuracy of shallow analysis. This controlled use of deep analysis minimizes the computational overhead of the approach. Furthermore, initial efforts have been made [51] to use machine learning techniques to acquire basic semantic entities, such as domain-specific terms, and parts of the extraction grammars, in particular domain-specific lexico-syntactic patterns.

This multi-strategy approach that attempts to combine the advantages of shallow and deep analysis, as well as the strengths of automated knowledge acquisition through learning with the use of rich syntactic and semantic resources, is indicative of the general trend towards optimal combination of methods at various levels. Recent research at the University of Sheffield, UK also follows that trend, starting from simple wrapper-type information extraction approaches [8] and moving towards learning methods that incorporate more linguistic information, as well as domain-specific ontologies [9]. However, the focus of this work is still on the minimization of human effort in producing linguistic resources and as a result the extraction grammars are much simpler than the HPSGs.

A related strand of work has been using conceptual graphs as the representation for the extracted information. Typically, these approaches require deep syntactic analysis of the text and some domain knowledge, e.g. in the form of an ontology, in order to construct a graph for each sentence that is analyzed (e.g. [21], [45], [33]). These approaches face similar difficulties in acquiring the required knowledge for the mapping between syntax and semantics. Machine learning has been proposed as a partial solution to the problem, e.g. for learning the mapping rules between syntactic parses and conceptual graphs [54].

Most of the generic information extraction platforms have been extended to use ontologies, instead of the simpler template schemata of the past. Reeve and Han [40] provide a brief survey of the use of ontologies in information extraction platforms. For the majority of these systems though, ontologies are only used as a rich indexing structure, adding possibly to the variety of entities and relations that are sought for in the text (e.g. [41]). A notable exception is the concept of 'information extraction ontologies' [15], where the ontology plays the role also of a simple extraction system. This is achieved by incorporating lexical constraints in the ontology, using the data frame approach [14]. Data frames associate regular expressions with the domain-specific concepts, in order to allow for their direct extraction from the text. In a similar manner, Patrick [35] uses Systemic Functional Grammars, which combine high-level conceptual descriptions with

---

[2] DFKI: Deutsches Forschungszentrum fur Kuenstliche Intelligenz; http://www.dfki.de/

low-level textual and linguistic features. This work shows initial signs of convergence of extraction grammars with ontologies, whereby a conceptual structure incorporates sufficient information to be used for the extraction of instances of its concepts from text. Whether this extended structure is a grammar, an ontology or a completely different representation is of less importance.

2.2   Ontology Learning

Machine learning methods have been used extensively to acquire various resources required for information extraction, in particular grammars and lexicons for part-of-speech tagging, sentence splitting, noun phrase chuncking, named-entity recognition, coreference resolution, sense disambiguation, etc. They have also been used to acquire the lexico-syntactic patterns or grammars that are used for information extraction (e.g. [53], [36], [44]), in particular the simpler regular expressions used in wrappers (e.g. [25], [29]). More recently this line of work has been extended to target the semantic model needed for information extraction, which is in most cases an ontology. Thus, the new research activity of ontology learning and population has emerged.

Ontology learning methods vary significantly according to their dependence on linguistic processing of the training data. At one extreme, the learning process is driven completely by the results of language processing [5]. Following this approach, the OntoLT toolkit allows the user to define or tune linguistic rules that are triggered by the data and result in the modification of the ontology. In other words, each rule defines linguistic preconditions, which, if satisfied by a sentence, lead to a change in the existing ontology, usually extending the ontology with new concepts and their properties. This is a deductive approach to concept learning, which has been the subject of dispute in the early days of machine learning, as it has been argued that it does not cause generalization and therefore is not learning at the knowledge level [13]. Nevertheless, it can be an effective method for enriching an ontology, although it requires significant expertise in defining the linguistic rules. The Ontology Express system [34] follows a similar linguistic approach to ontology learning, with two notable exceptions: (a) it concentrates on the discovery of new concepts, based on the identification of specific patterns in the text that usually denote particular relations, e.g. introduction of a new term as a subtype of an existing one, (b) it uses a frequency-based filter and non-monotonic reasoning to select the new concepts and add them to the ontology.

Term identification and taxonomic association of the discovered terms has been the most researched aspect of ontology learning. One of the earliest systems to adopt this approach was ASIUM [16], which uses lexico-syntactic patterns, in the form of subcategorization frames that are often used for information extraction, in order to identify interesting entities in text. Starting with a few generic patterns that are instantiated in the text, ASIUM uses syntactic-similarity in order to cluster terms into groups of similar functionality in the text. This process is repeated, building the taxonomy of an ontology in a bottom-up fashion. A similar approach is followed by the DOGMA system [42]. Verb-driven syntactic relations, similar to generic subcategorization frames, are used to cluster terms with syntactic similarity. Term clustering is also employed by the OntoLearn system [30]. However, OntoLearn differs from the other two systems in two ways: (a) it combines statistics about term occurrence with linguistic information for the identification of terms, and most importantly (b) clustering is based on semantic interpretation through a mapping of the terms onto an existing ontology, such as Wordnet. Thus, the resulting ontology is a domain-specific subset of the generic one. Wordnet is also used in [51] to identify an initial set of examples of the hyponymy relation in an untagged corpus. Given these examples, generic extraction patterns are learned. These patterns are combined with the results of a statistical term identification method and the collocation patterns learned by a different statistical method, to provide a set of candidate concepts for the new ontology. More recently, Wordnet and lexico-syntactic patterns have been combined in [7] using a simple voting strategy, in order to identify terms and organize them in a taxonomy. Despite the simplicity of the voting strategy, the combination of various evidence from different methods seems to provide added value to the ontology learning process.

Another promising approach to ontology learning is based on the use of Formal Concept Analysis for term clustering and concept identification. In [43] concept lattices are constructed from data with the use of a knowledge acquisition method known as 'ripple-down rules'. The acquired conceptual structures are then used to define domain ontologies, with the cooperation of a human expert. In a related approach, Corbett [10] represents ontologies with the use of Conceptual Graphs and uses Conceptual Graph Theory, in order to automate ontology learning through merging of conceptual graphs. Given the use of conceptual graphs in information extraction from text, as discussed in 2.1, this approach provides an interesting link between extraction and ontology learning. For instance, a clustering approach for conceptual graphs, such as the one presented in [52], could be used to learn ontologies, in the form of contextual graphs, from text.

The highlights of ontology learning research presented in this subsection indicate the close relation between extraction patterns and concept discovery. One usually learns the extraction patterns at the same time as identifying new terms and relations among them with the aim to construct or refine an ontology. The work of Hahn and Markó [19] emphasizes this interaction, providing a method to learn grammatical in parallel with conceptual knowledge. Adopting a deductive learning approach, like OntoLT, the proposed method refines a lexicalized dependency grammar and a KL-ONE-type conceptual model, through the analysis of text and the qualitative assessment of the results.

The interaction between information extraction and ontology learning has also been modelled at a methodological level as a bootstrapping process that aims to improve both the conceptual model and the extraction system through iterative refinement. In [27] the bootstrapping process starts with an information extraction system that uses a domain ontology. The system is used to extract information from text. This information is examined by an expert, who may decide to modify the ontology accordingly. The new ontology is used for further information extraction and ontology enrichment. Machine learning assists the expert by suggesting potentially interesting taxonomic and non-taxonomic relations between concepts. Brewster et al. [3] propose a slightly different approach to the bootstrapping process. Starting with a seed ontology, usually small, a number of concept instances are identified in the text. An expert separates these as examples and counter-examples which are then used to learn extraction patterns. These patterns are used to extract new concept instances and the expert is asked to re-assess these. When no new instances can be identified, the expert examines the extracted information and may decide to update the ontology and restart the process. The main difference between the two approaches is in the type of extraction system that is used, which is linguistically richer in the case of [27] and uses the ontology as a component.

3   Recent Research Results by SKEL

The Software and Knowledge Engineering Laboratory (SKEL) of the Institute of Informatics and Telecommunications in the National Center for Scientific Research "Demokritos" has set as its main goal for the past decade to advance knowledge technologies that are required for overcoming the obstacle of information overload on the Web. Towards that goal, it has produced innovative research results in the whole chain of technologies employed by intelligent information integration systems: information gathering (retrieval, Web crawling), information extraction (named entity recognition and classification, role identification, wrappers), personalization (user communities and stereotypes). The recent emphasis of our research has been on the automation of intelligent system development, customization and maintenance, which involves mainly the employment of machine learning methods for knowledge acquisition.

This section highlights SKEL's most recent research activity in the area of machine learning for information extraction and ontology enrichment. It starts by presenting briefly the CROSSMARC architecture for information integration, which is the main result of the European research project CROSSMARC and provides the framework for our research in this area. It then moves on to present briefly our meta-learning approach to information extraction from Web pages, an efficient learning method for context-free grammars and a bootstrapping methodology for ontology enrichment.

3.1   The CROSSMARC Approach to Web Information Integration

CROSSMARC (Cross-lingual Multi Agent Retail Comparison)[3] was a European research project that was completed at the end of 2003. The main result of CROSSMARC was an open, agent-based architecture for cross-lingual information integration, incorporating the full chain of technologies involved in the

---

[3] http://www.iit.demokritos.gr/skel/crossmarc/

process. Initially, CROSSMARC was meant to focus on retail comparison systems that collect product information from various suppliers and present it to customers in a localized and personalized manner. In addition to this application however, the CROSSMARC architecture has proven equally useful for other information integration tasks, such as employment search engines. Figure 1 presents the CROSSMARC architecture.

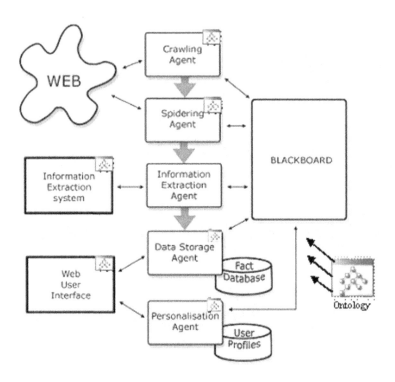

Fig. 1. CROSSMARC's agent based architecture.

As mentioned above, CROSSMARC implements the full information integration process, using independent agents that communicate via a blackboard and share the same domain ontology. The information gathering stage is separated into a crawling and a spidering step, collecting interesting sites from the Web and relevant pages from these sites, respectively. Machine learning is used to learn to identify relevant pages and the most promising paths to those pages. The information extraction agent serves as a controller for a number of different information extraction systems, each handling a different language (English, French, Italian and Greek are currently covered). The results of information extraction are stored into the fact database, which is accessed by the end-users through a personalized Web interface. The agent-based design of the

CROSSMARC architecture allows it to be open, distributed and customizable. The agents implementing each step of the process can be replaced by any other tool with the same functionality that respects the XML-based communication with the blackboard and the ontology. Furthermore, new information extraction systems, covering different languages can easily be connected to the information extraction agent. The ontology also plays an essential role in all stages, providing terms for modeling the relevance of Web pages, language-independent fact extraction, parameterization of the user models, etc. By collecting domain-specific knowledge in the ontology, the various agents become less dependent on the domain. More details about the CROSSMARC architecture and the prototype can be found in [24] and [47].

3.2   Meta-learning for Web Information Extraction

As we have seen in section 2 several approaches to information extraction and ontology learning attempt to combine the strengths of multiple methods in order to obtain better performance. Following this basic idea, our research in the area of Web information extraction has focussed on the combination of different learning methods in a meta-learning framework, aiming to improve recognition performance. For this reason, we have developed a stacked generalization framework that is suitable for information extraction, rather than classification which is the typical use of this approach. Figure 2 illustrates the use of the stacking framework proposed in [46], both at training and at run-time.

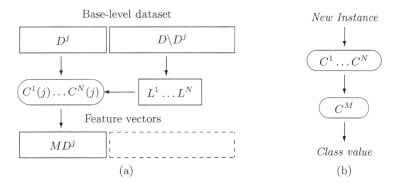

Fig. 2. (a) Illustration of the J-fold cross-validation process for creating the meta-level dataset. (b) The stacking framework at runtime.

At training time, the usual cross-validation approach of stacked generalization is followed, which trains all base-level learners $(L^1 \ldots L^N)$ on various subsets of the training dataset $(D \backslash D^j)$ and applies the learned systems $(C^1(j) \ldots C^N(j))$ on the unseen parts of the dataset (D^j), in order to construct the meta-level dataset (MD^j). However, in the case of information extraction the trained systems may extract different or contradictory information from the same subset of the data. Based on the confidence scores of the information extraction systems, the proposed framework combines their results into a common dataset that is suitable for training a classifier to choose whether to accept or reject an extracted piece of information and if accepted to recognize its type. This approach has led to considerable improvement in recognition performance, which is due to the complementarity of the trained base-level systems.

3.3   Grammar Induction

The importance of grammars for information extraction has become apparent in the description of relevant systems in section 2. With the exception of simple regular patterns, the acquisition of grammars using learning methods is limited to the refinement of specific parameters of hand-made grammars. This is due to the fact that the learning of more complex grammars from text is a hard task. Even harder is the induction of these grammars from positive only examples, which is practically the only type of example that a human annotator can provide. This is the reason why there are very few learning methods that deal with this problem, which are usually only applicable to datasets of small size and complexity. In an attempt to overcome this problem we have developed the e-GRIDS algorithm [37], which is based on the same principles as the GRIDS [26] and the SNPR [50] algorithms, but improves them substantially, in order to become applicable to realistic problems.

e-GRIDS performs a beam search in the space of grammars that cover the positive examples, guided by the MDL principle. In other words, it favors simpler grammars, in the sense that the sum of their code length and the code length of the data, assuming knowledge of these grammars, should be small. The starting state for the search is the most specific grammar, which covers only the training data. The search operators compress and generalize the grammars, by merging symbols and creating new ones. The latest version of e-GRIDS, called eg-GRIDS [38], replaces the beam search with a genetic one, within which the grammar-modification operators are treated as mutation operators. Figure 3 depicts graphically the eg-GRIDS architecture. The use of genetic search has provided a speed-up of an order of magnitude, facilitating the inclusion of more operators that allow the algorithm to search a larger part of the space and produce much better results. Thus, eg-GRIDS can handle larger datasets, and produce better estimates of the "optimal grammar".

3.4   Ontology Enrichment

Our approach to ontology enrichment [48] follows the basic bootstrapping methodology presented in section 2. Figure 4 illustrates the proposed methodology. The bootstrapping process commences with an existing domain ontology, which is used to annotate a corpus of raw documents. In this manner, a training corpus for information extraction is formed, without the need for a human annotator.

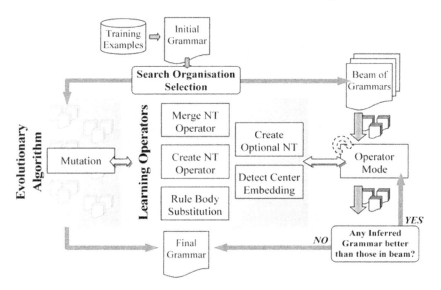

Fig. 3. The architecture of the eg-GRIDS algorithm. (NT: Non-Terminal symbol)

This can lead to a significant speed-up in the development of the information extraction system. The trained system usually generalizes beyond the annotated examples. Therefore, if applied again on the corpus it provides some new instances that do not appear in the initial ontology. These instances are screened by an expert who is responsible for maintaining the ontology. Once the ontology is updated, it can be used again to annotate a new training corpus that will lead to a new information extraction system. This process is repeated until no new instances are added to the ontology. Our initial experiments have shown that this approach works impressively well, even when the initial ontology is very sparsely populated.

As a further improvement of this method, COCLU [49], a novel compression-based clustering algorithm, was developed, which is responsible for identifying lexical variations of existing instances and clustering the lexical variations of new instances. This improvement minimizes the human effort in ontology enrichment, as the extracted instances can be treated in groups of lexical synonyms.

4 Discussion

The combination of ontology learning and information extraction under a bootstrapping framework of iterative refinement seems to be a promising path towards the Semantic Web vision. The use of machine learning for the (partial) automation of knowledge acquisition also seems to be a vital part of this process. However, there are various issues that arise under this framework and need to

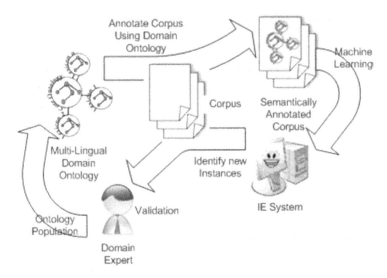

Fig. 4. Ontology enrichment methodology. (IE: Information Extraction)

be researched, in order to arrive at theoretically sound and practically effective solutions. This section raises some of these issues, together with some initial thoughts about them.

4.1 Knowledge Representation Issues

The most straightforward option in terms of knowledge representation is to keep the extraction grammars and the ontologies as separate entities, which will allow us to take full advantage of the work that has been done in each of the two research areas. This is the approach that is adopted in most of the work following the bootstrapping paradigm. However, in section 2 we have also seen some work on alternative representations that combine conceptual and syntactic knowledge under the same representation. This option would simplify the bootstrapping process, but may also have disadvantages, such as the fact that the combined representation needs to be task-specific, which limits the ability of the ontology to provide interoperability and knowledge sharing. Therefore, the combination of the ontology with the grammar remains an open issue to be studied.

If a combination is the preferred solution, a number of new questions arise, such as what type of grammars and what type of ontology one should use. A number of solutions already exist, as we have seen in section 2. However, the choice of an appropriate solution depends largely on the extraction task and the need for syntactic and conceptual support. Recently, the use of more complex grammars and ontologies has been proposed as a solution to the barrier in the performance of extraction systems. Nevertheless, there is still a number of problems that may be addressed with simpler solutions. Therefore, work on the typology of the extraction tasks and the need for resources is necessary.

Finally, in those cases where a combined representation is preferred, we should also study solutions that are inspired by the early work on knowledge representation, e.g. frames and semantic nets. Formal concept analysis and conceptual graphs, as well as probabilistic graphical models are examples of such representations, which have advanced considerably since their conception. Such solutions have started being studied in the context of the bootstrapping framework [6] and may prove very effective in practice.

4.2 Machine Learning Issues

The choice of knowledge representation affects directly the machine learning methods that will be used for knowledge acquisition. If grammars and ontologies are kept separate, the main question is which aspects of the ontology and the grammar will be learned and which will be provided by a human. So far, we have seen almost full automation for simple grammatical patterns and basic conceptual entities and relations. However, there is a host of other methods, such as those that induce context-free grammars, which have not been studied sufficiently in the context of information extraction. An additional issue is whether grammar learning can assist ontology learning and vice versa, i.e., can the elements of the representation acquired through learning be useful at both the conceptual and syntactic level? The answer to these questions depends very much on the type of training data that is available.

Supervised learning of complex representations requires data that may not be possible to acquire manually. Therefore, efforts to automate the generation of training data, such as in [48], are very interesting. Furthermore, unsupervised or partially supervised methods may prove particularly useful. Along these lines, we also need to find better ways to take into account existing background knowledge. Deductive learning methods are the extreme solution in that sense, but inductive learning methods can also benefit from existing knowledge resources.

If we opt for combined representations, it is more than likely that the learning methods will need to be extended, or existing methods will need to be combined in an intelligent way. Multi-strategy learning can prove particularly useful in this respect, as it aims to combine the strengths and special features of different learning methods in the best possible way.

4.3 Content Type Issues

Another major issue that affects directly the typology of extraction and learning tasks is what type of content we want to process. We have already seen that the semi-structured format of Web data can facilitate significantly the information extraction task and the learning of Web site wrappers. Further to that, some semantically annotated content has started appearing. We need to think about how we can make use of that, e.g. as training data for learning new extraction systems, or as background knowledge. An interesting alternative presented in [12] is to treat a set of resources, linked through RDF annotations, as a graph and construct a conceptual model from it.

Multimedia content is increasing on the Web. We need to examine more carefully the task of extracting information from such data, which is more demanding and less studied than text. Can we make assumptions that will allow us to produce practical extraction systems for multimedia data? At what conceptual level can we expect the extracted information to be placed? We need to go beyond the basic low-level features, but how feasible is object recognition within specific domains? Can multimedia ontologies assist in that process (e.g. [20]). There is even some initial work on enriching multimedia ontologies, through the processing of multimedia data [28]. Extraction grammars and the bootstrapping process advocated in this paper could be particularly useful in that respect. Initial ideas on how this can be achieved are presented in [23].

5   Summary

This paper advocates the need for a bootstrapping process, combining ontology learning and grammar learning, in order to semi-automate the construction of ontologies and information extraction systems. The aim of the paper was to present the most relevant work for this purpose, focussing on recent work at SKEL, the laboratory where the author belongs. Several related strands of SKEL research were presented: Web information integration, meta-learning for Web information extraction, induction of context-free grammars and ontology enrichment, through bootstrapping with information extraction learning. Finally, several research issues that need to be addressed towards the realization of the bootstrapping process were discussed. In summary, the main claim of this paper is that machine learning is the vehicle that could help integrate the construction of ontologies and extraction grammars and lead us closer to the Semantic Web.

Acknowledgments

This paper includes ideas and work that are not solely of the author. A number of current and past SKEL members have been involved in the presented work. This research was partially funded by the EC through the IST-FP5 project CROSSMARC (Cross-lingual Multi Agent Retail Comparison), contract number IST-2000-25366. The development of the CROSSMARC architecture has been carried out jointly by the partners of the project: National Center for Scientific Research "Demokritos" (GR), VeltiNet A.E. (GR), University of Edinburgh (UK), Universita di Roma Tor Vergata (IT), Lingway (FR).

References

1. Douglas E. Appelt, Jerry R. Hobbs, John Bear, David J. Israel, and Mabry Tyson. Fastus: A finite-state processor for information extraction from real-world text. In Ruzena Bajcsy, editor, *IJCAI*, pages 1172–1178, 1993.
2. Daniel M. Bikel, Scott Miller, Richard L. Schwartz, and Ralph M. Weischedel. Nymble: a high-performance learning name-finder. In *ANLP*, pages 194–201, 1997.

3. Christopher Brewster, Fabio Ciravegna, and Yorick Wilks. User-centred ontology learning for knowledge management. In Birger Andersson, Maria Bergholtz, and Paul Johannesson, editors, *NLDB*, volume 2553 of *Lecture Notes in Computer Science*, pages 203–207. Springer, 2002.

4. Paul Buitelaar, Siegfried Handschuh, and Bernardo Magnini, editors. *Proceedings of the ECAI Ontology Learning and Population Workshop, Valencia, Spain, 22-24 August.*, 2004.

5. Paul Buitelaar, Daniel Olejnik, and Michael Sintek. A protégé plug-in for ontology extraction from text based on linguistic analysis. In Christoph Bussler, John Davies, Dieter Fensel, and Rudi Studer, editors, *ESWS*, volume 3053 of *Lecture Notes in Computer Science*, pages 31–44. Springer, 2004.

6. Philipp Cimiano, Andreas Hotho, Gerd Stumme, and Julien Tane. Conceptual knowledge processing with formal concept analysis and ontologies. In Peter W. Eklund, editor, *ICFCA*, volume 2961 of *Lecture Notes in Computer Science*, pages 189–207. Springer, 2004.

7. Philipp Cimiano, Lars Schmidt-Thieme, Aleksander Pivk, and Steffen Staab. Learning taxonomic relations from heterogeneous evidence. In Buitelaar et al. [4].

8. Fabio Ciravegna. Adaptive information extraction from text by rule induction and generalisation. In Bernhard Nebel, editor, *IJCAI*, pages 1251–1256. Morgan Kaufmann, 2001.

9. Fabio Ciravegna, Alexiei Dingli, David Guthrie, and Yorick Wilks. Integrating information to bootstrap information extraction from web sites. In Subbarao Kambhampati and Craig A. Knoblock, editors, *IIWeb*, pages 9–14, 2003.

10. Dan Corbett. Interoperability of ontologies using conceptual graph theory. In *ICCS*, volume 3127 of *Lecture Notes in Computer Science*, pages 375–387. Springer, 2004.

11. Berthold Crysmann, Anette Frank, Bernd Kiefer, Stefan Mueller, Günter Neumann, Jakub Piskorski, Ulrich Schäfer, Melanie Siegel, Hans Uszkoreit, Feiyu Xu, Markus Becker, and Hans-Ulrich Krieger. An integrated archictecture for shallow and deep processing. In *ACL*, pages 441–448, 2002.

12. Alexandre Delteil, Catherine Faron, and Rose Dieng. Building concept lattices by learning concepts from rdf graphs annotating web documents. In Priss et al. [39], pages 191–204.

13. T. G. Dietterich. Learning at the Knowledge Level. *Machine Learning*, 1(3):287–316, 1986.

14. David W. Embley. Programming with data frames for everyday data items. In *NCC*, page 301305, 1980.

15. David W. Embley. Towards semantic understanding – an approach based on information extraction ontologies. In Klaus-Dieter Schewe and Hugh E. Williams, editors, *ADC*, volume 27 of *CRPIT*, page 3. Australian Computer Society, 2004.

16. David Faure and Claire Nedellec. Knowledge acquisition of predicate argument structures from technical texts using machine learning: The system asium. In Dieter Fensel and Rudi Studer, editors, *EKAW*, volume 1621 of *Lecture Notes in Computer Science*, pages 329–334. Springer, 1999.

17. Christiane Fellbaum, editor. *WordNet An Electronic Lexical Database*. Bradford Books, 1998.

18. R. Gaizauskas, T. Wakao, K. Humphreys, H. Cunningham, and Y. Wilks. University of sheffield: Description of the lasie system as used for muc-6. In *MUC-6*, pages 207–220, 1995.

19. Udo Hahn and Kornél G. Markó. An integrated, dual learner for grammars and ontologies. *Data Knowl. Eng.*, 42(3):273–291, 2002.
20. Asaad Hakeem, Yaser Sheikh, and Mubarak Shah. Casee: A hierarchical event representation for the analysis of videos. In Deborah L. McGuinness and George Ferguson, editors, *AAAI*, pages 263–268. AAAI Press / The MIT Press, 2004.
21. Jeff Hess and Walling R. Cyre. A cg-based behavior extraction system. In William M. Tepfenhart and Walling R. Cyre, editors, *ICCS*, volume 1640 of *Lecture Notes in Computer Science*, pages 127–139. Springer, 1999.
22. Paul S. Jacobs and Lisa F. Rau. Scisor: Extracting information from on-line news. *Communications of the ACM*, 33(11):88–97, 1990.
23. Vangelis Karkaletsis, Georgios Paliouras, and Constantine D. Spyropoulos. A bootstrapping approach to knowledge acquisition from multimedia content with ontology evolution. In Timo Honkela and Olli Simula, editors, *AKRR*. Helsinki University of Technology, 2005.
24. Vangelis Karkaletsis and Constantine D. Spyropoulos. Cross-lingual information management from web pages. In *PCI*, 2003.
25. Nicholas Kushmerick. Wrapper induction: Efficiency and expressiveness. *Artificial Intelligence*, 118(1-2):15–68, 2000.
26. Pat Langley and Sean Stromsten. Learning context-free grammars with a simplicity bias. In Ramon López de Mántaras and Enric Plaza, editors, *ECML*, volume 1810 of *Lecture Notes in Computer Science*, pages 220–228. Springer, 2000.
27. Alexander Maedche and Steffen Staab. Mining ontologies from text. In Rose Dieng and Olivier Corby, editors, *EKAW*, volume 1937 of *Lecture Notes in Computer Science*, pages 189–202. Springer, 2000.
28. Joseph Modayil and Benjamin Kuipers. Bootstrap learning for object discovery. In *IROS*. IEEE Press, 2004.
29. Ion Muslea, Steven Minton, and Craig A. Knoblock. A hierarchical approach to wrapper induction. In *Agents*, pages 190–197, 1999.
30. Roberto Navigli and Paola Velardi. Learning domain ontologies from document warehouses and dedicated websites. *Computational Linguistics*, 30(2), 2004.
31. Guenter Neumann and Feiyu Xu. Course on intelligent information extraction. In *ESSLI*, 2004.
32. Günter Neumann and Jakub Piskorski. A shallow text processing core engine. *Computational Intelligence*, 18(3):451–476, 2002.
33. Stéphane Nicolas, Bernard Moulin, and Guy W. Mineau. Sesei: A cg-based filter for internet search engines. In Aldo de Moor, Wilfried Lex, and Bernhard Ganter, editors, *ICCS*, volume 2746 of *Lecture Notes in Computer Science*, pages 362–377. Springer, 2003.
34. Norihiro Ogata and Nigel Collier. Ontology express: Statistical and non-monotonic learning of domain ontologies from text. In Buitelaar et al. [4], pages 19–24.
35. Jon Patrick. The scamseek project: Text mining for finanical scams on the internet. In S.J. Simoff and G.J. Williams, editors, *ADMC*, pages 33–38, 2004.
36. Georgios Petasis, Alessandro Cucchiarelli, Paola Velardi, Georgios Paliouras, Vangelis Karkaletsis, and Constantine D. Spyropoulos. Automatic adaptation of proper noun dictionaries through cooperation of machine learning and probabilistic methods. In Nicholas J. Belkin, Peter Ingwersen, and Mun-Kew Leong, editors, *SIGIR*, pages 128–135. ACM, 2000.
37. Georgios Petasis, Georgios Paliouras, Vangelis Karkaletsis, Constantine Halatsis, and Constantine D. Spyropoulos. e-grids: Computationally efficient grammatical inference from positive examples. *Grammars*, 2004.

38. Georgios Petasis, Georgios Paliouras, Constantine D. Spyropoulos, and Constantine Halatsis. eg-grids: Context-free grammatical inference from positive examples using genetic search. In Georgios Paliouras and Yasubumi Sakakibara, editors, *ICGI*, volume 3264 of *Lecture Notes in Computer Science*, pages 223–234. Springer, 2004.
39. Uta Priss, Dan Corbett, and Galia Angelova, editors. *Conceptual Structures: Integration and Interfaces, 10th International Conference on Conceptual Structures, ICCS 2002, Borovets, Bulgaria, July 15-19, 2002, Proceedings*, volume 2393 of *Lecture Notes in Computer Science*. Springer, 2002.
40. Lawrence Reeve and Hyoil Han. The survey of semantic annotation platforms. In *ACM/SAC*, 2005.
41. D. Reidsma, J. Kuper, T. Declerck, H. Saggion, and H. Cunningham. Cross document ontology based information extraction for multimedia retrieval. In *Supplementary proceedings of the ICCS03*, Dresden, 2003.
42. Marie-Laure Reinberger and Peter Spyns. Discovering knowledge in texts for the learning of dogma-inspired ontologies. In Buitelaar et al. [4], pages 19–24.
43. Debbie Richards. Addressing the ontology acquisition bottleneck through reverse ontological engineering. *Knowledge and Information Systems*, 6(4):402–427, 2004.
44. Ellen Riloff and Rosie Jones. Learning dictionaries for information extraction by multi-level bootstrapping. In *AAAI/IAAI*, pages 474–479, 1999.
45. G. Angelova S. Boytcheva, P. Dobrev. Cgextract: Towards extraction of conceptual graphs from controlled english. In *Supplementary proceedings of the ICCS01*, Stanford, USA, 2001.
46. Georgios Sigletos, Georgios Paliouras, Constantine D. Spyropoulos, and Takis Stamapoulos. Stacked generalization for information extraction. In Ramon López de Mántaras and Lorenza Saitta, editors, *ECAI*, pages 549–553. IOS Press, 2004.
47. Constantine D. Spyropoulos, Vangelis Karkaletsis, Claire Grover, Maria-Teresa Pazienza, Dimitris Souflis, and Jose Coch. Final report of the project crossmarc (cross-lingual multi agent retail comparison). Technical report, 2003.
48. Alexandros G. Valarakos, Georgios Paliouras, Vangelis Karkaletsis, and George A. Vouros. Enhancing ontological knowledge through ontology population and enrichment. In Enrico Motta, Nigel Shadbolt, Arthur Stutt, and Nicholas Gibbins, editors, *EKAW*, volume 3257 of *Lecture Notes in Computer Science*, pages 144–156. Springer, 2004.
49. Alexandros G. Valarakos, Georgios Paliouras, Vangelis Karkaletsis, and George A. Vouros. A name-matching algorithm for supporting ontology enrichment. In George A. Vouros and Themis Panayiotopoulos, editors, *SETN*, volume 3025 of *Lecture Notes in Computer Science*, pages 381–389. Springer, 2004.
50. Gerry Wolff. Grammar discovery as data compression. In *AISB/GI*, pages 375–379, 1978.
51. Feiyu Xu, Daniela Kurz, Jakub Piskorski, and Sven Schmeier. Term extraction and mining of term relations from unrestricted texts in the financial domain. In *BIS*, 2002.
52. Manuel Montes y Gómez, Alexander F. Gelbukh, and Aurelio López-López. Text mining at detail level using conceptual graphs. In Priss et al. [39], pages 122–136.
53. Roman Yangarber, Winston Lin, and Ralph Grishman. Unsupervised learning of generalized names. In *COLING*, 2002.
54. Lei Zhang and Yong Yu. Learning to generate cgs from domain specific sentences. In Harry S. Delugach and Gerd Stumme, editors, *ICCS*, volume 2120 of *Lecture Notes in Computer Science*, pages 44–57. Springer, 2001.

Conzilla — A Conceptual Interface to the Semantic Web

Matthias Palmér and Ambjörn Naeve

KMR group at NADA/KTH (Royal Institute of Technology)
Lindstedtsv. 5, 100 44 Stockholm, Sweden
{matthias,amb}@nada.kth.se
http://kmr.nada.kth.se/

Abstract This paper describes our approach to managing the complexity of semantic web-based information by creating conceptual information landscapes (context-maps) and connecting them into a structure called a *knowledge patch*. The aim is to support the creation of overview and clarity by providing mechanisms for presenting and hiding information in flexible ways. Moreover, we present our concept browser *Conzilla* as a knowledge management tool for navigating and editing knowledge patches, and we show how it provides a way to link the human-semantic and the machine-semantic perspectives.

1 Introduction

Today, it is well known that knowledge workers experience difficulties in managing complexity. This includes creating overview as well as supporting collaboration within the plethora of available information sources.

Our *concept browser*[10] *Conzilla*[12] originated from a need to build and present complicated knowledge structures that can enhance the learning process in various ways.

The concept browser approach can be summarized as "*content* in *context* through *concept*", meaning that concepts define an *inside* called *content* and a number of *outsides* called *contexts*. Conzilla context-maps capture "conceptual information landscapes" and connect them into a *knowledge patch*[8, 9] through their common concepts, making use of the *contextual neighbourhood* topology[10]. A context-map represents a focused selection in terms of highlighting some of the available concepts and concept-relations.

Conzilla context-maps are inspired by object-oriented modeling, most notably UML. Moreover, they include navigational support such as hyperlinks and occurrence relations similar to Topic Maps[15], as well as support for information filtering. In order to enhance the comprehensiveness and intuitiveness of the context-maps, we have used a more linguistically coherent modeling technique called *Unified Language Modeling* (ULM) [10], which focuses on depicting how we speak about concepts and their relations.

The technical basis for the semantic web is the description language RDF[5]. We have chosen to equip Conzilla with an RDF backend because we wanted to

F. Dau, M.-L. Mugnier, G. Stumme (Eds.): ICCS 2005, LNAI 3596, pp. 136–151, 2005.

strengthen the bridge between human- and machine-understandable semantics. Moreover, RDF represents an important step in the right direction with regard to scalability and extensibility. In general, with millions of users on the web - users with diverse and sometimes very conflicting opinions - it is crucial that RDF is designed to support knowledge representation in a scalable manner. It is also very important that items of knowledge can refer to other such items. Otherwise, questions like 'who said this' will not be answerable in a standardized manner. From a learning perspective these issues are equally important, especially since we believe that learning should not be regarded as an activity that is separated from other activities.

This paper presents Conzilla as a concept browser, which provides functionality such as hyperlinks between contex-maps and the specific separation between *context* and *content*. We believe that these features - beyond being useful in the context of a concept browser - are beneficial for the Semantic Web, since they contribute to providing a better overview and navigability of the rather verbose and obscure expressions in RDF.

The paper describes how context-maps are defined in RDF in order to enable generic user interfaces to edit and present knowledge expressed in RDF. Inspired by the needs of technology enhanced learning environments, we show how to design context-maps that allow parts of RDF expressions to be summarized, suppressed and presented in a form that is more comprehensible to humans. An effort has been made to design the management of context-maps in such a way that they can be kept in tune with a variety of different knowledge sources. In order to support flexible annotation and querying in Conzilla, we have used a framework called SHAME[4], which provides form-based presentations of information[1].

The paper is structured in the following way: In section 2 we discuss the Conzilla concept browser and present an overview of its different capabilities. In section 3 we compare Conzilla with similar tools, surveying the state of the art. In section 4 we discuss the design of context-maps, in section 5 we focus on their expression in RDF, and in section 6 we present our conclusions and outline some future work in this area. A few more technical issuses on how to handle RDF graphs are presented in an appendix.

2   Short Conzilla Interface Overview

In this section we introduce Conzilla as a conceptual browsing and editing tool and examplify the three aspects of presentation, *surf, view,* and *info*. We conclude by explaining the use of *sessions* in order to keep track of a knowledge patch, i.e. a set of context-maps that belong together.

2.1   Conceptual Browsing

Conzilla allows you to browse context-maps and inspect concepts and concept-relations. Figure 1 depicts two context-maps and a popup menu over the concept

---

[1] SHAME stands for Standardized Hyper-Adaptable Metadata Editor

'Brew Coffee' (in the map to the right) showing the alternatives *surf, view* and *info* on that concept.

Surf shows the contextual neighbourhood of the concept 'Brew Coffee', i.e. which other maps this concept occurs in. In this case 'Brew Coffee' occurs in precisely the two maps in the figure. Note that the 'Brew and serve Coffee' map is greyed out, since it is this context-map that is being navigated.

View allows the user to inspect content, or open up the 'inside' of a concept, creating a list of its *content-components* within its present context. For the 'Brew Coffee' concept, its list of content-components within the 'Brew and serve coffe' context is shown in the map to the right.

Info allows the user to bring up hidden information that is not directly shown in the map. The information is shown in a metadata inspector, similar to the form-based metadata editor shown to the right in figure 2. Part of the information is displayed as transparent popups in the context-maps as a result of the mouse hovering over something.

2.2   Conceptual Editing

Figure 2 depicts one of the context-maps from figure 1 - but this time in *edit* mode. It also shows a Dublin Core [1] metadata editor for the newly created concept of type 'Activity' with the English title 'Drink'. In order to edit other metadata fields than those provided by Dublin Core, you can change the metadata-form in the choice-box at the top, where it says 'Dublin Core'. Only a few metadata-forms are provided by default, e.g. Dublin Core, LOM and FOAF. However, via the *Formulator* application of the SHAME framework, you can create new metadata-forms or reuse parts from established standards / schemas. Which metadata-form that should be used in the browse mode can be specified in the style layer. The metadata can be filtered according to your language preferences, which are controlled from the 'settings' menu.

The menu shown in figure 2 is of the 'create' type, where you choose what kinds of concepts to create in the context-map.

2.3   Sessions for Adminstrating a Knowledge Patch

The context-maps of a knowledge patch are edited in *sessions*. Among other things, a session specifies *containers* [2] for storing concepts and concept-relations - as well as graphics and style information. From a collaboration perspective it is important to be able to use sessions with different containers for the same context-map. If these containers are loaded, the map shows everything combined. In editing mode, you can still see everything, but you can only edit according to the priviledges of your session. The *container-manager* can be used in order

---

[2] presently, containers are restricted to be regular RDF/XML files accessible locally or via ftp. We plan to support remote containers using RDF-stores like SCAM[13] or SESAME[3].

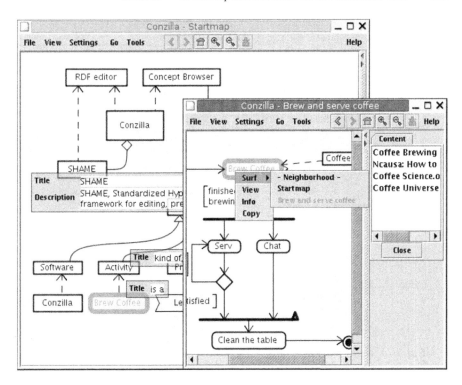

Fig. 1. The context-map in the back displays a ULM class/instance diagram with three transparent pop-ups showing Dublin Core information around the concept 'SHAME', a relation named 'kind of' (similar to rdfs:subclassof), and a relation named 'is a' (similar to rdf:type). The concept 'Brew Coffee' is highlighted (in both maps) in order to indicate that it has been selected. In the context-map in front we see a ULM activity diagram with a popup-menu showing the three alternatives, *Surf*, *View* and *Info* on the selected concept. The surf-submenu shows the conceptual neighborhood with the present context-map (Brew and Serve Coffee) grayed out. Since the 'Info' alternative was previously chosen, the list of content-components for the concept 'Brew Coffee' is still shown to the right.

to investigate context-maps that have mixed origin. The concepts and concept-relations can also be grouped into different *layers* in order to simplify the editing process or aid in a certain presentation perspective.

3   State of the Art

As a remedy for the web's lack of semantics - as well as to meet the rapidly increasing need to express metadata about web resources - the W3C has defined RDF[5]. At its basic level RDF has very little semantics. However, with the help of the RDF Vocabulary description language[2], people are encouraged to

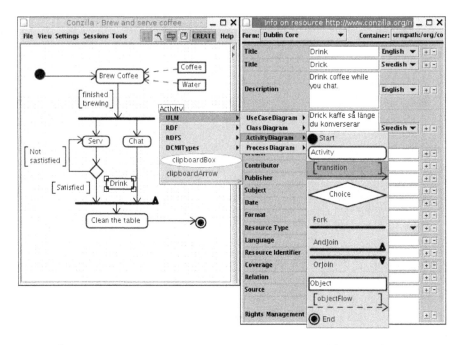

Fig. 2. A context-map showing how to brew coffe as a ULM Activity diagram. The activity 'Drink' has just been inserted, and the user is choosing the property 'transition' in the 'type' menu in order to connect the activity to the synchronization bar (which contains an "AND" sign to the right indicating the semantics of the join). To the right is a Dublin Core metadata editor for the 'Drink' activity, with Swedish translations of its 'Title' and 'Description'.

introduce new layers of semantics on top of the old ones. For more complicated data, the Web Ontology Language, OWL[14] is probably more suitable. It is important to notice that the base of RDF is agnostic to the kind of informtion that is actually expressed. There are many applications for working with RDF and the more specific languages expressed with the help of it. General frameworks such as RedLand [3], KAON [4], Jena2 [5], Sesame [6] and SCAM [7] for parsing, storing, querying and making connections to inference engines are sufficient for their respective purposes. However, generic authoring tools, especially visual, graph-based interfaces for end users, still seem to be lacking.

We will now compare a set of tools that aim towards filling this gap, qualifying by either being capable RDF editing tools (preferably graph-based) or by

---

[3] http://www.redland.opensource.ac.uk
[4] http://kaon.semanticweb.org/
[5] http://jena.sourceforge.net/
[6] http://www.openrdf.org/
[7] http://scam.sourceforge.net/

being modeling tools, especially if their functionality resembles that of a concept browser. In figure 3 we compare our tools Conzilla [8] and Meditor [9] with IsaViz [10], VUE [11] RDFAuthor [12], InferEd [13], and Protege [14].

We claim neither that this list of tools is complete, nor that the features considered are exhaustive with respect to the capabilities of the other tools. Instead, the features brought up reflect what we believe to be important in the design of a concept browser/RDF editor. Hence, it is no mystery that Conzilla stands out as the most feature-rich tool in this comparision.

Feature number:	Feature 1 to 3 requires that the tools can edit or export to RDF			4	Feature 5 to 9 requires that the tool supports a visual graph view				
	1	2	3		5	6	7	8	9
Feature description:	X = Works natively with RDF O = Some 'compatible' representation - = Not applicable	X = Edits any kind of RDF O = Editing restricted somehow - = Not applicable	X = Combines multiple RDF models O = Model centric - = Not applicable	X = Visual – graph like view O = Only text oriented view - = Not applicable	X = Has WYSIWYG editing O = Separated editing - = Not applicable	X = Can suppress information O = Presents everything - = Not applicable	X = Customizable layout O = Automatic layout only - = Not applicable	X = Appearance controlled by style O = Appearance fixed by the tool - = Not applicable	X = Navigation between views O = No navigation - = Not applicable
Tools:									
Conzilla	X	O	X	X	X	X	X	X	X
IsaViz	O	X	O	X	X	X	X	X	O
VUE	-	-	-	X	X	X	X	O	X
RDFAuthor	O	O	O	X	X	O	X	O	O
InferEd	X	X	O	O	-	-	-	-	-
Protege	O	O	X	X	O	O	O	O	O
Meditor	X	O	X	O	-	-	-	-	-

Fig. 3. A feature overview of RDF and/or conceptual modeling tools with respect to nine features ranging from representation format to apperance and navigation.

Let us perform a more qualitative investigation of IsaViz and VUE. We select IsaViz because it is a capable graph-based RDF-editor, and we select VUE because it is a capable knowledge-structuring and modeling tool.

VUE is a concept-mapping tool that connects to - and filters - the content of various digital repositories. The focus is on maps and the management of content in nodes and links. It has many nice features, including the filtering of content and presentation paths through maps. However, in comparison with

---

[8] http://www.conzilla.org/
[9] Meditor is an application of the SHAME framework, http://kmr.nada.kth.se/shame
[10] http://www.w3.org/2001/11/IsaViz/
[11] http://vue.tccs.tufts.edu/
[12] http://rdfweb.org/people/damian/RDFAuthor/
[13] http://www.intellidimension.com/pages/site/products/infered/default.rsp
[14] http://protege.stanford.edu/

Conzilla it lacks a visual language strong enough for modeling in e.g. UML and since nodes cannot occur in several maps the navigational aspects (feature 9) are rather weak compared to the capabilities of a concept browser. Furthermore the separation between information and presentation is done only for content, while the nodes and links appear in the presentation layer only. From the perspective of comparing VUE with Conzilla it would be interesting if the presentation file format used RDF. And provided machine understandable semantics for the concepts, concept-relations, navigational aspects and the relation to content. However, as it stands now, VUE is not an RDF tool (feature 1,2, and 3), even though it might integrate content from repositories that uses RDF internally.

IsaViz on the other hand has a strong focus on the information representation as is mainly an RDF editor. It has a lot of nice features like a zoomable interface and the ability to apply Graphical Stylesheets[16] to customize the presentation of individual graphs or schemas. However, because of how RDF is designed, there are problems with referring to parts of graphs from the outside. This has as a consequence that the IsaViz designers have chosen to use their own format (feature 1) for saving graphs whenever the appearance is customized (e.g. when doing your own layout or suppressing information). Unfortunately this has some serious drawbacks, e.g. you cannot customize the presentation of an RDF graph without making a snapshot of it, and therefore, independent subsequent changes cannot be incorporated without starting all over. IsaViz is oriented around RDF, not around maps and consequently has no navigational primitives beyond the navigation within the customized view of a single RDF-graph (feature 9).

4   Context-Map Design

Context-map should present concepts and concept relations in a manner that fullfills the requirement of a concept browser, basically supporting the three presentation modes of surf, view, and info. Moreover, from the perspective of authoring context-maps there are some additional design requirements:

1. Concepts and concept-relations are human expressions that may be more or less formal depending on the usage scenario. Still, it is important that the underlying expression has the potential for being machine readable. Hence, the expressibility of RDF and the formal languages that are available on top of it, makes RDF a good choice for expressing concepts and concept-relations.
2. Expressing information is often a task that requires collaboration or at least positioning within a larger setting. Hence, it is important that authors can work with separate information sources and still have access to a unified view. Therefore, we have to allow context-maps as - well as concepts and concept-relations - to be distributed in several RDF-graphs and to be separately editable.
3. Human expression is seldom uncontroversial, and it is important to allow the same information to be viewed from different perspective without loosing sight of the similarities. Reuse of information without making copies

requires precise referencing techniques, where the presentation of the originating context-maps can be left out. Hence, it is important to provide good separation between the information and presentation layers of a context-map.

Support for some of the reasoning behind these design issues can be found e.g in [17] and [6]. In the following subsections we will describe the context-map design in terms of three different layers, an *information*, a *presentation* and a *style* layer. This division in layers is inspired by how the regular web has evolved into a division between data storage (typically a database), HTML (typically generated via some template language) and a stylesheet (typically CSS or XSL).

4.1   Information Layer

The information in the information layer is made up of statements around resources expressed in RDF. In most cases the actual resources have an extent that cannot be captured in the information layer. This extent may include things with digital representations like documents or pictures or it might equally well include things like ideas, processes or people. A concept in a context-map is constituted of a subgraph of RDF-statements centered around a specific RDF-resource that represents it. In a similar manner a concept-relation in a context-map is constituted of a subgraph of RDF-statements centered around the reification of a specific RDF statement.

The expression of the information may follow a schema, a standard or be locally defined, it depends on what is suitable for the domain and the user.

As mentioned above, information presented in one context-map may reside in several RDF graphs.

4.2   Presentation Layer

The focus of the presentation is the idea of a context-map as introduced in [10]. In short it is a map of concepts and connecting concept-relations presented by boxes and lines. There might be text in the form of a label within the box and on the side of the line, both excavated from the information layer.

Every concept or concept-relation has its own integrity, i.e. its existence is not bound to a specific context-map. The concepts or concept-relations are included in a context-map via *layouts*. Layouts are intermediate resources holding information regarding position, size, and in-line styles such as text-alignment, visibility etc. Observe that the same concept or concept-relation may be inserted several times in the same map yielding different layouts. There is a special kind of layout, which groups other layouts. Such layouts are typically used for creating context-map layers.

In addition to the context-map presentation it is possible to interact with concepts and concept-relations in various ways. The three interactions presented below are central to the the idea of a concept browser and quite useful for an RDF editor as well:

Surfing First of all, there are hyperlinks on concepts and concept-relations leading to other context-maps. A hyperlink is stored on the layout, hence for a concept / concept-relation there might be different hyperlinks depending on where it is encountered. Second, since the same concept (or concept-relation) may occur in different context-maps it is possible to provide a *contextual neighborhood* [10] which is a list of all context-maps where the given concept (or concept-relation) occurs.

Since it is possible to include concepts / concept-relations in context-maps without anyone else's knowledge, it is impossible to be sure that you have a complete listing of contextual neighbourhoods. We have considered and tested to use the p2p network Edutella [11] for finding contextual neighbourhoods. However, in the long run it might be neccessary to develop a specific service which specializes in keeping track of contextualizations of concepts / concept-relations for efficiency reasons. The current implementation of Conzilla only checks all RDF-graphs that are loaded already.

Viewing Content One of the main design principles of a concept browser is to have content-components separated from their initial context. Content-components can then be assigned to relevant concepts or concept-relations as e.g. explanations, examples, motivations, discussions, or knowledgable people. It is possible to assign content-components to a concept or concept-relation within the scope of a context-map. These content-components will not be visible on that concept in any other context-map.

Just like a concept or a concept-relation a content-component might have an extent which goes beyond what is expressible in the information layer. Hence, it is only if a content-component is detected as a retrievable digital resource, it can be shown in a *content browser*. In most cases a regular web browser is suitable as a content browser, at least as an intermediate step in order to launch a better suited application.

Inspecting Further Information Presenting labels on concepts and concept-relations is nice but in general just represents "a scratch on the surface". For example, imagine that we are using the RDF version of the metadata standard Dublin Core [1] to express e.g. title, description, creator, creation date, subject, relations, and rights. The title is obviously suitable to use as a label and the relations can clearly be shown as conceptual relations. A context-map can, if needed, present the other fields in a graph manner as well, e.g. the date as a relation to a literal. However, in most cases it is more suitable to present metadata fields with string values as pure text in a form.

A set of fixed forms, to cover all possible situations, is not flexible enough. Instead we rely on the SHAME framework [15][4] to generate forms from small form-snippets called *formlets*. Much like how Conzilla separates information from its presentation SHAME formlets uses *queries* to capture elements of the RDF-graph and *form templates* to generate actual forms. Which formlets that should

---

[15] Documentation for SHAME can be found at http://kmr.nada.kth.se/shame

be used in a certain setting might be specified statically in the context-map or triggered via a type as specified in a stylesheet. Furthermore, by switching manually among available forms, more relevant information can be discovered. Another approach, yet to be perfected, is to let relevant formlets be automatically detected from the information itself, allowing all relevant information to be presented at once.

In the editing mode of Conzilla, SHAME is also used as the basis for a metadata editor, see figure 2.

4.3 Style Layer

A *style* describes the apperance of boxes and lines, typically their form, linetype, linewith, text alignment etc. A *local style* is a style applied to a specific layout of a given resource. A *global class style* is a style that is applied to an RDF Class or an RDF Property. A *global instance style* is a style that is applied to a specific resource, independently of context. More specifically, a global class style applies to *all* concepts or concept-relations that are expressed as instances of the RDF Class or RDF Property in this global class style. When detecting instances, we also take into account RDF Schema information, i.e. *subClassOf* and *subPropertyOf*. OWL ontologies are not yet taken into account because it is still unclear to us how they relate to the scope of context-maps, which relies heavily on an open-world assumption.

If a local style and a global style are simultaneously relevant, then the local style takes precedence. Observe that we have to override all parts of a style explicitly. As an example, if we provide a local style of a concept, which changes the label alignment but not the form of the box, this form will be determined by the global style - if one has been provided. Moreover, Conzilla offers a "fallback style" that applies if there is no other style that does so.

The present lack of an intermediate style level means that a context-maps cannot be associated with a specific and reusable set of styles. Of course you could force inline styles everywhere but that is not a recommended approach. Instead we plan to introduce a *style set* which would be the equivalence of a CSS style sheet. A style set would need to have a selector construction - similar to the ones in CSS or preferrably GSS [16]. A style set would override a global style, but be overridden by an local style.

Currently the style information makes use of a fixed set of hardcoded graphical primitives.

5   Context-Maps Expressed in RDF

In this section we will consider the RDF-expressions of the three layers in more detail. This will also include some nitty-gritty details of how RDF represents information in the information layer and how we can refer to that information via referring to RDF-constructs. In order to avoid confusion, we will refer to the RDF-expression of the information as *information triples*, to the RDF expression of context-maps as the *presentation triples* and to the RDF expression of styles as *style triples*.

5.1   General Thoughts on the Context-Map Construction in RDF

There are several reasons why we have chosen to express context-maps in RDF. First, this enables good integration with the information triples using internal referencing techniques such as URIs and the reification mechanism. Second, it allows inference engines to easily make use of the combination of information and presentation triples. Third, it allows context-maps to be extended and reused in other contexts. Fourth, it allows flexible authoring and annotation of the context-maps themselves, effectively allowing statements like, "I agree with what was said about that information".

There are several reasons why context-maps should have the ability to present information without changing it. The simplest one is that you may just have read-only access to the information triples. Hence, it is neccessary that the presentation triples can be located in different RDF-graphs than the information triples. Equally important is the condition that the presentation triples of the context-maps should not express anything that would change the semantics of the information triples. This is important if the two graphs are to be stored together or managed by tools without prior knowledge of context-maps. Observe that if the intention of the user is to express information that adds to - or changes - the semantics of existing information, he or she should of course be allowed to do this. But the presentation triples of context-maps should not do this automatically by their mere existence.

From this we clearly recognize the need for references to RDF-constructs across the borders of RDF-graphs.

5.2   Referring to Resources and Triples

As pointed out above, the layouts of a context-map should refer to the information triples. But we also need to reference the resources that are spoken about in the triples. Lets first note that a resource is something that is in general outside of RDF, it is merely referenced by a URI. You have to use a domain specific interpretation function[7] to get from the URI to the actual resource. This is quite natural since URIs can denote anything from e.g. a car to the idea of a perfect circle.

Hence, a layout-resource references a resource simply via its URI just like how the information triples references it. However, since we actually are interested in information around a resource as well as the resource itself we have to add a reference to the *container* where the information triples is stored. Currently the reference to the container is calculated [16] rather than explicitly stored on the layout or context-map.

---

[16] In some cases this is not possible, it remains to define an unambigous scheme for when it has to be expressed and when it can be calculated.

Triples, by default, have no identifiers, instead a layout-triple refers to a reification which in RDF is a standardized and identifiable representation of a specific triple [17]. Since a resource may be presented by several layout-resources, the layout-triple must indicate which layout-resources it refers to. Obviously, the layout-triples indicated layout-resources should match the ends of the reification referred to by the layout-triple, see Figure 4. If they do not match, the

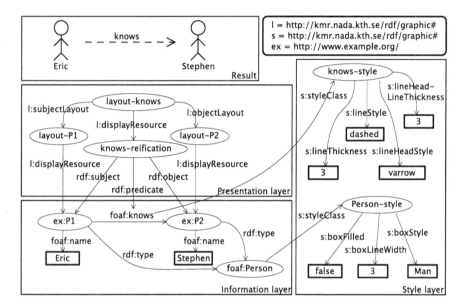

Fig. 4. An overview of the three layers (information, presentation and style) behind the fact that "Eric knows Stephen". The top left box shows the corresponding context-map, while the other three boxes show the underlying RDF representations. More specifically, the presentation layer (middle box to the left) shows three layouts: two layout-resources referring to their respective resources in the information layer (bottom left box), and one layout-triple (the one in the middle) referring indirectly to the triple in the information layer via a reification resource named 'knows-reification'.

layout-triple is incorrectly constructed. For a layout-triple we need to apply the interpretation function twice, since we first have to interpret the reification resource in order to get to the triple in the information triples, and then we must interpret the triple in order to get to the information that it expresses.

---

[17] In a given RDF graph there is only one triple with a given subject, predicate and object.

5.3   Referring Literals and Anonymous Resources from Layouts

It is not enough that context-maps can present resources and their connecting triples. Both literals and anonymous resources occur frequently and therefore we should be able to present them individually. Hence, we should try to find some way to refer to them from layouts.

Unfortunately, there is no perfect solution in standard RDF unless we force the presentation and information triples to be stored together. However, since the requirement to be able to keep them separate is vital, we choose to live with imperfect solutions and encourage people to use the integrated form based approach based on SHAME for displaying anonymous nodes and literals in connection with non anonymous resource. In the appendix we scetche an approach which have been partially implemented in Conzilla as of when this paper was published.

5.4   Style Expression

Figure 4 shows two different global class styles - one for the RDF Class foaf:Person and one for the RDF Property foaf:knows. The RDF-expressions for a global instance style and local style are almost identical. The only difference is that they are connected via the styleInstance property to the corresponding resources and layouts. The RDF-design of style sets remains to be developed.

6   Conclusions and Future Work

In this paper we have shown how context-maps can be designed and implemented (in Conzilla) in order to effectively present information expressed on the semantic web. The context-maps have been designed and implemented in a manner that allows WYSIWYG editing. Moreover, the context-maps have been expressed via a three-layered solution, where the lowest layer - the information layer - has been allowed to remain independent of the other two layers - the presentation layer and the style layer. In the presentation layer, the author is allowed to combine several sources into one or several maps, as well as to customize the layout and to specify the navigation through hyperlinks or contextual neighbourhoods. The style layer provides the final touch - together with the form-based metadata displays that are leveraged by the SHAME framework.

Future work will include the investigation of a better design for styles that are bound to maps, which we presently think of in terms of style sets. We will also think more about how to reference anonymous resources and literals. We also want to provide integration with RDF-based storage-and-access solutions like SCAM and Sesame. Also, in order to support inference-type business rules, the relations to OWL will be worked out, and in order to enable process management, a workflow engine will be interfaced. And of course, the interface of Conzilla needs improvements, which will be achieved through user-testing and feedback. We also plan to develop thin clients for use in browsers or mobiles. In fact, preparatory work in this direction is already under way.

References

[1] The Dublin Core Metadata Initiative. http://dublincore.org.

[2] D. Brickley and R.V. Guha. RDF Vocabulary Description Language 1.0: RDF Schema. http://www.w3.org/TR/2004/REC-rdf-schema-20040210/.

[3] J. Broekstra, A. Kampman, and F. van Harmelen. Sesame: A Generic Architecture for Storing and Querying RDF.

[4] H. Eriksson. Query Management For The Semantic Web. http://kmr.nada.kth.se/papers/SemanticWeb/CID-216.pdf.

[5] K. Graham and J. J. Carroll. Resource Description Framework (RDF): Concepts and Abstract Syntax. http://www.w3.org/TR/2004/REC-rdf-concepts-20040210/.

[6] T. Gruber. Every Ontology is a treaty - a social agreement - among people with some common motive in sharing. AIS Sigsemis Bulletin 1(3), October 2004.

[7] P. Hayes. RDF Semantics. http://www.w3.org/TR/2004/REC-rdf-mt-20040210/.

[8] A. Naeve. The Garden of Knowledge as a Knowledge Manifold - A Conceptual Framework for Computer Supported Subjective Education. CID17 project report: http://kmr.nada.kth.se/papers/KnowledgeManifolds/cid_17.pdf, 1997.

[9] A. Naeve. The knowledge manifold - an educational architecture that supports inquiry-based customizable forms of e-learning. Proc. of the 2nd European Web-based Learning Environments Conference (WBLE 2001), pp. 200-212, Lund, Sweden, 2001.

[10] A. Naeve. The Concept Browser: a new form of knowledge management tool. In Proceedings of the 2 nd European Web-based Learning Environments Conference (WBLE 2001), pp. 151-161, Lund, Sweden, 2001.

[11] W. Nejdl, B. Wolf, C. Qu, S. Decker, M. Sintek, A. Naeve, M. Nilsson, M. Palmér, and T. Risch. EDUTELLA: A P2P Networking Infrastructure Based on RDF. Proceedings of the 11th World Wide Web Conference, 2002.

[12] M. Nilsson. The conzilla design - the definitive reference. CID project report: http://kmr.nada.kth.se/papers/ConceptualBrowsing/conzilla-design.pdf, 2000.

[13] M. Palmér, A. Naeve, and F. Paulsson. The SCAM Framework: Helping Semantic Web Applications to Store and Access Metadata. In *ESWS*, pages 167–181, 2004.

[14] F. P. Patel-Schneider, P. Hayes, and I. Horrocks. OWL Web Ontology Language Semantics and Abstract Syntax. http://www.w3.org/TR/2004/REC-owl-semantics-20040210/.

[15] S. Pepper. The TAO of Topic Maps. http://www.ontopia.net/topicmaps/materials/tao.html.

[16] E. Pietriga. Graph Stylesheets (GSS) in IsaViz. http://www.w3.org/2001/11/IsaViz/gss/gssmanual.html.

[17] A. Sheth. The Informations Systems Perspective on Semantic Web Research. AIS Sigsemis Bulletin 1(1), April 2004.

Appendix — An Approach to Referencing Literals and Anonymous Nodes

By definition, *anonymous* resources have no identifiers outside of the RDF-graph where they occur. Therefore an anonymous resource cannot be referenced directly - except from triples in the same RDF-graph. If we want to refer to an anonymous resource, we need to invent an indirect referencing technique. Below we describe an approach called *graph-patterns* that captures anonymous resources and literals as a special case.

Referencing Anonymous Resources We will here define and state some findings without proof, a more formal treatment remains to be done.

Def: The *marked-graph* of a anonymous resources A consists of the RDF-graph where A occurs, and an extra marker-triple wherein A is the subject[18]. It follows that two anonymous resources are *indistinguishable* if their marked-graphs are *isomorphic* (defined in [7]). Even though it is not always possible, it is nevertheless interesting to consider the graph-pattern[19] that provides the most detailed matching of an anonymous resource.

Def: for an anonymous resource A, we say that a graph-pattern is *complete* relative to A, if the graph pattern captures the *anonymous closure*[20] of A. It can be shown that a complete graph pattern for A matches A - as well as each of its indistinguishable anonymous resources - which is the best you can expect under these circumstances.

Moreover, small independent changes to the surrounding RDF-graph should not invalidate the graph-pattern and break the references to its anonymous resources. From this perspecitve we can identify two inherently conflicting requirements on graph-patterns:

1. If we want to reference anonymous resources uniquely, the best we can do is use complete graph-patterns.
2. If we want to minimize the risk of broken references, a smaller graph-pattern reduces this risk.

It should be noted that in most practical situations complete graph-patterns are quite small and hence the conflict between the two requirements is neglible. A second somewhat weaker approach, which in most practical situations coincides with complete graph-patterns, is to use graph-patterns that capture all incoming and outgoing triples. A third - and even weaker - approach, is to use graph-patterns that only capture the incoming and outgoing triples that are shown in the context-map from where the anonymous resource is being referenced. In the last approach, the context-map itself can function as the graph-pattern, no secondary expression is neccessary. Whenever it is not enough to reuse context-maps as graph-patterns, an external query language is needed. The Edutella Query Language (QEL) [11] is a good alternative. In fact, the RDF-expression of QEL uses reifications that refer to *variables* instead of fixed resources, which is precisely how we will reference anonymous nodes in the triple- and resource-layouts described in section 5.2.

---

[18] the predicate and object should not have been introduced in the RDF-graph already.
[19] a graph-pattern is a query where the required anonymous resource is captured in a specific variable
[20] i.e. reachable graph from a node where only anonymous nodes may be traversed.

Referencing Literals Referencing literals constitutes a special case of referencing anonymous resources. In the case where you do not have an anonymous resource as the subject of the triple where the literal is expressed as object, the graph-pattern will be comparable to a reification. Moreover, if literals change often compared to resource URIs it is a bad idea to rely on exact string matching of the literal in the graph-pattern. The simplest solution is to replace the literal with a variable. However, this does not work when there are several triples that differ only in their literals. In such cases a constraint could be added on the variable in order to distinguish the literal in question. However, this approach would have to rely on heuristics and further investigation is needed to decide wether it would be worth the effort.

Variables in Concept Graphs

Frithjof Dau

Technische Universität Darmstadt, Fachbereich Mathematik
Schloßgartenstr. 7, D-64289 Darmstadt
`dau@mathematik.tu-darmstadt.de`

Abstract A main feature of many logics used in computer science is
a means to express quantification. Usually, syntactical devices like variables and quantifiers are used for this purpose. In contrast to that, in
conceptual graphs, a single syntactical item, the generic marker '∗' is
used. Nonetheless, sometimes conceptual graphs with variables have to
be considered. If the generic marker is replaced by variables, it has to be
investigated how this syntactical difference is reflected by the semantics
and transformation rules for conceptual graphs. In this paper, this task
is carried out for the system of concept graph with cuts (CGwCs). Two
different classes of CGwCs with variables are introduced, and for both,
a semantics and an adequate calculus for CGwCs is provided.

1 Introduction

A main feature of many logics used in computer science, like the different versions of description logic, first oder logic (FOL), conceptual graphs or the resource description framework (RDF), is a means to express quantification. In
any linear symbolic notion of logic (like the common notions for FOL or description logic), variables and quantifiers are the syntactical entities which are used
for this purpose. Even in RDF, which can be understood as a diagrammatic
reasoning system, so-called *blanks*, which are very much used like variables, are
used to express existential quantification.

The use of variables has some consequences for the handling of formulas. A
variable may occur several times in a formula. In order to grasp the meaning
of a formula, one has to keep track of these different occurrences, particularly,
whether they are in the same scope of an (existential or universal) quantifier.
Particularly, we have to distinguish between variables and their occurrences in a
formula. Moreover, one needs some means to rename variables, which has to be
captured either by a convention like the so-called *alpha conversion* of formulas,
or by some rules in the calculus.

For conceptual graphs, the situation is different. There exists only one syntactical element which is used to express existential quantification: the generic
marker '∗'. In a conceptual graph, different occurrences of '∗' as referent in different concept boxes refer to (not necessarily) different entities. As only one sign
for quantification is used, one does not have to keep track of generic markers.
Moreover, in the handling of conceptual graphs, there is no need to distinguish

F. Dau, M.-L. Mugnier, G. Stumme (Eds.): ICCS 2005, LNAI 3596, pp. 152–165, 2005.
© Springer-Verlag Berlin Heidelberg 2005

between the sign and its occurrences, and a convention like the alpha-conversion of symbolic logic is not needed. This makes conceptual graphs easier to read and handle by humans, which is one of their mains goals. Using '∗' instead of variables is a strong means to achieve this goal.

Nonetheless, sometimes it is reasonable to use variables in conceptual graphs as well. For example, finding a *linear* notation for conceptual graphs (the linear form) is much easier if generic markers are replaced by variables.[1] The use of variables instead of generic markers in any fragment of conceptual graphs has to be reflected by the semantics, as well by any calculus. Approaches for some specific forms of concept graphs with variables are investigated be Klinger in [7, 8] and Wille in [14, 15]. But, to the best of my knowledge, a comprehensive discussion between the differences of conceptual graphs with variables and conceptual graphs with generic markers has not been provided yet.

For simple conceptual graphs, which do not provide nestings or any means of negation, the use of variables instead of generic markers does not lead to major problems. The definition of their syntax and semantics is straightforward. Moreover, a calculus based on projections can easily provided. For example, the projections of [2] for conceptual graphs or the projection-like mappings for RDF (see [6, 1]) could be adopted for simple conceptual graphs, or the diagrammatic calculus of a work in preparation ([4]) could be used. But we run into problems if we add a means to express negation to conceptual graphs.

Recall that in conceptual graphs, a negation box ¬| 𝔊 | is used to negate the enclosed subgraph 𝔊. Let us first consider some simple examples to see some problems we have to cope with. Assume that we allow that arbitrary concept boxes are labeled with the same variable. Consider the following two graphs:

The intuitive meaning of the left graph is clear: It is 'it is not true that there exists an x with $S(x)$', i.e., no object has the property R. But what about the right graph? As we have no special sign like the quantifiers of symbolic logic for indicating the scope of variables, it is usually assumed that all variables in a conceptual graph denote the same object, thus the scope of a variable (i.e., the context where the existential quantification takes place) is the innermost context which contains all boxes which are labeled with x. For our right graph, this is the sheet of assertion. Realizing that the scope of a variable x can be a context where it does not occur (the sheet of assertion in our example) is counter-intuitive, and finding adaquate semantics and derivation rules for such graphs leads to unnecessary technical difficulties. Due to this problem, Sowa considers only conceptual graphs where each coreference set has a *defining label*.

---

[1] It is well known that conceptual graphs are based on Peirce's diagrammatical *existential graphs*, where *lines of identity* are used to express existential quantification and identity. But even Peirce replaced in some places the lines of identity by variables, which he called *selectives* (but he strongly recommended to avoid them).

The graph we consider is semantically equivalent to the graph below which has defining labels.

So, if we add a means to express negation, we have to be careful in the definition of the syntax and semantics of conceptual graphs with variables.

It is possible to consider conceptual graphs with variables where each variable occurs almost once. Or graph is equivalent to each of the two graphs below which satisfy this restriction.

Of course, the semantical equivalence of all the graphs we considered has to reflected by any calculus as well, i.e. in any calculus, we need rules which meet the specific properties of variables. For example, we need rules which allow to rearrange boxes which are labeled with the same variable, and we have to discuss how a renaming of variables has to he handled. A calculus like this goes beyond a simple one-to-one-translation of the calculus for CGwCs with generic markers.

In [3], the system of Concept Graphs with Cuts (CGwCs) is comprehensively investigated. CGwCs can be understood as a mathematical elaboration (in terms of mathematical graph theory) of conceptual graphs with negation boxes, where the negation boxes are replaced by a syntactical elements called *cuts*, and the coreference links are replaced by identity edges. Recall that cuts are graphically represented as bold ovals. In [3], a contextual semantics and a sound and complete calculus, based on Peirce's calculus for existential graphs, for CGwCs is provided.

In this paper, the differences between generic markers and variables are discussed. The scrutiny will be carried out on CGwCs, i.e. it will be shown how the results of [3] for CGwCs with generic markers can be transferred to CGwCs with variables. But although the remaining paper mainly focuses on CGwCs, the discussion should be helpful for other formalizations of conceptual graphs with variables as well.

In the next section, the basic definitions for CGwCs with variables are provided. In the third section, we start our investigation with a discussion of CGwCs with variables where each variable occurs almost once. These graphs will be called *purified* CGwCs with variables. This is no loss of expressivity, as each CGwCs with variables is semantically equivalent to a purified one. Purified CGwCs stand in one-to-one-correspondence to CGwCs with generic markers. Based on this correspondence, the adequate calculus for CGwCs with generic markers is translated into an adequate calculus for purified CGwCs with variables. This will be done in the next section of this paper. In the fourth section, we extend the class of purified CGwCs with variables to CGwCs with variables where variables may occur more than once. The calculus of the preceeding section is extended in order to obtain an adequate calculus for this bigger class of CGwCs with variables. Finally, a short discussion of the results is provided.

2   Basic Definitions and Semantics

We start with the definition of the underlying alphabet for CGwCs.

Definition 1 (Variables, Alphabet). *Let $Var := \{x_1, x_2, \ldots\}$ be a countably infinite set. The elements of Var are called* variables. *Let $*$ be a further sign, which is called the* generic marker.

An alphabet *is a triple $\mathcal{A} := (\mathcal{G}, (\mathcal{C}, \leq_{\mathcal{C}}), (\mathcal{R}, \leq_{\mathcal{R}}))$ where $\mathcal{G}$ is a finite set of object names, $(\mathcal{C}, \leq_{\mathcal{C}})$ is a finite ordered set of concept names with a greatest element $\top$, and $(\mathcal{R}, \leq_{\mathcal{R}})$ is a family of finite ordered sets $(\mathcal{R}_k, \leq_{\mathcal{R}_k})$, $k = 1, \ldots, n$ of relation names. Let $id \in \mathcal{R}_2$ be a special name which is called* identity. *We will sometimes write $(\mathcal{G}, \mathcal{G}, \mathcal{G})$ instead of $(\mathcal{G}, (\mathcal{C}, \leq_{\mathcal{C}}), (\mathcal{R}, \leq_{\mathcal{R}}))$.*

Next, we define two classes of CGwCs with variables. In [3], CGwCs with generic markers are provided as structures $(V, E, \nu, \top, Cut, area, \kappa, \rho)$, where $\rho : V \to \mathcal{G} \cup \{*\}$ is a mapping which assigns to each vertex $v \in V$ its referent $\rho(v)$, being an object name or the generic marker. Now we consider a modified version of these graphs with a label mapping $\rho : V \to \mathcal{G} \cup Var$. Let $\mathfrak{G}$ be such a graph. Let $Var^{\mathfrak{G}} := \{\alpha \in Var \mid \exists v \in V : \rho(v) = \alpha\}$ be the set of the variables which occur in $\mathfrak{G}$. In the introduction, we already discussed that the *scope* of a variable α is the innermost context which contains all vertices labeled with x. Mathematically, for $\alpha \in Var^{\mathfrak{G}}$, we set $scope(\alpha) := \bigvee \{c \in Cut \cup \{\top\} \mid \exists v \in area(c) : \rho(v) = \alpha\}$.[2] Finally, similar to the definition of V^* and $V^{\mathcal{G}}$ in [3], we set $V^{Var} := \{v \in V \mid \rho(v) \in Var\}$, and the vertices in V^{Var} are called *variable vertices* or *variable boxes*.

Similar to Sowa's defining labels of coreference sets, we define *dominating variable boxes* to be variable boxes $v \in V$ which satisfy $cut(v) = scope(\rho(v))$. For such a v, each $w \in V$ with $\rho(w) = \rho(v)$ is said to be *dominated by* v. If we otherwise have $\rho(w) \neq \rho(v)$ for each $w \in V$ with $w \neq v$, the dominating box v will be called *singular variable box*.

Now we can define the two classes of CGwCs with variables which will be discussed in this paper. A graph $\mathfrak{G} := (V, E, \nu, \top, Cut, area, \kappa, \rho)$ with $\rho[V] \subseteq \mathcal{G} \cup Var$ is called *concept graph with cuts (CGwC) and variables over $\mathcal{A}$*, if for each $\alpha \in Var^{\mathfrak{G}}$ there exists a dominating variable box v with $\rho(v) = \alpha$. If we have moreover $v_1 = v_2$ for all $v_1, v_2 \in V$ with $\rho(v_1) = \rho(v_2) \in Var$, then $\mathfrak{G}$ is called *purified concept graph with cuts (CGwC) and variables over $\mathcal{A}$*.

To provide a semantics, as in [3] we use power context families as model structures for concept graphs. The models are defined as follows:

Definition 2 (Models). *A power context family $\vec{\mathbb{K}} := (\mathbb{K}_i)_{k=0,\ldots,n}$ is a family of contexts $\mathbb{K}_k := (G_k, M_k, I_k)$ that satisfies $G_k \subseteq (G_0)^k$ for each $k = 1, \ldots, n$.*

For an alphabet $\mathcal{A} := (\mathcal{G}, \mathcal{C}, \mathcal{R})$ and a power context family $\vec{\mathbb{K}}$, we call the union $\lambda := \lambda_{\mathcal{G}} \dot{\cup} \lambda_{\mathcal{C}} \dot{\cup} \lambda_{\mathcal{R}}$ of the mappings $\lambda_{\mathcal{G}} : \mathcal{G} \to G_0$, $\lambda_{\mathcal{C}} : \mathcal{C} \to \mathfrak{B}(\mathbb{K}_0)$ and $\lambda_{\mathcal{R}} : \mathcal{R} \to \mathfrak{R}_{\vec{\mathbb{K}}}$ a $\vec{\mathbb{K}}$-interpretation of $\mathcal{A}$ if $\lambda_{\mathcal{C}}$ and $\lambda_{\mathcal{R}}$ are order-preserving, $\lambda_{\mathcal{C}}(\top) = \top$, $\lambda_{\mathcal{R}}(\mathcal{R}_k) \subseteq \mathfrak{B}(\mathbb{K}_k)$ for all $k = 1, \ldots, n$, and $(g_1, g_2) \in Ext(\lambda_{\mathcal{R}}(id)) \Leftrightarrow g_1 = g_2$ hold for all $g_1, g_2 \in \mathcal{G}$. The pair $(\vec{\mathbb{K}}, \lambda)$ is called $\mathcal{A}$-structure or $\mathcal{A}$-model.

---

[2] In [3], it is shown that $Cut \cup \{\top\}$ is a tree, so $\bigvee(\ldots)$ denotes the join in this tree.

Similar to [3], we have to define valuations for CGwCs with variables, and based on valuation, we can define how CGwCs with variables are evaluated in models. The next two definitions have in [3] counterparts for CGwCs with generic markers, which are slightly modified to encompass the fact that we allow different variable vertices which are labeled with the same variable.

Definition 3 (Partial and Total Valuations for CGwCs with Variables).
Let $\mathfrak{G} := (V, E, \nu, \top, Cut, area, \kappa, \rho)$ be a CGwC with variables and let $\mathcal{M}$ be a $\mathcal{A}$-structure. A mapping $ref : V' \to G_0$ with $V^{\mathcal{G}} \subseteq V' \subseteq V$, $ref(v) = \lambda_{\mathcal{G}}(\rho(v))$ for all $v \in V^{\mathcal{G}}$, and $ref(v_1) = ref(v_1)$ for all $v \in V^{Var}$ is called a partial valuation of $\mathfrak{G}$. If we moreover have $V' \supseteq \{v \in V^{Var} \,|\, scope(\rho(v)) > c\}$ and $V' \cap \{v \in V^{Var} \,|\, scope(\rho(v)) \leq c\} = \emptyset$, we say that ref is a partial valuation for the context c. If $V' = V$ holds, then ref is called (total) valuation of $\mathfrak{G}$.

Definition 4 (Endoporeutic Evaluation of Graphs).
Let $\mathfrak{G} := (V, E, \nu, \top, Cut, area, \kappa, \rho)$ be a CGwC and variables and let $\mathcal{M}$ be a $\mathcal{A}$-structure. Inductively over the tree $Cut \cup \{\top\}$, we define $(\vec{\mathbb{K}}, \lambda) \models \mathfrak{G}[c, ref]$ for each context $c \in Cut \cup \{\top\}$ and every partial valuation $ref : V' \subseteq V \to G_0$ for c. We set $(\vec{\mathbb{K}}, \lambda) \models \mathfrak{G}[c, ref] :\Longleftrightarrow$

> *ref can be extended to a partial valuation $\widetilde{ref} : \widetilde{V'} \to G_0$ with $\widetilde{V'} := V' \cup \{v \in V^{Var} \,|\, scope(\rho(v)) = c\}$ which satisfies:*

> - *$\widetilde{ref}(v) \in Ext(\lambda_{\mathcal{C}}(\kappa(v)))$ for each $v \in V \cap area(c)$ (vertex condition)*
> - *$\widetilde{ref}(e) \in Ext(\lambda_{\mathcal{R}}(\kappa(e)))$ for each $e \in E \cap area(c)$ (edge condition)*
> - *$(\vec{\mathbb{K}}, \lambda) \not\models \mathfrak{G}[d, \widetilde{ref}]$ for each $d \in Cut \cap area(c)$ (cut condition)*

For $(\vec{\mathbb{K}}, \lambda) \models \mathfrak{G}[\top, \emptyset]$ we write $(\vec{\mathbb{K}}, \lambda) \models \mathfrak{G}$. If we have two concept graphs $\mathfrak{G}_1$, $\mathfrak{G}_2$ such that $(\vec{\mathbb{K}}, \lambda) \models \mathfrak{G}_2$ for each contextual structure $(\vec{\mathbb{K}}, \lambda)$ with $(\vec{\mathbb{K}}, \lambda) \models \mathfrak{G}_1$, we write $\mathfrak{G}_1 \models \mathfrak{G}_2$.

We will consider CGwCs with variables only up to isomorphism and *renaming of the variables*. This idea is the well known *alpha-conversion* of formulas in linear and symbolic formalizations of FOL. Graphs which are identical up to different variable names will be called *equivalent* (this term is adopted from RDF).

Definition 5 (Equivalence of Graphs). *Let $\mathfrak{G} := (V, E, \nu, \top, Cut, area, \kappa, \rho)$, $\mathfrak{G}' := (V', E', \nu', \top', Cut', area', \kappa', \rho')$ be two CGwCs with variables. We will say that $\mathfrak{G}$ and $\mathfrak{G}'$ are equivalent, if $\mathfrak{G}'$ is isomorphic to a CGwC with variables $\mathfrak{G}'' := (V, E, \nu, \top, Cut, area, \kappa, \rho'')$ such that there is a bijective mapping $f : Var \to Var$ which satisfies $\rho''(v) = \rho(v)$ for each $v \in V$ with $\rho(v) \in \mathcal{G}$ and $\rho(v'') = f(\rho(v))$ for each $v \in V$ with $\rho(v) \in Var$.*

Obviously, equivalent graphs have same meaning, i.e. if $\mathcal{M}$ is a model and $\mathfrak{G}, \mathfrak{G}'$ are two equivalent CGwCs with variables over $\mathcal{A}$, we have $\mathcal{M} \models \mathfrak{G} \Longleftrightarrow \mathcal{M} \models \mathfrak{G}'$.

In the forthcoming calculus, we could employ a rule which allows to transform a CGwC with variables into an equivalent graph. In this work, we use a more convenient approach: To ease the handling of CGwCs with variables, we agree that CGwCs with variables are considered only up to equivalence.

3   Purified CGwCs with Variables

In the following sections, we will deal with different kinds of CGwCs. Before we proceed, a simple notational convention shall be introduced: We will sometimes use indices to denote which kind of CGwCs we use. An index ^{g} denotes CGwCs with generic markers, and the indices ^{v} and ^{pv} denotes CGwCs with variables and purified CGwCs with variables, respectively. Moreover, we will use upper indices in brackets to denote mappings between these different classes of CGwCs.

We start with the canonical translation from purified CGwCs with variables to CGwCs with generic markers, which is given by replacing each variable by a generic marker. Vice versa, if a CGwCs with generic markers is given, we can replace each generic marker by a fresh (i.e. a variable which is not used so far) variable. Of course, the assignment of variables to generic vertices is not uniquely given, but this poses no problem, as we consider CGwCs with variables only up to equivalence.

Definition 6 (Translations $^{(g)}$ and $^{(pv)}$). *Let $\mathfrak{G} := (V, E, \nu, \top, Cut, area, \kappa, \rho)$ be a purified CGwC with variables. Then let $\mathfrak{G}^{(g)} := (V, E, \nu, \top, Cut, area, \kappa, \rho^{(g)})$ the CGwC with generic markers with $\rho^{(g)}(v) = \rho(v)$, if $v \in \mathcal{G}$, and $\rho^{(g)}(v) = *$, if $v \in V^{Var}$. Vice versa, let $\mathfrak{G} := (V, E, \nu, \top, Cut, area, \kappa, \rho)$ be a CGwC with generic markers, and let $f : V^* \to Var$ be an injective mapping from the set of generic nodes into the set of variables. Then let $\mathfrak{G}^{(pv)}$ be (the equivalence class of) the purified CGwC with variables $\mathfrak{G}^{pv} := (V, E, \nu, \top, Cut, area, \kappa, \rho^{(pv)})$ with $\rho^{(pv)}(v) = \rho(v)$, if $v \in \mathcal{G}$, and $\rho^{(pv)}(v) = f(v)$, if $v \in V^*$.*

Obviously, the mappings $^{(pv)}$ and $^{(g)}$ are mutually inverse bijections between purified CGwCs with variables and CGwCs with generic markers. Moreover, entailment is respected by $^{(pv)}$ and $^{(g)}$, i.e. if $\mathcal{M}$ be is model and if $\mathfrak{G}_{pv}$ is a variable-purified CGwCs and $\mathfrak{G}_g$ is a CGwC with generic markers, we have

$$\mathcal{M} \models \mathfrak{G}_{pv} \iff \mathcal{M} \models \mathfrak{G}_{pv}^{(g)} \quad \text{and} \quad \mathcal{M} \models \mathfrak{G}_g \iff \mathcal{M} \models \mathfrak{G}_g^{(pv)} \qquad (1)$$

These results justify the use of the term 'translation' for $^{(pv)}$ and $^{(g)}$. Now we have to carry over the adequate calculus for CGwCs with generic markers to purified CGwCs with variables. The idea is straightforward: We will translate each rule for CGwCs with generic markers to a corresponding rule for purified CGwCs with variables. Let r be a rule of the calculus for generic CGwCs. We will write $\mathfrak{G}_a \vdash_r^g \mathfrak{G}_b$, if $\mathfrak{G}_a, \mathfrak{G}_b$ are two CGwCs with generic markers such that $\mathfrak{G}_b$ can be derived from $\mathfrak{G}_a$ by an application of the rule r. Now we will 'translate' each rule $\vdash_r^g$ to a rule $\vdash_r^{pv}$ for purified CGwCs with variables, i.e., the calculus $\vdash^{pv}$ will satisfy that

$$\mathfrak{G}_a \vdash_r^g \mathfrak{G}_b \iff \mathfrak{G}_a^{(pv)} \vdash_r^{pv} \mathfrak{G}_b^{(pv)} \qquad (2)$$

holds for all CGwCs with generic markers $\mathfrak{G}_a$ and $\mathfrak{G}_b$. Note that vice versa, as $^{(pv)}$ and $^{(g)}$ are mutually inverse bijections, from (2) we obtain $\mathfrak{G}_a \vdash_r^{pv} \mathfrak{G}_b \iff \mathfrak{G}_a^{(g)} \vdash_r^g \mathfrak{G}_b^{(g)}$ for all CGwCs with variables $\mathfrak{G}_a$ and $\mathfrak{G}_b$ as well.

If the calculus $\vdash^{pv}$ is designed this way, it is complete. In order to see this, let $\mathfrak{G}_a$, $\mathfrak{G}_b$ be two purified CGwCs with variables. Then we have:

$$\mathfrak{G}_a \models \mathfrak{G}_b \overset{(1)}{\Leftrightarrow} \mathfrak{G}_a^{(g)} \models \mathfrak{G}_b^{(g)} \Leftrightarrow \mathfrak{G}_a^{(g)} \vdash^g \mathfrak{G}_b^{(g)} \overset{(2)}{\Leftrightarrow} \mathfrak{G}_a^{(g)(pv)} \vdash^{pv} \mathfrak{G}_b^{(g)(pv)} \Leftrightarrow \mathfrak{G}_a \vdash^{pv} \mathfrak{G}_b$$

Now we can provide the translations of the calculus for CGwCs with generic markers to purified CGwCs with variables such which satisfies Eqn. (2). The differences to the calculus for CGwCs with generic markers are indicated by writing the changed phrases in a different text style.

Definition 7 (Calculus for Purified CGwCs with Variables).

The calculus for purified CGwCs with variables over the alphabet $\mathcal{A} := (\mathcal{G}, \mathcal{C}, \mathcal{R})$ *consists of the following rules:*

- **erasure:** *In positive contexts, any directly enclosed edge, isolated vertex, and closed subgraph may be erased.*
- **insertion:** *In negative contexts, any directly enclosed edge, isolated vertex, and closed subgraph whose variable vertices are labeled with fresh variables may be inserted.*
- **iteration:** *Let* $\mathfrak{G}_0 := (V_0, E_0, \nu_0, \top_0, Cut_0, area_0, \kappa_0, \rho_0)$ *be a (not necessarily closed) subgraph of* $\mathfrak{G}$ *and let* $c \leq cut(\mathfrak{G}_0)$ *be a context such that* $c \notin Cut_0$. *Then a copy of* $\mathfrak{G}_0$, *where each vertex* $v = \boxed{P : \alpha}$ *is replaced by* $v = \boxed{P : \alpha'}$ *for a fresh variable* α', *may be inserted into* c. *For every vertex* $v \in V_0^*$ *with* $cut(v) = cut(\mathfrak{G}_0)$, *an identity-link from* v *to its copy may be inserted.*
- **deiteration:** *If* $\mathfrak{G}_0$ *is a subgraph of* $\mathfrak{G}$ *which could have been inserted by rule of iteration, then it may be erased.*
- **double cuts:** *Double cuts (two cuts* c_1, c_2 *with* $area(c_1) = \{c_2\}$) *may be inserted or erased.*
- **generalization:** *For evenly enclosed vertices and edges, their concept names resp. their relation names may be generalized.* Moreover, for evenly enclosed vertices which carry an object name as reference, the object name may be replaced by a fresh variable α.
- **specialization:** *For oddly enclosed vertices and edges, their concept names resp. their relation names may be specialized.* Moreover, for oddly enclosed vertices which carry a variable as reference, the variable may be replaced by an object name.
- **exchanging references:** *Let* $e \in E^{id}$ *be an identity link with* $\rho(e|_1) = g_1$, $\rho(e|_2) = g_2$, $g_1, g_2 \in \mathcal{G} \cup Var$ *and* $cut(e) = cut(e|_1) = cut(e|_2)$. *Then the references of* v_1 *and* v_2 *may be exchanged, i.e., the following may be done: We can set* $\rho(e|_1) = g_2$ *and* $\rho(e|_2) = g_1$.[3]

---

[3] Note that we allow to exchange two variable references as well, but applying the rule this way to a purified CGwCs with variables yields simply an equivalent graph, and equivalent graphs are already considered to be identical. Nonetheless, in the next section we will use this rule for not necessarily purified CGwCs with variables as well, and for these graphs, the rule has indeed an effect.

- **merging two vertices:** *Let* $e \in E^{id}$ *be an identity link with* $\nu(e) = (v_1, v_2)$ *such that*
  - $cut(v_1) \geq cut(e) = cut(v_2)$,
  - $\rho(v_1) = \rho(v_2) \in \mathcal{G}$ *or* $\rho(v_1), \rho(v_2) \in \mathrm{Var}$, *and*
  - $\kappa(v_2) = \top$

  hold. Then v_1 may be merged into v_2, i.e., v_1 and e are erased and, for every edge $e \in E$, $e\big|_i = v_1$ is replaced by $e\big|_i = v_2$.

- **splitting a vertex:** *Let* $g \in \mathcal{G} \cup \mathrm{Var}$. *Let* $v = \boxed{P : g}$ *be a vertex in the context c_0 and incident with relation edges $R_1, \ldots, R_n$, placed in contexts $c_1, \ldots, c_n$, respectively. Let c be a context such that $c_1, \ldots, c_n \leq c \leq c_0$. Then the following may be done: In c, a new vertex $v' = \boxed{\top : g'}$, where $g' = g$, if $g \in \mathcal{G}$, or g' is a fresh variable, if $g \in \mathrm{Var}$, and a new identity-link between v and v' is inserted. On $R_1, \ldots, R_n$, arbitrary occurrences of v are substituted by v'.*

- **$\top$-erasure:** *For $g \in \mathcal{G} \cup \mathrm{Var}$, an isolated vertex* $\boxed{\top : g}$ *may be erased from arbitrary contexts.*

- **$\top$-insertion:** *For $g \in \mathcal{G} \cup \mathrm{Var}$, an isolated vertex* $\boxed{\top : g}$ *may be inserted in arbitrary contexts. Particularly, if $g \in \mathrm{Var}$, g has to be a fresh variable in order to obtain a well-formed purified CGwC with variables.*

- **identity-erasure:** *Let $g \in \mathcal{G}$, let $v_1 = \boxed{P_1 : g}$ and $v_2 = \boxed{P_2 : g}$ be two vertices. Then any identity-link between v_1 and v_2 may be erased.*

- **identity-insertion:** *Let $g \in \mathcal{G}$, let $v_1 = \boxed{P_1 : g}$, $v_2 = \boxed{P_2 : g}$ be two vertices in contexts c_1, c_2, resp. and let $c \leq c_1, c_2$ be a context. Then an identity-link between v_1 and v_2 may be inserted into c.*

4   CGwCs with Variables

In this section, we will extend the calculus $\vdash^{pv}$ of Sec. 3 to the system of (not necessarily purified) CGwCs with variables.

The calculus of Sec. 3 so far is defined only for purified CGwCs with variables. Particularly, it can only be applied to these graphs. Moreover, we had to design the rules of $\vdash^{pv}$ to make sure that an application of any rule to a purified CGwCs with variables yields a purified CGwCs with variables again. For example, the rule 'insertion' of $\vdash^{pv}$ can only be applied to a purified CGwC with variables, and we only allowed to insert subgraphs where all variable boxes are labeled with a *fresh* variable. As we now consider non-purified CGwCs with variables as well, this is now an unnecessary restriction.

It is reasonable not to add more rules to $\vdash^{pv}$ in order to obtain a sound and complete calculus $\vdash^{v}$ for CGwCs with variables. Instead, we will extend each rule of $\vdash^{pv}$ to a rule for CGwCs with variables, and if the rule had some restrictions which were needed to ensure that an application of the rule yields a purified CGwC with variables, these restrictions are now dismissed. On the other hand, when we extend a rule $\vdash^{pv}_r$ to a rule $\vdash^{v}_r$ for not necessarily purified CGwCs with variables, we have to take care that when $\vdash^{v}_r$ is applied, no scopes of variables are allowed to change. We exemplify this necessity with two examples.

We start with an example of extending the rule 'T-insertion'. Consider the following valid graph with the meaning 'every person is the child of a person'.

For CGwCs with variables, it is self-suggesting that we now allow to insert T-boxes into arbitrary contexts. Let us first consider an insertion of T-boxes such that no scopes of variables are changed, for example like this:

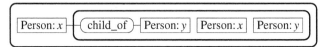

This is indeed a valid derivation. But if we insert a new box $\boxed{\top : x}$ which changes the scope of the variable x, like in the next graph,

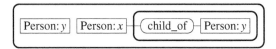

we obtain an invalid graph with the meaning 'every person is the child of every person'. Thus, for CGwCs with variables, we can insert a new box $\boxed{\top : \alpha}$ with $\alpha \in \mathrm{Var}$ only if this insertion does not change the scope of α.

In the rule 'generalization', we will not only allow to generalize an object name to a fresh variable, but we will allow to generalize a variable to a fresh variable as well. Again, we have to take care that no scope of a variable changes when this rule is applied. To see this, consider the following valid graph with the meaning 'there exists a married person, and if this person is a father, it is a male adult' (we assume that only adults are allowed to marry).

We can generalize the variable x of the innermost concept box to the fresh variable z. Then we obtain the following graph:

The meaning of this graph is 'there exists a married person, and if this person is a father, there exists a male adult'. Obviously, this derivation is valid. But if we generalize the variable x of the outermost concept box to the fresh variable z, we obtain the following graph where the scope of x has changed:

The meaning of this graph is 'there exists a married person, and every father is a male adult', which is, as there are fathers who are still a minor, not true.

The calculus $\vdash^{pv}$ had been designed to make sure that an application of any rule to a purified CGwCs with variables yields a purified CGwCs with variables again. This restriction can be dismissed now, but recall that we only consider CGwCs with variables which have dominating variable boxes. This has to be taken into account when the rules of $\vdash^{v}$ are introduced. To summarize: If we reformulate a rule $\vdash^{pv_r}$ to a rule $\vdash^{v}_r$, we have to make sure that no scope of any variable is changed, and applying a rule yields always a CGwC with variables having dominating variable boxes.

Now we are prepared to provide the calculus for CGwCs with variables. Most of the rules are direct extensions of the rules for purified CGwCs with variables, but there are two significant changes. First of all, in the generalization rule, we allow to replace the variable of a variable vertex by a fresh variable. Secondly, in the rule 'splitting a vertex', if the reference of the vertex is a variable, we now allow that the new copy is labeled with the same variable as well. The specialization rule and the rule 'merging two vertices' are extended in a way that they allows the reverse transformation in oddly enclosed contexts.

Definition 8 (Calculus for CGwCs with Variables).

The calculus for variable-purified CGwCs over the alphabet $\mathcal{A} := (\mathcal{G}, \mathcal{C}, \mathcal{R})$ consists of the following rules:

- **erasure:** *In positive contexts, any directly enclosed edge, isolated vertex, and closed subgraph $\mathfrak{G}'$, where each variable box of $\mathfrak{G}'$ is dominated by a variable box which does not belong to $\mathfrak{G}'$, may be erased.*
- **insertion:** *In negative contexts, any directly enclosed edge, isolated vertex, and closed subgraph $\mathfrak{G}'$, where each variable box of $\mathfrak{G}'$ is dominated by a variable box which does not belong to $\mathfrak{G}'$, may be inserted.*
- **iteration:** *Let $\mathfrak{G}_0 := (V_0, E_0, \nu_0, \top_0, Cut_0, area_0, \kappa_0, \rho_0)$ be a (not necessarily closed) subgraph of $\mathfrak{G}$ and let $c \leq cut(\mathfrak{G}_0)$ be a context such that $c \notin Cut_0$. Then a copy of $\mathfrak{G}_0$, where each vertex $v = \boxed{P : \alpha}$ is replaced by $v = \boxed{P : \alpha'}$ for a fresh variable α', may be inserted into c. For every vertex $v \in V_0^*$ with $cut(v) = cut(\mathfrak{G}_0)$, an identity-link from v to its copy may be inserted.*
- **deiteration:** *If $\mathfrak{G}_0$ is a subgraph of $\mathfrak{G}$ which could have been inserted by rule of iteration, then it may be erased.*
- **double cuts:** *Double cuts (two cuts c_1, c_2 with $area(c_1) = \{c_2\}$) may be inserted or erased.*
- **generalization:** *For evenly enclosed vertices and edges, their concept names resp. their relation names may be generalized. Moreover, for each evenly enclosed vertex, its reference may be replaced by a fresh variable α.*
- **specialization:** *For vertices v and edges e in the area of an odd cut c, their concept names resp. their relation names may be specialized. Moreover, if v is a singular variable vertex, this variable may be replaced by an object name or another variable α, provided we have $scope(\alpha) \geq c$ in $\mathfrak{G}$.*

- **exchanging references:** *Let $e \in E^{id}$ be an identity link with $\rho(e|_1) = g_1$, $\rho(e|_2) = g_2$, $g_1, g_2 \in \mathcal{G} \cup Var$ and $cut(e) = cut(e|_1) = cut(e|_2)$. Then the references of v_1 and v_2 may be exchanged, i.e., the following may be done: We can set $\rho(e|_1) = g_2$ and $\rho(e|_2) = g_1$.[4]*
- **merging two vertices:** *Let $e \in E^{id}$ be an identity link with $\nu(e) = (v_1, v_2)$ such that*
  - $cut(v_1) \geq cut(e) = cut(v_2)$,
  - $\rho(v_1) = \rho(v_2) \in \mathcal{G}$, *or* $\rho(v_1), \rho(v_2) \in Var$ *such that* $\rho(v_1) = \rho(v_2)$ *or* v_2 *is a singular variable vertex, and*
  - $\kappa(v_2) = \top$.

  hold. Then v_1 may be merged into v_2, i.e., v_1 and e are erased and, for every edge $e \in E$, $e|_i = v_1$ is replaced by $e|_i = v_2$.
- **splitting a vertex:** *Let $g \in \mathcal{G} \cup Var$. Let $v = \boxed{P : g}$ be a vertex in the context c_0 and incident with relation edges $R_1, \ldots, R_n$, placed in contexts $c_1, \ldots, c_n$, respectively. Let c be a context such that $c_1, \ldots, c_n \leq c \leq c_0$. Then the following may be done: In c, a new vertex $v' = \boxed{\top : g'}$, where $g' = g$, if $g \in \mathcal{G}$, and $g' = g$ or g' is a fresh variable, if $g \in Var$, and a new identity-link between v and v' is inserted. On $R_1, \ldots, R_n$, arbitrary occurrences of v are substituted by v'.*
- **$\top$-erasure:** *For $g \in \mathcal{G}$, an isolated vertex $\boxed{\top : g}$ may be erased from arbitrary contexts. For $\alpha \in Var$, an dominated isolated vertex $\boxed{\top : \alpha}$ may be erased from arbitrary contexts.*
- **$\top$-insertion:** *Let c be context. For $g \in \mathcal{G} \cup Var$, an isolated vertex $\boxed{\top : g}$ may be inserted into $area(c)$. For $\alpha \in Var$ with $scope(\alpha) \geq c$, an isolated vertex $\boxed{\top : \alpha}$ may be inserted into $area(c)$.*
- **identity-erasure:** *Let $g \in \mathcal{G} \cup Var$, let $v_1 = \boxed{P_1 : g}$ and $v_2 = \boxed{P_2 : g}$ be two vertices. Then any identity-link between v_1 and v_2 may be erased.*
- **identity-insertion:** *Let $g \in \mathcal{G} \cup Var$, let $v_1 = \boxed{P_1 : g}$, $v_2 = \boxed{P_2 : g}$ be two vertices in contexts c_1, c_2, resp. and let $c \leq c_1, c_2$ be a context. Then an identity-link between v_1 and v_2 may be inserted into c.*

in contrast to purified CGwCs with variables, as we extended the class of well-formed graphs, the soundness of these rules is not immediately clear. Nonetheless, to each rule of the calculus for CGwCs with variables corresponds a rule for CGwCs with generic markers. For each rule of the calculus for CGwCs with generic markers, a soundness-proof is provided in [3]. The underlying ideas for the rules are in both systems identical, and a closer observation of the proofs of [3] shows that they can be rewritten for the system of CGwCs with variables. Thus, the following lemma is given without a proof.

Lemma 1 (Soundness of $\vdash^v$). *The rules of $\vdash^v$ are sound, i.e., for two CGwCs with variables $\mathfrak{G}_a^v, \mathfrak{G}_b^v$, we have $\mathfrak{G}_a^v \vdash^v \mathfrak{G}_b^v \implies \mathfrak{G}_a^v \models \mathfrak{G}_b^v$.*

---

[4] Note that we allow to exchange two variable references as well, which has now, in contrast to purified CGwCs with variables has an effect.

Now we have to show that the rules of $\vdash^v$ are complete. We start with a lemma where we show that each CGwCs with variables can be transformed to a purified CGwCs with variables, and vice versa, with the rules of $\vdash^v$.

Lemma 2. *For each CGwC with variables $\mathfrak{G}^v$ exists an syntactically equivalent purified CGwC with variables $\mathfrak{G}^{pv}$, i.e., we have $\mathfrak{G}^v \vdash^v \mathfrak{G}^{pv}$ and $\mathfrak{G}^{pv} \vdash^v \mathfrak{G}^v$.*

Proof: We will transform $\mathfrak{G}^v$ into a purified CGwC with variables $\mathfrak{G}^{pv}$. The procedure of the proof shall be exemplified with the graph on the right.

For each variable vertex v, we do the following: First, v is split such that the copy v' of v is labeled with a fresh variable, and *all* occurrences of v on an edge are replaced by v'. Then, the references of v and v' are exchanged.

After this, each variable box which is not a singular variable box is labeled with the concept name $\top$.

Then, for each variable α, we choose an dominating variable box v_α, and insert an identity-link between v_α and all remaining vertices labeled with α.

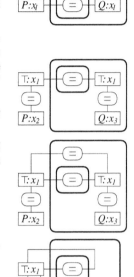

Finally, for each variable α, each variable box $w \neq v_\alpha$ labeled with α is merged into v_α.

The resulting graph $\mathfrak{G}^{pv}$ is purified. As each step in the proof can be carried out in both directions, $\mathfrak{G}^{pv}$ is provably equivalent to $\mathfrak{G}^v$, thus we are done. □

Each rule $\vdash^v_r$ of $\vdash^v$ is an extension of the rule $\vdash^{pv}_r$. Thus, $\vdash^v$ is a complete calculus for purified CGwCs. Together with the last lemma, we immediately obtain the completeness of $\vdash^v$, i.e., we get:

Corollary 1 (Completeness of $\vdash^v$). *The rules of $\vdash^v$ are complete, i.e., for two CGwCs with variables $\mathfrak{G}_a, \mathfrak{G}_b$, we have $\mathfrak{G}_a \models \mathfrak{G}_b \Longrightarrow \mathfrak{G}_a \vdash^v \mathfrak{G}_b$.*

5 Conclusion

In this paper, we investigated how the results of [3] for CGwCs with generic markers can be transferred to CGwCs with variables. At a first glance, the difference between these two systems is of minor syntactical nature. But a closer observation shows that one has to take care of several technical details, mostly in the formalization of the transformation rules for CGwCs with variables. Particularly, we had to take care that an application of a transformation rule to a

CGwCs yields a well-formed graph again, and we had to ensure that no application of an transformation rule changes the scope of a variable. These restrictions result in transformation rules CGwCs with variables which are technically more complex than their counterparts for CGwCs with generic markers.

The discussion of the preceeding sections can be applied to other formalizations of conceptual graphs as well. So we see that on the one hand, it is necessary to investigate conceptual graphs. with variables on its own. On the other hand, this work shows that one main goal of conceptual graphs, namely that humans can better handle them than any symbolic notation for logic, is better achieved if generic markers instead of variables are used for existential quantification. For this reason, this conclusion advocates the use of generic markers.

References

[1] J. F. Baget: *Homomorphismes d'hypergraphes pour la subsumption en RDF/RDFS*. RSTI L'objet, LMO 04, 2004, p. 203–216.

[2] M.-L. Mugnier: *Concept Types and Coreference in Simple Conceptual Graphs*. In: Pfeiffer, H. D.; Wolff, K. E. (Eds): Conceptual Structures at Work, LNAI, Vol. 3127, Springer-Verlag, Berlin – Heidelberg – New York, 2004, p 303–318.

[3] F. Dau: *The Logic System of Concept Graphs with Negation (And Its Relationship to Predicate Logic)*. LNAI, Vol. 2892, Springer, Berlin–Heidelberg–New York, 2003.

[4] F. Dau: *RDF as Graph-Based Diagrammatic Reasoning System: Syntax, Semantics, Calculus* Submitted to the 2nd European Semantic Web Conference.

[5] F. Dau: *Rhetorical Structures in Diagrammatic Reasoning Systems*. Submitted to the symposium on Visual Languages and Human Centric Computing 05.

[6] P. Hayes: *RDF Semantics: W3C Recommendation*. 10 February 2004. http://www.w3.org/TR/rdf-mt/

[7] J. Klinger: *Semiconcept Graphs with Variables*. In: U. Priss, D. Corbett, and G. Angelova (Eds.): Conceptual Structures: Integration and Interfaces. LNAI 2393. Springer-Verlag, Heidelberg-Berlin, 2002.

[8] J. Klinger: *The Logic System of Protoconcept Graphs*. PhD-thesis. To appear.

[9] F. Manola, E. Miller: *RDF Primer*. http://www.w3.org/TR/rdf-primer/

[10] C. S. Peirce: Collected Papers. Harvard University Press, Cambridge, Massachusetts, 1931–1935.

[11] J. F. Sowa: *Conceptual Structures: Information Processing in Mind and Machine*. The System Programming Series. Adison-Wesley, Reading 1984.

[12] J. F. Sowa: *Conceptual Graphs Summary*. in: T. E. Nagle, J. A. Nagle, L. L. Gerholz, P. W. Eklund (Eds.): Conceptual Structures: current research and practice, Ellis Horwood, 1992, 3–51.

[13] J. F. Sowa: Logic: Graphical and Algebraic, Manuscript, Croton-on-Hudson 1997.

[14] R. Wille: *Existential Concept Graphs of Power Context Families*. In: U. Priss, D. Corbett and G. Angelova (Eds.): Conceptual Structures: Integration and Interfaces. LNAI 2393, Springer Verlag, Berlin–New York, 2002.

[15] R. Wille: *Implicational Concept Graphs*. In: H. D.. Pfeiffer, K. E. Wolff (Eds.): Conceptual Struchtures at Work. LNAI 3127, Springer Verlag, Berlin–New York, 2004.

6 Appendix: The Calculi for Both Systems

In this appendix, a short overview on the handling of variables in for the calculi $\vdash^{pv}$ for purified CGwCs with variables and $\vdash^{v}$ for CGwCs with variables is provided.

	purified CGwCs with variables	CGwCs with variables
erasure insertion	Each closed subgraph (which can be an isolated vertex) may be erased. Arbitrary closed subgraphs (which can be isolated vertices), if they contain only fresh variables, may be inserted.	Only subgraphs 𝔊' (which can be a single isolated vertex, where all variable boxes of 𝔊' are dominated by other variable boxes which do not belong to 𝔊', subgraph may be erased or inserted.
iteration deiteration	If a subgraph 𝔊' is iterated, then in the copy of 𝔊', each variable has to be replaced by a fresh variable.	If a subgraph 𝔊' is iterated, then in the copy of 𝔊', each variable has to be replaced by a fresh variable. Only subgraphs with singular variable boxes can be deiterated.
double cuts	no problems	no problems
general. special.	For the rule 'generalization', object names can be replaced by fresh variables. (Note that generalizing variable boxes to variable boxes with a fresh variable has no effect). Vice versa, for the rule 'specialization', variables can be specialized to object names.	For the rule 'generalization', object names can be generalized to fresh variables. If a variable box is dominated, its variable can be generalized to a fresh variable. Vice versa, for the rule 'specialization', variables in singular variable boxes can be specialized to object names. Moreover, variables in singular variable boxes can be replaced by a variable α, if the box is -after the transformation- dominated by another variable box.
exchanging references	No restrictions, but due to the equivalence of graphs, exchanging the references of two variable boxes has no effect.	No restrictions. Note that exchanging the references of two variable boxes now has an effect.
split/merge vertices	If a variable box is split, the new box has to be labelled with fresh variable.	If a variable box is split, the new box has to be labelled with same or fresh variable.
⊤-erasure ⊤-insertion	An isolated ⊤-box can be erased. An isolated ⊤-box which is labelled with an object name or a fresh variable can be inserted.	An isolated ⊤-box, if it is labelled with an object name, or if is a variable box which is dominated by another variable box, of if it is a singular variable box, can be erased or inserted.
id-era. id-ins.	id-erasure/id-insertion is possible only between two object boxes with the same references.	id-erasure/id-insertion is possible between two object boxes or two variable boxes with the same references.

Arbitrary Relations in Formal Concept Analysis and Logical Information Systems

Sébastien Ferré, Olivier Ridoux, and Benjamin Sigonneau

IRISA/Université de Rennes 1
Campus de Beaulieu, 35042 Rennes cedex, France
Firstname.Lastname@irisa.fr

Abstract A logical view of formal concept analysis considers attributes of a formal context as unary predicates. In a first part, we propose an augmented definition that handles *binary relations* between objects. A Galois connection is defined on augmented contexts. It represents concept inheritance as usual, but also relations between concepts. As usual, labeling operators are also defined. In particular, concepts and relations are visible and labeled in a single structure. In a second part, we show how relations can be used for navigating in an augmented concept lattice. This part augments the theory of Logical Information Systems. An implementation is sketched, and first experimental results are presented.

1   Motivation

Previous works have shown how FCA can serve as a basis for navigating in a set of objects [1,2,3], automated learning [4], and other operations like querying, datamining, and context updating [5]. All these works show the versatility of FCA and of its basic schema of a Galois connection between sets of objects and their descriptions. The Galois connection yields a concept lattice structure that is the formal foundation for navigating, infering, approximating, etc. However in most applications of FCA, and Logical Information Systems (LIS) [5] in particular, objects are described in isolation; no explicit relation between objects can be described. Still, many applications are better modelled by arbitrary relations between objects rather than by atomic objects only: e.g., in software engineering and geographical information systems.

Power context families [6] introduce arbitrary relations in formal contexts but not in the concept lattice of objects, which is used as a basis for navigating, querying, and datamining. We want to extend the definitions of Galois connection, intent, concept lattice, and labeling to a power context family as a whole, instead to each context of the family in isolation, while focusing on concepts whose extent is a set of objects. Another way to say it is that we want to incorporate relations in the description of objects as well as in the concept lattice. A direct application will be to augment navigation in LIS by allowing to follow relations of the formal context between concepts in addition to hierarchical relations.

F. Dau, M.-L. Mugnier, G. Stumme (Eds.): ICCS 2005, LNAI 3596, pp. 166–180, 2005.

Though this work applies equally well to standard FCA, we express it in terms of logical concept analysis [3] because intentions of relations will take the form of quantified formulas, which fits better a framework in which logical formulas are native.

The structure of the sequel is as follows. Section 2 presents other attempts to introduce relations in FCA, and logical formalisms that have inspired this work. Section 3 describes our proposal. Section 4 presents its application to navigation in a logical information system, and Section 5 sketches its implementation as a file system, and an example.

2   Related Work

The main attempt to incorporate relations in FCA is the *power context family* [6]. It consists of a vector of formal contexts $(K_1, \ldots, K_n)$ $(n \geq 2)$ with $K_i = (\mathcal{O}_i, \mathcal{A}_i, I_i)$ $(i = 1, \ldots, n)$ such that $\mathcal{O}_i \subseteq (\mathcal{O}_1)^i$. It encompasses well arbitrary relations at the context level, including n-ary relations. However, a different concept lattice is generated for each context. On the contrary, we seek to define a single concept lattice, where the concepts combine information about objects, the relations they have, the objects accessible from these relations, and so on. This is because we use concept lattices as a navigation structure in order to retrieve sets of objects depending on their properties (including relations).

The handling of arbitrary relations in conceptual graphs [7] and description logics [8] is just natural, since this is precisely what they are built for. However in both formalisms, concepts are given *a priori*, and not generated from a formal context. A conceptual graph and an ABox (set of objects and relations labeled by logical properties) can be used to build a power context family, but both formalisms lack the ability to exhibit collections of objects as FCA do.

Conceptual graphs have been combined with FCA by defining the concept and relation types as the concepts of a power context family [9]. However, this does not put relations in the concept lattice.

Description logics are languages of unary predicates whose most relevant feature for this article is that they handle relations (called *roles* in the realm of description logics) via quantifications. The theory of description logic tells how to express classes and test whether a class entails another or whether an individual belongs to a class. Prediger and Stumme [10] have used description logic to defined logical scales, and build a formal context over a relational database, but this context contains no relation. Baader et al. have also combined FCA and description logics [11]. Besides the fact they have different objectives, a key difference is that they use relations inside the representation of each object (description logic concepts), whereas we here consider explicit relations between objects of a context. In the terms of description logics, their context is a TBox (set of terminological definitions), whereas our context is an ABox.

In this paper we adopt the choice of description logics to consider only binary relations, knowing this is not really a restriction as n-ary relations can always be converted to binary relations through reification.

3   Relations in Logical Concept Analysis

3.1   Adding Relations to the Formal Context

Firstly, we define an *object context* that is identical to logical contexts previously defined in LCA [3].

Definition 1 (object context).
An object context *is a triple* $K_1 = (\mathcal{O}, \mathcal{L}_1, d_1)$, *where:*
- $\mathcal{O}$ *is a finite set of objects,*
- $\mathcal{L}_1 =_{\text{def}} (L_1, \sqcup_1, \perp_1)$ *is a 0-sup-semilattice.* $\mathcal{L}_1$ *can be tought as a logic: elements of* L_1 *are called formulas,* $\sqcup_1$ *is called disjunction,* $\perp_1$ *is the neutral element for disjunction (i.e., false) and the order* $\sqsubseteq_1$, *defined by* $x \sqsubseteq_1 y =_{\text{def}} x \sqcup_1 y = y$, *is called subsumption.*
- $d_1 \in \mathcal{O} \to L_1$ *is a mapping from objects to their logical description.*

Lemma 1. *Let* K_1 *be an object context. The pair* (ext_1, int_1), *defined by*
$$ext_1(f_1) =_{\text{def}} \{o \in \mathcal{O} \mid d_1(o) \sqsubseteq_1 f_1\} \qquad \text{for } f_1 \in L_1$$
$$int_1(O) =_{\text{def}} \sqcup_1\{d_1(o) \mid o \in O\} \qquad \text{for } O \subseteq \mathcal{O}$$

is a Galois connection between $\mathcal{P}(\mathcal{O})$ *and* L_1: $O \subseteq ext_1(f_1) \Leftrightarrow int_1(O) \sqsubseteq_1 f_1$.

Note 1. $\sqcup_1$ is well-defined because $\mathcal{O}$ is finite, and $\sqcup_1 \emptyset = \perp_1$.

Secondly, we define a *relation context* that logically describes binary relations between objects of an object context.

Definition 2 (relation context). *Let* $\mathcal{O}$ *be a set of objects, as in Definition 1. A* relation context *is a triple* $(\mathcal{R}, \mathcal{L}_2, d_2)$, *where:*
- $\mathcal{R}$ *is a binary relation, i.e. a set of pairs in* $\mathcal{O} \times \mathcal{O}$, *equiped with two mappings* start *and* end, *s.t.* $start((o, o')) =_{\text{def}} o$ *and* $end((o, o')) =_{\text{def}} o'$. *Moreover,* $\mathcal{R}$ *is closed w.r.t. an* inverse operation $^{-1}$ *such that for every relation* $r \in \mathcal{R}$, $start(r^{-1}) = end(r)$ *and* $end(r^{-1}) = start(r)$.

  Note 2. In concrete contexts, r *and* r^{-1} *may have proper names, like* parent *and* child. *Moreover,* start *and* end *are generalized to sets of pairs, i.e. relations, as follows:* $start(R) = \bigcup_{r \in R} start(r)$ *and* $end(R) = \bigcup_{r \in R} end(r)$.

- $\mathcal{L}_2 =_{\text{def}} (L_2, \sqcup_2, \perp_2, \cdot^{-1})$, *s.t.* $(L_2, \sqcup_2, \perp_2)$ *is a 0-sup-semilattice. Again,* $\mathcal{L}_2$ *can be thougth as a logic. It must support an inverse operation* $(^{-1})$ *over formulas reflecting the inverse of relations (see below), and such that for every* $f_2, g_2 \in L_2$, *the following axioms are satisfied:*
  - $(f_2^{-1})^{-1} \equiv_2 f_2$ *(where* $f \equiv_2 g \quad =_{\text{def}} \quad f \sqsubseteq_2 g \wedge g \sqsubseteq_2 f$*)*
  - $f_2 \sqsubseteq_2 g_2 \Leftrightarrow f_2^{-1} \sqsubseteq_2 g_2^{-1}$ *(*$\sqsubseteq_2$ *is invariant to reading direction).*
- $d_2 \in \mathcal{R} \to L_2$ *is a mapping from binary relations to logical formulas that is compatible with the inverse operation on relations, i.e.,* $d_2(r^{-1}) \equiv_2 d_2(r)^{-1}$, *for all* $r \in \mathcal{R}$.

	male	female	dead	old	grown-up	young	30s	40s	90s	none	Carl Barks	Don Rosa	Taliaferro	Osborne	Walt Disney
downy		x	x						x			x			
fergus	x		x						x			x	x		
matilda		x	x						x			x	x		
scrooge	x			x		x						x			
hortense		x	x						x			x	x		
quackmore	x		x						x			x	x		
unknown	x									x					
della		x			x						x	x	x	x	
donald	x				x										x
huey	x						x	x						x	x
dewey	x						x	x						x	x
louie	x						x	x						x	x

	husband	mother
(downy, fergus)	x	
(hortense, quackmore)	x	
(della, unknown)	x	
(hortense, downy)		x
(scrooge, downy)		x
(mathilda, downy)		x
(donald, hortense)		x
(della, hortense)		x
(huey, della)		x
(dewey, della)		x
(louie, della)		x

Fig. 1. Object and relation context for the Duck's family example.

Example 1. Figure 1 represents the object and relation context for the members of Donald Duck's family. This example takes place in classical FCA, which is a special case of LCA where $L_1 = \mathcal{P}(A_1), \perp_1 = A_1$ and $\sqcup_1 = \cap$; e.g., $d_1(\text{downy}) = \{\text{female}, \text{dead}, 90\text{s}, \text{Don Rosa}\}$. $\mathcal{L}_2$ is defined in a similar fashion. Note that according to Definition 2 the second table should actually be closed by $^{-1}$ to hold a relation context.

Lemma 2. *Let K_2 be a relation context. The pair (ext_2, int_2), defined by*

$$ext_2(f_2) =_{\text{def}} \{r \in \mathcal{R} \mid d_2(r) \sqsubseteq_2 f_2\} \quad \text{for } f_2 \in L_2$$
$$int_2(R) =_{\text{def}} \bigsqcup_2 \{d_2(r) \mid r \in R\} \quad \text{for } R \subseteq \mathcal{R}$$

is a Galois connection between $\mathcal{P}(\mathcal{R})$ and L_2: $R \subseteq ext_2(f_2) \Leftrightarrow int_2(R) \sqsubseteq_2 f_2$.

Note 3. $\bigsqcup_2$ is well-defined because $\mathcal{O}$ is finite, and $\bigsqcup_2 \emptyset = \perp_2$.

The idea is now to combine contexts k_1 and K_2 in a context K, and logics $\mathcal{L}_1$ and $\mathcal{L}_2$ in a logic $\mathcal{L}$ similar to description logics in that relations can be used to retrieve objects depending on their relationships to other objects.

Definition 3 (combined context and logic). *Let K_1 be an object context, and K_2 be a relation context. The combined context is the pair (K_1, K_2) gathering objects, relations, and their logical descriptions. The combined logic $\mathcal{L}$ is the couple $(L, \sqsubseteq)$, where:*

$$- L \longrightarrow \top \mid L_1 \mid \exists L_2.L$$

$$- f \sqsubseteq g =_{\text{def}} \begin{cases} true & \text{if } g = \top \\ f \sqsubseteq_1 g & \text{if } f, g \in L_1 \\ (f_2 \sqsubseteq_2 g_2) \wedge (f' \sqsubseteq g') & \text{if } f = \exists f_2.f', g = \exists g_2.g' \text{ (where } f', g' \in L) \\ false & \text{otherwise} \end{cases}$$

Note 4. Formulas of L can be expressed in first-order predicate logic in the following manner. L_1 and L are sets of unary predicates, and L_2 is a set of binary predicates such that $\exists f_2.f$ is defined by $(\exists f_2.f)(x) =_{\text{def}} \exists x'.(f_2(x, x') \wedge f(x))$.

Note 5. For convenience, we extend the subsumption $\sqsubseteq$ to sets of formulas $G, F \subseteq L$ by defining $G \sqsubseteq f =_{\text{def}} \exists g \in G : g \sqsubseteq f$, and $G \sqsubseteq F =_{\text{def}} \forall f \in F : G \sqsubseteq f$.

We now prove there exists a Galois connection between sets of objects and sets of formulas from $\mathcal{L}$, and we give a definition of it as a couple (ext, int). Then it is well known from FCA [12] that a complete concept lattice can be defined.

Definition 4 (extent). *Let K and $\mathcal{L}$ be the context and logic combined from K_1 and K_2. The* extent *in K of a set of formulas $F \subseteq L$ is a set of objects defined by $ext(F) =_{\text{def}} \bigcap_{f \in F} ext'(f)$, where*

$$ext'(f) =_{\text{def}} \begin{cases} \mathcal{O} & \text{if } f = \top \\ ext_1(f) & \text{if } f \in L_1 \\ \{start(r) \mid r \in ext_2(f_2), end(r) \in ext'(f')\} & \text{if } f = \exists f_2.f'. \end{cases}$$

Definition 5 (intent). *Let K and $\mathcal{L}$ be the context and logic combined from K_1 and K_2. The* intent *in K of a set of objects $O \subseteq \mathcal{O}$ is defined by*

$$int(O) = \{f \in L \mid O \subseteq ext'(f)\}.$$

Theorem 1 (relational Galois connection). *Given a combined context K, the pair (ext, int) defined above is a Galois connection between $(\mathcal{P}(\mathcal{O}), \subseteq)$ and $(\mathcal{P}(L), \supseteq)$, i.e. for every $O \subseteq \mathcal{O}$ and $F \subseteq \mathcal{L}$: $O \subseteq ext(F) \Leftrightarrow int(O) \supseteq F$.*

Theorem 2 (concept lattice). *Let K be a combined context. Let the set of concepts $\mathcal{C}$ be defined as the set of all pairs $(O, F) \in \mathcal{P}(\mathcal{O}) \times \mathcal{P}(L)$ such that $O = ext(F)$ and $F = int(O)$: O is called the* extent *of the concept, and F is called its* intent. *The partial ordering $(\mathcal{C}, \leq)$, where $(O, F) \leq (O', F') =_{\text{def}} O \subseteq O'$, is a complete lattice, the* concept lattice, *as a direct consequence of Theorem 1.*

It is important here to note that concept intents gather properties about objects as formulas in $\mathcal{L}_1$, properties about existing relations as formulas like $\exists f_2.\top$, and recursively properties about related objects as formulas like $\exists f_2.f'$.

Definition 5 suggests that intents can be computed only by testing every formula in L. In the following we show that approximations of these intents can be effectively computed to an arbitrary accuracy.

Firstly, we define $L(n)$ as the subset of L containing exactly all formulas that have no more than n times the existential quantifier $\exists$. This corresponds to restricting the depth of relation paths to n.

Definition 6 (depth-n intent). *Let K and $\mathcal{L}$ be the context and logic combined from K_1 and K_2. The depth-n intent $int(n)(O)$ in K for $O \subseteq \mathcal{O}$ is defined by*

$$int(0)(O) \quad = int_1(O)$$
$$int(n+1)(O) = int(n)(O)$$
$$\cup \; \{\exists int_2(R).f' \mid \exists R \subseteq \mathcal{R}, O = start(R), int(n)(end(R)) \sqsubseteq f'\}.$$

This definition is well-founded, and because of the finiteness of objects and relations every approximate intent is also finite. However they can represent an infinite set of formulas thanks to subsumption $\sqsubseteq$ in L: if $f \notin int(n)(O)$ but $int(n)(O) \sqsubseteq f$, then f implicitly belongs to the intent of O at depth n.

Theorem 3 (depth-n relational Galois connection). *Given a combined context K and a depth n, the pair of mappings $(ext, int(n))$ is a Galois connection between $(\mathcal{P}(\mathcal{O}), \subseteq)$ and $(\mathcal{P}(L(n)), \sqsubseteq)$, i.e. for every $O \subseteq \mathcal{O}$ and $F \subseteq L(n)$, $O \subseteq ext(F) \Leftrightarrow int(n)(O) \sqsubseteq F$.*

Proof. We split proof in three parts.

1. Firstly, we prove that for all $n \in \mathbb{N}$, $O \subseteq \mathcal{O}$ and $f \in L(n)$, $O \subseteq ext'(f) \Rightarrow int(n)(O) \sqsubseteq f$. The proof works by recurrence on the depth n, and by induction on the syntax of formulas. The case where $n = 0$ follows from Lemma 1, so we only show the general case $n+1$, when $f = \exists f_2.f'$, i.e., $f \in L(n+1)$, and so $f' \in L(n)$.
   $O \subseteq ext'(\exists f_2.f') \Longrightarrow O \subseteq \{start(r) \mid r \in ext_2(f_2), end(r) \in ext'(f')\}$
   $\Longrightarrow \forall o \in O : \exists r \in ext_2(f_2) : o = start(r), end(r) \in ext'(f')$
   $\Longrightarrow \exists R \subseteq ext_2(f_2) : O = start(R), end(R) \subseteq ext'(f')$
   $\Longrightarrow \exists R \subseteq \mathcal{R} : O = start(R), R \subseteq ext_2(f_2), end(R) \subseteq ext'(f')$
   $\Longrightarrow \exists R \subseteq \mathcal{R} : O = start(R), int_2(R) \sqsubseteq_2 f_2, int(n)(end(R)) \sqsubseteq f'$
   $\qquad\qquad$ (Lemma 2, recurrence hypothesis because $f' \in L(n)$)
   $\Longrightarrow \exists R \subseteq \mathcal{R} : O = start(R), int(n)(end(R)) \sqsubseteq f', \exists int_2(R).f' \sqsubseteq \exists f_2.f'$
   $\Longrightarrow \exists g \in int(n+1)(O) : g \sqsubseteq f$ $\qquad\qquad\qquad (g = \exists int_2(R).f')$
   $\Longrightarrow int(n+1)(O) \sqsubseteq f.$

2. Secondly, we prove in the same way the reciprocal lemma, i.e. $int(n)(O) \sqsubseteq f \Rightarrow O \subseteq ext'(f)$.
   Suppose $int(n+1)(O) \sqsubseteq \exists f_2.f'$
   Either $int(n)(O) \sqsubseteq \exists f_2.f' \Longrightarrow O \subseteq ext'(\exists f_2.f')$ $\quad$ (recurrence hypothesis)
   or $\{\exists int_2(R).f' \mid R \subseteq \mathcal{R}, O = start(R), int(n)(end(R)) \sqsubseteq f'\} \sqsubseteq \exists f_2.f'$
   $\Longrightarrow \exists R \subseteq \mathcal{R} : O = start(R), int(n)(end(R)) \sqsubseteq f', \exists int_2(R).f' \sqsubseteq \exists f_2.f'$
   $\Longrightarrow \exists R \subseteq \mathcal{R} : O = start(R), int_2(R) \sqsubseteq_2 f_2, int(end(R)) \sqsubseteq f'$
   $\Longrightarrow \exists R \subseteq \mathcal{R} : O = start(R), R \subseteq ext_2(f_2), end(R) \subseteq ext'(f')$
   $\qquad\qquad$ (Lemma 2, recurrence hypothesis because $f' \in L(n)$)
   $\Longrightarrow \forall o \in O : \exists r \in ext_2(f_2) : o = start(r), end(r) \in ext'(f')$
   $\Longrightarrow O \subseteq ext'(\exists f_2.f') \Longrightarrow O \subseteq ext'(f).$

3. Finally, we prove the theorem for all $n \in \mathbb{N}$, $O \subseteq \mathcal{O}$ and $f \in L(n)$.
   $O \subseteq ext(F) \Longleftrightarrow \forall f \in F : O \subseteq ext'(f) \Longleftrightarrow \forall f \in F : int(n)(O) \sqsubseteq f$ (first and second part of this proof) $\Longleftrightarrow int(n)(O) \sqsubseteq F.$ $\qquad\qquad\square$

This last result entails that for every depth n a Galois connection is defined, and so, a depth-n concept lattice can be derived from it. We complete this by showing that when the depth tends to infinity the depth-n intent is equivalent to the *full* intent of Definition 5.

Theorem 4 (limit relational Galois connection). *Let K be a combined context. For every set of objects O, when the depth n tends to infinity, the set of formulas in L that are subsumed by the depth-n intent tends to be equal to the full intent $int(O)$, i.e.*

$$\forall O \subseteq \mathcal{O} : \forall f \in L : f \in int(O) \iff \exists n \in \mathbb{N} : int(n)(O) \sqsubseteq f.$$

Proof. $f \in int(O) \iff O \subseteq ext'(f) \iff O \subseteq ext(\{f\})$
$\iff \exists n \in \mathbb{N} : int(n)(O) \sqsubseteq \{f\}$ \hfill (Theorem 3 because $f \in L(n)$)
$\iff \exists n \in \mathbb{N} : int(n)(O) \sqsubseteq f.$ \hfill $\Box$

The exact full intents and concept lattice cannot be computed, especially if there is a cycle between objects related by $\exists f_2.f'$ formulas. But this is not really a problem since Theorems 3 and 4 show that we have a series of finite depth-n intents and related concept lattices, which can be made as close as possible to *full* intents and concept lattice.

3.2   Adding Relations to the Concept Lattice Labeling

We show in this section that all the information contained in the binary relation context is present in the concept lattice $\mathcal{C}$, and can be made explicit by adding a relational labeling to the concept lattice, in addition to the usual labeling by objects and formulas.

Definition 7 (labeling). *Let $\mathcal{C}$ be a concept lattice. The labeling of $\mathcal{C}$ by formulas, noted μ, and by objects, noted γ, are defined as follows:*
$$\mu \in L \to \mathcal{C}, \quad \mu(f) =_{\text{def}} (ext(\{f\}), int(ext(\{f\}))),$$
$$\gamma \in \mathcal{O} \to \mathcal{C}, \quad \gamma(o) =_{\text{def}} (ext(int(\{o\})), int(\{o\})).$$

It is well-known that a concept lattice contains the same information as the object context from which it derives. We show now that the concept lattice derived from a combined context also contains its relational information.

A way of showing that $\mathcal{C}$ contains in some way the relation concept lattice $\mathcal{C}_2$ is to build an order-preserving mapping from the latter to pairs of concepts of the former. So we go on defining a relational version of μ and γ, applying to pairs of concepts.

Definition 8 (relational labeling). *Let $c, c' \in \mathcal{C}$. We define:*
$$\mu^2 \in L_2 \to \mathcal{C}^2, \mu^2(f_2) =_{\text{def}} (\mu(\exists f_2.\top), \mu(\exists f_2^{-1}.\top)),$$
$$\gamma^2 \in \mathcal{R} \to \mathcal{C}^2, \gamma^2(r) =_{\text{def}} (\gamma(start(r)), \gamma(end(r))).$$

This implies that in addition to subsumption links between concepts ($\leq$), there are relation links between concepts (either individual relations between object-labeled concepts, or relational formulas between formula-labeled concepts). Moreover, most properties on labeling functions are kept [12,5]. In the following, the ordering on pairs of concepts is defined by $(c_1, c_1') \leq (c_2, c_2')$ iff $c_1 \leq c_2$ and $c_1' \leq c_2'$; and the inverse of a pair of concept is defined by $(c, c')^{-1} = (c', c)$.

There exists an order-preserving mapping from the concept lattice of the relation context K_2 into the concept lattice of the combined context $K = (K_1, K_2)$.

Theorem 5 (order-preserving mapping). *The mapping $\varphi \in \mathcal{C}_2 \to \mathcal{C}^2$, $\varphi(c_2) = \mu^2(int_2(c_2))$ is order-preserving, i.e.,*

$$\forall c_2, c_2' \in \mathcal{C}_2 : c_2 \leq c_2' \Rightarrow \varphi(c_2) \leq \varphi(c_2').$$

Corollary 1. *A corollary of Theorem 5 is that the ordering between formula-labels is preserved:* $\forall f_2, f_2' \in L_2, \mu_2(f_2) \leq \mu_2(f_2') \Rightarrow \mu^2(f_2) \leq \mu^2(f_2').$

So, each time two relation labels are ordered in the relation concept lattice $\mathcal{C}_2$, they are also ordered in the combined concept lattice $\mathcal{C}$. However the reverse does not hold. For example, consider the two relations `parent` and `grand-parent`, whose inverse are respectively `child` and `grand-child`. It is true that anyone who has a grand-parent also has a parent ($\mu(\exists grand\text{-}parent.\top) \leq \mu(\exists parent.\top)$); reciprocally, anyone who has a grand-child also has a child ($\mu(\exists grand\text{-}child.\top) \leq \mu(\exists child.\top)$). This implies the label `grand-parent/grand-child` is lower than the label `parent/child` in the combined concept lattice. But this is certainly not the case in the relation concept lattice as a relation can never have both properties `parent/child` and `grand-parent/grand-child`. In conclusion the combined concept lattice can add useful implications compared to the relation concept lattice.

Example 2. Figure 2 shows the depth-1 concept lattice built from the combined context of Example 1, and its labeling by attributes, objects, and relational properties $\exists r.\top$. Grey circles represent the concepts introduced by relations, dashed arrows and italic labels stand for labeling by relations.

It shows that every mother has a husband, and reciprocally that every married female is a mother. Hence, every duck who has a mother also has a father.

4 Querying and Navigating with Relations in LIS

In previous work about Logical Information Systems (LIS) [5], querying and navigation are defined on an object context $K_1 = (\mathcal{O}, \mathcal{L}_1, d_1)$. Queries are formulas in $\mathcal{L}_1$, and for every query $q \in \mathcal{L}_1$, $ext_1(q)$ is the set of answers to the query. In order to help users building queries, even without knowledge about both the context and the logic, a set of navigation links can be computed for any query q in order to refine it. Navigation links are not searched in the whole space of logical formulas. They are searched in a subset of L_1 which we call navigation features. It is computed by application of a user-defined function $feat_1$

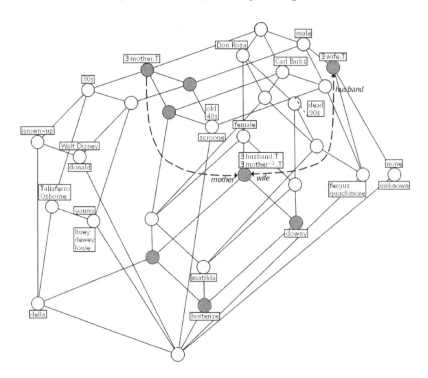

Fig. 2. Concept lattice labeled by formulas, objects and some relations

to context K_1. The smaller and simpler those features are, the simpler and more efficient the navigation is.

The set of navigation links $dirs(q)$ that can refine a query q is defined by:

$$dirs(q) =_{\text{def}} Max_{\sqsubseteq_1}\{x \in feat_1(K_1) \mid \emptyset \subsetneq ext_1(q \sqcap_1 x) \subsetneq ext_1(q)\}.$$

In order to have an efficient and progressive navigation, links computed by $dirs(q)$ must carry information, i.e. $ext_1(q \sqcap_1 x) \subsetneq ext_1(q)$, not lead to a dead-end, i.e. $\emptyset \subsetneq ext_1(q \sqcap_1 x)$, and be as general as possible w.r.t. subsumption, i.e. $Max_{\sqsubseteq_1}$.

Each link x enables the user to move from the query q to the query $q \sqcap_1 x$. Finally individual objects have to be found in some place. Hence the definition of local objects for any query q:

$$locals(q) =_{\text{def}} ext_1(q) \setminus \bigcup_{x \in dirs(q)} ext_1(x).$$

Note that the definition of *dirs* relies on a conjunction operation $\sqcap_1$ on queries. However, such a connective was not required on the development of logical concept analysis. Similarly, navigation is supposed to start from the top element of the logic, *true* or $\top_1$, whereas it is the bottom element, *false* or $\bot_1$

that is required in logical concept analysis. This is not a contradiction, but this must be examined in the actual definition of a querying and navigation system.

In the following we present the relational extension of querying and navigation in formal contexts.

4.1   Querying: Query Language and Extent

In order to have boolean operators in the *query language* L_q, we extend the combined logic L in the following way:

$$L_q \rightarrow \top \mid \bot \mid L_q \wedge L_q \mid L_q \vee L_q \mid \neg L_q \mid L_1 \mid \exists L_2.L_q.$$

It is not necessary to extend the subsumption $\sqsubseteq$ to the query language as will be made clear later. However we do need to define the extent of queries, given some context $K = (K_1, K_2)$ in order to answer queries.

Definition 9 (query extent). *Let K be a combined context from K_1 and K_2.*

$$
\begin{aligned}
ext_q(\top) &= \mathcal{O} & ext_q(q_1 \wedge q_2) &= ext_q(q_1) \cap ext_q(q_2) \\
ext_q(\bot) &= \emptyset & ext_q(q_1 \vee q_2) &= ext_q(q_1) \cup ext_q(q_2) \\
ext_q(f_1 \in L_1) &= ext_1(f_1) & ext_q(\neg q) &= \mathcal{O} \setminus ext_q(q) \\
ext_q(\exists f_2.q) &= \{start(r) \mid r \in ext_2(f_2), end(r) \in ext_q(q)\}
\end{aligned}
$$

The last 2 lines are taken from Definition 4, given that in these cases, queries are in the logic $\mathcal{L}$.

4.2   Features: Useful Features for Navigation

In this section we show that not all formulas in L_q need to be considered as candidates for navigation links. This leads to a definition of $feats_q(K)$ that is only a small subset of the full query language itself.

Firstly we define $X(q)$ as the set of possible navigation links for some query q, i.e., formulas satisfying the "strictly refining and relevant" property:

$$X(q) =_{\mathrm{def}} \{x \in L_q \mid \emptyset \subsetneq ext_q(q \wedge x) \subsetneq ext_q(q)\}.$$

This definition differs from the definition of *dirs* by the fact that the selection of subsumption-maximal elements is not applied, and that the full language L_q is considered instead of a subset of features $feat_q(K)$. Our purpose is precisely to characterize the latter.

Lemma 3 (elimination of connectives). *For every query $q \in L_q$,*

$$
\begin{aligned}
q_1 \wedge q_2 \in X(q) &\Rightarrow q_1 \in X(q) \vee q_2 \in X(q), & \top \notin X(q) \\
q_1 \vee q_2 \in X(q) &\Rightarrow q_1 \in X(q) \vee q_2 \in X(q), & \bot \notin X(q) \\
\neg q_1 \in X(q) &\Rightarrow q_1 \in X(q)
\end{aligned}
$$

This indicates that boolean connectors can all be ignored in features. For example, consider the proposition about conjunction, and let $q_1 \wedge q_2 \in X(q)$. Either both q_1 and q_2 are in $X(q)$, and so $q_1 \wedge q_2$ can be obtained by successively selecting q_1 and q_2; or only one subquery, say q_1, is in $X(q)$, which means that $q_1 \wedge q_2$ is in fact equivalent to q_1 for navigation (they reach the same concept).

Lemma 4. *For all queries* $q, q' \in L_q$, *and all relation formula* $x_2 \in L_2$,
1. $\exists x_2.q' \in X(q) \Leftrightarrow q' \in X(\exists x_2^{-1}.q)$,
2. $\exists x_2.q' \in X(q) \Rightarrow ext_q(q \wedge \exists x_2.\top) \neq \emptyset$.

Lemma 4 allows to recursively decompose the search for relational links. It shows this can be achieved by first looking for $\exists x_2.\top$ formulas that make the new query non-empty (second proposition), and then by replacing the $\top$ by navigation links among the images of objects in q through relation x_2 (first proposition).

Hence, the set of useful navigation features for the language L_q is defined as follows, given user-defined feature extraction functions $feat_1$ and $feat_2$.

Definition 10 (features).
$$feat_q(K) = feat_1(K_1) \cup \{\exists x_2.x \mid x_2 \in feat_2(K_2), x \in feat_q(K)\}$$

This shows that the useful part of the query language for navigation is included in L, even if general queries do not belong to L. Note that if the query language in contexts K_1 and K_2 is based on propositional calculus, a variant of Lemma 3 will hold; $feat_1(K_1)$ and $feat_2(K_2)$ are just the atomic formulas of the logic of their context.

4.3   Navigation: Links and Local Objects, Selection and Traversal

We can now define a version of *dirs* that is extended with relations.

Definition 11 (Navigation links).

$$dirs(q) =_{\text{def}} Max_{\sqsubseteq}\{x \in feat_q(K) \mid \emptyset \subsetneq ext_q(q \wedge x) \subsetneq ext_q(q)\}.$$

There is no problem with using the subsumption $\sqsubseteq$ *because* $feat_q(K) \subseteq L$.

Local objects are defined as usual: an object is local if it not accessible from any navigation link. Navigation links in a combined context are of two kinds: object formulas, and relation formulas. Object formulas are used in logical queries, but relation formulas can be used either in logical queries, or as paths to go through relations. In each case following a link transforms the current query q as follows (assignement is denoted by :=).

1. A link $x_1 \in L_1$ is used as usual to refine a query: $q := q \wedge x_1$. For instance, property *old* can be such a refinement.
2. A link $\exists x_2.x \in L_2$ can also be used for refining a query, except that the refinement applies to objects in relation to current objects instead of the current objects themselves: $q := q \wedge \exists x_2.x$. For instance, property $\exists parent.old$ can be used to select individuals that have an old parent.
3. A link $\exists x_2.x \in L_2$ also means that relations of type x_2 can be traversed to reach objects of type x: $q := x \wedge \exists x_2^{-1}.q$. Only those objects of type x that can be reached from q through x_2 are considered. For instance, property $\exists parent.old$ can be used to select the old parents of the curently selected individuals.

Cases 1 and 2 are kinds of *conceptual navigation* (from some concept to a subconcept), whereas case 3 is a kind of *relational navigation* (from some concept to a related concept).

One may want to consider a query like $\exists x_2.(x \wedge x')$ (consider, say, $\exists parent.(old \wedge loving)$ for old and loving parents). Lemma 3 shows that it is not necessary to produce navigation links of this form. However, if we successively select links $\exists x_2.x$ and $\exists x_2.x'$, then the resulting query is $\exists x_2.x \wedge \exists x_2.x'$, which is not equivalent to $\exists x_2.(x \wedge x')$. This is where relational navigation comes in for help: first, traverse x_2 ($q := \exists x_2^{-1}.\top$), then select x ($q := \exists x_2^{-1}.\top \wedge x$), and select x' ($q := \exists x_2^{-1}.\top \wedge x \wedge x'$), and finally, traverse x_2^{-1} ($q := \exists x_2.(\exists x_2^{-1}.\top \wedge x \wedge x')$), i.e., $q := \exists x_2.(x \wedge x')$).

5   Implementation as a File System

Though a LIS can be implemented as a stand-alone application, a parallel between LIS notions and file system notions makes it natural to implement it as a file system. This is the purpose of LISFS (LIS File System [13]) which is a file system that offers the operations of a LIS at the operating system level. This is achieved by considering files as objects, directories as formulas and paths as conjunctions of directories. The root directory (/ under UNIX) plays the role of the $\top$ formula. Thus, the absolute name of a file stands as its logical description.

Commands have essentially the same effects as UNIX shell commands, w.r.t. this change. For instance, the shell command `cd a` changes the current query q to $q \wedge a$, and the command `ls` is used to list the navigation links (computed by *dirs*) and the local objects (computed by *objects*) of the current query.

5.1   Adding Arbitrary Relations to LISFS

From a file-system point of view, arbitrary relations can somewhat be understood as symbolic links. As a matter of fact, having a relation r between two objects o and o' declares that o and o' are linked and gives a description r to the link. Regular UNIX links are just a special case where r can only take up one value: the "synonym" relation.

Relations also extend the notion of symbolic link in that they can play two roles in navigation. Indeed, they can be used for conceptual navigation, in which case they behave as normal properties, and for relational navigation. In the case of relational navigation, traversing a navigation link in LISFS is the counterpart of following a symbolic link in UNIX, with the additional benefit that links in LISFS have an inverse and apply to sets of objects.

Regarding navigation and querying, the following concrete syntax is adopted (q denotes the current query):

- `cd r>x` selects objects whose image by r satisfies x: $q := q \wedge \exists r.x$,
- `cd r<x` selects objects whose antecedent by r satisfies x: $q := q \wedge \exists r^{-1}.x$,
- `cd r>>` traverses relation r: $q := \exists r^{-1}.q$,
- `cd r<<` traverses relation r^{-1}: $q := \exists r.q$.

```
[1]# ls                                    3        young/
total 12                                   4        first_appearance:30s/
1        first_appearance:40s/             5        mother>true/
1        first_appearance:none/            7        creator:/
1        old/                              [3]# cd husband<true; ls
2        grown-up/                         total 3
3        husband<true/                     1        first_appearance:none/
3        husband>true/                     1        husband<creator:osborne/
3        mother<true/                      1        husband<creator:taliaferro/
3        young/                            1        husband<first_appearance:30s/
4        female/                           1        husband<grown-up/
5        dead/                             2        creator:/
5        first_appearance:30s/             2        dead/
5        first_appearance:90s/             2        first_appearance:90s/
8        male/                             2        husband<creator:carl_barks/
8        mother>true/                      2        husband<dead/
11       creator:/                         2        husband<first_appearance:90s/
[2]# cd male; ls                           [4]# cd !creator: ; ls
total 8                                    total 1
1        first_appearance:40s/                      unknown
1        first_appearance:none/            [5]# cat unknown
1        grown-up/                         The husband of Della Duck
1        old/                              [6]# cd husband<< ; ls
2        dead/                             total 1
2        first_appearance:90s/                      della
3        husband<true/
```

Fig. 3. A Running Example with the Duck Family Context

We use a slightly modified version of the shell command `ln` to create a relation between two files that takes care of the name of the relation. E.g., we write `ln r o o'` to add a relation r between o and o'.

5.2   Example

Figure 3 shows LISFS augmented with arbitrary relations in action. Navigation and querying take place in the Duck family context presented in Figure 1.

Starting from the $\top$ formula, command 1 asks for available navigation links. The number of objects in the current query and in its possible refinements are given. Quantified formulas appear under their concrete syntax: `mother>true` for $\exists mother.\top$, `mother<true` for $\exists mother^{-1}.\top$, etc.

From there, command 2 lets the user select the males from the members of the Duck family by setting the current query to $\top \wedge male \equiv male$. New increments are listed. Then, command 3 restricts the query to married males. Thus, increments concerning relation $husband^{-1}$ become more precise: two ducks have a wife that is dead ($\exists husband^{-1}.dead$, shown as `husband<dead`) and so on.

Moreover, three ducks in the family satisfy the current query (see `total 3`), among whom only two of them have a creator (see `2 creator:`). This seems odd, so the user asks to see ducks that does not have a creator with command 4 (`!` denotes negation). This duck is `unknown`. As objects are files in LISFS, they have a content which command 5 lists. Here, the user sees that `unknown` is supposed to be the husband of Della. To verify that this really holds in the context, the user then traverses relation $husband^{-1}$ with command 6.

6   Conclusion

In this paper we have shown that relations can be smoothly introduced not only in formal contexts as is done by power context families, but also in the definition of intents, and hence in concept lattices and their labeling. The advantage over previous approaches is that information from both object and relation contexts is combined in a single concept lattice. This enables a natural but powerful extension of LIS navigation and querying [5]. Relational features express properties over objects w.r.t. their related objects, and can be used both for refining a set of objects, as usual, and for traversing some relation from a set of objects to another. This relational navigation has been implemented as an extension to an existing LIS file system (LISFS).

Although relations in contexts are arbitrary, including non tree-like structures, our query language allows only tree-like queries, as in description logics, but unlike conceptual graphs. We plan to extend the query language so as to remove this tree-like constraint. This could be done by inserting variables in queries, like in $\exists r_1.\exists r_2.X : f' \wedge \exists r_3.\exists r_4.X$ where X must refer to a same object in both occurrences. We also plan to extend the logic so as to handle more complex path patterns (e.g., regular expressions), such as $\exists parent^+.famous$ meaning "has a famous ancestor".

References

1. Godin, R., Missaoui, R., April, A.: Experimental comparison of navigation in a galois lattice with conventional information retrieval methods. International Journal of Man-Machine Studies **38** (1993) 747–767
2. Lindig, C.: Concept. In Köhler, J., Giunchiglia, F., Green, C., Walther, C., eds.: IJCAI95 Workshop on Formal Approaches to the Reuse of Plans, Proofs, and Programs, Montreal, Canada (1995)
3. Ferré, S., Ridoux, O.: A logical generalization of formal concept analysis. In Mineau, G., Ganter, B., eds.: Int. Conf. Conceptual Structures. Number 1867 in Lecture Notes in Computer Science, Darmstadt, Germany, Springer (2000) 371–384
4. Ganter, B., Kuznetsov, S.: Formalizing hypotheses with concepts. In Mineau, G., Ganter, B., eds.: Int. Conf. Conceptual Structures. Number 1867 in Lecture Notes in Computer Science, Darmstadt, Germany, Springer (2000) 342–356
5. Ferré, S., Ridoux, O.: An introduction to logical information systems. Information Processing & Management **40** (2004) 383–419
6. Wille, R.: Conceptual graphs and formal concept analysis. In: Int. Conf. Conceptual Structures. Volume 1257 of LNCS., Seattle, Washington, USA, Springer (1997) 290–303
7. Sowa, J.F.: Conceptual structures. Information processing in man and machine. Addison-Wesley, Reading, MA, USA (1984)
8. Donini, F.M., Lenzerini, M., Nardi, D., Nutt, W.: The complexity of concept languages. Information and Computation **134** (1997) 1–58
9. Mineau, G., Stumme, G., Wille, R.: Conceptual structures represented by conceptual graphs and formal concept analysis. In Tepfenhart, W.M., Cyre, W.R., eds.: Int. Conf. Conceptual Structures. Volume 1640 of LNCS., Blacksburg, Virginia, USA, Springer (1999) 423–441

10. Prediger, S., Stumme, G.: Theory-driven logical scaling. In: International Workshop on Description Logics. Volume 22., Sweden (1999)
11. Baader, F., Sertkaya, B.: Applying formal concept analysis to description logics. In Eklund, P.W., ed.: Int. Conf. Formal Concept Analysis. Volume 2961 of LNCS., Sydney, Australia, Springer (2004) 261–286
12. Ganter, B., Wille, R.: Formal Concept Analysis — Mathematical Foundations. Springer (1999)
13. Padioleau, Y., Ridoux, O.: A logic file system. In: USENIX Annual Technical Conference, General Track, San Antonio, Texas, USA, USENIX (2003) 99–112

Merge-Based Computation of Minimal Generators

Céline Frambourg[1], Petko Valtchev[2], and Robert Godin[1]

[1] Département d'informatique, UQAM, Montréal (Qc), Canada
[2] DIRO, Université de Montréal, Montréal (Qc), Canada

Abstract. Minimal generators (mingens) of concept intents are valuable elements of the Formal Concept Analysis (FCA) landscape, which are widely used in the database field, for data mining but also for database design purposes. The volatility of many real-world datasets has motivated the study of the evolution in the concept set under various modifications of the initial context. We believe this should be extended to the evolution of mingens. In the present paper, we build up on previous work about the incremental maintenance of the mingen family of a context to investigate the case of lattice merge upon context subposition. We first recall the theory underlying the singleton increment and show how it generalizes to lattice merge. Then we present the design of an effective merge procedure for concepts and mingens together with some preliminary experimental results about its performance.

1 Introduction

Formal Concept Analysis (FCA) has been proved to be a suitable tool for representing the knowledge contained in a database. It is also used as a basis for association rule mining (ARM).

ARM from a transaction database is a classical data mining topic, whereby the most challenging problem is the detection of informative patterns in the transaction sets. A major difficulty with association rules is the prohibitive number of itemsets (and hence association rules) that can be generated even from a reasonably large data set. Moreover, this approach generates a large number of redundant rules. Formal concept analysis (FCA) has helped to solve this problem as it introduces closed itemsets (CIs) , which are a promising solution to the problem of reducing the number of reported association rules. A further step in this direction is the construction of association rule bases from CIs: an operation which largely relies on the notion of closed itemset mingens. Yet another difficulty arises with dynamic databases where the transaction set is frequently updated. Although the necessity of processing volatile data in an incremental manner has been repeatedly emphasized in the general data mining literature, few incremental algorithms for association rule generation (and hence frequent itemset detection) have been reported so far.

CIs mingens are a lossless and concise representation of knowledge in databases. In [8], Kryszkiewicz and Gajek explain the construction of the mingens representation of itemsets and they also show that it is sufficient to determine all the itemsets and their supports. The growth of databases may induce another problem in the data mining task, which is the lack of space. To solve this problem, we propose to distribute the

F. Dau, M.-L. Mugnier, G. Stumme (Eds.): ICCS 2005, LNAI 3596, pp. 181–194, 2005.

computation of the lattices prior to a final merging of the results at the end. Some assembly algorithms have been published recently [18, 16], but most of the methods do not take care of the mingens. That is why we study the evolution of the mingen family during the lattice assembly. To answer that question, we made an exhaustive study of the structures used during CIs family assembly. This study is based on the results of the incremental case that can be found in [19].

We first recall the theoretical framework underlying the lattice assembly and the mingens (Section 2). We will then present the algorithmic aspects of the dynamic computation (Section 3) and finally we will present recent algorithmic results of the problems (Section 4).

2   Closures and Mingens

2.1   FCA Basics

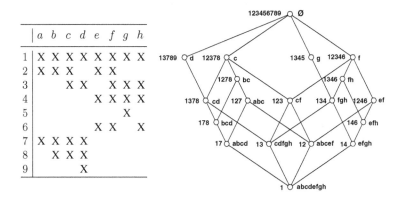

Fig. 1. Left: Context $\mathcal{K}$ (adapted from [4]) with $O = \{1, 2, ..., 9\}$ and $A = \{a, b, ..., h\}$. **Right:** The Hasse diagram of the concept (Galois) lattice derived from $\mathcal{K}$.

FCA studies the way lattices emerge out of data. It considers an incidence relation I over a pair of sets O and A, of objects and attributes, respectively. The relation is given by the matrix which is called a *(formal) context* $\mathcal{K} = (O, A, I)$. Moreover, I gives rise to two $'$ mappings and the composite operators $''$ define *closures* on $\mathcal{P}(O)$ and $\mathcal{P}(A)$, hence each of them induces a family of *closed* subsets, denoted $\mathcal{C}_\mathcal{K}^o$ and $\mathcal{C}_\mathcal{K}^a$, respectively. Those two families, provided with $\subseteq$ become two *complete lattices* which are dually isomorphic via $'$. These lattices overlap perfectly, thus giving rise to the *concept lattice*[3]. A pair (X, Y) of mutually corresponding subsets, i.e. $X = Y'$ and $Y = X'$, is called a *(formal) concept* whereby X is the *extent* and Y is the *intent*. An itemset Z is called a generator of a closed set X if $Z'' = X$. It is called a minimal generator (mingen) if

---

[3] Also known as the *Galois lattice*

$T \subset Z$ implies $T'' \subset Z'' = X$. The closure operator on $\mathcal{C}_{\mathcal{K}}^{a}$ defines an equivalence relation[4] where intents represent the maximum of each equivalence class, and mingens their minimum.

2.2 Mingens in the Literature

In the literature, mingens have been given different names and various properties thereof have been exploited with or without an explicit mention of the concept. For instance, in the database field, *keysets* (see [9]) or *minimal keys* of the sub-relations obtained by decomposing a given relation into 3NF, represent mingens of the attribute sets corresponding to the sub-relations. In early literature on closures and implications, mingens are referred to as *irreducible gaps* thus alluding at their status of minima among all non-closed elements ("gaps") [6] within the same closure class. In a different branch of the same closure field (see [12]), mingens are termed *minimal blockers*, a notion that closely follows that of a minimal transversal of a hypergraph [1].

Remarkable properties of the mingen family include its ideal-shaped structure. In fact, in the Boolean lattice of all the parts of the ground set of a closure family, the mingens represent an order ideal since the family is downwards closed by the subset-of relation. On the cardinality side, it is known that the size of the mingen family can grow up to exponential in the dimensions of the context [10].

Finally, the mingens have been used as target but also as auxiliary structures for lattice algorithms. For instance, the calculation of concepts in Close [11], AClose [11] or Titanic [13] relies on the construction of all mingens. In contrast, in NextClosure [3], the notion of mingen is not explicitly mentioned, but is nevertheless present: a concept intent is canonically generated by its smallest prefix containing a mingen.

3 Summary of the Dynamics of the Closed and the Mingens Computation

So far, three different paradigms of lattice construction have been proposed: batch, incremental and divide-and-conquer. Here we only present the incremental and the merge one.

3.1 Incremental Lattice Construction

The incremental lattice construction paradigm emerged as a response to evolving datasets. Indeed, when a new object is inserted into a dataset, the intent families of the two underlying contexts, i.e., the one "before" ($\mathcal{K}$) and the one "after" ($\mathcal{K}^{+}$), are tightly related: the later is the closure by intersection of the former augmented by the new object description, o'. The later operation is a computationally less expensive basis for the construction of the lattice of $\mathcal{K}^{+}$ than the straightforward construction from scratch. Moreover, as shown in [5], instead of computing all possible intersections of subsets from $\mathcal{C}^{a} \cup \{o'\}$, only pair-wise intersections of o' with an intent from $\mathcal{C}^{a}$ need to be

---

[4] $X, Y \subseteq A$ are equivalent iff $X'' = Y''$.

considered. Among those, some are already existing intents while others are specific to $\mathcal{K}^+$. Concepts from the $\mathcal{L}$ are divided into three categories with respect to the intersections produced by their intents. First, some concepts serve as canonical representatives for their intersections: These are the concepts $c = (X, Y)$ whose intents are the closures of the respective intersections, i.e., $Y = (Y \cap o')''$. They are further divided into *modified* (denoted $\mathbf{M}(o)$) and *genitors* (denoted $\mathbf{G}(o)$), meaning that the intersection is itself an intent in $\mathcal{L}$ (i.e., $(Y \cap o')'' = Y \cap o'$) or that this is not the case (i.e., $(Y \cap o')'' \supset Y \cap o'$), respectively. The remaining concepts in $\mathcal{L}$ are called old or unchanged (denoted by $\mathbf{U}(o)$) and have little importance in the approach. As genitor and modified intents are the closures of the corresponding intersections with o', each concept $c = (X, Y)$ from $\mathbf{G}(o) \cup \mathbf{M}(o)$ is the unique maximum in $\mathcal{L}$ among all those generating the intersection $E = Y \cap o'$.

Given $\mathcal{L}$ (concept set and precedence relation) and o, the incremental lattice construction problem amounts to restructuring and completing the data structure representing $\mathcal{L}$ up to reaching a structure that represents $\mathcal{L}^+$. For convenience reasons, the homologous concepts[5] in $\mathcal{L}^+$ of modified and genitors from $\mathcal{L}$ will be denoted by $\mathbf{M}^+(o)$ and $\mathbf{G}^+(o)$, respectively. New concepts in $\mathcal{L}^+$ with respect to $\mathcal{L}$ will be denoted by $\mathbf{N}^+(o)$. The restructuring of $\mathcal{L}$ into $\mathcal{L}^+$ is summarized by the following facts (see [15] for a detailed description):

- $\mathcal{L}$ is isomorphic to a join-sub-semi-lattice of $\mathcal{L}^+$, made of the homologous concepts of those in $\mathcal{L}$,
- the suborder of $\mathcal{L}^+$ induced by the set of new concepts $\mathbf{N}^+(o)$ is isomorphic to the suborder induced by $\mathbf{G}^+(o)$, the homologous concepts of the genitors in $\mathcal{L}$.

3.2   Computation of the Mingens

In a way similar to lattice construction, the incremental mingen computation amounts to transforming the mingen family of an initial context to the one of the augmented context (see [17]). The reasoning underlying the transformation is based on the Boolean lattice 2^A and the equivalence relation induced by the intent family $\mathcal{C}^a$.

First, the equivalence relation of the augmented intent family $\mathcal{C}^{a+}$ is a refinement of the one corresponding to $\mathcal{C}^a$. The classes of the initial relation either remain stable in the new one or are split into two new classes. Second, split classes correspond to genitors, while modified and old have their classes stable. Thus, given a genitor c of intent Y, the resulting two classes for $\mathcal{C}^{a+}$ correspond to the intent of the homologous concept in $\mathcal{L}^+$, i.e., Y, and to the respective new intent $Y \cap o'$, respectively. Recall that $Y \cap o'$ is a former non-closed subset that becomes closed and hence its respective class for $\mathcal{C}^{a+}$ is made of non-closed elements that lay in the class of Y whenever $\mathcal{C}^a$ is considered. For instance, Fig. 5 shows the evolution of the equivalence class of the intent $cdfgh$ in 2^A (diagram on the top) after the insertion of a new closed set d (diagram on the bottom left).

The above fact explains why only generators of genitor concepts need to be examined in the transformation. Indeed, on the one hand, part of those become generators of

---

[5] A concept from a context is homologous to another one from a different context if both have the same intent.

the respective new intent, while on the other hand, new generators are emerging for the genitor intent. The new generators are minimal sets in the newly formed equivalence class for $\mathcal{C}^{a+}$ that were not minimal in the larger class from $\mathcal{C}^{a}$. In the above example, the initial set of generators of $cdfgh$ is cg, d, ch. After d becomes closed, its equivalence class, here consisting of d itself, is split from the former class of $cdfgh$. The new set of generators of $cdfgh$, i.e., minima in the new equivalence class is much larger: cg, cd, df, dg, dh, ch.

Formally, let us denote by $gen_{\mathcal{K}}()$ the set of mingens of a concept (intent) within the context $\mathcal{K}$ (the subscript will be skipped whenever confusion is excluded). The following property summarizes the evolution in the mingen family between $\mathcal{K}$ and $\mathcal{K}^{+}$:

Proposition 1 *For any $c = (X, Y)$ in $\mathbf{G}(o)$, let c_+ and $\bar{c}$ be its homologous concept in $\mathcal{L}^{+}$ and the generated new concept, respectively ($\bar{c} = (X \cup \{o\}, Y \cap o')$ and $c_+ = (X, Y)$). The sets of generators of $\bar{c}$ and c_+ are as follows:*

- $gen_{\mathcal{K}^+}(\bar{c}) = gen_{\mathcal{K}}(c) \cap 2^{Y \cap o'}$,
- $gen_{\mathcal{K}^+}(c_+) = \min((gen_{\mathcal{K}}(c) - 2^{Y \cap o'}) \cup (gen_{\mathcal{K}}(c) \cap 2^{Y \cap o'} \times \{Y - o'\}))$.

The only non-trivial aspect of the above proposition is that newly occurring mingens in the class of the homologous concept c_+ are composed of a former mingen that "went" to the new concept $\bar{c}$ to which a single attribute is added which stems from the difference between the genitor intent and the new one ($Y \cap o'$).

```
 1: procedure COMPUTE-MINGENS(In/Out: c, c̄ concepts)
 2:
 3: for all g in c.gens do
 4:    if g ⊆ c̄.Intent then
 5:       c̄.gens ← c̄.gens ∪ {g}
 6: c.gens ← c.gens - c̄.gens
 7: SORT(c̄.gens)
 8: for all ḡ in c̄.gens do
 9:    new-gens ← ∅
10:    for all a in (c.Intent - c̄.Intent) do
11:       gen-cond ← true
12:       for all g in c.gens do
13:          if g ⊆ ḡ ∪ {a} then
14:             gen-cond ← false
15:       if gen-cond then
16:          new-gens ← new-gens ∪ {ḡ ∪ {a}}
17:    c.gens ← c.gens ∪ new-gens
```

Algorithm 1: Computation of the mingens of a new concept and of its genitor.

Algorithm 1 embodies the computation of the mingens for a pair of corresponding concepts, genitor and new. Thus, given a genitor concept c and its corresponding new concept $\bar{c}$, it updates the set of mingens associated with the intent of c and identifies the set of mingens for the intent of $\bar{c}$.

3.3   Generalization of the Incremental Case

The restructuring of the lattice upon insertion of a single new object has been generalized to the case of n such objects. The problem has been reformulated as the one
of merging two lattices corresponding to contexts that share their attribute sets. This
section describes the basic facts about lattice merge.

Product of Lattices Along Context Subposition   Subposition is the horizontal assembly of contexts sharing a same set of attributes [4]. Let $\mathcal{K}_1 = (O_1, A, I_1)$ and
$\mathcal{K}_2 = (O_2, A, I_2)$ be two contexts sharing the attribute set A. Their *subposition* is the
context $\mathcal{K}_3 = (O_1 \dot\cup O_2, A, I_1 \dot\cup I_2)$ denoted $\mathcal{K}_3 = \frac{\mathcal{K}_1}{\mathcal{K}_2}$. For example, for the context
$\mathcal{K} = (O, A, I)$ given in Fig. 1, let $O_1 = \{1, 2, 3, 4\}$ and $O_2 = \{5, 6, 7, 8, 9\}$. The par-
tial lattices corresponding to $\mathcal{K}_1$ and $\mathcal{K}_2$, say $\mathcal{L}_1$ and $\mathcal{L}_2$, are given in Fig. 2, on the left.
In the remainder, to avoid confusion, the derivation operators ' for the various contexts
will be replaced by the respective indexes i, e.g., ^{2} will stand for ' for the context $\mathcal{K}_2$.

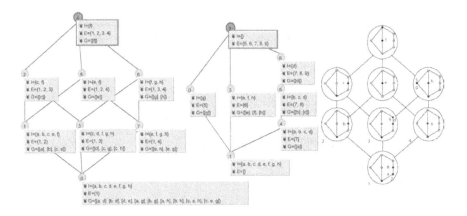

Fig. 2. Left: Factor lattices $\mathcal{L}_1$ and $\mathcal{L}_2$ of the context in Fig. 1. **Right**: The NLD of $\mathcal{L}_3$.

The lattices $\mathcal{L}_1$ and $\mathcal{L}_2$, further called the *factor* lattices, are related to the lattice
of the subposed context by two order morphisms. For convenience reasons, the direct
product of $\mathcal{L}_1$ and $\mathcal{L}_2$, denoted $\mathcal{L}_{1,2}$ is used in the definition of those morphisms:

Definition 1 *The order morphism* $\varphi : \mathcal{L}_3 \to \mathcal{L}_{1,2}$ *maps a concept from the global
lattice to a pair of concepts from the partial lattices by splitting its extent over the
partial context attribute sets* O_1 *and* O_2:

$$\varphi((X, Y)) = ((X \cap O_1, (X \cap O_1)'), ((X \cap O_2, (X \cap O_2)'))$$

The morphism $\psi : \mathcal{L}_{1,2} \to \mathcal{L}_3$ *maps a pair of concepts over partial contexts into a
global concept by the intersection over their respective intents:*

$$\psi(((X_1, Y_1), (X_2, Y_2))) = ((Y_1 \cap Y_2)', Y_1 \cap Y_2).$$

In other terms, the composition of factor concepts into a global one is made along the intent dimension shared by $\mathcal{K}_j$ ($j = 1, 2, 3$): The corresponding operation may be seen as the merge of two closure spaces on A. Each node $((X_1, Y_1), (X_2, Y_2))$ from $\mathcal{L}_{1,2}$ is sent to a concept (X, Y) from $\mathcal{L}_3$ such that $Y = Y_1 \cap Y_2$ (e.g., in Fig. 2, $(c_{\#7}, c_{\#3})$ is sent to $(146, efh)$). The underlying mapping ψ is a surjective order morphism that preserves lattice joins (see [18] for details). Conversely, $\mathcal{L}_3$ is mapped onto $\mathcal{L}_j$ ($j = 1, 2$) by simply projecting concept intents on A_j (e.g., $(127, abc)$ is projected to the node $(c_{\#5}, c_{\#6})$). It is noteworthy that $\mathcal{L}_3$ is in general only a meet-sub-semi-lattice of $\mathcal{L}_{1,2}$.

Merge of Factor Lattices Following [18, 16], the factor merge process filters $\mathcal{L}_{1,2}$, and keeps the nodes from the meet-sub-semi-lattice isomorphic to $\mathcal{L}_3$. These are the maximal nodes in the equivalence classes induced by the homomorphism ψ on $\mathcal{L}_{1,2}$ (i.e., equivalence means nodes are sent to the same concept from $\mathcal{L}_3$). More specifically, the maximum node in such a class, say $((X_1, Y_1), (X_2, Y_2))$, is such that if $\psi((X_1, Y_1), (X_2, Y_2)) = (X, Y)$, then $X = X_1 \cup X_2$ and $Y = Y_1 \cap Y_2$. The canonical pairs of concepts can be compared to genitors in the incremental case, hence the concepts from such a pair will be called the *i-genitors* ($i = 1, 2$) of the respective concept from $\mathcal{L}_3$.

The straightforward procedure for lattice merge illustrated by Algorithm 2, presented in the next section, follows the i-genitor definition together with a characterization of the precedence relation in $\mathcal{L}_3$. The procedure, when applied to the lattices on the left of Fig. 2 yields the result presented in Fig. 3.

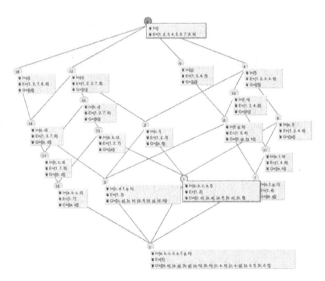

Fig. 3. Result lattice from the merge of $\mathcal{L}_1$ and $\mathcal{L}_2$ in Fig. 2.

4 Generalization of the Incremental Case for the Assembly

4.1 Theoretical Results

Mingen computation can easily be extended to the construction of the subposition-based product of lattices $\mathcal{L}_1$ and $\mathcal{L}_2$. We chose a straightforward approach which consists in applying the incremental paradigm to lattice merging. First, recall that any concept in the product $\mathcal{L}_3$ is created by a pair of factor concepts that play a symmetric role (the genitors) in the generalization of the incremental paradigm. We nevertheless adopt an asymmetric view of the factors and set $\mathcal{L}_1$ to the initial lattice where mingens already exist whereas $\mathcal{L}_2$ is seen as a surrogate for "new" concepts that constitute $\mathcal{L}_3$. Consequently, when such a new concept is detected by the assembly algorithm, its mingens will be computed with respect to its genitor in $\mathcal{L}_1$, i.e., the respective component of the canonical representative in $\mathcal{L}_{1,2}$. Obviously, unlike the specific case of the incremental update, there can be several new concepts per genitor. The challenge will be to determine their mingens without interference between those.

The new concepts created during the merge respect the proposition below.

Proposition 2   *The new concepts corresponding to a 1-genitor c_1, $\{c_3 \in C_3 | \Pi_1(\varphi(c_3)) = c_1\}$ where Π_1 is the projection over C_1 operator, have intents that lay in the equivalence class $[Intent(c_1)]_{11}$ in 2^A. Those new concepts will be denoted $prod_1(c_1)$:*

$$\forall c_3 \in prod_1(c_1), Intent(c_3)^{11} = Intent(c_1)$$

Proof. (sketch)

1. $Intent(c_3) \subseteq Intent(c_1)$
2. Let (c_1, c_2) be the genitor pair. $Intent(c_1) \cap Intent(c_2) = Intent(c_3)$ and (c_1, c_2) is maximal. This means that $(Intent(c_1), Intent(c_2))$ is minimal in $(2^A \times 2^A, \subseteq \times \subseteq)$. The canonicity property proves that $Intent(c_1)$ and $Intent(c_2)$ are also minimal. As $Intent(c_1)$ is closed, we have $Intent(c_3)^{11} = Intent(c_1)$.

This equivalence class is partitioned into finer classes according to the [33] closure and a unique new concept has the same intent as c_1.

Our aim is to simulate the incremental computation of the mingen family in the merge process. We assume that it is possible but we have to choose a strategy for the concepts insertion. We follow some criteria for this choice, as our main goal was the effectiveness. First, we tried to define a program that will compute the mingens with a minimal number of computation operations, but we also wanted to have no redundancies during the computation.

Three strategies may be suitable. The first one amounts to computing the concepts during a "*top-down*" lattice $\mathcal{L}_3$ exploration (that also means a "*bottom-up*" exploration of $[Intent(c_1)]_{11}$ in the 2^A lattice), the second one is a "*bottom-up*" lattice $\mathcal{L}_3$ exploration ("*top-down*" exploration of $[Intent(c_1)]_{11}$) and the last one is a direct strategy, which means that we make no use of incrementality, but that all the mingens will have to be computed once.

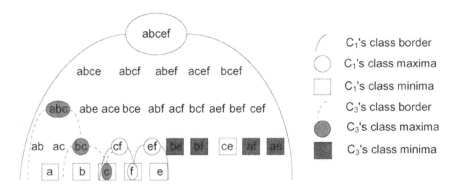

Fig. 4. Example of the partitioning.

The last strategy is the ideal one as it computes the mingens with no redundancies, but we do not have enough information to use the last strategy yet but it could be the object of further study.

The second strategy is that whenever a new concept c_3 is inserted, all its successors of $[Intent(c_1)]_{11}$ in $\mathcal{L}_3$ have to be computed. This means that the $[Intent(c_3)]_{33}$ is only really accessible at the end of all the $prod_1(c_1)$ concept insertions. So, $gen_3(c_3)$ may only be established once all insertions have been made. Therefore, the second strategy is not a suitable one except if the mingens are computed via a batch process.

The first strategy is then the only one which is suitable. We can compute the $gen(c_3)$ set from a temporal mingens list associated to c_1, that reflects the current state of the insertion process in the $[Intent(c_1)]_{11}$ class. We can also say that it is as if each c_3 was inserted separately in an incremental way and in order. The $gen^{inc}(c_1)$ list, representing the mingens of c_1 which evolve during the computation, represents after each insertion of a $c_{3,i}$, the minima of $[Intent(c_1)]_{11} - \bigcup_{j \leqslant i} [Intent\,(c_{3,j})]_{33}$. At the end, the mingens lists are up to date. Once the $gen_3(c_{3,i})$ list is computed, it will never be recomputed again.

The mingens list of a new concept of $\mathcal{L}_3$ may be described as follows:

Proposition 3 *Given an order* $(c_{3,1}, ..., c_{3,n})$ *such that* $\forall i, j | 1 \leqslant i \leqslant j \leqslant n,\ c_{3,j} \not\leqslant_3 c_{3,i}$:

$$gen_3\,(c_{3,i}) = \left\{ \begin{array}{l} Y | Y \subseteq Intent\,(c_{3,i}) \\[2ex] Y \in \min \left([Intent\,(c_1)]_{11} - \bigcup_{j \leqslant i} [Intent\,(c_{3,j})]_{33} \right) \end{array} \right\}$$

This strategy was chosen for consecutive insertions of the new concepts. It may be explained by the fact that there exist an equivalence between the merge, restricted to the concept c_1, and the insertion of n new concepts in $\mathcal{L}_1$ where c_1 is a genitor. This approximatively amounts to inserting n new objects having the respective intents of each $c_{3,i}$. By this strategy, we insure that whenever a new $c_{3[c_1]}$ is inserted, all the concepts with a smaller intent are computed and that the mingens can be computed with the mingens of c_1. With this strategy, we are sure that the mingens of every new concept from the class $[Intent(c_1)]_{11}$ can be computed by looking only at the mingens of one concept, c_1. These mingens will have to be updated after each insertion. Given the set of the intents of the new concepts generated by c_1, say $C_3^a \cap [Intent(c_1)]_{11}$, the previous condition imposes that at any moment the set of already inserted concepts corresponds to an order ideal of that set, provided with inclusion order. A noteworthy fact about the concrete computation method is that it perfectly fits the merge algorithm and Algorithm 1 (with parameters c_1 and c_3). The resulting algorithm is described below (Algorithm 2). Moreover, along all the insertions, the temporary set of mingens that are to be considered for the next insertion is stored at the genitor node within $\mathcal{L}_1$. Indeed, the concept corresponding to c_1 in $\mathcal{L}_3$ (i.e., with the same intent) will be the last one to be created since its intent is the greatest element of the equivalence class.

4.2   Implementation

The COMPUTE-MINGENS method has been associated to the merge process ([17]). The resulting algorithm is given in Algorithm 2.

1: **procedure** MERGE(**In:** $\mathcal{L}_1$, $\mathcal{L}_2$ lattices; **Out:** $\mathcal{L}_3$ a lattice)
2:
3: $\mathcal{L} \leftarrow \emptyset$
4: SORT($\mathcal{C}_1$); SORT($\mathcal{C}_2$) {Decreasing order}
5: **for all** (c_i, c_j) **in** $\mathcal{L}_1 \times \mathcal{L}_2$ **do**
6:      $I \leftarrow Intent(c_i) \cap Intent(c_j)$
7:      **if** CANONICAL($(c_i, c_j), I$) **then**
8:          $c \leftarrow$ NEW-CONCEPT($Extent(c_i) \cup Extent(c_j), I$)
9:          $\mathcal{L}_3 \leftarrow \mathcal{L}_3 \cup \{c\}$
10:         UPDATE-ORDER($c, \mathcal{L}_3$)
11:         COMPUTE-MINGENS(c_i, c)

Algorithm 2: Assembling the global Concept lattice from a pair of partial ones.

The concept c_i from the lattice $\mathcal{L}_1$ is used as a buffer for the intermediate computation. The COMPUTE-MINGENS method modifies c_i's mingens in a destructive way, i.e. when all the concepts created from c_i have been inserted in $\mathcal{L}_3$, the family of mingens of c_i in $\mathcal{L}_1$ is empty.

For example, the evolution of the equivalence class associated to $cdfgh$ (from $\mathcal{L}_1$) is depicted in Fig. 5. Indeed, the inner loop of Algorithm 2 discovers three new concepts with intents d, cd, and $cdfgh$, respectively, and in this order. These are gradually "inserted" in the class of $cdfgh$: the mingens of the new intent are computed and those of $c_1 = (13, cdfgh)$ are updated in $\mathcal{L}_1$. Thus, the new intent d which corresponds to an initial mingen of the class forces the creation of four new mingens (cd, df, dg, dh). In contrast, the closed cd merely converts a former mingen into a closure.

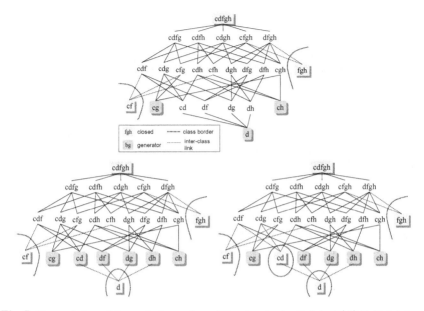

Fig. 5. The evolution of the equivalence class of the closed set $cdfgh$ in $\mathcal{P}(A)$ (initial state up) during the assembly process: after the creation of the new closed d (*left*) and after the creation of cd (*right*).

$Intent(c_1)$	$gen_{i-1}^{inc}(c_1)$	$Intent(c_i)$	$gen(c_i)$	$gen_i^{inc}(c_1)$
$cdfgh$	$\{d, cg, ch\}$	d	$\{d\}$	$\{cd, cg, ch, df, dg, dh\}$
$cdfgh$	$\{cd, cg, ch, df, dg, dh\}$	cd	$\{cd\}$	$\{cg, ch, df, dg, dh\}$
$cdfgh$	$\{cg, ch, df, dg, dh\}$	$cdfgh$	$\{cg, ch, df, dg, dh\}$	$\{\}$

Table 1. The evolution of the equivalence class of the closed set $cdfgh$ in $\mathcal{P}(A)$.

4.3 Performance Tests

The MERGE algorithm was implemented in Java, within the 2.0 version of the Galicia platform[6]. The method has been tested as a stand-alone application and its performance was compared with two incremental methods, one with the incremental computation of the mingens, and the other with a batch computation of the mingens (computed after

---

[6] See the website at: http://www.iro.umontreal.ca/~galicia.

each object insertion). The experiments were done on a Windows PC station (Pentium Xeon 3.06 GHz with 1.2 GB of RAM) using various subsets of the IBM transaction database T25I10D10K. This dataset is made out of 10 000 transactions over a set of 10 000 items. It is known to be a sparse one, with an average of 28 items per transaction. We did not use a dense dataset, as the lattices to merge contains too many concepts and require too much resources.

In order to improve the results, we combined the COMPUTE-MINGENS algorithm to a batch method called JEN (in [2]). In fact, when a concept is created by the bottom node, its mingens will be computed by JEN. In all the other cases, the COMPUTE-MINGENS algorithm is used. This combination has been motivated by the fact that JEN computes the mingens using the information provided by the successors rather then by the predecessors. This is particularly beneficial for large attribute sets, where the concepts created by the bottom, may imply a lot of computation operation. For example, for the context described previously, that contains 10 000 attributes, suppose the first object contains 28 attributes, the mingens computation of that concept will produce 279 216 operations. The way JEN works has helped solve this problem. This method has also been generalized in the merge case, as the partial order insure that a node is only computed after all its successors have been computed.

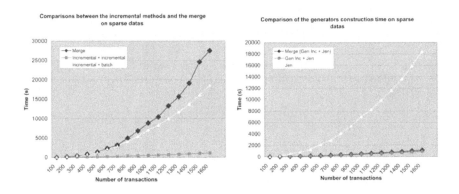

Fig. 6. Cumulative CPU-time for all three algorithms on transaction batches up to 1600 drawn from the T25I10D10K dataset.(*Left*) Lattice construction and computation of the mingens (*Right*) computation of the mingens only.

The graphs drawn in Fig. 6 summarize our findings so far. They clearly show that the incremental method gives currently better results than the merge process. This fact is not surprising given the large number of concepts that the merge algorithm must examine on each merge operation ($l_1 \cdot l_2$, where $l_1 = |\mathcal{L}_1|$ and $l_2 = |\mathcal{L}_2|$) and the even larger number of mingens. However, when we extracted the mingen computation time from the global execution time, we can see that the merge method is about as good as the incremental method and that they are 10 times faster than the batch method used in an incremental way.

The results show that the incremental mingen computation method and its generalization seem promising. It means that provided a good merge algorithm for lattice construction, where the mingens computation can be applied, we would be able to have a better execution time than batch algorithms.

5   Conclusion

The work presented here builds on a previous study of the incremental maintenance of the mingens family of a context. We investigated the case of lattice merge upon context subposition and showed a way to extend the incremental update of the mingen family to this more general case. To that end, first, a precise characterisation of the lattice structures involved in mingen computation/update was provided. The characterization was then embedded into a concrete update method for both concepts and mingens. The performances of the new method were compared to those of two other methods performing concept and mingen construction: an incremental lattice builder (ADD-OBJECT) coupled to a batch procedure for mingen computation (JEN), and a purely incremental lattice and mingen extraction. Although the overall performances of the later method has proven superior to the other two, when only mingen construction cost is considered, merging is as good as one-by-one incremental reconstruction. This fact rises the hopes that with an efficient lattice merge technique, e.g., one that takes advantage of effective decentralization of both data and computation resources, there can be a significant improvement in the speed of the global method computing both concepts and mingens. Such efficient methods have already been designed for lattice merge upon context apposition, i.e., in the dual case of the one considered here, in both sequential [14] and parallel [7] algorithmic settings.

Acknowledgments

This research was supported by the authors' individual NSERC grants as well as by the FQRNT team grant.

References

[1] C. Berge. *Hypergraphs: Combinatorics of Finite Sets*. North Holland, Amsterdam, 1989.

[2] A. Le Floc'h, C.Fisette, R. Missaoui, P. Valtchev, and R. Godin. Jen : un algorithme efficace de construction de générateurs pour l'identification des règles d'association. *numéro spécial de la revue des Nouvelles Technologies de l'Information*, 1(1):135–146, 2003.

[3] B. Ganter. Two basic algorithms in concept analysis. preprint 831, Technische Hochschule, Darmstadt, 1984.

[4] B. Ganter and R. Wille. *Formal Concept Analysis, Mathematical Foundations*. Springer-Verlag, 1999.

[5] R. Godin, R. Missaoui, and H. Alaoui. Incremental Concept Formation Algorithms Based on Galois (Concept) Lattices. *Computational Intelligence*, 11(2):246–267, 1995.

[6] J.L. Guigues and V. Duquenne. Familles minimales d'implications informatives résultant d'un tableau de données binaires. *Mathématiques et Sciences Humaines*, 95:5–18, 1986.

[7] Jean François Djoufak Kengue, Petko Valtchev, and Clémentin Tayou Djamegni. A parallel algorithm for lattice construction. In B. Ganter and R. Godin, editors, *roceedings of the 3rd Intl. Conference on Formal Concept Analysis (ICFCA'05), Lens (FR), (14-18 February) 2005*, pages 248–263, 2005.

[8] Marzena Kryszkiewicz and Marcin Gajek. Concise representation of frequent patterns based on generalized disjunction-free generators. In *PAKDD*, pages 159–171, 2002.

[9] D. Maier. *The theory of Relational Databases*. Computer Science Press, 1983.

[10] H. Mannila and K.-J. Räihä. On the complexity of inferring functional dependencies. *Discrete Applied Mathematics*, 40(2):237–243, 1992.

[11] N. Pasquier. *Data Mining : Algorithmes d'extraction et de réduction des règles d'association dans les bases de données*. Ph. d. thesis, Université Blaise Pascal,Clermont-Ferrand II, 2000.

[12] J. Pfaltz and C. Taylor. Scientific discovery through iterative transformations of concept lattices. In *Proceedings of the 1st International Workshop on Discrete Mathematics and Data Mining*, pages 65–74, Washington (DC), USA, April 2002.

[13] G. Stumme, R. Taouil, Y. Bastide, N. Pasquier, and L. Lakhal. Computing Iceberg Concept Lattices with Titanic. *Data and Knowledge Engineering*, 42(2):189–222, 2002.

[14] P. Valtchev and V. Duquenne. Towards scalable divide-and-conquer methods for computing concepts and implications. In E. SanJuan, A. Berry, A. Sigayret, and A. Napoli, editors, *Proceedings of the 4th Intl. Conference Journées de l'Informatique Messine (JIM'03): Knowledge Discovery and Discrete Mathematics, Metz (FR), 3-6 September*, pages 3–14. INRIA, 2003.

[15] P. Valtchev, M. Rouane Hacene, and R. Missaoui. A generic scheme for the design of efficient on-line algorithms for lattices. In B. Ganter A. de Moor, W. Lex, editor, *Proceedings of the 11th Intl. Conference on Conceptual Structures (ICCS'03)*, volume 2746 of *Lecture Notes in Computer Science*, pages 282–295, Berlin (DE), 2003. Springer-Verlag.

[16] P. Valtchev and R. Missaoui. Building concept (Galois) lattices from parts: generalizing the incremental methods. In H. Delugach and G. Stumme, editors, *Proceedings of the ICCS'01*, volume 2120 of *Lecture Notes in Computer Science*, pages 290–303, 2001.

[17] P. Valtchev, R. Missaoui, and R. Godin. Formal Concept Analysis for Knowledge Discovery and Data Mining: The New Challenges. In P. Eklund, editor, *Concept Lattices: Proceedings of the 2nd Int. Conf. on Formal Concept Analysis (FCA'04)*, volume 2961 of *Lecture Notes in Computer Science*, pages 352–371. Springer-Verlag, 2004.

[18] P. Valtchev, R. Missaoui, and P. Lebrun. A partition-based approach towards building Galois (concept) lattices. *Discrete Mathematics*, 256(3):801–829, 2002.

[19] P. Valtchev, R. Missaoui, M. Rouane-Hacene, and R. Godin. Incremental maintenance of association rule bases. In *Proceedings of the 2nd Workshop on Discrete Mathematics and Data Mining*, San Francisco (CA), USA, May 2003.

Representation of Data Contexts and Their Concept Lattices in General Geometric Spaces

Tim B. Kaiser

Darmstadt University of Technology, 64287 Darmstadt, Germany
`tkaiser@mathematik.tu-darmstadt.de`

Abstract. We present a possibility for coordinatizing many-valued contexts and their concept lattices, i.e. we investigate when an algebra (in the sense of universal algebra) can be assigned to the object set of a many-valued context such that the extents can be described by the congruence classes of the algebra. Since congruence class spaces have a natural geometric nature the outlined approach can be interpreted as a geometric representation of concept lattices.

1   Introduction

Data contexts can be seen as an essential bridge between reality and mathematics, because they can be represented mathematically but are still very close to raw observations. To activate the potential of mathematics in supporting humans in their aim to understand reality (realities) it is from this viewpoint important to develop mathematical methods which help exhibiting structure hidden in data contexts. This can be done via *representing* the data context in a meaningful way in a geometrical space since

> [...] the geometric nature of those representations make relationships graphic, intelligible, and workable in multiple ways and therefore may successfully support human thought and action.[...] ([WW02], p. 6)

A framework for investigations of this type is given by *measurement theory*, which is concerned with the problem of representing given *empirical relational structures* in *numerical relational structures*. In this paper we use an extended version of this paradigm, the four-level model (cf. [Wi96], see Fig. 1), where an additional level, the *synthetic* level, is added between the level of relational structures and the level of numerical (algebraic) structures. In [Wi96], [Wi97], and [WW02] it is argued that the extended paradigm is more adequate for representations in geometrical spaces, since firstly the step from the synthetic level to the analytic (numerical) level reflects classical coordinatization theorems in geometry, f.e. the theorem that a three-dimensional affine geometry can always be coordinatized by some vector space over a skew field. And secondly the step from the formal level to the synthetic level allows for developing structure theory for qualitative representations.

F. Dau, M.-L. Mugnier, G. Stumme (Eds.): ICCS 2005, LNAI 3596, pp. 195–208, 2005.

In [Wi96] this approach is applied to the problem of representing ordinal data contexts in vector spaces over ordered skew fields, where strong conditions are required which limit the area of possible applications. Already in [WW92] conditions for ordinal structures to be representable by *ordered n-loops* are given. This result is generalized slightly in [Wi96]. So the most general, well-known structure for representing ordinal data is an ordered n-loop. This representation still enforces strong conditions on the data, called ordinal dependency conditions.

In this paper, we try to establish a new way of representing data contexts geometrically by considering *congruence class spaces* of algebras in the sense of universal algebra as representing structures. This implements the four level model as shown in Fig. 1. On the formal level we will work with *many-valued contexts*, on the synthetic level we use *closure structures* fulfilling certain conditions, and on the analytic level *congruence class spaces* derived from an algebra in the sense of universal algebra serve as coordinatizing structures.

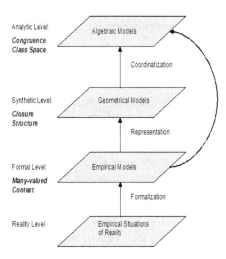

Fig. 1. The four level model

2   Closure Structures with Weak Parallelism

Closure structures allow a very general approach to geometry where the basic entities are *points* and an *operator* assigning to any set of points the smallest subspace they belong to [Jo59].

Definition 1 (closure structure, restriction). *We call* $\Gamma := (G, [\cdot])$ *a closure structure if* G *is a set and* $[\cdot] : \mathfrak{P}(G) \to \mathfrak{P}(G)$ *an operator satisfying the following properties (let* $X, Y \subseteq G$*):*

(C1) $X \subseteq [X]$ *(extensitivity)*
(C2) $[X] = [[X]]$ *(idempotency)*
(C3) $X \subseteq Y \Rightarrow [X] \subseteq [Y]$ *(isotony)*

If $H \subseteq G$ we call $\Gamma_{|H} := (H, [\cdot]_{|H})$ a restriction of Γ if for $J \subseteq H$ the closure is defined as $[J]_{|H} := [J] \cap H$.

Obviously, a restriction of a closure structure is again a closure structure. The closed sets $\mathfrak{C}(\Gamma) := \{ X \in \mathfrak{P}(G) \mid X = [X] \}$ of any closure structure form a complete lattice via set inclusion. Often we call the elements of G *points* and the closed sets *subspaces*. If $[X] = Y$ we call the set X a *generating subset* for Y. The *rank* of a subspace V is defined as $|X|$ if $[X] = V$ and there does not exist a X' with $[X'] = V$ and $|X'| < |X|$.

For closure structures the notion of an embedding is given by:

Definition 2 (embedding). *Let $\Gamma := (G, [\cdot])$ and $\Gamma' := (G', [\cdot]')$ be closure structures. An injection $\phi : G \rightarrow G'$ is called an* embedding *if for $H \subseteq G$*

$$\phi([H]) = [\phi(H)]' \cap \phi(G).$$

Note that – if $\Gamma := (G, [\cdot])$ is a closure structure and $H \subseteq G$ – a restriction $\Gamma_{|H}$ is embeddable into Γ via $\phi : H \rightarrow G$ with $\phi(h) := h$. The next definition makes explicit what *(weak) parallelity* means in a closure structure:

Definition 3 (weak parallelism). *Let $\Gamma := (G, [\cdot])$ be a closure structure. We call a relation $\|$ on $\mathfrak{C}^+(\Gamma) := \mathfrak{C}(\Gamma) \setminus \{\emptyset\}$* weak parallelism *if the following conditions hold:*

(P1) *For $R \in \mathfrak{C}^+(\Gamma)$ we have $R \parallel R$.*
(P2) *For $R, S, T, U \in \mathfrak{C}^+(\Gamma)$ the conditions $R \parallel S$, $T \subseteq S$, and $T \parallel U$ imply the existence of $V \in \mathfrak{C}^+(\Gamma)$ with $U \subseteq V$ and $R \parallel V$.*
(P3) *For $R, S \in \mathfrak{C}^+(\Gamma)$ and $p \in S$ with $R \parallel S$ we have $S \subseteq [\{p\} \cup R]$.*
(P4) *For $R \in \mathfrak{C}^+(\Gamma)$ and $p \in G$ there exists one and only one $S \in \mathfrak{C}^+(\Gamma)$ with $R \parallel S$ and $p \in S$.*

If $\|$ is an equivalence relation it is called parallelism.

Let $\Gamma := (G, [\cdot], \|)$ be a closure structure with weak parallelism, $p \in G$, and $V \in \mathfrak{C}^+(\Gamma)$. We denote by $\pi(p|V)$ the unique subspace W containing p and satisfying $V \parallel W$ (this definition is justified by (P4)). If $\|$ is a parallelism we get the following result:

Proposition 1. *Let $\Gamma := (G, [\cdot], \|)$ be a closure structure with parallelism and $A, B \in \mathfrak{C}(\Gamma)$. Then $A \parallel B$ and $rank(A) = 1$ implies $rank(B) = 1$.*

Proof. Let A be of rank 1. Then $A = [a]$ for some $a \in G$. We assume $A \parallel B$ with $rank(B) > rank(A)$. Then $B = [b_1, b_2, ..., b_n]$ with $n > 1$. But $C := \pi(a|[b_1])$ has to be a proper superset of A, because otherwise (P3) would be violated. This leads to the following situation: $A \parallel B$, $[b_1] \subseteq B$, and $C \parallel [b_1]$. Using (P2) we can deduce the existence of a closed set D with $C \subseteq D$ and $A \parallel D$. Since $a \in C$, we have $D = A$, but $D \subsetneq A$, a contradiction. $\square$

We think of a *dilation* as a mapping $\delta : G \to G$ with $\delta(p) \in \pi(\delta(q)|p, q)$ for any points $p, q \in G$. The set of all dilations is denoted by $\Delta(\Gamma)$. Now, we can assign an algebra $\mathbb{A}(\Gamma) := (G, \Delta(\Gamma))$ to a closure structure with weak parallelism.

Following [Ma52], we define a *geometric space* as a special closure structure. By $A \subseteq_{fin} B$ we mean that A is finite and a subset of B.

Definition 4 (geometric space). *We call a closure structure $(G, [\cdot])$ geometric space if $[\cdot]$ satisfies the additional properties:*

(G1) $[p] = \{p\}$ *for* $p \in G$,
(G2) $[\emptyset] = \emptyset$
(G3) $[X] = \bigcup_{Y \subseteq_{fin} X} [Y]$ *for* $X \subseteq G$.

So a closure structure is a geometric space if the empty set and the one-element sets are closed and the closure operator is algebraic (G3). In a geometric space, subspaces of rank 2 are called *lines* and subspaces of rank 3 are called *planes*. A geometric space which satisfies for $X \subseteq G$

$$X = [X] \iff [p, q, r] \subseteq X \text{ for } p, q, r \in X$$

is called *planar*.

Now, we turn to a special class of examples for closure structures with weak parallelism which are even geometric spaces satisfying certain additional properties. Let $\mathbb{A} := (A, (f)_I)$ be an algebra. The congruence classes of $\mathbb{A}$ together with the empty set form a closure system. Therefore every set $M \subseteq A, M \neq \emptyset$ is contained in a smallest congruence class $[M]_\theta := \bigcap\{[a]\theta \mid a \in A, \theta \in Con(\mathbb{A}), M \subseteq [a]\theta\}$ and $[\cdot]_\theta$ is a closure operator on A if we define $[\emptyset]_\theta := \emptyset$. We denote the closure structure $(A, [\cdot]_\theta)$ by $\Gamma\mathbb{A}$ and the closed sets of this closure structure by $\Gamma_c\mathbb{A}$. The closure structure $\Gamma\mathbb{A}$ is called the *congruence class space* of the algebra $\mathbb{A}$. We call an algebra *regular* if every congruence relation can be recovered by any of its congruence classes, i.e. $[a]\theta = [a]\theta' \implies \theta = \theta'$. Also we call a set $R \subseteq \mathfrak{P}(M \times M)$ of relations on a set M regular if for $\theta, \theta' \in R$ and $m \in M$ we have $[m]\theta = [m]\theta' \implies \theta = \theta'$.

Theorem 1. *Let $\mathbb{A} := (A, (f)_I)$ be an algebra. Then the structure $\Gamma(\mathbb{A}) := (A, [\cdot]_\theta)$ is a geometric space and the relation $\|_\mathbb{A}$ defined by*

$$R \parallel_\mathbb{A} S : \iff S \text{ is congruence class of } \theta(R)$$

is a weak parallelism on $\Gamma(\mathbb{A})$.

Example 1. We will use the smallest non-modular lattice N_5 to give an example of an congruence class space. The Hasse diagram of N_5 is shown in figure 2. To build the congruence class space of an algebra one has to determine all its congruence classes. Figure 3 displays the congruence lattice of N_5. Note that N_5 is not regular. Figure 4 shows the lattice of the congruence classes of N_5, the closed sets of its congruence class space.

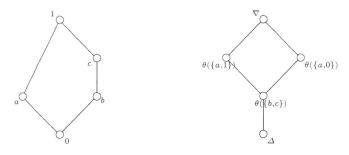

Fig. 2. N_5 **Fig. 3.** The congruence lattice of N_5

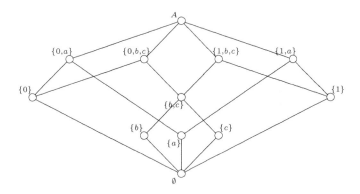

Fig. 4. Lattice of the congruence classes of N_5

The regularity of the algebra from which one derives the congruence class space and the weak parallelism $\|_{\mathbb{A}}$ are tightly coupled:

Proposition 2. *Let $\mathbb{A}$ be an algebra. Then the weak parallelism $\|_{\mathbb{A}}$ in $\Gamma_c(\mathbb{A})$ is a parallelism if and only if $\mathbb{A}$ is regular.*

Proof. "$\Rightarrow$": *Let $\|_{\mathbb{A}}$ be a parallelism. We consider parallel pencils $\Pi_C := \{B \in \Gamma_c(\mathbb{A}) \mid B \|_{\mathbb{A}} C, C \in \Gamma_c(\mathbb{A})\}$. Such a parallel pencil is simultaenously a partition on the carrier set A of the algebra as well as a block of a partition on the congruence classes. The set of all parallel pencils is the set of all blocks of the partition induced by $\|_{\mathbb{A}}$. For $A, B \in \Pi_C$ by symmetry of $\|_{\mathbb{A}}$ we have $\theta(A) = \theta(B)$. That means all blocks in Π_C exhaust exactly one congruence relation. Since every closed set is in exactly one parallel pencil we can deduce the regularity of $\mathbb{A}$.*
"$\Leftarrow$": *Let $\mathbb{A}$ be regular. Assume $B \|_{\mathbb{A}} C$ which is equivalent to C is a class of $\theta(B)$. First, we show symmetry. Since $\mathbb{A}$ is regular we have $\theta(B) = \theta(C)$, hence $C \|_{\mathbb{A}} B$. Transitivity holds using an analogous argument.* □

If for a given geometric space Γ there exists an algebra $\mathbb{A}$ with $\Gamma \cong \Gamma(\mathbb{A})$ we say $\mathbb{A}$ *coordinatizes* Γ. The following theorem from [Wi70] generalizes the result that an affine geometry can be coordinatized by a vector space over a skew field if and only if the Condition of Desargue holds. In that sense it is natural to ask under which conditions such a coordinatization is possible for a geometric space in our more general setting. Before presenting the characterization we need one more definition:

Definition 5. *Let A be a set, let $B \subseteq A$, and let Δ be a set of maps from A to A. Then we define a relation $\equiv \subseteq A \times A$ where for $a, b \in A$ we have $a \equiv b \bmod (B, \Delta)$ if and only if there exist $\delta_i \in \Delta$ for $i = 0, 1, ..., n$ with $a \in \delta_0(B)$ & $b \in \delta_n(B)$ & $\delta_i(B) \cap \delta_{i+1}(B) \neq \emptyset$ for $i \in \{1, 2, .., n-1\}$.*

Theorem 2. *Let $\Gamma := (G, [\cdot])$ be a geometric space. Then the following conditions are equivalent:*

(C1) *There exists an algebra $\mathbb{A}$ with $\Gamma \cong \Gamma(\mathbb{A})$.*
(C2) *Γ is planar and there exists a weak parallelism $\|$ on $\mathfrak{C}(\Gamma)$ where*

$$a, b, c, d \in G \text{ and } d \in [a, b, c] \Rightarrow d \equiv a \bmod (\{a, b, c\}, \Delta(\|)). \qquad (1)$$

We will employ Theorem 2 for answering the question under which conditions a closure structure is representable in a congruence class space. In the following we collect properties a closure structure representable in a congruence class space automatically possesses.

Proposition 3. *Let ϕ be an embedding of the closure structure $\Gamma := (G, [\cdot])$ into the congruence class space $\Gamma' := (A, [\cdot]_\theta)$ of some algebra $\mathbb{A}$. Then Γ is a geometric space inheriting a weak parallelism from Γ'. We will refer to this parallelism as* trace parallelism.

Proof. Let $p \in G$ and $X \subseteq G$. Note that since ϕ is an embedding and therefore injective we have

$$[X] = \phi^{-1}([\phi(X)]_\theta \cap \phi(G)) = \phi^{-1}([\phi(X)]_\theta) \qquad (2)$$

So we get $[p] = \phi^{-1}([\phi(p)]_\theta) = \phi^{-1}(\phi(p)) = p$ which shows (G1). Similarly, we get $[\emptyset] = \phi^{-1}([\phi(\emptyset)]_\theta) = \phi^{-1}(\emptyset) = \emptyset$ which shows (G2). For verifying (G3) we use again equation 2 and that $[\cdot]_\theta$ is algebraic. We get

$$
\begin{aligned}
[X] &\overset{(2)}{=} \phi^{-1}([\phi(X)]_\theta) \\
&\overset{[\cdot]_\theta \ algebraic}{=} \phi^{-1}\left(\bigcup_{Y \subseteq_{fin} X} [\phi(Y)]_\theta \right) \\
&= \bigcup_{Y \subseteq_{fin} X} \phi^{-1}([\phi(Y)]_\theta) \\
&\overset{(2)}{=} \bigcup_{Y \subseteq_{fin} X} [Y].
\end{aligned}
$$

This makes explicit that Γ is a geometric space.

Now, we show that Γ naturally inherits a weak parallelism $\parallel$, the trace parallelism, *from Γ' if we define $\parallel$ for $R, S \in \Gamma_c$ as*

$$R \parallel S : \Longleftrightarrow \exists W \in \Gamma'_c : [\phi(R)]_\theta \parallel_\mathbb{A} W \ with \ \phi^{-1}W = S.$$

*We have to show that $\parallel$ is indeed a weak parallelism. In the following proof let
R, S, T, U and V denote closed sets of Γ. For showing (P1) we have to assure
that $H \parallel H$ for all $H \in \mathfrak{C}\Gamma$. But $H \parallel H$ is equivalent to $[\phi(H)]_\theta \parallel_\mathbb{A} K$ where
$\phi^{-1}K = H$. With $K := [\phi(H)]_\theta$ (P1) obviously holds.*

*For verifying (P2) let $R \parallel S$ and $T \subseteq S$. Now assume $T \parallel U$. We have to
show that there exists a closed set V with $U \subseteq V$ and $R \parallel V$. But $R \parallel S$ is
equivalent to $[\phi(R)]_\theta \parallel_\mathbb{A} W$ with $\phi^{-1}W = S$ and $T \subseteq S$ yields $[\phi(T)]_\theta \subseteq [\phi(S)]_\theta$.
Again $T \parallel U$ gives $[\phi(T)]_\theta \parallel_\mathbb{A} W'$ with $\phi^{-1}W' = U$. Therefore we can use (P2)
for $\parallel'$ to deduce the existence of some Γ'-closed set V' with $[\phi(U)]_\theta \subseteq V'$ and
$[\phi(R)]_\theta \parallel_\mathbb{A} V'$. Let $V := \phi^{-1}(V')$. Obviously, we have $U \subseteq V$ and $R \parallel V$.*

*For verifying property (P3) let $R \parallel S$ and $p \in S$. We get $[\phi(R)]_\theta \parallel_\mathbb{A} W$ where
$\phi^{-1}W = S$. That implies $\phi(p) \in W$. Now we can use (P3) for $\parallel_\mathbb{A}$ to deduce that
$[\{\phi(p)\} \cup [\phi(R)]_\theta]_\theta W$. Now it remains to show that $\phi^{-1}[\{\phi(p)\} \cup [\phi(R)]_\theta]_\theta =
[\{p\} \cup R]$. But using equation 2 we get the desired result since for any closure
system we have $[\{a\} \cup [A]] = [\{a\} \cup A]$.*

*As the last one, we have to show (P4) which is an analogy of Euclids Parallel
Postulate. Let $p \in G$ and $R \in \mathfrak{C}\Gamma$. For $[\phi(R)]_\theta]$ we get the existence of one and
only one subspace $W \in \mathfrak{C}(\Gamma')$ with $\phi(p) \in W$ and $R \parallel_\mathbb{A} W$. Let $S := \phi^{-1}(W)$. By
definition, we have $R \parallel S$ and $p \in S$. If there is another subspace T with $R \parallel T$
and $p \in T$ the closure of its image would generate a suspace of Γ' fulfilling the
conditions. That leads to a violation of (P4) regarding $\parallel_\mathbb{A}$. This completes our
proof that the trace parallelism is indeed a weak parallelism.* $\square$

In the following we provide examples of geometric spaces representable in a
congruence class space where the first example is a non-planar geometric space
and the second example is a geometric space violating condition (1) to show
that indeed more general geometric spaces can be represented in a congruence
class spaces, i.e. that embeddability into a congruence class space is definitely a
weaker condition than isomorphy to a congruence class space.

Example 2. The language of universal algebra allows to consider a vector space
over a (skew) field V_K as an algebra $\mathbb{G} := (V, +, -, 0, (*)_K)$ of type $(2, 1, 0, \underbrace{1, ..., 1}_{|K|})$.
In that sense our example is a restriction of the congruence class space of V_K
where $V = (\mathbb{Z}_3)^3$ and $K = \mathbb{Z}_3$ on the set of points

$$R := \{(1, 0, 0), (2, 0, 0), (2, 1, 0), (2, 1, 1), (0, 0, 2)\}.$$

The closure operator of the induced closure structure $\mathfrak{G} := \Gamma(\mathbb{G}) = ((\mathbb{Z}_3)^3,
< \cdot >)$ assigns to any set of points the smallest affine space they are contained
in.

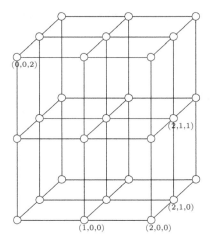

Fig. 5. $(\mathbb{Z}_3)^3$, labelled points define non-planar restriction

Figure 5 visualizes the restriction as labelled circles. We investigate the restriction $\mathfrak{G}_{|R}$. Clearly, $\mathfrak{G}_{|R}$ is embeddable into $\mathfrak{G}$ via identification. First we argue that $\mathfrak{G}_{|R}$ is not planar. Regard the set $L := \{(1,0,0),(2,0,0),(2,1,0),(2,1,1)\}$. As restricted closure we get $< L >_{|R} = R$. So L is not closed. But for any closure of three points $p_1, p_2, p_3 \in L$ we get $< \{p_1, p_2, p_3\} >_{|R} \subseteq L$. That means L is three-closed but not a closed set, hence $\mathfrak{G}_{|R}$ is not planar.

Example 3. For violating condition (1) in a planar geometric space embeddable into a congruence class space we pick a flat $\mathfrak{F}$ in $\mathfrak{G}$ and regard the restriction on the set of points $S := \{(0,0),(1,0),(2,1),(1,2)\}$ as shown in Figure 6.

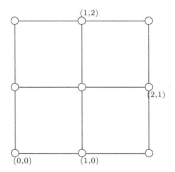

Fig. 6. $(\mathbb{Z}_3)^2$ where labelled points define a critical configuration

Clearly, $\mathfrak{F}_{|S}$ is planar. We will show that the restriction $\mathfrak{F}_{|S}$ does not fulfill condition (1). Let $D := \{(0,0),(1,0),(2,1)\}$. Then we have $(1,2) \in< D >_{|S}$. So condition (1) requires the existence of dilations $\delta_i, i = 0,...,n$ which satisfy $(1,2) \in \delta_0(\{a,b,c\})$, $\delta_i \cap \delta_{i+1} \neq \emptyset$ for $i \in \{0,...,n-1\}$, and $(0,0) \in \delta_n(\{a,b,c\})$. There are two possibilities for a dilation to act on D: $\delta(D) \subseteq D$ and $\delta(D) \nsubseteq D$. In the first case we do not win anything for reaching $(1,2)$. So let us assume $\delta(D) \nsubseteq D$. Then $(1,2) \in \delta(D)$ which implies that $\delta((0,0)) = (1,2), \delta((1,0)) = (1,2)$, or $\delta((2,1)) = (1,2)$. Assume $\delta((0,0)) = (1,2)$. Since δ is a dilation we get $\delta((1,0)) = (1,2)$ and also $\delta((2,1)) = (1,2)$. Therefore we have $\delta(D) \cap D = \emptyset$. The two other cases work analogously. This shows that the point $(1,2)$ which is in the closure of D can not be reached via a dilation construction.

We can use the result of the above proposition together with the examples to reformulate our initial question: *Which closure structures can be represented in congruence class spaces?* into *Which geometric spaces carry a weak parallelism such that they are extendible to planar geometric spaces satisfying condition (1).* Since we want to apply our results to data contexts later on and data contexts naturally carry a weak parallelism, in the following we will focus on geometric spaces with a given weak parallelism. So again our question changes to: *Which geometric spaces with weak parallelism are extendible to planar geometric spaces satisfying condition (1)?*

To answer this question we need a notion of embeddability for closure structures with weak parallelism:

Definition 6. *Let $\Gamma := (G,[\cdot],\Pi)$ and $\Gamma' := (H,[\cdot]',\Pi')$ be closure structures with weak parallelism. Then we call a mapping $\epsilon : G \to H$ embedding if*

1. *ϵ is an embedding of $(G,[\cdot])$ into $(H,[\cdot]')$.*
2. *For $A,B \in \Gamma^c : A\Pi B \implies [\epsilon(A)]'\Pi'[\epsilon(B)]'$.*

For this notion of embeddability it is easy to find a closure structure with weak parallelism which cannot be represented in a congruence class space. In the following we give an infinite sequence of congruence class spaces not representable in congruence class spaces.

Example 4. Let $R_n := \{1,2,3,4,...,n\}$ and let exactly all one-element sets, all two-element sets the empty set and R_n be closed. Let $\Psi_n := (R_n,[\cdot])$ denote the induced closure structure. Furthermore, define a weak parallelism on Ψ_n^c as $A\Pi B$ if and only if B is a one-element set with $A \cap B = \emptyset$ or $A = B$. So $(R_n)_{n \in \mathbb{N}}$ forms an infinite sequence of geometric spaces with weak parallelism which are not satisfying condition (1), since every dilation δ with $\delta(\{a,b,c\}) \cap \{a,b,c\} \neq \emptyset$ is constant. Moreover, we have exactly $\binom{n}{3}$ point configurations violating the condition. This remains true for every extension of Ψ_n, therefore Ψ_n cannot be represented in a congruence class space.

3   Data Contexts and Closure Systems with Weak Parallelism

In section 1 we have pointed out that the main goal of measurement theory is to find a *numerical relational structure* in which a given *relational structure* can be represented homomorphically. Here, we modify this idea. We generalize this approach allowing not only numerical relational structures as representing entities but every kind of algebra. As relational structure we use many-valued contexts. To activate our previous results concerning closure structures with weak parallelism we use the idea of *conceptual scaling* to derive a closure structure/system from the many-valued context.

Definition 7 (many-valued context). *A many-valued context is a structure* (G, M, W, I) *where* G, M, *and* W *are sets and* I *is a subset of* $G \times M \times W$ *with* $(g, m, w_1) \in I$ *and* $(g, m, w_2) \in I \Rightarrow w_1 = w_2$. *Every attribute* $m \in M$ *can be regarded as a partial function* $m : G \to W$, *dually every object* $g \in G$ *can be regarded as a partial function* $g : M \to W$. *A many-valued context is called* complete *if every attribute* $m \in M$ *is a function (equivalently, every* $g \in G$ *is a function).*

For a map $f : A \to B$ the *kernel* of f is defined as the equivalence relation $\ker(f) := \{(a, b) \in A^2 \mid f(a) = f(b)\}$. For every complete many-valued context $\mathbb{K} := (G, M, W, I)$ we can define two sets of equivalence relations $\mathbf{Eq}^G(\mathbb{K}) := \{ker(m) \mid m \in M\}$ and $\mathbf{Eq}^M(\mathbb{K}) := \{ker(g) \mid g \in G\}$. Also we define the set of arbitrary meets of equivalence relations of a complete many-valued context as $\overline{\mathbf{Eq}}^M(\mathbb{K}) := \{\theta \in Eq(G) \mid \exists N \subseteq M : \theta = \cap_{m \in N} ker(m)\}$, dually for $\overline{\mathbf{Eq}}^G(\mathbb{K})$. In the following we restrict our considerations to complete many-valued contexts. This is no serious restriction since every non-complete many-valued context (G, M, W, I) can be completed to (G, M, W', I') with $W' := W \cup \{\square\}$ and $I' := I \cup \{(g, m, \square) \mid \text{if } m(g) \text{ is undefined}\}$.

Formal Concept Analysis provides an instrument to assign a closure system to a many-valued context through conceptual scaling. For our purposes we use a special case of plain scaling:

Definition 8 (derived context via nominal scaling). *Let* $\mathbb{K} := (G, M, W, I)$ *be a complete many-valued context. Then the formal context* $\mathbb{K}^n := (G, N, J)$ *is called* derived context via nominal scaling *of* $\mathbb{K}$ *if* $N := \{(m, w) \in M \times W \mid \exists g \in G : m(g) = w\}$ *and* $J := \{(g, (m, w)) \in G \times N \mid (g, m, w) \in I\}$.

Definition 9 (object closure system of a many-valued context). *Let* $\mathbb{K} := (G, M, W, I)$ *be a complete many-valued context. Then its (nominal) object closure system is defined as* $\mathfrak{U}(\mathbb{K}) := \{Ext(\mathfrak{c}) \mid \mathfrak{c} \in \mathfrak{B}(\mathbb{K}^n)\}$.

We can assign a weak parallelism to a closure system derived from a many-valued context using its equivalence relations. For $A \in \mathfrak{U}(\mathbb{K})$ we denote by $\overline{\mathbf{Eq}}(A)$

the smallest equivalence relation θ in $\overline{\mathbf{Eq}}^G(\mathbb{K})$ with $A \times A \subseteq \theta$. Now we can define a relation $\pi(\mathbb{K}) \subseteq \mathfrak{U}(\mathbb{K}) \times \mathfrak{U}(\mathbb{K})$ as follows:

$$(A, B) \in \pi(\mathbb{K}) \iff B \text{ is a block of } \overline{\mathbf{Eq}}^G(A).$$

It turns out that $\pi(\mathbb{K})$ is a weak parallelism.

Theorem 3. *Let* $\mathbb{K} := (G, M, W, I)$ *be a complete many-valued context. Then* $\pi(\mathbb{K})$ *is a weak parallelism on* $\mathfrak{U}(\mathbb{K})$.

Proof. Obviously, $\pi(\mathbb{K})$ is reflexive.
Let $A, B, C, D, E \in \mathfrak{U}(\mathbb{K})$, let $A\pi(\mathbb{K})B$, let $C \subseteq B$, and let $C\pi(\mathbb{K})D$. This means that B is a class of $\overline{\mathbf{Eq}}^G(A)$. Since C is contained B we have $\overline{\mathbf{Eq}}^G(C) \subseteq \overline{\mathbf{Eq}}^G(A)$. Using that D is a class of $\overline{\mathbf{Eq}}^G(C)$ we get that $D \times D \subseteq \overline{\mathbf{Eq}}^G(A)$ which implies the existence of a class E of $\overline{\mathbf{Eq}}^G(A)$ with $D \subseteq E$.
Now, let $A, B \in \mathfrak{U}(\mathbb{K})$ with $A\pi(\mathbb{K})B$ and let $b \in B$. We want to show that $B \subseteq (A \cup \{b\})^{\mathfrak{U}}$. We have that B is a block of $\overline{\mathbf{Eq}}^G(A)$. That means that any relation containing $A \times A$ also contains $B \times B$. Therefore $\overline{\mathbf{Eq}}^G(A \cup \{b\})$ contains both $A \times A$ and $B \times B$ which implies by transitivity of the equivalence relation that $A \cup B \times A \cup B \subseteq \overline{\mathbf{Eq}}^G(A \cup \{b\})$. This yields $B \subseteq (A \cup \{b\})^{\mathfrak{U}}$.
For the last one, let $A \in \mathfrak{U}(\mathbb{K})$ and let $g \in G$. Clearly there is exactly one class of $\overline{\mathbf{Eq}}^G(A)$ containing g, since $\overline{\mathbf{Eq}}^G(A)$ is an equivalence relation.

$\square$

A first naive way of subscribing a weak parallelism to an object closure system of a complete many-valued context could have been considering $A, B \in \mathfrak{U}(\mathbb{K})$ as related, denoted by $A\theta(\mathbb{K})B$, if there exist $v, w \in W$ and $m \in M$ such that $A = (m, w)^J$ and $B = (m, v)^J$. But this does not yield a weak parallelism for arbitrary many-valued contexts. Clearly the properties of the mentioned relation are determined by the equivalence relations $\mathbf{Eq}^G(\mathbb{K})$. The next proposition explains the connection between $\pi(\mathbb{K})$ and $\theta(\mathbb{K})$.

Proposition 4. *Let* $\mathbb{K} := (G, M, W, I)$ *be a complete many-valued context. Then we have* $\pi(\mathbb{K}) = \theta(\mathbb{K})$ *if and only if* $\mathbf{Eq}^G(\mathbb{K})$ *is regular and* $\mathbf{Eq}^G(\mathbb{K}) = \overline{\mathbf{Eq}}^G(\mathbb{K})$, *i.e.* $\mathbf{Eq}^G(\mathbb{K})$ *is regular and meet-closed.*

Proof. "$\Rightarrow$": Assume $\pi(\mathbb{K}) = \theta(\mathbb{K})$. First we have to show that $\mathbf{Eq}^G(\mathbb{K})$ is regular, i.e. that any class of a relation $\psi \in \mathbf{Eq}^G(\mathbb{K})$ determines its relation. Suppose that there exist different relations $\psi, \psi' \in \mathbf{Eq}^G(\mathbb{K})$ having $A \subseteq G$ as a class. Then there are blocks B of ψ and C of ψ' with $B \neq C$ and $B \cap C \neq \emptyset$. Clearly, this violates axiom (P4), a contradiction. Suppose $\mathbf{Eq}^G(\mathbb{K})$ is not meet-closed. Then the relation $\theta(\mathbb{K})$ is not even reflexive on $\mathfrak{U}(\mathbb{K})$.
"$\Leftarrow$": Assume $\mathbf{Eq}^G(\mathbb{K})$ is regular and meet-closed. Let $A\pi(\mathbb{K})B$. Then B is a class of $\overline{\mathbf{Eq}}^G(A)$. Since $\mathbf{Eq}^G(\mathbb{K})$ is regular and meet-closed there is exactly one relation $\psi \in \mathbf{Eq}^G(\mathbb{K})$ with $\psi = \ker(m) = \overline{\mathbf{Eq}}^G(A)$ for one $m \in M$. This shows

that $A\theta(\mathbb{K})B$. Now assume $A\theta(\mathbb{K})B$. This means that there exists an $m \in M$ with A and B blocks of $\ker(m)$. Since $\mathbf{Eq}^G(\mathbb{K})$ is meet-closed $\overline{\mathbf{Eq}}^G(A) \in \mathbf{Eq}^G(\mathbb{K})$ and since $\mathbf{Eq}^G(\mathbb{K})$ is regular $\ker(m) = \overline{\mathbf{Eq}}^G(A)$, which implies $A\pi(\mathbb{K})B$. □

Now, we have a way to assign to a complete many-valued context a closure system with weak parallelism $\mathfrak{C}(\mathbb{K}) := (\mathfrak{U}(\mathbb{K}), \pi(\mathbb{K}))$ or a closure structure with weak parallelism $\Gamma(\mathbb{K}) := (G, \mathfrak{U}, \pi(\mathbb{K}))$.

In the next section we apply our proposed technique to an example.

4   Example

Consider the many-valued context $\mathbb{K} := (G, M, W, I)$ represented by the following data table:

	1	2	3
a	A	C	E
b	B	C	F
c	A	D	F
d	B	D	E

Fig. 7. Sample context

If we consider the meet-closure of its equivalence relations we see that

$$\overline{\mathbf{Eq}}^G(\mathbb{K}) = \mathbf{Eq}^G(\mathbb{K}) \cup \Delta_G \cup \nabla_G.$$

The set of relations $\overline{\mathbf{Eq}}^G(\mathbb{K})$ is regular, therefore we can use proposition 4 to argue that $\pi(\mathbb{K})$ is a parallelism.

Now, we have to look at the set of dilations $\Delta(\pi(\mathbb{K}))$ induced by the parallelism. It turns out that it contains the following non-constant mappings:

1. $\delta_1 : a \mapsto b, c \mapsto d, b \mapsto a$, and $d \mapsto c$
2. $\delta_2 : a \mapsto c, b \mapsto d, c \mapsto a$, and $d \mapsto b$
3. $\delta_3 : a \mapsto d, b \mapsto c, d \mapsto a$, and $c \mapsto b$
4. $\delta_4 = id_G$

Now we have to check if all conditions for a coordinatization are satisfied. First of all $\mathfrak{U}(\mathbb{K})$ is a geometry since the empty set and its points are closed and it is algebraically since it is finite. Furthermore $\mathfrak{U}(\mathbb{K})$ is planar because there is no closure whith a rank higher than three. Now, only condition 1 remains to be checked. For this we only have to consider the case where $d \in [a, b, c]$, because of symmetry. But clearly, δ_3 does the job. we have $d \in \delta_3(\{a, b, c\}) = \{b, c, d\} \cap \{a, b, c\} \neq \emptyset$ which fulfills the condition. Now we know that we can coordinatize the derived closure system with parallelism. If we look closer at the constructed algebra $(G, (\delta_i)_{i=1}^4)$ we see that it is isomorphic to the vector space $\mathbb{Z}_2^2$, which is the smallest finite affine geometry. Therefore by applying our technique we found another way of representing our closure system:

Fig. 8. $\mathfrak{U}(\mathbb{K})$ represented as $\mathbb{Z}_2^2$

5   Conclusion

In the this paper we have shown how to assign a closure structure with weak parallelism to a many-valued context. This closure structure can be investigated to see if it matches the conditions listed in Theorem 2. If it fulfills these conditions we know that we can assign an algebra to it gaining another way of describing its extents, not only be applying the derivation operator to attribute sets but also by forming the smallest congruence class an object set is contained in. As in this paper only the question of isomorphic representation is answered sufficiently using the results from [Wi70] the future task in this line of research clearly is to find "nice" conditions for embeddability were some leverage points are indicated by our examples. Since a coordinatizing algebra gives us a way to introduce (partial) operations on the set of concepts of the scaled context – if we consider the concepts as elements of suitable factor algebras of the coordinatizing algebra – it might also be interesting to investigate if these operations can yield supportive insights or meaningful interpretations for "real world data".

References

[GW99]   B. Ganter, R. Wille: Formal Concept Analysis, Mathematical Foundations. Springer-Verlag, Berlin Heidelberg New York (1999)

[Jo59]   B. Jonsson: Lattice-theoretic approach to affine and projective geometry. In L. Henkin, P. Suppes, A. Tarski (eds.): The axiomatic method with special references to geometry and physics. Amsterdam, North-Holland (1959), 188-203.

[KLST71] D. H. Krantz, R. D. Luce, P. Suppes, A. Tversky: Foundations of Measurement, Volume I. Academic Press, Inc., New York (1971)

[Ma52]   F. Maeda: Lattice-theoretic characterization of abstract geometries. Jour. of Sci. of Hiroshima Univ. Ser. A 15 (1951/52), p. 87-96.

[Wi70]   R. Wille: Kongruenzklassengeometrien. Springer-Verlag, Berlin Heidelberg New York (1970)

[Wi96]   U. Wille: Geometric Representation of Ordinal Contexts. Shaker Verlag, Aachen (1996)

[Wi97]   U. Wille: The role of synthetic geometry in representational measurement theory. J. Math. Psych. 42 (1997), 71-78.

[WW92]   R. Wille, U. Wille: Coordinatization of ordinal structures. Discrete Mathematics. (1992)

[WW02]   R. Wille, U. Wille: Restructuring General Geometry: Measurement and Visualization of Spatial Structures. Preprint (2002)

Local Negation in Concept Graphs

Julia Klinger

Technische Universität Darmstadt, Fachbereich Mathematik
Schloßgartenstr. 7, D-64289 Darmstadt,
`jklinger@mathematik.tu-darmstadt.de`

Abstract. The aim of this paper is to present several examples illustrating challenges (local) negation in concept graphs yields with respect to semantical entailment. The examples are mainly concerned with inequality and with the impact missing information can have on relationships. The problems described here arose during the development of a logic system of protoconcept graphs (cf. [Kl05]), hence they will serve as exemplification.

1 Introduction

The theory of protoconcept graphs is part of a program called 'Contextual Logic' which can be understood as a formalization of the traditional philosophical logic with its doctrine of concepts, judgments, and conclusions (cf. [Wi96], [Pr98a], [Wi00b], [DK03]). As semantical structures, protoconcept graphs have been introduced in [Wi02a]. Complementing the semantic theory, in [Kl05] a logic approach to protoconcept graphs based on a separation of syntax and semantics was developed. Roughly spoken, protoconcept graphs are simply concept graphs with a negation on the level of concepts and relations (thus a limited negation) and with equality as syntactical element. However, the investigation of the semantical entailment relation for protoconcept graphs resulted in several interesting challenges. The aim of the present paper is to report on these problems. As shown in [Kl05], the examples provided here are indeed representative for all non-standard problems occurring with respect to semantical entailment.

This paper consists of four more sections. In Section 2, syntax and semantics for protoconcept graphs are introduced and explained by examples. Both the third and the fourth section deal with the challenges mentioned above. In particular, Section 3 is devoted to problems which arise when dealing with inequality, and Section 4 focuses on so-called 'hidden relationships'. Section 5 finishes this paper with some concluding remarks.

2 Basic Definitions

We assume that the reader is familiar with the basic notions of Formal Concept Analysis. We will proceed as follows: First, the notion of an alphabet is introduced, then protoconcept graphs are defined as 'formulas' over this alphabet. For the semantics, these syntactic graphs are then interpreted in power context families, and, based on this, models are defined.

F. Dau, M.-L. Mugnier, G. Stumme (Eds.): ICCS 2005, LNAI 3596, pp. 209–222, 2005.

2.1 Syntax

We start with the definition of an alphabet.

Definition 1. An *alphabet* for protoconcept graphs is a triple $\mathcal{A} := (\mathcal{G}, \mathcal{P}, \mathcal{R})$ of non-empty disjoint sets. The set $\mathcal{G}$ is a finite set whose elements are called *object names*, $(\mathcal{P}, \leq_{\mathcal{P}})$ is a finite ordered set whose elements are called *protoconcept names*, and $(\mathcal{R}, \leq_{\mathcal{R}})$ is the union of a finite family $(\mathcal{R}_k, \leq_{\mathcal{R}_k})_{k=1,2,\ldots,n}$ of finite ordered sets such that $\mathcal{R} := \dot{\bigcup}_{k=1}^{n} \mathcal{R}_k$ and $\leq_{\mathcal{R}} := \dot{\bigcup}_{k=1}^{n} \leq_{\mathcal{R}_k}$ whose elements are called *relation names*. The set of protoconcept names is supplied with a distinguished greatest element $\top_0$ and a distinguished smallest element $\bot_0$, and each set $\mathcal{R}_k$ $(k = 1, 2, \ldots, n)$ of relation names contains a distinguished greatest element $\top_k$ and a distinguished smallest element $\bot_k$. Moreover, the set of relation names includes a special element $\doteq \in \mathcal{R}_2$ which is called *identity*. Furthermore, let $\mathrm{Var} := \{x_1, x_2, \ldots\}$ be an infinite set of signs, the elements of which are called *variables*. For every finite subset $X \subseteq \mathrm{Var}$ we set $\mathcal{G}_X := \mathcal{G} \dot{\cup} X$ and $\mathcal{A}_X := (\mathcal{G}_X, \mathcal{P}, \mathcal{R})$. Finally, we set $\dot{\mathcal{G}} := \mathcal{G} \dot{\cup} \mathrm{Var}$.

$\diamond$

As an example, consider the alphabet in Figure 1. All sample graphs depicted in the next sections are defined over this alphabet.

Let $X := \{x, y, z, u, v\} \subseteq \mathrm{Var}$ and $\mathcal{A} := (\mathcal{G}, \mathcal{P}, \mathcal{R})$ with

$\mathcal{G} := \{g, h, g_1, g_2, g_3, g_4, g_5, g_6\}$

$\mathcal{P} := \{\bot_0, \top_0, P, Q\}$

$\mathcal{R} := \mathcal{R}_1 \cup \mathcal{R}_2 \cup \mathcal{R}_3$

$\mathcal{R}_1 := \{\top_1, \bot_1, R_1\}$

$\mathcal{R}_2 := \{\top_2, \bot_2, \doteq, R_2, R_2'\}$

$\mathcal{R}_3 := \{\top_3, \bot_3, R_3\}$

$\leq_{\mathcal{R}} := \leq_{\mathcal{R}_1} \dot{\cup} \leq_{\mathcal{R}_2} \dot{\cup} \leq_{\mathcal{R}_3}$

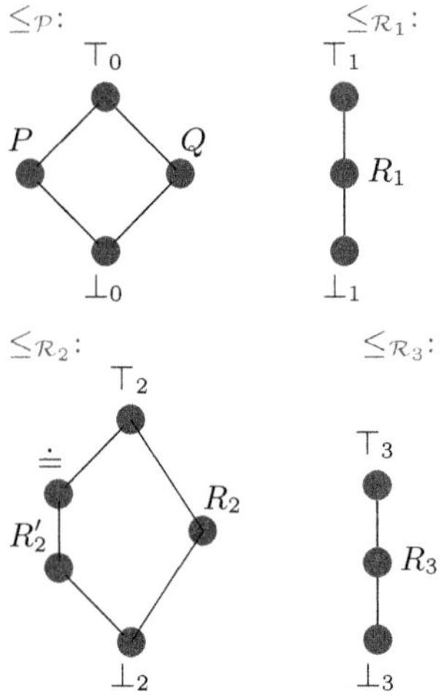

Then $\mathcal{A}_X := (\{g, h, g_1, g_2, g_3, g_4, g_5, x, y, z, u, v\}, \mathcal{P}, \mathcal{R})$.

Fig. 1. An alphabet for protoconcept graphs

Protoconcept graphs structure information rhetorically. This structure is coded in relational graphs.

Definition 2. A *relational graph* is a triple (V, E, ν) with a finite set V of vertices, a finite set E of edges and a map $\nu\colon E \to \bigcup_{k \in \mathbb{N}} V^k$. If $\nu(e) = (v_1, \ldots, v_k)$ for some $e \in E$ then $v_1, \ldots, v_k$ are called the *adjacent vertices* of the k-*ary* edge e, and $|e| := k$ is called the arity of e. Let $E^{(k)} := \{e \in V \cup E \mid |e| = k\}$ be the set of all elements of $V \cup E$ of arity k $(k \in \mathbb{N})$, and $E^{(0)} := V$. $\diamond$

An example for a relational graph is shown in Figure 2 (with the relational graph $(\{v, w\}, \{e\}, \nu)$ and $\nu(e) = (v, w)$).

Fig. 2. Example for a relational graph

Definition 3. A *protoconcept graph* over the alphabet $\mathcal{A}$ is a structure $\mathfrak{G} := (V, E, \nu, \kappa, \varrho)$ for which

- (V, E, ν) is a relational graph,
- $\kappa\colon V \cup E \to \mathcal{P} \cup \mathcal{R}$ is a mapping such that $\kappa(V) \subseteq \mathcal{P}$ and $\kappa(E) \subseteq \mathcal{R}$ and all $e \in E^{(k)}$ satisfy $\kappa(e) \in \mathcal{R}_k$ for all $k \geq 1$.
- $\varrho\colon V \to \mathfrak{P}(\dot{\mathcal{G}}) \times \mathfrak{P}(\dot{\mathcal{G}})$, $\quad \varrho\colon v \mapsto (\varrho^+(v), \varrho^-(v))$ is given by two functions $\varrho^+\colon V \to \mathfrak{P}(\dot{\mathcal{G}})$ and $\varrho^-\colon V \to \mathfrak{P}(\dot{\mathcal{G}})$ such that all $u \in E^{(k)}$ $(k \in \mathbb{N}_0)$ satisfy $\varrho^+(u) \cup \varrho^-(u) \neq \emptyset$ (where for $\nu(e) = (v_1, \ldots, v_n)$ we set $\varrho^+(e) := \varrho^+(v_1) \times \cdots \times \varrho^+(v_n)$ and $\varrho^-(e) := \varrho^-(v_1) \times \cdots \times \varrho^-(v_n)$) .

$\diamond$

Figure 3 shows an example, namely the graph $\mathfrak{G} := (V, E, \nu, \kappa, \varrho)$, with the relational graph (V, E, ν) as in Figure 2, and with $\kappa(v) = P$, $\kappa(w) = Q$, $\kappa(e) + R_2$, $\varrho(v) = (\{x\}, \{g_1\})$, and $\varrho(w) = (\{g_2\}, \{y\})$ (if $\varrho^+(u)$ or $\varrho^-(u)$ are singleton sets, then in the graphical representation of the graph we omit the set-brackets around them). Anticipating the next section, we describe how the protoconcept graph in Figure 3 is read: "there is something (namely x) which is P, and g_1 is not P and g_2 is Q and there is something (namely y) which is not Q and (x, g_2) are in R_2 and (g_1, y) are not in R_2". Note that the variables are never free variables, but always existentially quantified. This follows from the definition of the semantics as introduced in the next section.

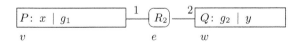

Fig. 3. A protoconcept graph

Protoconcept graphs extend the theory of concept graphs as introduced in [Pr98b] by including a negation on the level of concepts and relations. In particular, every concept graph can be understood as a protoconcept graph. On the other hand, every protoconcept graph corresponds to a concept graph with cuts (see [Da03a]). This logic system contains a negation on the level of propositions. In contrast to Dau's system, however, the theory of protoconcept graphs is decidable with respect to satisfiability and derivability. For a thorough discussion of the inter-relationships between the three theories we refer to [KV03].

Furthermore, it is important to note that there is a correspondance between protoconcept graphs and polarised simple conceptual graphs as described in [Ke01]. These graphs are conceptual graphs with an atomic negation on the relations. While there are certain differences in the alphabet and the definition of the semantics, the main distinction beween the two theories is that protoconcept graphs with relation $\doteq$ to identify objects and variables while polarized concept graphs employ coreference links. Since the latter are not relation edges, they may not be negated. However, in the theory of protoconcept graphs, we can set objects or variables unequal. As discussed in the following sections, this inequaltiy yields several difficulties. Additionally, polarised simple graphs use projections in order to study derivability, while protoconcept graphs have a sound and complete calculus.

2.2   Semantics

For the semantics, we first briefly recall the definition of protoconcepts (see [Wi00a]). They are chosen over concepts for our theory, since in contrast to concepts they are closed under extensional negation. Thus, let $\mathbb{K} = (G, M, I)$ be a formal context with a set G of *objects*, a set M of *attributes* and is a binary relation $I \subseteq G \times M$ called the *incidence raltion*. For a subset $A \subseteq G$ we set $A^I := \{m \in M \mid gIm \text{ for all } g \in A\}$. Dually, for $B \subseteq M$ we set $B^I := \{g \in G \mid gIm \text{ for all } m \in B\}$. A *protoconcept* of $\mathbb{K}$ is a pair (A, B) with $A \subseteq G$ and $B \subseteq M$ such that $A^I = B^{II}$ (which is equivalent to $A^{II} = B^I$). The set $\mathfrak{P}(\mathbb{K})$ of all protoconcepts of $\mathbb{K}$ is structured by the *generalization order* $\sqsubseteq$, defined by

$$(A_1, B_1) \sqsubseteq (A_2, B_2) :\Leftrightarrow A_1 \subseteq A_2 \text{ and } B_1 \supseteq B_2,$$

and by logical operations defined as follows:

$$(A_1, B_1) \sqcap (A_2, B_2) := (A_1 \cap A_2, (A_1 \cap A_2)^I)$$
$$(A_1, B_1) \sqcup (A_2, B_2) := ((B_1 \cap B_2)^I, B_1 \cap B_2)$$
$$\neg(A, B) := (G \backslash A, (G \backslash A)^I)$$
$$\top := (G, \emptyset)$$
$$\bot := (\emptyset, M).$$

As underlying structure for the semantics, families of formal contexts are defined.

Definition 4. A *power context family* $\overrightarrow{\mathbb{K}} := (\mathbb{K}_k)_{k=0,1,2,\dots,n}$ is a family of contexts $\mathbb{K}_k := (G_k, M_k, I_k)$ such that $G_k \subseteq G_0^k$ for $k \in \{1, \dots, n\}$.   ◇

As an example for a power context family we consider the left hand side of Figure 4 which contains the power context family $\overrightarrow{\mathbb{K}} := (\mathbb{K}_0, \mathbb{K}_1, \mathbb{K}_2, \mathbb{K}_3)$ with $I_1 = \emptyset = I_3$. Note that the object set of the kth context consists of the $k-$tuples of objects of $\mathbb{K}_0$.

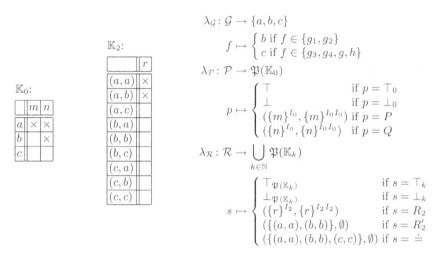

Fig. 4. A model for the graph in Figure 3

Next the syntactical elements of the alphabet are interpreted in a power context family via an interpretation. The conditions 1-3 in the following Definition guarantee that the $k-$ary relation names are indeed interpreted as protoconcepts of the kth context and that all special elements are treated according to their intended meaning.

Definition 5. For an alphabet $\mathcal{A} := (\mathcal{G}, \mathcal{P}, \mathcal{R})$ of type n and a power context family $\overrightarrow{\mathbb{K}} := (G_k, M_k, I_k)_{k=0,1,\dots,n}$, we call the triple $\lambda := (\lambda_{\mathcal{G}}, \lambda_{\mathcal{P}}, \lambda_{\mathcal{R}})$ consisting of the mappings $\lambda_{\mathcal{G}} : \mathcal{G} \to G_0$, $\lambda_{\mathcal{P}} : P \to \underline{\mathfrak{P}}(\mathbb{K}_0)$ and $\lambda_{\mathcal{R}} : \mathcal{R} \to \bigcup_{k \in \mathbb{N}} \underline{\mathfrak{P}}(\mathbb{K}_k)$ a $\overrightarrow{\mathbb{K}}-interpretation$ of $\mathcal{A}$ if $\lambda_{\mathcal{P}}$ and $\lambda_{\mathcal{R}}$ are both order-preserving and if they satisfy

1. $\lambda_{\mathcal{P}}(\top_0) = \top \in \underline{\mathfrak{P}}(\mathbb{K}_0)$ and $\lambda_{\mathcal{P}}(\bot_0) = \bot \in \underline{\mathfrak{P}}(\mathbb{K}_0)$ and $\lambda_{\mathcal{R}}(\top_k) = \top \in \underline{\mathfrak{P}}(\mathbb{K}_k)$ and $\lambda_{\mathcal{R}}(\bot_k) = \bot \in \underline{\mathfrak{P}}(\mathbb{K}_k)$ for all $k = 1, 2, \dots, n$,
2. $(g_1, g_2) \in \text{Ext}(\lambda_{\mathcal{R}}(\doteq)) \Leftrightarrow g_1 = g_2$ for all $g_1, g_2 \in G_0$,
3. $\lambda_{\mathcal{R}}(\mathcal{R}_k) \subseteq \underline{\mathfrak{P}}(\mathbb{K}_k)$ for all $k = 1, 2, \dots, n$.

The pair $(\overrightarrow{\mathbb{K}}, \lambda)$ is called a *contextual structure* for $\mathcal{A}$.   ◇

The triple $\lambda := (\lambda_{\mathcal{G}}, \lambda_{\mathcal{P}}, \lambda_{\mathcal{R}})$ with $\lambda_{\mathcal{G}}, \lambda_{\mathcal{P}}$ and $\lambda_{\mathcal{R}}$ as in the right hand side of Figure 4 is a $\overrightarrow{\mathbb{K}}$–interpretation of $\mathcal{A}$ (with $\overrightarrow{\mathbb{K}}$ as in Figure 4 and $\mathcal{A}$ as in Figure 1). Moreover, the variables need to be evaluated in the power context family:

Definition 6. Let $\mathcal{A}$ be an alphabet for protoconcept graphs and let $(\overrightarrow{\mathbb{K}}, \lambda)$ with $\mathbb{K}_k := (G_k, M_k, I_k)$ $(k = \{0, \ldots, n\})$ be a contextual structure for $\mathcal{A}$. Then each function $\beta \colon \mathrm{Var} \to G_0$ is called a *valuation* of Var in $\overrightarrow{\mathbb{K}}$. Valuations are naturally extended to mappings

$$\dot{\beta} \colon \dot{\mathcal{G}} \to G_0$$
$$a \mapsto \begin{cases} \beta(a) & \text{if } a \in \mathrm{Var} \\ \lambda_{\mathcal{G}}(a) & \text{if } a \in \mathcal{G}. \end{cases}$$

$\diamond$

Now we are finally in a position to define if a contextual structure is a model for a protoconcept graph:

Definition 7. Let $\mathfrak{G} := (V, E, \nu, \kappa, \varrho)$ be a protoconcept graph over $(\mathcal{A}, \mathrm{Var})$ and let $(\overrightarrow{\mathbb{K}}, \lambda)$ be a contextual structure for $\mathcal{A}$. Moreover, let $\beta \colon \mathrm{Var} \to G_0$ be a valuation. We say that $\mathfrak{G}$ is an *existential protoconcept graph of $(\overrightarrow{\mathbb{K}}, \lambda)$ under the valuation β* (and write $(\overrightarrow{\mathbb{K}}, \lambda) \models \mathfrak{G}[\beta]$) if

- all $v \in V$ satisfy
$$\dot{\beta}\varrho^+(v) \subseteq \mathrm{Ext}(\lambda_{\mathcal{P}} \kappa(v)) \qquad \text{(vertex condition)}$$
$$\dot{\beta}\varrho^-(v) \subseteq G_0 \setminus \mathrm{Ext}(\lambda_{\mathcal{P}} \kappa(v))$$
- all $e \in E$ satisfy
$$\dot{\beta}\varrho^+(e) \subseteq \mathrm{Ext}(\lambda_{\mathcal{R}} \kappa(e)) \qquad \text{(edge condition)}$$
$$\dot{\beta}\varrho^-(e) \subseteq (G_0)^k \setminus \mathrm{Ext}(\lambda_{\mathcal{R}} \kappa(e))$$

(where for $\nu(e) = (v_1, \ldots, v_n)$ we set $\dot{\beta}\varrho^+(e) := \dot{\beta}\varrho^+(v_1) \times \cdots \times \dot{\beta}\varrho^+(v_n)$ and $\dot{\beta}\varrho^-(e) := \dot{\beta}\varrho^-(v_1) \times \cdots \times \dot{\beta}\varrho^-(v_n)$). If there is a valuation β with $(\overrightarrow{\mathbb{K}}, \lambda) \models \mathfrak{G}[\beta]$, then we call $(\overrightarrow{\mathbb{K}}, \lambda)$ a *model* for $\mathfrak{G}$. $\diamond$

For the contextual structure in Figure 4 we set $\beta \colon \mathrm{Var} \to \{a, b, c\}$, $\beta(x) := a$, $\beta(y) := c$ and $\beta(w) = a$ for all $w \in \mathrm{Var} \setminus \{x, y\}$. Then $(\overrightarrow{\mathbb{K}}, \lambda)$ is a model for the graph in Figure 3 under the valuation β. We only check one vertex condition, namely that of v: We have $\dot{\beta}(\varrho^+(v)) = \dot{\beta}(\{x\}) = \{a\} \subseteq \{a\} = \mathrm{Ext}(\lambda_{\mathcal{P}} P)$ and $\dot{\beta}(\varrho^-(v)) = \dot{\beta}(\{g_1\}) = \{b\} \subseteq G_0 \setminus \{a\} = G_0 \setminus \mathrm{Ext}(\lambda_{\mathcal{P}} P)$. With respect to the edge condition for e we find $\dot{\beta}(\varrho^+(e)) = \dot{\beta}(\{(x, g_2)\}) = \{(a, b)\} \subseteq \{(a, a), (a, b)\} = \mathrm{Ext}(\lambda_{\mathcal{R}} R_2)$ and $\dot{\beta}(\varrho^-(e)) = \dot{\beta}(\{(g_1, y)\}) = \{(b, c)\} \subseteq G_2 \setminus \{(a, a), (a, b)\} = G_2 \setminus \mathrm{Ext}(\lambda_{\mathcal{R}} R_2)$.

Finally, we recall the definition of semantical entailment.

Definition 8. Let $\mathfrak{G}_1$ and $\mathfrak{G}_2$ be two protoconcept graphs over the same alphabet. We say $\mathfrak{G}_1$ *entails* $\mathfrak{G}_2$ if $\mathfrak{G}_2$ is a protoconcept graph of every model for $\mathfrak{G}_1$. We denote this by $\mathfrak{G}_1 \models \mathfrak{G}_2$.   ◇

3   Inequality

In this section, we describe three types of examples which are all concerned with the consequences of inequality. First we investigate how two object names can be set unequal explicitly and implicitly in a graph. Next we argue that the fact that the syntactical element $\doteq$ is integrated in the order $\leq_{\mathcal{R}_2}$ like any other relation name has to be especially taken into account. Finally, we illustrate that the combination of inequality with existential quantification needs a special treatment.

3.1   Inequality of Object Names

Inequality of object names can be coded explicitly or implicitly in a protoconcept graph. As examples consider the graphs in Figure 5. These graphs state explicitly or implicitly that in any model, the object names g_1 and g_2 cannot be interpreted as the same object: The first graph obviously states that g_1 and g_2 are unequal. The protoconcept graph $\mathfrak{G}_2$ says that g_1 is P and g_2 is not P. Hence, g_1 and g_2 cannot be interpreted as the same object. In the third graph (consisting of three edges) we find that g and h are syntactically equated while g_1 is in the relation R_2 to g and g_2 is not in the relation R_2 to h. Again, the interpretations of g_1 and g_2 may never be equal. In particular, we find that $\mathfrak{G}_2 \models \mathfrak{G}_1$ and $\mathfrak{G}_3 \models \mathfrak{G}_1$.

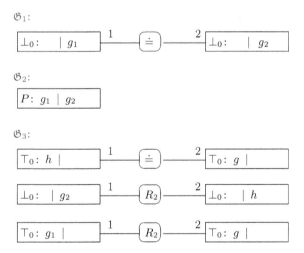

Fig. 5. How inequality can be coded

3.2 Integration of $\doteq$ in $\leq_{\mathcal{R}_2}$

Now we argue that the integration of $\doteq$ in $\leq_{\mathcal{R}_2}$, together with a local negation can have rather unexpected results.

Let $\mathfrak{G}_1$, $\mathfrak{G}_2$ as in Figure 6 with $R_2' <\doteq$. We prove that $\mathfrak{G}_1 \models \mathfrak{G}_2$. Note that this can only be shown by activating the order $\leq_{\mathcal{R}_2}$. In particular, the information in $\mathfrak{G}_2$ is not stated explicitly in $\mathfrak{G}_1$. Thus, let $(\overrightarrow{\mathbb{K}}, \lambda)$ be a model for $\mathfrak{G}_1$. Then this model either satisfies $\lambda_{\mathcal{G}}(g) \neq \lambda_{\mathcal{G}}(h)$, which implies that $\lambda_{\mathcal{G}}((g,h)) \in G_2 \setminus \mathrm{Ext}(\lambda_{\mathcal{R}} \doteq) \subseteq G_2 \setminus \mathrm{Ext}(\lambda_{\mathcal{R}} R_2')$. Or we have $\lambda_{\mathcal{G}}(g) = \lambda_{\mathcal{G}}(h)$, but then $\lambda_{\mathcal{G}}((g,h)) = \lambda_{\mathcal{G}}((g,g)) \in G_2 \setminus \mathrm{Ext}(\lambda_{\mathcal{R}} R_2')$, because $(\overrightarrow{\mathbb{K}}, \lambda)$ satisfies the edge condition for $\mathfrak{G}_1$. Hence, $\mathfrak{G}_1 \models \mathfrak{G}_2$.

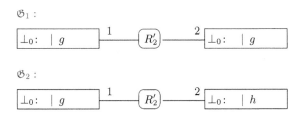

Fig. 6. Example

3.3 Existential Quantification and Inequality

In Section 3.1 we have illustrated that object names can be set unequal explicitly and implicitly. This investigation is now extended to variables. We find that there are even more possibilities to express inequality, due to a local 'or':

We consider the graphs $\mathfrak{G}_1, \mathfrak{G}_2$ as in Figure 7. Although g is neither set unequal to g_1 nor to g_2, we find that $\mathfrak{G}_1 \models \mathfrak{G}_2$. For if $(\overrightarrow{\mathbb{K}}, \lambda) \models \mathfrak{G}_1$, we define $\beta \colon \mathrm{Var} \to G_0$ as follows: Depending on whether $\lambda_{\mathcal{G}}(g_1) \neq \lambda_{\mathcal{G}}(g)$ or $\lambda_{\mathcal{G}}(g_2) \neq \lambda_{\mathcal{G}}(g)$ (and at least one of these conditions has to be fulfilled) we set $\beta(x) = \lambda_{\mathcal{G}}(g_1)$ or $\beta(x) = \lambda_{\mathcal{G}}(g_2)$, respectively. Then $(\overrightarrow{\mathbb{K}}, \lambda) \models \mathfrak{G}_2[\beta]$.

Thus, Figure 7 shows that with variables as references and the possibility to set two names unequal via equality and local negation, we obtain a disjunction: we have that in every model for $\mathfrak{G}_1$, the interpretations of g and g_1 *or* the interpretations of g and g_2 have to be different. As an additional example of the same type consider the two graphs in Figure 8, where $g_1, g_2, g_3 \in \dot{\mathcal{G}}$ are pairwise set unequal in the graph $\mathfrak{G}_1$. Then for any $g, h \in \dot{\mathcal{G}}$ and any $x \in \mathrm{Var}$ we obtain $\mathfrak{G}_1 \models \mathfrak{G}_2$: Every model for $\mathfrak{G}_1$ contains at least three objects, hence if we have g, h, then in every model for $\mathfrak{G}_1$ there is something, which is different from them.

Note, that in both examples we used the direct approach of setting the object names respectively variables unequal. Alternatively, any one of the possibilities

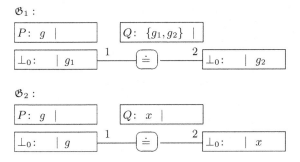

Fig. 7. Example

of setting them unequal implicitly (see the lower two examples in Figure 5) could have been employed.

The main argumentation for proving that $\mathfrak{G}_1 \models \mathfrak{G}_2$ for $\mathfrak{G}_1, \mathfrak{G}_2$ in Figure 7 was that g, g_1 and g_2 cannot be mapped to the same element in a model. This can be generalized to more than three elements: Consider the two graphs in Figure 9. For g we obtain $\mathfrak{G}_1 \models \mathfrak{G}_2$. For let $(\overrightarrow{\mathbb{K}}, \lambda) \models \mathfrak{G}_1[\beta]$. Then $\lambda_{\mathcal{G}}((g_1, g_2)) \in \text{Ext}(\lambda_{\mathcal{P}}(R_2))$ and $\lambda_{\mathcal{G}}((g_3, g_4)) \in G_2 \setminus \text{Ext}(\lambda_{\mathcal{P}}(R_2))$, thus g_1, g_2, g_3, g_4 can not all be mapped to the same element of G_0 by $\lambda_{\mathcal{G}}$. Hence $|\lambda_{\mathcal{G}}(\{g_1, g_2, g_3, g_4\})| > 1$ and for g there is an $a \in \lambda_{\mathcal{G}}(\{g_1, g_2, g_3, g_4\})$ with $\lambda_{\mathcal{G}}(g) \neq a$. This a certainly satisfies $a \in \text{Ext}(\lambda_{\mathcal{P}}(P))$, thus we simply set $\beta'(x) = a$, $\dot{\beta}'(g) = \dot{\beta}(g)$ and find $(\overrightarrow{\mathbb{K}}, \lambda) \models \mathfrak{G}_2[\beta']$.

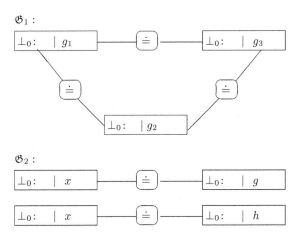

Fig. 8. Example

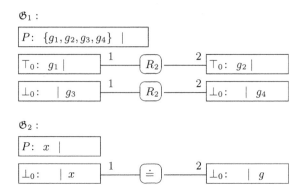

Fig. 9. Example

4 Hidden Relationships

The aim of this section is to investigate so-called 'hidden relationships', which are pieces of information about relations that can only be obtained from a graph via case differentiation. The examples described in this sections somehow generalize the examples in Section 3.1 (for instance to edges with relation names other than $\doteq$). As an example, consider the graphs in Figure 10. Note that the graph $\mathfrak{G}_1$ does not directly code all the information given in $\mathfrak{G}_2$, i.e., that there are three things x, y, z (which are not necesseraly distinct) such that (x, y) is in R_2 and (y, z) is not in R_2. Still, we find $\mathfrak{G}_1 \models \mathfrak{G}_2$. Let $(\overrightarrow{\mathbb{K}}, \lambda)$ be a model for $\mathfrak{G}_1$. We distinguish two cases for the definition of a valuation $\beta \colon \mathrm{Var} \to G_0$ with $(\overrightarrow{\mathbb{K}}, \lambda) \models \mathfrak{G}_2[\beta]$:

(i) If $(\lambda_{\mathcal{G}}(g_2), \lambda_{\mathcal{G}}(g_2)) \notin \mathrm{Ext}(\lambda_{\mathcal{R}}\, R_2)$, then we set $\beta(x) := \lambda_{\mathcal{G}}(g_1)$, $\beta(y) := \lambda_{\mathcal{G}}(g_2)$ and $\beta(z) = \lambda_{\mathcal{G}}(g_1)$ (and $\beta(a) := a$ for all $a \in \mathrm{Var} \setminus \{x, y, z\}$).

(ii) If $(\lambda_{\mathcal{G}}(g_2), \lambda_{\mathcal{G}}(g_1)) \in \mathrm{Ext}(\lambda_{\mathcal{R}}\, R_2)$, then we set $\beta(x) := \lambda_{\mathcal{G}}(g_1)$, $\beta(y) := \lambda_{\mathcal{G}}(g_2)$ and $\beta(z) = \lambda_{\mathcal{G}}(g_3)$ (and $\beta(a) := a$ for all $a \in \mathrm{Var} \setminus \{x, y, z\}$).

Next let us investigate in which different ways 'hidden relationships' like the one visualized in Figure 10 can occur. The graphs in Figure 10 consist of two edges of the same label R_2. The next example shows that such implicitly coded information can also be obtained from edges of different arity (e.g. by combining information coded in vertices and edges):

We take the graphs $\mathfrak{G}_1, \mathfrak{G}_2$ from Figure 11. Let $(\overrightarrow{\mathbb{K}}, \lambda)$ be a model for $\mathfrak{G}_1$. Depending on whether or not $\lambda_{\mathcal{G}}(g_1) \in \mathrm{Ext}(\lambda_{\mathcal{P}}(Q))$, we can define a valuation β for $\mathfrak{G}_2$ in $(\overrightarrow{\mathbb{K}}, \lambda)$: If $\lambda_{\mathcal{G}}(g_1) \in \mathrm{Ext}(\lambda_{\mathcal{P}}(Q))$, then the valuation $\beta \colon \mathrm{Var} \to G_0$ with $\beta(x) := \lambda_{\mathcal{G}}(g_2)$, $\beta(y) := \lambda_{\mathcal{G}}(g_1)$ is a valuation for $\mathfrak{G}_2$ in $(\overrightarrow{\mathbb{K}}, \lambda)$. If $\lambda_{\mathcal{G}}(g_1) \notin \mathrm{Ext}(\lambda_{\mathcal{P}}(Q))$, then $\beta \colon \mathrm{Var} \to G_0$, $\beta(x) := \lambda_{\mathcal{G}}(g_1)$, $\beta(y) := \lambda_{\mathcal{G}}(g_3)$ is a valuation for $\mathfrak{G}_2$ in $(\overrightarrow{\mathbb{K}}, \lambda)$. Thus, $\mathfrak{G}_1 \models \mathfrak{G}_2$.

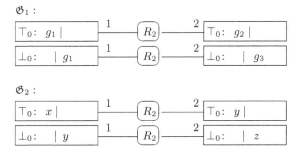

Fig. 10. Example

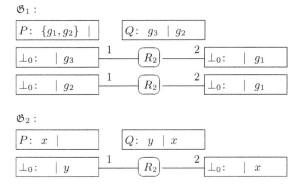

Fig. 11. Example

Moreover, the references in the second graph need not all be variables. Via a case differentiation with respect to whether $\lambda_\mathcal{G}((g_2, g_1, g_3))$ in $\mathrm{Ext}(\lambda_\mathcal{R}(R_3))$ or not we see that the graphs in Figure 12 satisfy $\mathfrak{G}_1 \models \mathfrak{G}_2$.

Finally, we note that it is not enough to consider inequality and implicit relationships separately. For this, consider the example described in Figure 13. We find that $\mathfrak{G}_1 \models \mathfrak{G}_2$, which follows similarly to the argumentation for the example in Figure 10, however, the argumentation has an additional twist in it: Let $(\overrightarrow{\mathbb{K}}, \lambda)$ be a model for $\mathfrak{G}_1$. If $\lambda_\mathcal{G}((g_2, g_1, g_6)) \in \mathrm{Ext}(\lambda_\mathcal{R} R_3)$, *then the model needs to satisfy* $\lambda_\mathcal{G}(g_2) \neq \lambda_\mathcal{G}(g_5)$ *(otherwise the edge condition for the bottommost edge of $\mathfrak{G}_1$ would not be satisfied).* Hence, $\beta\colon \mathrm{Var} \to G_0$ with $\beta(x) = \lambda_\mathcal{G}(g_2), \beta(y) = \lambda_\mathcal{G}(g_1), \beta(z) = \lambda_\mathcal{G}(g_3), \beta(g_6) = \lambda_\mathcal{G}(a), \beta(v) = \lambda_\mathcal{G}(g_4)$ and $\beta|_{\mathrm{Var}\setminus\{x,y,z,u,v\}}(d) = \lambda_\mathcal{G}(g_1)$ is a valuation of $\mathfrak{G}_2$ in $(\overrightarrow{\mathbb{K}}, \lambda)$. On the other hand, if $\lambda_\mathcal{G}((g_2, g_1, g_6)) \notin \mathrm{Ext}(\lambda_\mathcal{R} R_3)$, then $\beta\colon \mathrm{Var} \to G_0$ with $\beta(x) = \lambda_\mathcal{G}(g_1), \beta(y) = \lambda_\mathcal{G}(g_2), \beta(z) = \lambda_\mathcal{G}(g_1), \beta(u) = \lambda_\mathcal{G}(g_4), \beta(v) = \lambda_\mathcal{G}(g_6)$ and $\beta|_{\mathrm{Var}\setminus\{x,y,z,u,v\}}(d) = \lambda_\mathcal{G}(g_1)$ is a valuation of $\mathfrak{G}_2$ in $(\overrightarrow{\mathbb{K}}, \lambda)$. Hence, $\mathfrak{G}_1 \models \mathfrak{G}_2$. Note that within the argumentation we had to *infer* that g_2 and g_1 have to be mapped to different objects; without drawing this conclusion we would not have been able to deduce $\mathfrak{G}_1 \models \mathfrak{G}_2$.

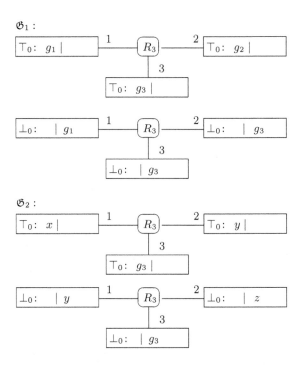

Fig. 12. Example

5   Conclusion

The aim of this paper was to illustrate and classify the problems arising from the introducton of a limited negation. Although the examples described above can be combined with each other, we find that (beside such combinations) there is a characterization of semantical entailment in [Kl05] which shows that the examples described above can indeed be considered representative for all non-standard problems.

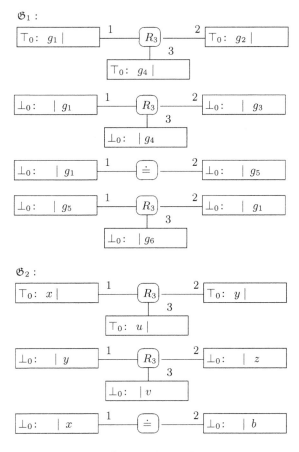

Fig. 13. Example

References

[Da03a] F. Dau: The Logic System of Concept Graphs with Negations (and its Relationship to Predicate Logic). Springer Verlag, Berlin–New York 2003.

[DK03] F. Dau, J. Klinger: From Formal Concept Analysis to Contextual Logic. FB4–Preprint, TU Darmstadt 2003.

[Ke01] G.N. Kerdiles: Saying It with Pictures: a logical landscape of conceptual graphs. http://staff.science.uva.nl/ kerdiles/.

[Kl05] J. Klinger: The Logic System of Protoconcept Graphs. Preprint, FB Mathematik, TU Darmstadt 2005.

[KV03] J. Klinger, B. Vormbrock: Contextual Boolean Logic: How did it develop? In: B. Ganter, A. de Moor (Eds.): Using Conceptual Structures. Contributions to ICCS 2003. Shaker Verlag, Aachen 2003, 143–156.

[Pr98a] S. Prediger: Kontextuelle Urteilslogik mit Begriffsgraphen, Shaker Verlag, Aachen 1998.

[Pr98b] S. Prediger: Simple Concept Graphs: A Logic Approach. In: M.-L. Mugnier, M. Chein (Eds): Conceptual Structures: Theory, Tools and Application, Springer Verlag, Berlin–New York 1998, 225–239.

[Wi96] R. Wille: Restructuring Mathematical Logic: An Approach Based on Peirce's Pragmatism. In: A. Ursini, P. Agliano (Eds.): Logic and Algebra. Marcel Dekker, New York 1996, 267–281.

[Wi00a] R. Wille: Boolean Concept Logic. In: B. Ganter, G.W. Mineau (Eds.): Conceptual Structures: Logical, Linguistic, and Computational Issues, Springer Verlag, Berlin–New York 2000, 317–331.

[Wi00b] R. Wille: Contextual Logic Summary. In: G. Stumme (Ed.): Working with Conceptual Structures. Contributions to ICCS 2000. Shaker Verlag, Aachen 2000, 256–276.

[Wi02a] R. Wille: Existential Graphs of Power Context Families. In: U. Priss, D. Corbett, G. Angelova (Eds.): Conceptual Structures: Integration and Interfaces, Springer Verlag, Berlin–New York 2002, 382–396.

Morphisms in Context

Markus Krötzsch[1]*, Pascal Hitzler[1]**, and Guo-Qiang Zhang[2]

[1] AIFB, Universität Karlsruhe, Germany
[2] Department of Electrical Engineering and Computer Science, Case Western Reserve
University, Cleveland, Ohio, U.S.A.

Abstract. Morphisms constitute a general tool for modelling complex relationships between mathematical objects in a disciplined fashion. In Formal Concept
Analysis (FCA), morphisms can be used for the study of structural properties of
knowledge represented in formal contexts, with applications to data transformation and merging. In this paper we present a comprehensive treatment of some
of the most important morphisms in FCA and their relationships, including dual
bonds, scale measures, infomorphisms, and their respective relations to Galois
connections. We summarize our results in a concept lattice that cumulates the relationships among the considered morphisms. The purpose of this work is to lay a
foundation for applications of FCA in ontology research and similar areas, where
morphisms help formalize the interplay among distributed knowledge bases.

1 Introduction

Formal Concept Analysis (FCA) [1] provides a fundamental mathematical methodology for the creation, analysis, and manipulation of data and knowledge. Its field of application ranges from social and natural sciences to most prominently computer science.
The automated processing of knowledge necessitates an understanding of its structural
properties in order to develop sound transformation algorithms, ontology merging procedures, and other operations needed for practical applications. FCA is ideally suited
for such an understanding due to its sound mathematical and philosophical base, rooted
in algebra and logic.

Fundamental structural properties can be captured by category-theoretical treatments [2], the heart of which are *morphisms* as structure-preserving mappings. In turn,
morphisms provide abstract means for the modelling of data translation, communication, and distributed reasoning, to give a few examples. Thus the theory and application
of morphisms between formal contexts have recently become a focal point in FCA.

Institution theory [3], developed in the 80's, uses formal contexts and appropriate
morphisms to represent a broad class of logics. The resulting mathematical theory has
been applied as a basis for various programming languages. More recently, similar ideas

---
* This author acknowledges support by Case Western Reserve University, Cleveland, Ohio,
where part of his work was carried out, sponsorship by the German Academic Exchange Service (DAAD) and by the Gesellschaft von Freunden und Förderern der TU Dresden e.V., and
support by the EU KnowledgeWeb network of excellence.
** This author acknowledges support by the German Federal Ministry of Education and Research
under the SmartWeb project, and by the EU under the KnowledgeWeb network of excellence.

F. Dau, M.-L. Mugnier, G. Stumme (Eds.): ICCS 2005, LNAI 3596, pp. 223–237, 2005.

have been used as a foundation of a theory of *information flow* [4], recently considered in the context of *ontology research* [5, 6]. While the focus of the above is more on communication and transport of information, research on *Chu spaces* [7] (a special case of which are formal contexts) considers similar morphisms from a categorical viewpoint in order to obtain categories with certain specific properties.

Although morphisms between contexts have been investigated in all of the above research areas, they mostly study the same kind of morphisms, which today are typically called *infomorphisms*. These, however, are only a special choice for morphisms in FCA, and it is unclear whether they are in general preferable to other possible notions. At least two other kinds of morphisms known in FCA deserve particular attention. One is the so-called *(dual) bond*, a specific kind of relation between contexts which is of importance due to its close relationship to Galois connections. The other is *scale measure*, characterized by certain functional continuity properties.

In order to develop concrete applications to knowledge processing from structural analysis based on category theory, it is of fundamental importance to understand the properties of and relationships between different notions of morphisms. So far, only a few order- and category-theoretic treatments of morphisms in FCA are available, either studying one kind of morphism in isolation or focusing on further specific types of morphisms (an overview is given at the end of the following section). The purpose of this paper is thus to present a comprehensive study of the relations among the three types of major morphisms in FCA mentioned above. We explicate the rich interrelationships and dependencies as a step stone for further developments.

The paper is structured as follows. After explaining some preliminaries in Section 2, we study dual bonds and their relationships to direct products of formal contexts and Galois connections in Section 3. In Section 4, dual bonds featuring certain continuity properties are identified as an important subclass. Section 5 deals with the relationship between scale measures, functional types of dual bonds, and Galois connections, while Section 6 is devoted to infomorphisms. Finally, in Section 7 we summarize our results in the form of a concept lattice obtained by attribute exploration, and discuss possible directions for future research. All proofs are omitted for space restrictions. They can be found in [8].

2   Preliminaries

Our notation basically follows [1], with a few exceptions to enhance readability for our purposes. Especially, we avoid the use of the symbol $'$ to denote the operations that are induced by a context. We shortly review the main terminology using our notation, but we assume that the reader is familiar with the notation and terminology from [1]. Our treatment also requires some basic knowledge of *(antitone) Galois connections* and their *monotone* variant (a.k.a. *residuated maps*), which can also be found in [1].

A *(formal) context* $\mathbb{K}$ is a triple (G, M, I) where G is a set of *objects*, M is a set of *attributes*, and $I \subseteq G \times M$ is an *incidence relation*. Given $O \subseteq G$ and $A \subseteq M$, we define:

$$O^I := \{m \in M \mid g \, I \, m \text{ for all } g \in O\}, \qquad I(O) := \{m \in M \mid g \, I \, m \text{ for some } g \in O\},$$
$$A^I := \{g \in G \mid g \, I \, m \text{ for all } m \in A\}, \qquad I^{-1}(A) := \{g \in G \mid g \, I \, m \text{ for some } m \in A\}.$$

For singleton sets we use the common abbreviations $g^I := \{g\}^I$, $I(g) := I(\{g\})$, etc. The notation X^I can be ambiguous if it is not clear whether X is considered a set of objects or a set of attributes, so we will be careful to avoid such situations. We refer to $I(O)$ as the *image* of O and to $I^{-1}(A)$ as the *preimage* of A with respect to I. We use these notations for arbitrary binary relations.

A subset $O \subseteq G$ is an *extent* of $\mathbb{K}$ whenever $O = O^{II}$. O is an *attribute extent (object extent)* if there is some attribute n (object g) such that $O = n^I$ ($O = g^{II}$). *Intents, object intents* and *attribute intents* are defined dually. A *concept* of $\mathbb{K}$ is an extent-intent pair (O, A) such that $O = A^I$ (or, equivalently, $A = O^I$).

Since the extent and intent of a concept determine each other uniquely, we will usually prefer to consider only one of them. Our use of the terms *object extent* and *attribute intent* constitutes a slight deviation from standard terminology.

The central result of FCA is that contexts can be used to represent complete lattices.

Theorem 1 ([1, Theorem 3]). *For any context $\mathbb{K} = (G, M, I)$, the mapping $(\cdot)^{II} : 2^G \to 2^G$ constitutes a closure operator on the powerset 2^G. The corresponding closure system (in the sense of [1]) is the set $\mathsf{B}_o(\mathbb{K}) := \{O \subseteq G \mid O = O^{II}\}$ of all extents of $\mathbb{K}$.*

Similar statements are true for the mapping $(\cdot)^{II} : 2^M \to 2^M$, which induces a closure system $\mathsf{B}_a(\mathbb{K})$. Under set inclusion, $\mathsf{B}_o(\mathbb{K})$ and $\mathsf{B}_a(\mathbb{K})$ are dually order-isomorphic, with $(\cdot)^I : 2^G \to 2^A$ and $(\cdot)^I : 2^A \to 2^G$ as the according isomorphisms.

We refer to $\mathsf{B}_o(\mathbb{K})$ and $\mathsf{B}_a(\mathbb{K})$ ordered by set inclusion as the *object-* and *attribute-concept lattices*.

An important aspect of FCA is that contexts can be dualized and complemented to obtain new structures. These operations turn out to be vital for our subsequent studies. Given a context $\mathbb{K} = (G, M, I)$, the context *dual* to $\mathbb{K}$ is $\mathbb{K}^d := (M, G, I^{-1})$. It is easy to see that dualizing a context merely changes the roles of extent and intent. Thus, with respect to the order of the concept lattices we have $\mathsf{B}_o(\mathbb{K}^d) = \mathsf{B}_a(\mathbb{K})$ and $\mathsf{B}_a(\mathbb{K}^d) = \mathsf{B}_o(\mathbb{K})$. The situation for complement, defined as $\mathbb{K}^c = (G, M, \bar{I})$ with $\bar{I} := (G \times M) \setminus I$, is more involved since the concept lattices of $\mathbb{K}$ and $\mathbb{K}^c$ are in general not (dually) isomorphic to each other. We can observe immediately that dualization and complementation commute: $\mathbb{K}^{cd} = \mathbb{K}^{dc}$. Furthermore, the following lemma will be helpful.

Lemma 1. *Given a context $\mathbb{K} = (G, M, I)$ with objects $g, h \in G$, we find that $g \in h^{II}$ if and only if $h \in g^{\bar{I}\bar{I}}$.*

Definitions of the relevant context-morphisms will be introduced in the subsequent sections. An overview of the existing results on morphisms in FCA is given in [1, Chapter 7], which incorporates much information from [9], though the latter contains further details from a more category-theoretic viewpoint. Bonds and infomorphisms, as well as several other kinds of morphisms that we shall not consider in this paper, have been studied in greater detail in [10]. Some newer results on dual bonds and *relational Galois connections* between contexts can be found in [11]. Further related investigations can be found in [12], where infomorphisms are studied in conjunction with monotone Galois connections, *complete homomorphisms*, and the so-called *concept lattice morphisms*. Morphisms relating FCA, domain theory, and logic have been studied in [13].

3   Dual Bonds and Direct Product

The construction of concept lattices exploits the fact that the derivation operators $(\cdot)^I$ form an antitone Galois connection. Naturally, Galois connections are also of interest when one looks for suitable morphisms for concept lattices. To represent Galois connections on the level of contexts, functions between the sets of attributes or objects turn out to be too specific. Instead, one uses certain relations called *dual bonds* which we study in this section. Most of the materials before Lemma 3 can be found in [1, 10, 11].

Definition 1. *A dual bond between formal contexts* $\mathbb{K} = (G, M, I)$ *and* $\mathbb{L} = (H, N, J)$ *is a relation* $R \subseteq G \times H$ *for which the following hold:*

- *for every object* $g \in G$, g^R *(which is equal to* $R(g)$*) is an extent of* $\mathbb{L}$ *and*
- *for every object* $h \in H$, h^R *(which is equal to* $R^{-1}(h)$*) is an extent of* $\mathbb{K}$.

   This definition is motivated by the following result:

Theorem 2 ([1, Theorem 53]). *Consider a dual bond R between contexts* $\mathbb{K}$ *and* $\mathbb{L}$ *as above. The mappings*

$$\vec{\phi}_R : B_o(\mathbb{K}) \to B_o(\mathbb{L}) : X \mapsto X^R \quad and \quad \overleftarrow{\phi}_R : B_o(\mathbb{L}) \to B_o(\mathbb{K}) : Y \mapsto Y^R$$

form an antitone Galois connection between the (object) concept lattices of $\mathbb{K}$ *and* $\mathbb{L}$.
   Conversely, given such an antitone Galois connection $(\vec{\phi}, \overleftarrow{\phi})$*, the relation* $R_{(\vec{\phi},\overleftarrow{\phi})} = \left\{(g,h) \mid h \in \vec{\phi}(g^{II})\right\} = \left\{(g,h) \mid g \in \overleftarrow{\phi}(h^{JJ})\right\}$ *is a dual bond, and these constructions are mutually inverse in the following sense:*

$$\vec{\phi} = \vec{\phi}_{R_{(\vec{\phi},\overleftarrow{\phi})}} \qquad \overleftarrow{\phi} = \overleftarrow{\phi}_{R_{(\vec{\phi},\overleftarrow{\phi})}} \qquad R = R_{\vec{\phi}_R,\overleftarrow{\phi}_R}$$

   Hence, formal contexts with dual bonds are "equivalent" to complete lattices with antitone Galois connections. Referring to dual bonds as morphisms might be somewhat misleading, since they do not immediately satisfy the necessary axioms for category theoretic morphisms. However, we will adhere to this terminology since it is indeed possible to use dual bonds in a categorical fashion, provided that objects, homsets and composition are chosen appropriately (see [14] for details).
   Before proceeding, let us note the following consequence of Lemma 1.

Lemma 2. *Consider a dual bond R between contexts* $\mathbb{K} = (G, M, I)$ *and* $\mathbb{L} = (H, N, J)$. *Then* $R(g^{\star\star}) = R(g)$ *and* $R^{-1}(h^{\star\star}) = R^{-1}(h)$ *holds for any* $g \in G$, $h \in H$. *Especially,* $R(g^{\star\star})$ *and* $R^{-1}(h^{\star\star})$ *are extents.*

   Now we ask how the dual bonds between two contexts can be represented. Since extents are closed under intersections, the same is true for the set of all dual bonds between two contexts. Thus the dual bonds form a closure system and one might ask for a way to cast this into a formal context which has dual bonds as concepts. An immediate candidate for this purpose is the direct product of the contexts.

Definition 2. *Given contexts* $\mathbb{K} = (G, M, I)$ *and* $\mathbb{L} = (H, N, J)$*, the* direct product *of* $\mathbb{K}$ *and* $\mathbb{L}$ *is the context* $\mathbb{K} \times \mathbb{L} = (G \times H, M \times N, \nabla)$*, where* $(g, h) \nabla (m, n)$ *iff* $g \, I \, m$ *or* $h \, J \, n$.

Proposition 1 ([11]). *Extents of a direct product* $\mathbb{K} \times \mathbb{L}$ *are dual bonds from* $\mathbb{K}$ *to* $\mathbb{L}$.

However, it is known that the converse of this result is false, i.e. there are dual bonds which are not extents of the direct product.

As a consequence, the direct product does only represent a distinguished subset of all dual bonds. In order to find additional characterizations for these relations, we use the following result.

Lemma 3. *Given a binary relation R between objects, let R^{∇} denote the intent associated with R when viewed as a set of objects of the direct product. Consider the contexts* $\mathbb{K} = (G, M, I)$ *and* $\mathbb{L} = (H, N, J)$ *and a relation* $R \subseteq G \times H$. *For any attribute* $m \in M$, *the following sets are equal:*

- $X_1 := R^{\nabla}(m) = \{n \in N \mid (m, n) \in R^{\nabla}\}$
- $X_2 := R(m^{\mathit{1}})^J = \{h \in H \mid \text{there is } g \in G \text{ with } g \not{I} m \text{ and } (g, h) \in R\}^J$
- $X_3 := \bigcap_{g \in m^{\mathit{1}}} R(g)^J$

Furthermore, for any object $g \in G$, *we find that* $R^{\nabla\nabla}(g) = R^{\nabla}(g^{\mathit{1}})^J = \bigcap_{m \in g^{\mathit{1}}} R(m^{\mathit{1}})^{JJ}$.

Now we can state a characterization theorem for dual bonds in the direct product.

Theorem 3. *Consider the contexts* $\mathbb{K} = (G, M, I)$ *and* $\mathbb{L} = (H, N, J)$ *and a relation* $R \subseteq G \times H$. *The following are equivalent:*

(i) *R is an extent of the direct product* $\mathbb{K} \times \mathbb{L}$.
(ii) *For all* $g \in G$, $R(g) = R^{\nabla}(g^{\mathit{1}})^J \; \left(= \bigcap_{m \in g^{\mathit{1}}} R(m^{\mathit{1}})^{JJ}\right)$.
(iii) *R is a dual bond and, for all* $g \in G$, $\bigcap_{m \in g^{\mathit{1}}} R(m^{\mathit{1}})^{JJ} = R(g^{\mathit{1}\mathit{1}})$.

Another feature of dual bonds in the direct product allows for the construction of Galois connections other than those considered in Theorem 2. Given a dual bond R in $\mathbb{K} \times \mathbb{L}$, its intent R^{∇} is a dual bond from $\mathbb{K}^d$ to $\mathbb{L}^d$, which induces another antitone Galois connection between the dual concept lattices. This Galois connection appears to have no simple further relationship to the antitone Galois connection derived from R.

Corollary 1. *Consider the contexts* $\mathbb{K} = (G, M, I)$ *and* $\mathbb{L} = (H, N, J)$ *and an extent R of the direct product* $\mathbb{K} \times \mathbb{L}$. *There are two distinguished Galois connections* $\phi_R : B_o(\mathbb{K}) \to B_o(\mathbb{L})$ *and* $\phi_{R^{\nabla}} : B_o(\mathbb{K})^{op} \to B_o(\mathbb{L})^{op}$ *and each of R, R^{∇}, ϕ_R and $\phi_{R^{\nabla}}$ uniquely determines the others (using $(\cdot)^{op}$ to denote the order duals of the respective concept lattices).*

Of course any antitone Galois connection between two posets contravariantly induces another antitone Galois connection, obtained by exchanging both adjoints. But there appears to be no general way to construct an additional antitone Galois connection between the order duals of the original posets. Some of our results, like Proposition 3 and Theorem 7 below, can be extended to account for this second Galois connection, but we will usually prefer to save space and refrain from stating this explicitly.

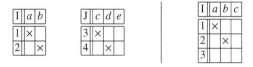

Fig. 1. Formal contexts for Counterexamples 1 (left) and 2 (right).

4 Continuity for Dual Bonds

Continuity is a central concept in many branches of mathematics. It is also of importance for formal concept analysis. However, we will generally not be dealing with functions but with relations such as dual bonds, so the notion of continuity will be lifted accordingly (the following is partially taken from [1]).

Definition 3. *Consider contexts* $\mathbb{K} = (G, M, I)$ *and* $\mathbb{L} = (H, N, J)$*. A relation* $R \subseteq G \times H$ *is* extensionally continuous *if it reflects* extents *of* $\mathbb{L}$*, i.e. if for every extent* O *of* $\mathbb{L}$ *the preimage* $R^{-1}(O)$ *is an extent of* $\mathbb{K}$*.*

R is extensionally object-continuous *(attribute-continuous) if it reflects all object extents (attribute extents) of* $\mathbb{L}$*, i.e. if for every object extent* $O = h^{JJ}$ *(attribute extent* $O = n^J$*) the preimage* $R^{-1}(O)$ *is an extent of* $\mathbb{K}$ *(but not necessarily an object extent).*

A relation is extensionally closed *from* $\mathbb{K}$ *to* $\mathbb{L}$ *if it preserves* extents *of* $\mathbb{K}$*, i.e. if its inverse is extensionally continuous from* $\mathbb{L}$ *to* $\mathbb{K}$*. Extensional object- and attribute-closure are defined accordingly.*

The dual definitions give rise to intensional *continuity and closure properties.*

Lemma 2 earlier shows that extensional object-continuity and -closure are properties of any dual bond when considered as a relation between one context and the complement of the other. We thus focus on extensional attribute-continuity and -closure in this section. The other notions will however become important later on in Section 5.

Whenever it is clear whether we are dealing with a relation on attributes or on objects, we will tend to omit the additional qualifications "extensionally" and "intensionally." We also remark that neither object- nor attribute-continuity is sufficient to obtain full continuity in the general case, as can be seen from R^∇ in Counterexample 1.

Now we can investigate the interaction between continuity and the representation of dual bonds.

Theorem 4. *Consider a dual bond R from* $\mathbb{K} = (G, M, I)$ *to* $\mathbb{L} = (H, N, J)$*. If R is extensionally attribute-continuous from* $\mathbb{K}$ *to* $\mathbb{L}^c$*, then R is an extent of* $\mathbb{K} \times \mathbb{L}$ *and* R^∇ *is intensionally object-closed from* $\mathbb{K}^c$ *to* $\mathbb{L}$*.*

Of course, analogous results can be obtained for closure by exchanging the roles of $\mathbb{K}$ and $\mathbb{L}$. One may wonder whether similar statements can be proven for dual bonds which are fully continuous and/or closed. However, this is not the case:

Counterexample 1. Consider the contexts $\mathbb{K} = (\{1, 2\}, \{a, b\}, I)$ and $\mathbb{L} = (\{3, 4\}, \{c, d, e\}, J)$ depicted in Fig. 1 (left). Define $R = \{(1, 3), (2, 4)\}$. All subsets of $\{1, 2\}$ are extents of both $\mathbb{K}$ and $\mathbb{K}^c$. Likewise, all subsets of $\{3, 4\}$ are extents of $\mathbb{L}$ and $\mathbb{L}^c$. Thus R is trivially closed and continuous in every sense. However, we find that $R^\triangledown = \{(a,d),(b,c)\}$ is not closed from $\mathbb{K}^c$ to $\mathbb{L}$. Indeed, $\{a,b\}$ is an intent of $\mathbb{K}^c$ but $R^\triangledown(\{a,b\}) = \{c,d\}$ is not an intent of $\mathbb{L}$, since $\{c,d\}^{JJ} = \{c,d,e\}$.

Another false assumption one might have is that the conditions given in Theorem 4 for being an extent of the direct product are not just sufficient but also necessary. However, neither closure nor continuity is needed for a dual bond to be represented in the direct product.

Counterexample 2. Consider the context $\mathbb{K} = (\{1,2,3\},\{a,b,c\},I)$ depicted in Fig. 1 (right). Define $R = \{(1,1),(2,2)\}$. We find that $R^\triangledown = \{(a,b),(b,a)\}$. Thus $R = R^{\triangledown\triangledown}$ and R is a dual bond which is an extent of the direct product $\mathbb{K} \times \mathbb{K}$. However, R is not even attribute-continuous from $\mathbb{K}$ to $\mathbb{K}^c$, since $R^{-1}(c^I) = R^{-1}(\{1,2,3\}) = \{1,2\}$ is not closed in $\mathbb{K}$. On the other hand, using that $R = R^{-1}$, we find that R is not attribute-closed from $\mathbb{K}^c$ to $\mathbb{K}$ either.

Although this shows that continuity is not a characteristic feature of all dual bonds in the direct product, we still find that there are many situations where there is a wealth of continuous dual bonds. This is the content of the following theorem.

Theorem 5. *Consider the contexts* $\mathbb{K} = (G,M,I)$ *and* $\mathbb{L} = (H,N,J)$. *If*

$$\emptyset \text{ is an extent of } \mathbb{K} \qquad or \qquad \emptyset \text{ is not an extent of } \mathbb{L}^c$$

then the set of all dual bonds which are continuous from $\mathbb{K}$ *to* $\mathbb{L}^c$ *is* $\bigcap$-*dense in* $\mathsf{B}_o(\mathbb{K}\times\mathbb{L})$ *and thus forms a basis for the closure system of all dual bonds in the direct product.*

If the assumptions also hold with $\mathbb{K}$ *and* $\mathbb{L}$ *exchanged, then the set of all dual bonds which are both continuous from* $\mathbb{K}$ *to* $\mathbb{L}^c$ *and closed from* $\mathbb{K}^c$ *to* $\mathbb{L}$ *is* $\bigcap$-*dense as well.*

Note that the previous theorem could of course also be stated using closure in place of continuity. Furthermore it is evident that dual bonds of the form $(m,n)^\triangledown$ are such that the (pre)image of almost any set is an extent. The only exception is the empty set, which is why we needed to add the given preconditions. We remark that these conditions are indeed very weak. By removing or adding full rows, any context can be modified in such a way that the empty set either is an extent or not. Since the concept lattices of the context and its complement are not affected by this procedure, one can enforce the necessary conditions without loosing generality.

5   Functional Bonds and Scale Measures

In FCA, (extensionally) continuous functions have been studied under the name *scale measures*, the importance of which stems from the fact that they can be regarded as a model for concept scaling and data abstraction. Topology provides additional interpretations for continuous functions in the context of knowledge representation and reasoning, but we will not give further details here.[3] We merely remark that continuity between topological spaces coincides with continuity between appropriate contexts.

---

[3] Roughly speaking, the potential of topology for our purposes resides in its well-known connections to FCA (data representation), formal logic (reasoning), and domain theory (computation/approximation), all of which are based on essentially the same mechanisms of *Stone duality* (see [14] for further details).

Continuity for functions constitutes a special case of continuity in the relational case as defined above.

Definition 4. *Consider contexts* $\mathbb{K} = (G, M, I)$ *and* $\mathbb{L} = (H, N, J)$. *A function* $f : G \to H$ *is* extensionally continuous *whenever its graph* $\{(x, f(x)) \mid x \in G\}$ *is an extensionally continuous relation, i.e. if* $f^{-1}(O)$ *is an extent of* $\mathbb{K}$ *for any extent* O *of* $\mathbb{L}$.

Extensional attribute- *and* object-continuity, *as well as the according intensional properties and closures are defined similarly based on the graph of the function.*

This definition agrees with [1, Definition 89], where extensionally continuous maps have also been called *scale measures*. Extensional attribute-continuity (and thus intensional object-continuity) is of course redundant, as the following lemma shows.

Lemma 4. *Given contexts* $\mathbb{K} = (G, M, I)$ *and* $\mathbb{L} = (H, N, J)$, *a function* $f : G \to H$ *is extensionally continuous iff it is extensionally attribute-continuous.*

This statement relies on the fact that attribute extents are $\bigcap$-dense in the object concept lattice and that preimage commutes with intersection. On the one hand, this is not true for images of functions, and hence extensional attribute-closure does not yield full closure. On the other hand, though object extents are supremum-dense, the respective suprema are not the set-theoretical unions. Hence extensional object-continuity and -closure are reasonable notions as well.

The link from functions to our earlier studies of dual bonds is established through a specific class of dual bonds which can be represented by functions.

Definition 5. *Consider a dual bond R between contexts (G, M, I) and (H, N, J). Then R is* functional *whenever, for any $g \in G$, the extent $R(g)$ is generated by a unique object $f_R(g) \in H$:*

$$R(g) = f_R(g)^{JJ}.$$

In this case R is said to induce *the corresponding function $f_R : G \to H$.*

It is obvious that functional dual bonds are uniquely determined by the function they induce. In fact, it is easy to see that R is the least dual bond that contains the graph of the function f_R. However, not for every function will this construction yield a dual bond that is functional. The next result characterizes the functions that are of the form f_R for some functional dual bond R.

Proposition 2. *Consider a context $\mathbb{K} = (G, M, I)$ and a context $\mathbb{L} = (H, N, J)$ for which the map $h \mapsto h^J$ is injective. There is a bijective correspondence between*

- *the set of all functional dual bonds from $\mathbb{K}$ to $\mathbb{L}$ and*
- *the set of all extensionally object-continuous functions from $\mathbb{K}$ to $\mathbb{L}^c$.*

The required bijections consist of the functions

- *$R \mapsto f_R$ mapping each functional dual bond to the induced function and*
- *$f \mapsto R_f$ mapping each object-continuous function to the least dual bond which contains its graph $\{(g, f(g)) \mid g \in G\}$.*

Object-continuity of the functions f_R is not too much of a surprise in the light of Lemma 2. The fact that this property suffices for the above result demonstrates how specific functional dual bonds really are. In contrast, the properties established in Lemma 2 are generally not sufficient for a relation to be a dual bond.

Also note that the additional requirements for $\mathbb{L}$, which guarantee that no two functions induce the same dual bond, are again rather weak. Indeed, they are implied by the common assumption that the contexts under consideration are clarified.

We can now go further and characterize the antitone Galois connections obtained from functional dual bonds.

Proposition 3. *Consider a context $\mathbb{K} = (G, M, I)$ and a context $\mathbb{L} = (H, N, J)$ for which the map $h \mapsto h^J$ is injective. The bijection between dual bonds and antitone Galois connections given in Theorem 2 restricts to a bijective correspondence between*

- *the set of all functional dual bonds from $\mathbb{K}$ to $\mathbb{L}$ and*
- *the set of all antitone Galois connections from $\mathsf{B}_o(\mathbb{K})$ to $\mathsf{B}_o(\mathbb{L})$ which map object extents of $\mathbb{K}$ to object extents of $\mathbb{L}$.*

In the light of the previous proposition we give a definition for the corresponding property of Galois connections.

Definition 6. *Consider contexts $\mathbb{K} = (G, M, I)$ and $\mathbb{L} = (H, N, J)$ and a (monotone or antitone) Galois connection $\phi = (\vec{\phi}, \overleftarrow{\phi})$ between $\mathsf{B}_o(\mathbb{K})$ and $\mathsf{B}_o(\mathbb{L})$.*

Then ϕ is functional *(from $\mathbb{K}$ to $\mathbb{L}$) if $\vec{\phi}$ maps object extents to object extents and, for any $g \in G$ there is a unique object $f_{\vec{\phi}}(g)$ such that*

$$\vec{\phi}(g^{II}) = f_{\vec{\phi}}(g)^{JJ}.$$

In this case, ϕ is said to induce *the function $f_{\vec{\phi}} : G \to H$.*

Proposition 3 shows rather natural classes of dual bonds and Galois connections, respectively. However, functional dual bonds do not generally arise as extents of the direct product. Moreover, the corresponding class of extensionally object-continuous functions as described in Proposition 2 appears to be unidentified. As Theorem 6 below shows, the more common class of extensionally continuous functions still allows for a nice characterization in terms of dual bonds.

Theorem 6. *Consider a context $\mathbb{K} = (G, M, I)$ and a context $\mathbb{L} = (H, N, J)$ for which the map $h \mapsto h^J$ is injective. The bijection given in Proposition 2 restricts to a bijective correspondence between*

- *the set of all extensionally continuous functions from $\mathbb{K}$ to $\mathbb{L}^c$ and*
- *the set of all functional dual bonds from $\mathbb{K}$ to $\mathbb{L}$ that are continuous from $\mathbb{K}$ to $\mathbb{L}^c$.*

Especially, every dual bond R_f induced by a continuous function from $\mathbb{K}$ to $\mathbb{L}^c$ is an extent of the direct product $\mathbb{K} \times \mathbb{L}$.

Thus we find that extensionally continuous functions, or scale measures, are a rather specific kind of dual bonds. Again we must be careful: It is certainly not the case that all functional dual bonds which are extents in the direct product are continuous. Just consider the context $\mathbb{K} = (\{g\}, \{m\}, \{(g, m)\})$. The relation $R = \{(g, g)\}$ is an extent of the direct product $\mathbb{K} \times \mathbb{K}$ and it is functional with f_R being the identity. However, the preimage of the empty set (which is closed in $\mathbb{K}^c$) is not an extent of $\mathbb{K}$.

As a dual bond, every continuous function naturally induces an antitone Galois connection – Propositions 2 and 3 discussed the according constructions for object-continuous functions. Due to their special structure, continuous functions can additionally be used to derive another monotone Galois connection. It should not come as a surprise that these entities determine each other uniquely under some mild assumptions.

Theorem 7. *Consider contexts $\mathbb{K} = (G, M, I)$ and $\mathbb{L} = (H, N, J)$, and a function $f : G \rightarrow H$ which is continuous from $\mathbb{K}$ to $\mathbb{L}^c$.*

(i) An antitone Galois connection $\phi_f : \mathsf{B}_o(\mathbb{K}) \rightarrow \mathsf{B}_o(\mathbb{L})$ is given by the mappings

$$\vec{\phi}_f : \mathsf{B}_o(\mathbb{K}) \rightarrow \mathsf{B}_o(\mathbb{L}) : X \mapsto \bigcap \{f(x)^{JJ} \mid x \in X\} \quad and$$

$$\overleftarrow{\phi}_f : \mathsf{B}_o(\mathbb{L}) \rightarrow \mathsf{B}_o(\mathbb{K}) : Y \mapsto \bigcap \{f^{-1}(y^{JJ}) \mid y \in Y\}.$$

(ii) A monotone Galois connection $\psi_f : \mathsf{B}_o(\mathbb{K}) \rightarrow \mathsf{B}_o(\mathbb{L}^c)$ is given by the mappings

$$\vec{\psi}_f : \mathsf{B}_o(\mathbb{K}) \rightarrow \mathsf{B}_o(\mathbb{L}^c) : X \mapsto f(X)^{JJ} \quad and$$

$$\overleftarrow{\psi}_f : \mathsf{B}_o(\mathbb{L}^c) \rightarrow \mathsf{B}_o(\mathbb{K}) : Y \mapsto f^{-1}(Y).$$

Moreover, if $\mathbb{L}$ is such that $h \mapsto h^J$ is injective, the above mappings provide bijective correspondences between

- *the set of all extensionally continuous functions from $\mathbb{K}$ to $\mathbb{L}^c$,*
- *the set of all antitone Galois connections $\mathsf{B}_o(\mathbb{K})$ to $\mathsf{B}_o(\mathbb{L})$ that are functional (from $\mathbb{K}$ to $\mathbb{L}$) and for which the induced function is continuous from $\mathbb{K}$ to $\mathbb{L}^c$,*
- *the set of all monotone Galois connections $\mathsf{B}_o(\mathbb{K})$ to $\mathsf{B}_o(\mathbb{L}^c)$ that are functional (from $\mathbb{K}$ to $\mathbb{L}^c$).*

Part (ii) of the theorem and the corresponding bijections are known (see [1, Propositions 118 and 119]). Note that the two Galois connections from the preceding result are not obtained from each other by some simple dualizing. This is also evident when comparing the different side conditions in both cases: functional monotone Galois connections always relate to continuous functions, while continuity has to be required explicitly for functional antitone Galois connections.

6   Infomorphisms

Infomorphisms are a special kind of morphism between formal contexts that have been considered quite independently in rather different research disciplines. The name "infomorphism" we use here has been coined in the context of *information flow theory* [4]. Literature on Chu spaces means the same when speaking about "Chu mappings";

institution theory [3] refers the corresponding definition as the "Satisfaction condition" without naming the emerging morphisms at all. In FCA, the antitone version of these morphisms has been studied under the name *(context-)Galois connection* [10, 11].

Probably the most decisive feature of informorphisms is self-duality, an immediate consequence of their symmetry. Some of the relationships between infomorphisms and Galois connections are known, but our results in earlier sections reveal a more complete picture.

Definition 7. *Given contexts* $\mathbb{K} = (G, M, I)$ *and* $\mathbb{L} = (H, N, J)$, *an* infomorphism *from* $\mathbb{K}$ *to* $\mathbb{L}$ *is a pair of mappings* $\vec{f} : G \rightarrow H$ *and* $\overleftarrow{f} : N \rightarrow M$ *such that*

$$g \ I \ \overleftarrow{f}(n) \qquad \text{if and only if} \qquad \vec{f}(g) \ J \ n$$

holds for arbitrary $g \in G, n \in N$.

We first establish the following basic facts.

Lemma 5. *Consider contexts* $\mathbb{K} = (G, M, I)$ *and* $\mathbb{L} = (H, N, J)$. *The infomorphisms from* $\mathbb{K}$ *to* $\mathbb{L}$ *are exactly the infomorphisms from* $\mathbb{K}^c$ *to* $\mathbb{L}^c$.

Given such an infomorphism $(\vec{f}, \overleftarrow{f})$ *and sets* $O \subseteq G, A \subseteq N$, *we find that*

$$\vec{f}^{-1}(A^J) = \overleftarrow{f}(A)^I, \quad \vec{f}^{-1}(A^{\not{J}}) = \overleftarrow{f}(A)^{\not{I}}, \quad \overleftarrow{f}^{-1}(O^I) = \vec{f}(O)^J \quad \text{and} \quad \overleftarrow{f}^{-1}(O^{\not{I}}) = \vec{f}(O)^{\not{J}}.$$

Especially, $\vec{f}$ *is extensionally continuous from* $\mathbb{K}^{(c)}$ *to* $\mathbb{L}^{(c)}$ *and* $\overleftarrow{f}$ *is intensionally continuous from* $\mathbb{L}^{(c)}$ *to* $\mathbb{K}^{(c)}$.

Using these continuity properties, we can already specify a number of possible Galois connections constructed from infomorphisms. We remark that continuity between two contexts is in general not equivalent to continuity between the respective complements, such that Theorem 7 can be applied to one part of an infomorphism in two different ways, whereas this is not possible for arbitrary continuous functions.

From Theorem 6, we know that we can obtain continuous dual bonds from both $\vec{f}$ and $\overleftarrow{f}$. Since these relations are extents and intents, respectively, in the direct product, one may ask whether they belong to the same concepts or not. The following proposition shows the expected result.

Proposition 4. *Consider contexts* $\mathbb{K} = (G, M, I)$ *and* $\mathbb{L} = (H, N, J)$ *and an infomorphism* $(\vec{f}, \overleftarrow{f})$ *from* $\mathbb{K}$ *to* $\mathbb{L}$. *Define relations* $R \subseteq G \times H$ *and* $S \subseteq M \times N$ *by setting*

$$R(g) = \vec{f}(g)^{JJ} \quad \text{and} \quad S^{-1}(n) = \overleftarrow{f}(n)^{\not{I}\not{I}}.$$

Then R *is a dual bond from* $\mathbb{K}^c$ *to* $\mathbb{L}$ *which is an extent of* $\mathbb{K}^c \times \mathbb{L}$ *with* $R^\nabla = S$.

Furthermore, R *is extensionally continuous from* $\mathbb{K}^c$ *to* $\mathbb{L}^c$ *and* S^{-1} *is intensionally continuous from* $\mathbb{L}^c$ *to* $\mathbb{K}$.

Observe that the above construction of R (and S) relies only on the continuity of $\vec{f}$ from $\mathbb{K}^c$ to $\mathbb{L}^c$ (and the corresponding continuity of $\overleftarrow{f}$). One can also construct a dual bond based on the continuity properties of these functions between the non-complemented contexts. However, Proposition 4 does not imply any relationship between these two dual bonds beyond the obvious fact that they induce the same infomorphism.

We already know that the dual bonds induced by (one part of) an infomorphism have rather specific properties. The next result shows that these features are sufficient for characterizing the respective dual bonds.

Proposition 5. *Consider contexts* $\mathbb{K} = (G, M, I)$ *and* $\mathbb{L} = (H, N, J)$ *and let R be a dual bond from* $\mathbb{K}^c$ *to* $\mathbb{L}$ *such that both R and* $R^{\nabla - 1}$ *are functional. If R is extensionally continuous then the functions induced by R and* $R^{\nabla - 1}$ *constitute an infomorphism from* $\mathbb{K}$ *to* $\mathbb{L}$.

Note that, according to Lemma 4, extensional continuity of a functional dual bond R is equivalent to extensional attribute-continuity. This in turn implies intensional object-closure of R^{∇} (Theorem 4) which, since $R^{\nabla - 1}$ is also functional, implies the closure of R^{∇}. Thus our assumptions are perfectly symmetrical. Furthermore, Propositions 4 and 5 induce a bijection between infomorphisms and the described class of dual bonds.

Having understood how infomorphisms are characterized in terms of dual bonds, we can specify their relationship with Galois connections.

Theorem 8. *Consider contexts* $\mathbb{K} = (G, M, I)$ *and* $\mathbb{L} = (H, N, J)$, *and let $f = (\vec{f}, \overleftarrow{f})$ be an infomorphism from* $\mathbb{K}$ *to* $\mathbb{L}$.

– *An antitone Galois connection* $\phi_f : B_o(\mathbb{K}) \to B_o(\mathbb{L}^c)$ *is given by the mappings*

$$\vec{\phi}_f : B_o(\mathbb{K}) \to B_o(\mathbb{L}^c) : X \mapsto \bigcap\{\vec{f}(x)^{\lambda\lambda} \mid x \in X\} \quad = \quad \bigcap\{\overleftarrow{f}^{-1}(x^{\lambda})^{\lambda} \mid x \in X\} \quad and$$

$$\overleftarrow{\phi}_f : B_o(\mathbb{L}^c) \to B_o(\mathbb{K}) : Y \mapsto \bigcap\{\vec{f}^{-1}(y^{JJ}) \mid y \in Y\} \quad = \quad \bigcap\{\overleftarrow{f}(y^J)^I \mid y \in Y\}.$$

Further, three antitone Galois connections $\phi_f^c : B_o(\mathbb{K}^c) \to B_o(\mathbb{L})$, $\phi_f^d : B_o(\mathbb{K}^d) \to B_o(\mathbb{L}^{cd})$ *and* $\phi_f^{cd} : B_o(\mathbb{K}^{cd}) \to B_o(\mathbb{L}^d)$ *are defined similarly, using the complemented incidence relations (for $(\cdot)^c$) and exchanging $\vec{f}$ and $\overleftarrow{f}$ (for $(\cdot)^d$), respectively.*

– *A monotone Galois connection* $\psi_f : B_o(\mathbb{K}) \to B_o(\mathbb{L})$ *is given by the mappings*

$$\vec{\psi}_f : B_o(\mathbb{K}) \to B_o(\mathbb{L}) : X \mapsto \vec{f}(X)^{JJ} \quad = \quad \overleftarrow{f}^{-1}(X^I)^J \quad and$$

$$\overleftarrow{\psi}_f : B_o(\mathbb{L}) \to B_o(\mathbb{K}) : Y \mapsto \vec{f}^{-1}(Y) \quad = \quad \overleftarrow{f}(Y^J)^I.$$

Another monotone Galois connection $\psi_f^c : B_o(\mathbb{K}^c) \to B_o(\mathbb{L}^c)$ *is defined similarly, but with all incidence relations complemented.*

Note that Proposition 4 shows that the antitone Galois connections ϕ_f^d and ϕ_f^{cd} can also be constructed as in Corollary 1 from the two dual bonds induced by the function $\vec{f}$. Especially, Corollary 1 does not yield any further Galois connections.

7   Summary and Future Work

The above considerations show that scale measures and infomorphisms can be identified with special types of dual bonds, and thus that part of this work can also be regarded

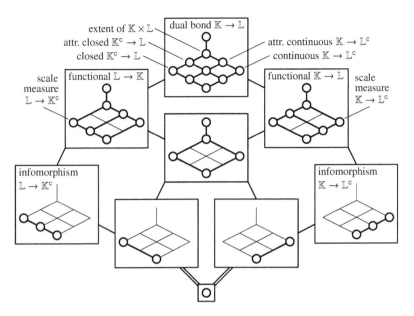

Fig. 2. The concept lattice of the discussed properties of dual bonds, displayed as a nested line diagram. The included attributes are defined in Definition 1 (dual bond), 2 ($\mathbb{K} \times \mathbb{L}$), 3 (continuity and closure), and 5 (functionality). The attributes "scale measure" and "infomorphism" refer to the dual bonds described in Theorem 6 and Proposition 5, respectively, and thus imply functionality. Further nontrivial implications are those established in Theorem 3 (attribute-continuity or -closure implies being an extent of $\mathbb{K} \times \mathbb{L}$) and Lemma 5 (infomorphisms are continuous).

as a study of various attributes of dual bonds and of the implications between them. The resulting concept lattice of context-morphisms is represented by the *nested line diagram*[4] in Fig. 2. This diagram indeed gives a complete picture of the situation: not only have all of the entailed implications between attributes been proven within this article, but there are also no other implications. In fact, one can obtain Fig. 2 by *attribute exploration*. We skip the details for lack of space.[5]

In spite of this rather complete picture, there are many aspects of the theory of morphisms in FCA which could not be considered within this article; they are left as possible directions for future research. As mentioned in the introduction, the use of morphisms to model knowledge transfer and information sharing may employ methods from category theory [15]. But not all of the above morphisms immediately yield categories of contexts, especially since antitone Galois connections cannot be composed in

---

[4] The concept lattice represented by a nested line diagram consists of the boldfaced nodes, where connections between boxes represent parallel connections between boldfaced nodes at corresponding positions wrt. the background structure. See [1, pp. 75].

[5] ConExp (http://sourceforge.net/projects/conexp) helped us a lot.

an obvious way. As a solution, one can dualize one context and consider *bonds* which yield monotone Galois connections that can be composed easily [1]. One can also restrict to special classes of dual bonds: scale measures, infomorphisms, and dual bonds that are both closed and continuous suggest obvious composition mechanisms.

The next step after identifying possible categories is to investigate the properties of these structures. What are their natural interpretations in terms of knowledge representation? Do they support all of the constructions that one may be interested in? How are they related to other known categories, e.g. from formal logic, order theory, or topology? This does also involve comparisons to the usage of context-morphisms in institution theory and information flow, where a relaxation of the rather strict definition of infomorphisms may yield advantages for certain applications.

In institution theory, many specific collections of formal contexts have been introduced in order to handle given logics, basically by considering the consequence relation between the models and the formulae of a logic as a formal context. In this setting, dual bonds allow for a proof theoretic interpretation as *consequence relations* and may have special properties due to the additional (logical) restrictions on contexts. For example, *compactness*[6] of classical propositional logic yields additional continuity and closure properties of dual bonds between (appropriate complements of) the respective contexts. Furthermore, extensionally continuous functions between such contexts are continuous in the usual topological sense with respect to the associated Stone spaces (see [14]).

Besides the mentioned (onto-)logical and categorical investigations, there are also further questions related to lattice theory. We characterized the Galois connections that are induced by certain types of dual bonds, especially in the functional case (Proposition 3, Theorems 7 and 8). For many other types of dual bonds, the corresponding descriptions are missing. Likewise, although dual bonds are closed under intersections, we are aware of no (non-canonical) context that has all dual bonds as extents.

In FCA, the concept lattice of the direct product $\mathbb{K} \times \mathbb{L}$ is known as the *tensor product* of the lattices $B_o(\mathbb{K})$ and $B_o(\mathbb{L})$. Theorem 5 showed that the study of dual bonds can also yield additional results on the tensor product, but further relationships between both subjects have not been investigated yet. As shown in [10, Satz 15], infomorphisms can be represented by a concept lattice as well, but the role of this structure in the light of our present investigations still needs to be explored.

Many other results from [1, 10, 11, 12] could not be discussed due to space limitations.

Acknowledgement We very much apprechiated some comments by the referees, which helped us to improve our presentation.

References

[1] Ganter, B., Wille, R.: Formal Concept Analysis: Mathematical Foundations. Springer (1999)
[2] Lawvere, F.W., Rosebrugh, R.: Sets for mathematics. Cambridge University Press (2003)
[3] Goguen, J., Burstall, R.: Institutions: abstract model theory for specification and programming. Journal of the ACM **39** (1992)

---

[6] This basically amounts to saying that the induced concept lattice is *(co-)algebraic*, see [13].

[4] Barwise, J., Seligman, J.: Information flow: the logic of distributed systems. Volume 44 of Cambridge tracts in theoretical computer science. Cambridge University Press (1997)

[5] Kent, R.E.: The information flow foundation for conceptual knowledge organization. In: Proc. of the 6th Int. Conf. of the International Society for Knowledge Organization. (2000)

[6] Kent, R.E.: Semantic integration in the Information Flow Framework. In Kalfoglou, Y., et al, eds.: Semantic Interoperability and Integration. Dagstuhl Seminar Proceedings 04391 (2005)

[7] Pratt, V.: Chu spaces as a semantic bridge between linear logic and mathematics. Theoretical Computer Science **294** (2003) 439–471

[8] Krötzsch, M., Hitzler, P., Zhang, G.Q.: Morphisms in context. Technical report, AIFB, Universität Karlsruhe (2005) www.aifb.uni-karlsruhe.de/WBS/phi/pub/KHZ05tr.pdf.

[9] Erné, M.: Categories of contexts. *Unpublished* (2005) Rewritten version.

[10] Xia, W.: Morphismen als formale Begriffe, Darstellung und Erzeugung. PhD thesis, TH Darmstadt (1993)

[11] Ganter, B.: Relational Galois connections. *Unpublished manuscript* (2004)

[12] Kent, R.E.: Distributed conceptual structures. In de Swart, H., ed.: Sixth International Workshop on Relational Methods in Computer Science. Volume 2561 of Lecture Notes in Computer Science., Springer (2002) 104–123

[13] Hitzler, P., Krötzsch, M., Zhang, G.Q.: A categorical view on algebraic lattices in formal concept analysis. Technical report, AIFB, Universität Karlsruhe (2004)

[14] Krötzsch, M.: Morphisms in logic, topology, and formal concept analysis. Master's thesis, Dresden University of Technology (2005)

[15] Goguen, J.: Three perspectives on information integration. In Kalfoglou, Y., et al., eds.: Semantic Interoperability and Integration. Dagstuhl Seminar Proceedings 04391 (2005)

Contextual Logic and Aristotle's Syllogistic

Rudolf Wille

Technische Universität Darmstadt, Fachbereich Mathematik,
Schloßgartenstr. 7, D–64289 Darmstadt
wille@mathematik.tu-darmstadt.de

Abstract. This paper is concerned with incorporating *negational relationships* into the semantics of Contextual Logic by mathematizing Aristotle's Syllogistic in contextual-logic terms. For preparing this, a short sketch of *Aristotle's Syllogistic* is presented. Then it is shown how a *contextual semantics* for Aristotle's syllogisms can be developed on the basis of so-called syllogistic contexts. This semantic approach is used to determine *implication bases* for elementary judgments within syllogistic contexts. Finally, directions of further research are mentioned.

Contents

1 Introduction

Contextual Logic has been developed as a mathematization of the traditional philosophical logic with its doctrines of concepts, judgments, and conclusions (cf. [Wi96], [Wi00b]). For mathematizing the doctrine of concepts, *Formal Concept Analysis* [GW99a] has been proven useful and, for mathematizing the doctrines of judgments and conclusions, Sowa's *Theory of Conceptual Graphs* [So84] has given essential stimulations. In general, the development of Contextual Logic is guided by the aim to support humans in creating, communicating, and processing knowledge (cf. [DK03]).

This paper is concerned with incorporating *negational relationships* into the semantics of Contextual Logic. Because of its conceptual nature, Contextual Logic cannot treat *negations* as they are treated in Predicate Logic, since the complement of the extension of a concept needs not to be a concept extension again (e.g., non-piano is usually not considered as a concept). Therefore, if negations shall be understood as unary operations like in Predicate Logic, concepts have to be generalized, for instance, to protoconcepts as discussed in [Wi00a]. In this paper, we want to mathematize negational relationships – like *implications* and *incompatibilities* – without using negation operations.

F. Dau, M.-L. Mugnier, G. Stumme (Eds.): ICCS 2005, LNAI 3596, pp. 238–249, 2005.

Implications as judgments have already been mathematized in [Wi04] as *implicational concept graphs*. But the logical relationships of implications and incompatibilities, as they are basic for instance for Brandom's theory of discursive practice [Br94], cannot be studied within Contextual Logic in its present state. A first step to overcome this deficiency may be the incorporation of *Aristotle's Syllogistic* [Ar75] into Contextual Logic. Such an incorporation will be elaborated in this paper.

Aristotle's syllogistic is based on four types of *elementary judgments* which combine two concepts $\mathfrak{a}$ and $\mathfrak{b}$, respectively. Those judgment types, denoted by a, e, i, and o, are defined in Fig. 1 in modern mathematical terms ($\underline{\mathfrak{a}}$ and $\underline{\mathfrak{b}}$ denote the extents of the concepts $\mathfrak{a}$ and $\mathfrak{b}$, respectively). The *graphical representa-*

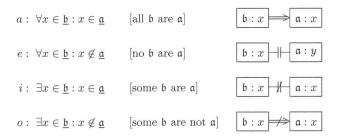

$$a : \forall x \in \underline{\mathfrak{b}} : x \in \underline{\mathfrak{a}} \qquad [\text{all } \mathfrak{b} \text{ are } \mathfrak{a}]$$

$$e : \forall x \in \underline{\mathfrak{b}} : x \notin \underline{\mathfrak{a}} \qquad [\text{no } \mathfrak{b} \text{ are } \mathfrak{a}]$$

$$i : \exists x \in \underline{\mathfrak{b}} : x \in \underline{\mathfrak{a}} \qquad [\text{some } \mathfrak{b} \text{ are } \mathfrak{a}]$$

$$o : \exists x \in \underline{\mathfrak{b}} : x \notin \underline{\mathfrak{a}} \qquad [\text{some } \mathfrak{b} \text{ are not } \mathfrak{a}]$$

Fig. 1. The four types of Aristotle's elementary judgments

tions of the elementary judgments on the right side of Fig. 1 are proposed for an extented Contextual Judgment Logic which also covers the contextual logic of implications and incompatibilities. The diagram for the type a was already introduced in greater generality for implicational concept graphs in [Wi04]. It should also be mentioned that, for the elementary judgments of type i, the shown diagram is common in the Theory of Conceptual Graphs without the graphical connection between the two boxes. The added connection in Fig. 1 shall indicate the contradictoriness of the types e and i, analogously to the indicated contradictoriness of the types a and o.

In the following, first a short sketch of *Aristotle's Syllogistic* is presented in Section 2 to give the necessary background knowledge for understanding the approached incorporation of Aristotle's Syllogistic into Contextual Logic. In Section 3, it is then shown how a *contextual semantics* for Aristotle's syllogisms can be developed on the basis of so-called syllogistic contexts. In Section 4, this semantic approach is used to determine implication bases for elementary judgments within syllogistic contexts.

2   Aristotle's Syllogistic

In his syllogistic (his doctrine of syllogisms), Aristotle understood a *syllogism* as "a discourse in which, a certain thing being stated, something other what is stated follows of necessity from being so" ([Au95], p. 780; [Ar75], p. 2). In this paper we concentrate on the categorical syllogisms which are the most important ones. A *categorical syllogism* is an argument consisting of three elementary judgments, two serving as premises and one serving as conclusion. As already shown in Fig. 1, Aristotle distinguished four types of *elementary judgments* (formed by two concepts A and B):

a-judgments (universal affirmative): AaB (A covers B)
e-judgments (universal negative): AeB (A covers no part of B)
i-judgments (particular affirmative): AiB (A covers some part of B)
o-judgments (particular negative): AoB (A does not cover B)

The possible forms of *logic diagrams* of Aristotle's four judgments types are depicted in Fig. 2 (cf. [St86], p. 74ff.). In the diagrams, all proper areas bounded by circular arcs represent genuine parts of the corresponding concepts. The region outside the two circles might however disappear, which is the case when the universe of discourse is represented just by the union of the areas of the two circles.

The three elementary judgments of a syllogism are always formed by three concepts as in the following example:
"humans are not robots", "children are humans" $\Rightarrow$ "children are not robots"

- The first judgment is of the type AeB
  where A is the predicate concept "robot" and B is the subject concept "human".

- The second judgment is of the type BaC
  where B is the predicate concept "human" and C is the subject concept "child".

- The third judgment is of the type AeC
  where A is the predicate concept "robot" and C is the subject concept "child".

Thus, the syllogism has the form AeB, $BaC \Rightarrow AeC$, which is named "Celarent" for retaining the vowel sequence $e - a - e$. In general, the common concept of the two premise judgments is called the *middle term*, the predicate concept of the conclusion judgment is called the *major term* and the subject concept of the conclusion judgment is called the *minor term*. Every syllogism belongs to one of three *figures* depending on how the middle term functions: (1.) once as predicate concept and once as subject concept, (2.) twice as predicate concept or (3.) twice as subject concept. Our example is a syllogism of figure 1.

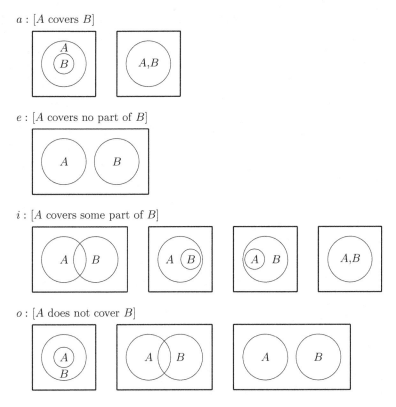

Fig. 2. The logic diagrams of Aristotle's elementary judgments

The basic task of Aristotle's Syllogistic is to determine all valid and invalid categorical syllogisms. This needs a semantics which, in particular, serves with counterexamples. Such a semantics is given by the logic diagrams shown in Fig. 2 which functions more or less the same as the *semantics of sailclothes* invented already by Platon (cf. [St86], p. 41). A survey on the many attemps to improve and extend Aristotle's Syllogistic can be found in [Th98].

3   A Contextual Semantics for Syllogisms

The aimed mathematical semantics for syllogisms will be based on *formal contexts* and their *formal concepts* as they are defined in Formal Concept Analysis [GW99a]. They will be used to mathematize the logic diagrams of Aristotle's elementary judgments (see Fig. 2). From the two concepts of an elementary judgment, such mathematizations abstract two formal concepts $\mathfrak{a}$ and $\mathfrak{b}$ of a suitably chosen formal context $\mathbb{K} := (G, M, I)$. The extents $\underline{\mathfrak{a}}$ and $\underline{\mathfrak{b}}$ of those formal concepts allow a mathematical representation of the partition formed by the circular areas of the two concepts in the corresponding logic diagram. The general form of such representations is shown in Fig. 3. It should be mentioned that the object sets $\underline{a} \setminus \underline{b}$, $\underline{b} \setminus \underline{a}$, and $(G \setminus \underline{a}) \setminus \underline{b}$ need not to be concept extents and might even be empty, but the object set $\underline{a} \cap \underline{b}$, which might also be empty, is always an extent of a formal concept.

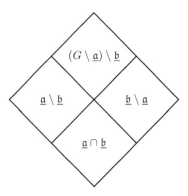

Fig. 3. The general partition pattern of two formal concepts a and b

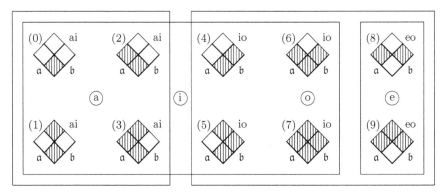

Fig. 4. The ten partition patterns derived from two formal concepts a and b

In a logic diagram of Aristotle's elementary judgments (Fig. 2), the circular areas and their parts are always non-empty. Therefore we have to assume that the extents of the corresponding formal concepts are also non-empty. This has as consequence that, derived from two formal concepts, there are ten partition patterns as visualized in Fig. 4 within the general pattern of Fig. 3 by different hatchings, which represent the underlying object set G, respectively. From the hatchings one can easily determine Aristotle's judgment types for the ten partition patterns by using the following characterizations:

$$\begin{aligned}
\text{type } a &\iff \underline{\mathfrak{b}} \setminus \underline{\mathfrak{a}} = \emptyset && [\text{patterns: } (0), (1), (2), (3)] \\
\text{type } e &\iff \underline{\mathfrak{a}} \cap \underline{\mathfrak{b}} = \emptyset && [\text{patterns: } (8), (9)] \\
\text{type } i &\iff \underline{\mathfrak{a}} \cap \underline{\mathfrak{b}} \neq \emptyset && [\text{patterns: } (0), (1), (2), (3), (4), (5), (6), (7)] \\
\text{type } o &\iff \underline{\mathfrak{b}} \setminus \underline{\mathfrak{a}} \neq \emptyset && [\text{patterns: } (4), (5), (6), (7), (8), (9)]
\end{aligned}$$

Now, we are ready to introduce a contextual semantics for syllogisms. Let $\mathbb{K} := (G, M, I)$ be a formal context with $G \neq \emptyset$ and $M^I = \emptyset$; furthermore, let $\mathfrak{B}_\triangle(\mathbb{K}) := \mathfrak{B}(\mathbb{K}) \setminus \{(\emptyset, M)\}$. For determining the desired semantics, the basic idea is to derive a new formal context $\mathbb{K}^\Sigma$ which has as formal attributes the pairs of formal concepts of $\mathbb{K}$ understood as mathematizations of Aristotle's elementary judgments. More precisely, we define the *"syllogistic derivative"* of $\mathbb{K}$ as the formal context $\mathbb{K}^\Sigma := (G^2, \mathfrak{B}_\triangle(\mathbb{K})^2, I_\Sigma)$ with

$$((g, h), (\mathfrak{a}, \mathfrak{b})) \in I_\Sigma : \iff \{g, h\} \subseteq \underline{\mathfrak{a}} \cap \underline{\mathfrak{b}} \text{ or } \{g, h\} \subseteq \underline{\mathfrak{a}} \setminus \underline{\mathfrak{b}} \text{ or }$$
$$\{g, h\} \subseteq \underline{\mathfrak{b}} \setminus \underline{\mathfrak{a}} \text{ or } \{g, h\} \subseteq (G \setminus \underline{\mathfrak{a}}) \setminus \underline{\mathfrak{b}}.$$

For $(\mathfrak{a}, \mathfrak{b}) \in \mathfrak{B}_\triangle(\mathbb{K})^2$, we obtain

$$\{(\mathfrak{a}, \mathfrak{b})\}^{I_\Sigma} = (\underline{\mathfrak{a}} \cap \underline{\mathfrak{b}})^2 \cup (\underline{\mathfrak{a}} \setminus \underline{\mathfrak{b}})^2 \cup (\underline{\mathfrak{b}} \setminus \underline{\mathfrak{a}})^2 \cup ((G \setminus \underline{\mathfrak{a}}) \setminus \underline{\mathfrak{b}})^2;$$

i.e., the derivation $\{(\mathfrak{a}, \mathfrak{b})\}^{I_\Sigma}$ is an equivalence relation on the object set G with the equivalence classes out of $\underline{\mathfrak{a}} \cap \underline{\mathfrak{b}}$, $\underline{\mathfrak{a}} \setminus \underline{\mathfrak{b}}$, $\underline{\mathfrak{b}} \setminus \underline{\mathfrak{a}}$, and $(G \setminus \underline{\mathfrak{a}}) \setminus \underline{\mathfrak{b}}$. In general, we can prove the following proposition by using Proposition 11 in [GW99a]:

Proposition 1 *For every non-empty subset $\mathfrak{A}$ of $\mathfrak{B}_\triangle(\mathbb{K})^2$, the derivation $\mathfrak{A}^{I_\Sigma}$ is an equivalence relation on G determined by $\mathfrak{A}^{I_\Sigma} = \bigcap_{(\mathfrak{a}, \mathfrak{b}) \in \mathfrak{A}} \{(\mathfrak{a}, \mathfrak{b})\}^{I_\Sigma}$.*

In the following we will show that the logic of Aristotle's categorical syllogisms can be based on the *Contextual Attribute Logic* [GW99b] of the syllogistic derivatives $\mathbb{K}^\Sigma$. For understanding this connection between Contextual Attribute Logic and Aristotle's logic, it is helpful to study in Fig. 5 the represensions of the ten different *pairs of equivalence relations* derived from two formal concepts $\mathfrak{a}$ and $\mathfrak{b}$ of formal contexts $\mathbb{K}$, respectively (cf. Fig. 4):

- The areas of the ten large squares in Fig. 5 represent the set G^2, respectively.
- The areas of the horizontal hatchings in Fig. 5 represent the equivalence relations
  $\{(\mathfrak{a}, \mathfrak{a})\}^{I_\Sigma}$ $(= \underline{\mathfrak{a}}^2 \cup (G \setminus \underline{\mathfrak{a}})^2)$, respectively.
- The areas of the vertical hatchings in Fig. 5 represent the equivalence relations
  $\{(\mathfrak{b}, \mathfrak{b})\}^{I_\Sigma}$ $(= \underline{\mathfrak{b}}^2 \cup (G \setminus \underline{\mathfrak{b}})^2)$, respectively.
- The areas of horizontal and vertical hatchings in Fig. 5 represent the equivalence relations $\{(\mathfrak{a}, \mathfrak{b})\}^{I_\Sigma}$ $(= \{\mathfrak{a}, \mathfrak{b}\}^{I_\Sigma} = \{(\mathfrak{a}, \mathfrak{a})\}^{I_\Sigma} \cap \{(\mathfrak{b}, \mathfrak{b})\}^{I_\Sigma})$, respectively.

The *categorical syllogisms* can be analogously founded on *triples of equivalence relations* derived from three formal concepts $\mathfrak{a}$, $\mathfrak{b}$, and $\mathfrak{c}$ of formal contexts $\mathbb{K}$, respectively. Fig. 6 shows representations of the syllogisms *Barbara* and *Celarent*:

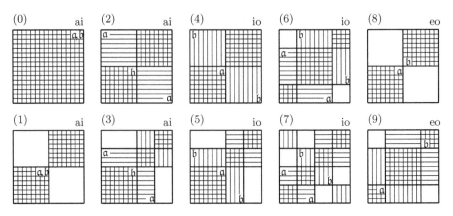

Fig. 5. Representations of the ten pairs of equivalence relations in G^2 which correspond to the partition patterns in Fig. 4

- The areas of the two large squares in Fig. 6 represent the set G^2, respectively.
- The areas of the horizontal hatchings in Fig. 6 represent the equivalence relations $\{(a, b), (b, c)\}^{I_\Sigma}$ $(= \{(a, b)\}^{I_\Sigma} \cap \{(b, c)\}^{I_\Sigma} = \{(a, a)\}^{I_\Sigma} \cap \{(b, b)\}^{I_\Sigma} \cap \{(c, c)\}^{I_\Sigma})$, respectively.
- The areas of the vertical hatchings in Fig. 6 represent the equivalence relations
  $\{(a, c)\}^{I_\Sigma}$ $(= \{(a, a)\}^{I_\Sigma} \cap \{(c, c)\}^{I_\Sigma})$, respectively.

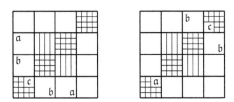

Fig. 6. Representations of the syllogisms *Barbara* and *Celarent*

In both representations we have $\{(a, b), (b, c)\}^{I_\Sigma} \subseteq \{(a, c)\}^{I_\Sigma}$, i.e., $\{(a, b), (b, c)\} \to (a, c)$ is an attribute implication of the syllogistic derivative $\mathbb{K}^\Sigma$.

Since $(\underline{b} \setminus \underline{a})^2 = \emptyset$ ($\iff \underline{b} \setminus \underline{a} = \emptyset$) and $(\underline{c} \setminus \underline{b})^2 = \emptyset$ ($\iff \underline{c} \setminus \underline{b} = \emptyset$) implies $(\underline{c} \setminus \underline{a})^2 = \emptyset$ ($\iff \underline{c} \setminus \underline{a} = \emptyset$),

the implication $\{(a, b), (b, c)\} \to (a, c)$ represents indeed the syllogism *Barbara*.

Accordingly,

since $(\underline{a} \cap \underline{b})^2 = \emptyset$ ($\Longleftrightarrow$ $\underline{a} \cap \underline{b} = \emptyset$) and $(\underline{c} \setminus \underline{b})^2 = \emptyset$ ($\Longleftrightarrow$ $\underline{c} \setminus \underline{b} = \emptyset$)

implies $(\underline{a} \cap \underline{c})^2 = \emptyset$ ($\Longleftrightarrow$ $\underline{a} \cap \underline{c} = \emptyset$),

the implication $\{(a, b), (b, c)\} \rightarrow (a, c)$ represents indeed the syllogism *Celarent*.

In the same way, the representations of all other valid categorical syllogisms can be justified. But it is easier and more substantial to use Aristotle's *syllogistic deduction* showing that the valid categorical syllogisms are exactly the syllogisms which can be deduced by the following *basic principles* (cf. [St86], p. 94ff.):

1. The inversion rules: (a, b) of type $a \Rightarrow (b, a)$ of type i
   (a, b) of type $e \Rightarrow (b, a)$ of type e
   (a, b) of type $i \Rightarrow (b, a)$ of type i

2. The categorical syllogisms *Barbara* and *Celarent*

3. The contradictoriness of the types a and o and of the types e and i

4   Implication Bases of Syllogistic Contexts

The introduced contextual semantics for Aristotle's elementary judgments and categorical syllogisms based on syllogistic contexts $\mathbb{K}^\Sigma$ may open further insights into logical coherences. In particular, it seems interesting to investigate useful bases of elementary judgments within the contexts $\mathbb{K}^\Sigma$. To facilitate this investigation, we define a *syllogistic Θ-context* of a formal context $\mathbb{K} := (G, M, I)$ with $|G| > 1$ and $M^I = \emptyset$ by $\mathbb{K}^\Theta := (G^2, \{\{(a, a)\}^{I_\Sigma} \mid a \in \mathfrak{B}_\triangle(\mathbb{K}) \setminus \{(G, \emptyset)\}\}, \in)$. Because of $\{(a, b)\}^{I_\Sigma} = \{(a, a)\}^{I_\Sigma} \cap \{(b, b)\}^{I_\Sigma}$ in $\mathbb{K}^\Sigma$, $\mathfrak{B}(\mathbb{K}^\Theta) \cong \mathfrak{B}(\mathbb{K}^\Sigma)$ and $\mathbb{K}^\Theta$ is isomorphic to the attribute reduction of $\mathbb{K}^\Sigma$ with the attribute set $\{\{(a, a)\}^{I_\Sigma} \mid a \in \mathfrak{B}_\triangle, a \neq G\}$. Since the *stem basis* of the attribute implications of a formal context is the most useful implication basis, we concentrate our investigation on determining the *pseudo-intents* of syllogistic Θ-contexts which are the premises of the implications of the corresponding stem bases, respectively (cf. [GW99a], p. 83).

Let us first consider a small example of a syllogistic Θ-context derived from the formal context $\mathbb{K}_4 := (G_4, G_4, \neq)$ with $G_4 := \{1, 2, 3, 4\}$. Its syllogistic Θ-context $\mathbb{K}_4^\Theta$ is represented by the cross-table in Fig. 7 (brackets and commas of the common descriptions of sets are deleted). Fig. 8 shows the concept lattice $\mathfrak{B}(\mathbb{K}_4^\Theta)$. The pseudo-intents of $\mathbb{K}_4^\Theta$ are just the sets consisting of two attributes. The pairs of attributes out of $\{[12, 34], [13, 24], [14, 23]\}$ imply all other attributes because their corresponding concept pairs have the smallest concept (the identity) as meet. The other pairs of attributes imply just one further attribute because the corresponding concept pairs have an atom of the concept lattice as meet (each atom is below exactly three attribute concepts).

Suprisingly, an analogous result as for $\mathbb{K}_4$ can be proved for all formal contexts $\mathbb{K}_n := (G_n, G_n, \neq)$ with $G_n := \{1, 2, \ldots, n\}$ for arbitrary natural numbers $n > 1$. This is stated in the following proposition:

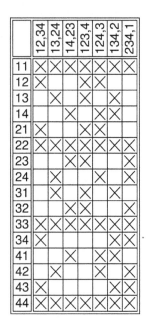

Fig. 7. The syllogistic Θ-context $\mathbb{K}_4^{\Theta}$

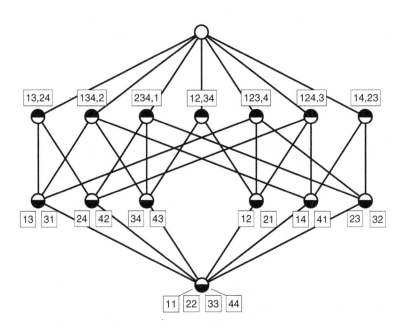

Fig. 8. Concept lattice of the formal context in Fig. 7

Proposition 2 *The pseudo-intents of the syllogistic Θ-context $\mathbb{K}_n^\Theta$ are exactly the sets consisting of two attributes.*

For proving Proposition 2, we need the following lemma about equivalence relations on a finite set:

Lemma 1 *Let S be be a finite set and let $\mathfrak{C}(S)$ be the set of all equivalence relations on S with exactly two equivalence classes. Then a subset $\mathfrak{E}$ of $\mathfrak{C}(S)$ contains the sets $\{F \in \mathfrak{C}(S) \mid F \supseteq D \cap E\}$ for all $D, E \in \mathfrak{E}$ if and only if $\mathfrak{E} = \{F \in \mathfrak{C}(S) \mid F \supseteq \bigcap \mathfrak{E}\}$.*

Proof Obviously, $\mathfrak{E} = \{F \in \mathfrak{C}(S) \mid F \supseteq \bigcap \mathfrak{E}\}$ implies $\mathfrak{E} = \bigcup_{D, E \in \mathfrak{E}}\{F \in \mathfrak{C}(S) \mid F \supseteq D \cap E\}$. Conversely, let $\mathfrak{E} = \bigcup_{D, E \in \mathfrak{E}}\{F \in \mathfrak{C}(S) \mid F \supseteq D \cap E\}$. Now, let C be an equivalence class of the equivalence relation $\bigcap \mathfrak{E}$ and let C_E be the equivalence class of $E \in \mathfrak{E}$ with $C_E \supseteq C$. Then $C = \bigcap_{E \in \mathfrak{E}} C_E$. Let $\mathfrak{E}$ consist of m equivalence relations $E_1, E_2, \ldots, E_m$. Then $E_{12} := (C_{E_1} \cap C_{E_2})^2 \cup (G_n \setminus (C_{E_1} \cap C_{E_2}))^2$ is an equivalence relation having $C_{E_1} \cap C_{E_2}$ as equivalence class and containing $E_1 \cap E_2$. Analogously, we can construct equivalence relations $E_{123}, \ldots, E_{12\ldots m}$ with equivalence classes $C_{E_1} \cap C_{E_2} \cap C_{E_3}, \ldots, C_{E_1} \cap C_{E_2} \cap \cdots \cap C_{E_m}$ satisfying $E_{123} \supseteq E_{12} \cap E_3, \ldots, E_{12\ldots m} \supseteq E_{12\ldots(m-1)} \cap E_m$. In this way we obtain an equivalence relation $E_C := E_{12\ldots m}$ in $\bigcup_{D, E \in \mathfrak{E}}\{F \in \mathfrak{C}(S) \mid F \supseteq D \cap E\}$ having C and $S \setminus C$ as its equivalence classes. This construction can be performed with any equivalence class C of $\bigcap \mathfrak{E}$. Now, let $E_{\hat{C}}$ be any equivalence relation in $\mathfrak{C}(S)$ with $E_{\hat{C}} \supseteq \bigcap \mathfrak{E}$ having $\hat{C}$ as one of its two equivalence classes. Then

$$\hat{C} = \bigcap\{S \setminus C \mid C \text{ is an equivalence class of } \bigcap \mathfrak{E} \text{ with } C \cap \hat{C} = \emptyset\}.$$

Analogously to the previous construction, we can iterately derive the equivalence relation $E_{\hat{C}}$ within $\bigcup_{D, E \in \mathfrak{E}}\{F \in \mathfrak{C}(S) \mid F \supseteq D \cap E\}$. This finally proves that $\mathfrak{E} = \bigcup_{D, E \in \mathfrak{E}}\{F \in \mathfrak{C}(S) \mid F \supseteq D \cap E\}$ implies $\mathfrak{E} = \{F \in \mathfrak{C}(S) \mid F \supseteq \bigcap \mathfrak{E}\}$ too.
$\square$

Proof of Proposition 2 Since the concept extents of $\mathbb{K}_n$ are exactly the subsets of G_n, the concept extents of $\mathbb{K}_n^\Theta$ are exactly the equivalence relations on G_n. The attributes of $\mathbb{K}_n^\Theta$ are exactly the maximal equivalence relations on G_n unequal G_n^2, i.e., each of them has two equivalence classes. An arbitary equivalence relation on G_n is always the intersection of suitable attributes of $\mathbb{K}_n^\Theta$.

Now, let $\{(\mathfrak{a}, \mathfrak{a})\}^{I_\Sigma}$ and $\{(\mathfrak{b}, \mathfrak{b})\}^{I_\Sigma}$ be two different attributes of $\mathbb{K}_n^\Theta$. Then the equivalence relation $\{(\mathfrak{a}, \mathfrak{a})\}^{I_\Sigma} \cap \{(\mathfrak{b}, \mathfrak{b})\}^{I_\Sigma}$ has either three or four equivalence classes. In the case of three, the interval $[\{(\mathfrak{a}, \mathfrak{a})\}^{I_\Sigma} \cap \{(\mathfrak{b}, \mathfrak{b})\}^{I_\Sigma}, G_n^2]$ of the lattice of equivalence relations on G_n contains exactly three attribute extents; hence each two of the corresponding three attributes form a pseudo-intent. In the case of four equivalence classes, the interval $[\{(\mathfrak{a}, \mathfrak{a})\}^{I_\Sigma} \cap \{(\mathfrak{b}, \mathfrak{b})\}^{I_\Sigma}, G_n^2]$ is isomorhic to $\underline{\mathfrak{B}}(\mathbb{K}_4^\Theta)$; hence each two attributes in that interval form a pseudo-intent as we have seen by the preceding example. Together we obtain that all sets of two attributes of $\mathbb{K}_n^\Theta$ are pseudo-intents. Finally, we obtain with Lemma 1 that there are no other pseudo-intents of $\mathbb{K}_n^\Theta$. $\square$

Proposition 2 confirms Aristotle's conception that (categorical) syllogisms are sufficient for concluding elementary judgments from (finitely many) other elementary judgments. However, for Proposition 2, it is essential that all subsets of the object set G_n are concept extents of the formal context $\mathbb{K}_n$. If an arbitrary finite formal context $\mathbb{K}$ is taken as a basis, then the syllogistic Θ-context $\mathbb{K}^\Theta$ might have quite a number of pseudo-intents consisting of more than two attributes. Those pseudo-intents are caused by pseudo-intents of the underlying formal context K.

5   Further Research

The main direction of further research shall be concerned with generalizing the semantics of syllogistic contexts to a semantics based on generalized power context families. It is the hope that in this way a contextual-logic theory of *implications and incompatibilities* can be developed. A theory of implicational concept graphs has already been successfully invented; that suggests to generalize this theory to something like *"syllogistic concept graphs"* of which the diagrams in Fig. 1 represent the elementary cases. A serious test of such extented theory would be how well it can be used for mathematizing *Brandom's theory of discursive practice*. More stimulation for extending Contextual Logic can be expected by further studies and mathematizations of Aristotle's Syllogistic and its continuations; in particular, the *hypothetical and disjunctive syllogisms* are challenging.

References

[Ar75]   Aristoteles: *Lehre vom Schluß oder Erste Analytik*. Felix Meiner, Hamburg 1975.

[Au95]   R. Audi (ed.): *The Cambridge dictionary of philosophy*. Cambridge University Press, Cambridge 1995.

[Br94]   R. B. Brandom: *Making it explicit. Reasoning, representing, and discursive commitment*. Harvard University Press, Cambridge 1994.

[DK03]   F. Dau, J. Klinger: From Formal Concept Analysis to Contextual Logic. FB4-Preprint, TU Darmstadt 2003.

[GW99a] B. Ganter, R. Wille: *Formal Concept Analysis: mathematical foundations*. Springer, Heidelberg 1999; German version: Formale Begriffsanalyse: Mathematische Grundlagen. Springer, Heidelberg 1996.

[GW99b] B. Ganter, R. Wille: Contextual Attribute Logic. In: W. Tepfenhart, W. Cyre (eds.): *Conceptual structures: standards and practices*. LNAI **1640**. Springer, Heidelberg 1999, 377–388.

[So84]   J. F. Sowa: *Conceptual structures: Information processing in mind and machine.* Adison-Wesley, Reading 1984.

[St86]   P. Stekeler-Weithofer: Grundprobleme der Logik: Elemente einer Kritik der formalen Vernunft. de Gruyter, Berlin 1986.

[Th98]   P. Thom: Syllogismus; Syllogistik. In: K. Gründer (Hrsg.): *Historisches Wörterbuch der Philosophie.* Bd.10. Schwabe, Basel 1998, 687–707.

[Wi96]   R. Wille: Restructuring mathematical logic: an approach based on Peirce's pragmatism. In: A. Ursini, P. Agliano (eds.): *Logic and algebra.* Marcel Dekker, New York 1996, 267–281.

[Wi00a]   R. Wille: Boolean Concept Logic. In: B. Ganter, G. W. Mineau (eds.): *Conceptual structures: logical, linguistic, and computational issues.* LNAI **1867**. Springer, Heidelberg 2000, 317–331.

[Wi00b]   R. Wille: Contextual Logic summary. In: G. Stumme (ed.): *Working with conceptual structures: Contributions to ICCS 2000.* Shaker-Verlag, Aachen 2000, 265–276.

[Wi04]   R. Wille: Implicational concept graphs. In: K. E. Wolff, H. D. Pfeiffer, H. S. Delugach (eds.): *Conceptual structures at work.* LNAI **3127**. Springer, Heidelberg 2004, 52–61.

States of Distributed Objects in Conceptual Semantic Systems

Karl Erich Wolff

Mathematics and Science Faculty
Darmstadt University of Applied Sciences
Schoefferstr. 3, D-64295 Darmstadt, Germany
karl.erich.wolff@t-online.de
http://www.fbmn.fh-darmstadt.de/home/wolff

Abstract. Our classical understanding of objects in spatiotemporal systems is based on the idea that such an object is at each moment at exactly one place. As long as the notions of "moment" and "place" are not made explicit in their granularity the meaning of that idea is not clear. It became clear by the introduction of Conceptual Time Systems with Actual Objects and a Time Relation (CTSOT) using an explicit granularity description for space and time and an object representation such that each object is at each moment in exactly one state - where the states are formal concepts of the CTSOT.

For the purpose of introducing also a granularity tool for the objects the author has defined Conceptual Semantic Systems where relational information is combined with the granularity tool of conceptual scales. That led to a mathematical definition of particles and waves such that the usual notions of particles and waves in physics are covered. Waves and wave packets are "distributed objects" in the sense that they may appear simultaneously at several places.

Now the question arises how to introduce a mathematical notion for the "state of a distributed object", as for example the state of an electron or the state of an institution, in the general framework of Conceptual Semantic Systems. That question is answered in this paper by the introduction of the notion of the "aspect of a concept **c** with respect to some view Q", in short "the Q-aspect of **c**" which is defined as a suitable set of formal concepts. For spatiotemporal Conceptual Semantic Systems the state of an object **p** at a time granule **t** is defined as the spatial aspect of the infimum of "realizations" of **p** and **t**. The one-element states of "actual objects" in a CTSOT are special cases of these states which may have many elements.

The information units (instances) of a Conceptual Semantic System connect the concepts of different semantical scales, for example scales for objects, space, and time. That allows for defining the information distribution of the Q-aspect of a distributed object **c** which leads to a mathematical definition of the "BORN-frequency"; that is defined as a relative frequency of information units which can be understood as a very meaningful mathematical representation of the famous "probability distribution of a quantum mechanical system".

F. Dau, M.-L. Mugnier, G. Stumme (Eds.): ICCS 2005, LNAI 3596, pp. 250–266, 2005.
© Springer-Verlag Berlin Heidelberg 2005

1   Objects and Their Representation

In this introductory section we discuss the notion of an "object" and some representations of objects from a conceptual point of view which developed during the last twenty years when the author was working in the field of Conceptual Knowledge Processing.

1.1   Objects and Systems

Examples of objects in the sense of our intuitive understanding of "objects" are single stones, tables, or persons. Obviously we are using the notion of an "object" in a much more general way, for example, when we say that an institution is an object. Then we focus on the institution as a whole, while calling it a system we concentrate on the fact that it is decomposable into parts. Since parts of a system can be treated again as new systems certain "basic" or "elementary" or "atomic" subsystems are often called "objects".

We give a simple example which shows our freedom in focussing on a system and selecting its parts depending on our purpose. Let us consider a shepherd counting his sheep with the purpose to be sure that no sheep of his herd is lost. Here, each sheep is understood as an object in the sense that it is "atomic", that is, it is not necessary for the given purpose to subdivide it into parts. For another purpose, namely for investigating an ill sheep, that sheep plays the role of a subsystem, and its broken leg may be understood as an "object" in that subsystem.

1.2   Objects in Spatiotemporal Systems

Clearly, not all parts of a system are usually understood as "objects". For example, in spatiotemporal systems the objects are usually clearly distinguished from the spatial parts, the "places", or from the temporal parts, say "days" or "weeks", of the system. That spatiotemporal distinction between "objects" and other "entities" like "places" or "days" is basic for applications in physics.

For the discussion of classical and quantum objects in modern physics the reader is referred to the fine collection of important papers on that subject by Elena Castellani [Ca98]. We quote from the beginning of her introduction:

> In the philosophical literature, as well as in ordinary usage, physical objects are most frequently referred to as "bodies", "material things", "material beings", "objects in space and time". With respect to the totality of existing objects, they are usually (but not always) distinguished from "persons". Whichever are the particular descriptions employed, *having mass*, *being located in space and time*, and *persisting through time* seem to constitute the fundamental features required for something to qualify as a "physical object". ...
>
> Now what about the objects of physics? Are those entities, whose nature and behavior it is the aim of physical theories to describe, "physical

objects" in the preceding sense? Let us consider, for example, the entities at the very center of the developments of contemporary physics: the so-called elementary particles, the microscopic "objects" supposed to be the ultimate constituents of the physical world. ...

Are such entities to be called *objects* after all?

From the other papers collected in [Ca98] the following problems around physical objects seem to be closely related to our approach: Objects and Individuality; Part/Whole Relation; Vagueness of Objects; Objects as Sums of Properties; Identity through Time; Genidentity; Indistinguishability; Objects and Measurement.

It is impossible to mention in this paper the broad discussion about problems around the notion of "objects" in the literature. Before discussing some of these problems from our point of view we shortly mention the connections to classical mathematical tools in physics.

1.3   Connections to Classical Mathematical Tools in Physics

Most readers who are interested in the formal representation of movements of objects in space and time might expect to find in any paper on that subject after a few general remarks the classical notions of real numbers, vector spaces, metrics and analytical notions like differentiation and integration. But we do not use any of these notions in our present investigation. All these basic mathematical notions can be included into the theory, but we like to do that step by step - checking the conceptual role of the actually included classical notion at each step.

That led us to a conceptual understanding of basic "physical notions" like for example "life tracks", "particles", "waves", "wave packets". In the present paper we develop a conceptual way of understanding the "probability distribution of a quantum object (or system)" based on a conceptual representation of "objects" and "aspects of objects" as discussed in the next sections.

1.4   Conceptual Representation of Objects

To make our ideas about "objects" clear we first mention that we do not try to define what an "object in reality" is. Instead, we describe the main steps which led us from simple conceptual notions designed for the formal description of "objects in reality" to the object representation in the definition of spatiotemporal Conceptual Semantic Systems and discuss the advantages and disadvantages of these formalizations. We assume that the reader is familiar with the basic notions in Formal Concept Analysis as described in [GW99].

First Step: Objects as Formal Objects. The most simple representation of "real objects" is to label different real objects by different real marks, such that each object is uniquely determined by its mark; that can be done easily for "human sized objects", but it seems to be impossible for some physical objects like photons. A corresponding formal description for n "real objects" is the formal context $(G, G, =)$ where $G := \{1, ..., n\}$ is the set of formal objects. That is a suitable conceptual description of n distinguishable objects. To describe indistinguishable objects we can take a formal context of the form $(G, M, G \times M)$. Its concept lattice has exactly one concept which is clearly the object concept of each of the objects. An arbitrary formal context (G, M, I) can be used as a formal description of "real objects" some of which can be distinguished by attributes while others remain indistinguishable. Clearly, indistinguishability is here understood as "not distinguishable by the attributes in the given context". Two formal objects g, h of a formal context (G, M, I) are called *indistinguishable* if they have the same attributes; that is equivalent to the property that their object concepts are equal: $\gamma(g) = \gamma(h)$.

The disadvantage of the representation of "real objects" as formal objects is the following. If we wish to represent "real temporal objects" which change their attributes in time then we have to cope with three kinds of things, namely objects, time, and attributes which are connected by a ternary relation, for example: "an object g has at time t an attribute m". A simple and very successful, but not yet general representation of such ternary relations is described in the following.

Second Step: Temporal Objects and Life Tracks. The ternary relation "an object g has at time t an attribute m" can be represented by a binary relation if we consider "actual objects (g, t)" as new entities and build a formal context with the binary relation "(g, t) has the attribute m". Clearly, we could also use the binary relation "g has the (time-dependent) attribute (m, t)". The first choice with actual objects is very successful in all applications since the object concept $\gamma(g, t)$ has the meaning of a *state* of the object g at time granule t. That led to a very fruitful development of so-called "Conceptual Time Systems with Actual Objects and a Time Relation", in short CTSOT, which allow for representing states, transitions, and life tracks of such objects. For a detailed discussion the reader is referred to [Wo02a, Wo02b]. In a CTSOT each object g is represented by its life track or equivalently by the subsystem "generated by the rows (g, t)" where t is a time granule of g. The CTSOTs are well-suited for representing those "real objects" which are at each time granule in exactly one state, or at exactly one place if we interpret the attributes as "local attributes" - as for example as x- and y-coordinates or as attributes like "was on the market place".

Third Step: Tuples of Objects as Formal Objects. For the conceptual representation of k-ary relations (where k is an integer) on a set G_0 of "objects" one can use a formal context $\mathbb{K}_k := (G_k, M_k, I_k)$ where $G_k \subseteq (G_0)^k$. Then the formal concepts of $\mathbb{K}_k$ are called *relation concepts* since their extents are as subsets of $(G_0)^k$ k-ary relations on the set G_0. That conceptual representation of k-ary relations is used in *power context families* [Wi97, PW99] in connection with the relational representation of knowledge in concept graphs [Wi00, Wi02,

Wi03] which are contextual representations of conceptual graphs which have been introduced by John Sowa [So84].

In the following we represent relational knowledge not by power context families but by Conceptual Semantic Systems as introduced by the author [Wo04c]. A comparison between these two similar formal methods will become possible by applying them in practice.

2   Basic Notions in Conceptual Semantic Systems

In the following we continue developing the investigation of Conceptual Semantic Systems, which combine the simplicity of relational data tables with the method of conceptual scaling and the knowledge representation in form of semantics (often also called "ontologies"). It contains all three forms of conceptual object representation mentioned in the previous section. It allows for a representation of "real objects" like clouds, institutions, epidemics and other concepts which are usually not understood as objects in the sense that they are at each moment at exactly one place - for example, an institution may be located in several houses. It is obvious that the notions of "place" and "moment" can be understood also as such general "objects" like clouds or institutions which are in a certain sense "distributed" over some "space" and not necessarily "located at a single point". To describe all that we combine the conceptual representation of granularity, the representation of life tracks of objects in a CTSOT and the conceptual representation of relational structures.

2.1   Definition of Conceptual Semantic Systems

The main idea behind the formal definition of a Conceptual Semantic System (CSS) is that of a data table which represents in its rows more or less short forms of relational expressions combining concepts. The meaning of these concepts is represented by formal contexts, called semantic scales. With each column of the data table we associate a semantic scale representing the meaning of the values of that column. The values in each column are formal concepts of the semantic scale of that column.

Since the formal concepts in each semantic scale are ordered by their conceptual hierarchy the data tables of a CSS form an "ordinal context" as introduced and investigated in [SWi92]. For convenience we repeat the general definition of an ordinal context.

Definition 1. *"Ordinal Context"*
Let G and M be sets, and $(W_m, \leq_m)_{m \in M}$ be a family of ordered sets; let I be a ternary relation with $I \subseteq \bigcup_{m \in M}(G \times \{m\} \times W_m)$ such that for each $g \in G$ and $m \in M$ there is exactly one $w \in W_m$ with $(g, m, w) \in I$, denoted by $m(g) := w$. Then the tuple $(G, M, (W_m, \leq_m)_{m \in M}, I)$ is called an ordinal context. The many-valued context $(G, M, \bigcup_{m \in M} W_m, I)$ is called the underlying many-valued context of the given ordinal context.

Instead of arbitrary ordered sets $(W_m, \leq_m)_{m \in M}$, we choose the concept lattices of the semantic scales for defining "Conceptual Semantic Systems" in Definition 2.

Definition 2. *"Conceptual Semantic System"*

Let M be a set and, for each $m \in M$, let $\mathbb{S}_m := (G_m, N_m, I_m)$ be a formal context. Let $\underline{\mathfrak{B}}(\mathbb{S}_m)$ be the concept lattice of $\mathbb{S}_m$. Then any ordinal context

$$\mathfrak{K} := (G, M, (\underline{\mathfrak{B}}(\mathbb{S}_m))_{m \in M}, I)$$

is called a Conceptual Semantic System (CSS) with semantic scales $\mathbb{S}_m$ ($m \in M$). The elements of G are called instances or information units. We interpret the concepts of the semantic scales as "types" and the concept lattice of a semantic scale as a "type hierarchy". A triple $(g, m, \mathbf{c}) \in I$ is interpreted as "instance g tells something about the concept $\mathbf{c} \in \mathfrak{B}(\mathbb{S}_m)$". For any instance g the tuple $(m(g)|m \in M)$ is interpreted as a short description of a statement connecting the concepts $m(g)$ where $m \in M$.

The purpose for the introduction of Conceptual Semantic Systems is twofold: first, by calling the formal objects in G "instances" or "information units" we suggest to use the formal objects as "chunks of information which connect concepts" and not for the representation of these concepts. The instances of a CSS play the role of the rows (or the tuples) in a database table; an instance $g \in G$ is, roughly speaking, just a formal connection of the "meaningful" concepts $m(g)$ where $m \in M$. That allows for the representation of arbitrary relations among the chosen concepts of the semantic scales. A special example is the parametric representation of the unit circle by triples $(t, \cos(t), \sin(t))$. In that sense a CSS is a parametric representation of relational conceptual knowledge.

Therefore we are no longer forced to decide which kinds of concepts in the sense of persons or places or time granules have to be selected for the special role of formal objects of a many-valued context. For example, in a CTSOT the actual objects (p, t) play the role of the formal objects. As a consequence, any actual object is "located" at exactly one place in the concept lattice, namely at its object concept. That is very convenient for many purposes but not general enough for the representation of "distributed objects" which are "located at many places".

The second reason for the introduction of Conceptual Semantic Systems is to make explicit the meaning of basic concepts used in a statement by describing these concepts as formal concepts. That has the advantage that the "original" attributes of the intents of these concepts are also used in the following *semantically derived context*.

Definition 3. *"Semantically Derived Context"*

Let $(G, M, (\underline{\mathfrak{B}}(\mathbb{S}_m))_{m \in M}, I)$ be a Conceptual Semantic System with semantic scales $\mathbb{S}_m = (G_m, N_m, I_m)$ ($m \in M$) and let $int(c)$ denote the intent of a concept c. Then the formal context

$$\mathbb{K} := (G, \{(m,n)|m \in M, n \in N_m\}, J) \; where$$
$$gJ(m,n) :\Longleftrightarrow n \in int(m(g))$$

is called the semantically derived context of $(G, M, (\underline{\mathfrak{B}}(\mathbb{S}_m))_{m \in M}, I)$.

It is easy to see that the semantically derived context of a CSS can be obtained also by plain scaling as the usual derived context. Therefore we write in the following only "derived context" instead of "semantically derived context". The formal concepts of the semantic scales yield "realized concepts" in the derived context in the sense of the following definition.

Definition 4. "Realization of a concept of a semantic scale"
Let $(G, M, (\underline{\mathfrak{B}}(\mathbb{S}_m))_{m \in M}, I)$ be a Conceptual Semantic System with derived context $\mathbb{K}$. Then for $m \in M$ the following mapping

$$r_m : \mathfrak{B}(\mathbb{S}_m) \to \mathfrak{B}(\mathbb{K})$$
$$\boldsymbol{c} = (A_{\boldsymbol{c}}, B_{\boldsymbol{c}}) \mapsto r_m(\boldsymbol{c}) := ((\{m\} \times B_{\boldsymbol{c}})^{\downarrow}, (\{m\} \times B_{\boldsymbol{c}})^{\downarrow\uparrow})$$

is called the m-realization of $\boldsymbol{c}$ in $\mathfrak{B}(\mathbb{K})$.

If different many-valued attributes k and m have the same scales $\mathbb{S}_k = \mathbb{S}_m$, then the k-realization of a concept $\boldsymbol{c} \in \mathfrak{B}(\mathbb{S}_k)$ may be different from its m-realization.

2.2 Spatiotemporal Conceptual Semantic Systems

For spatiotemporal applications we specify in the set M of attributes of a CSS three special attributes P,T,L where P serves for the description of "general objects" like "real persons" or "real particles" or other "real objects", T for temporal concepts like "days", and L for "localities" like "places". Mathematically these special attributes are not distinguished from the other attributes. Instead of three different attributes we could also take three different subsets of attributes for the description of three different kinds of "categorically different concepts".

Definition 5. "Spatiotemporal Conceptual Semantic System"
Let $\mathfrak{K} := (G, M, (\underline{\mathfrak{B}}(\mathbb{S}_m))_{m \in M}, I)$ be a Conceptual Semantic System with semantic scales $\mathbb{S}_m = (G_m, N_m, I_m)$ $(m \in M)$ and let $P, T, L \in M$. Then the pair $(\mathfrak{K}, (P, T, L))$ is called a spatiotemporal Conceptual Semantic System.
The formal concepts of $\mathbb{S}_P, \mathbb{S}_T, \mathbb{S}_L$ are called types of "general objects", types of "time granules", and types of "space granules" respectively.
Let $\mathbb{K}$ denote the derived context of $\mathfrak{K}$, $\underline{\mathfrak{B}}(\mathbb{K})$ its concept lattice and γ its object concept mapping. For each $m \in M$ the context

$$\mathbb{K}_m := (G, \{(m,n)|n \in N_m\}, J \cap (G \times \{(m,n)|n \in N_m\}))$$

is called the m-part of $\mathbb{K}$. Its concept lattice is denoted by $\underline{\mathfrak{B}}_m$, its object concept mapping by γ_m.
The formal concepts of $\underline{\mathfrak{B}}_P, \underline{\mathfrak{B}}_T, \underline{\mathfrak{B}}_L$ are called "general objects" or "packets", "time granules", and "space granules" of $(\mathfrak{K}, (P, T, L))$ respectively.

For $m \in M$, the concept lattice $\underline{\mathfrak{B}}_m$ can be supremum-embedded into $\underline{\mathfrak{B}}(\mathbb{S}_m)$ which follows from Proposition 31 in [GW99], p. 98.

For the definition of "particles" and "waves" as special packets the author has used in [Wo04c] the following definition of the *location of a packet at a time granule*.

Definition 6. *"Location of a Packet at a Time Granule"*
Let $\mathfrak{C} := (\mathfrak{K}, (P, T, L))$ *be a spatiotemporal CSS. Let* $\boldsymbol{p} := (A_p, B_p)$ *be a packet of* $\mathfrak{C}$, *and* $\boldsymbol{t} := (A_t, B_t)$ *be a time granule of* $\mathfrak{C}$. *Then the pair* $(\boldsymbol{p}, \boldsymbol{t})$ *is called an actual packet of* $\mathfrak{C}$ *and the set*

$$L(\boldsymbol{p}, \boldsymbol{t}) := \{\gamma_L(g) | g \in A_p \cap A_t\}$$

is called the location of $(\boldsymbol{p}, \boldsymbol{t})$ *in* $\mathfrak{C}$ *or the location of the packet* $\boldsymbol{p}$ *at time granule* $\boldsymbol{t}$.

In biology, the habitat of a species can be understood as the location of an actual packet representing the species as a general object at some time granule. In the next section we generalize that successful notion to the notion of an "aspect of a concept with respect to a given view". That will lead us to the definition of the information distribution of an aspect of a concept. Its application to physics yields a conceptual interpretation of the "probability distribution of a quantum object".

3 Aspects of Concepts

In colloquial speech an "aspect of a thing" contains some special information about a thing. An example of an aspect of an actual person, say of *my father in his youth*, might be the set of towns where he lived during his youth. That is a local aspect, another one is his educational aspect, describing which schools he visited during his youth. Obviously, the locality of an actual packet of a spatiotemporal CSS is also such an aspect of that actual packet. The following definition introduces the notion of an "aspect of a concept with respect to some view" in an arbitrary given formal context.

Definition 7. *"Q-aspect of a concept and distributed concepts"*
Let $\mathbb{K} := (G, M, I)$ *be a formal context and* $Q \subseteq M$. *Then the Q-part of* $\mathbb{K}$ *is the formal context* $\mathbb{K}_Q := (G, Q, I \cap (G \times Q))$, *its concept lattice is denoted by* $\underline{\mathfrak{B}}_Q$, *and its object concept mapping by* γ_Q. *For any concept* $\boldsymbol{c} := (A_c, B_c) \in \underline{\mathfrak{B}}(\mathbb{K})$ *the set*

$$\alpha_Q(\boldsymbol{c}) := \{\gamma_Q(g) | g \in A_c\}$$

is called the aspect of the concept $\boldsymbol{c}$ *with respect to the view* Q, *in short the Q-aspect of* $\boldsymbol{c}$. *If* $|\alpha_Q(\boldsymbol{c})| \geq 2$, *then* $\boldsymbol{c}$ *is called distributed in* $\underline{\mathfrak{B}}_Q$.

In this paper we are interested in the Q-aspects of concepts of the derived context $\mathbb{K} = (G, N, J)$ of a CSS $\mathfrak{K} := (G, M, (\underline{\mathfrak{B}}(\mathbb{S}_m))_{m \in M}, I)$ with semantic scales $\mathbb{S}_m := (G_m, N_m, I_m)$ $(m \in M)$. Then the set of attributes of $\mathbb{K}$ is $N = \{(m, n) | m \in M, n \in N_m\}$. For a many-valued attribute $m \in M$ we call

$$\alpha_m(\mathbf{c}) := \alpha_{\{m\} \times N_m}(\mathbf{c})$$

the *m-aspect of $\mathbf{c}$* or *the aspect of $\mathbf{c}$ in the m-part of $\mathbb{K}$*.

The following proposition shows that the location $L(\mathbf{p},\mathbf{t})$ of an actual packet of a spatiotemporal CSS can be described as an aspect.

Proposition 1 *Let $\mathfrak{K} := (G, M, (\mathfrak{B}(\mathbb{S}_m))_{m \in M}, I)$ be a CSS with semantic scales $\mathbb{S}_m$ ($m \in M$). Let $\mathfrak{C} := (\mathfrak{K}, (P, T, L))$ be a spatiotemporal CSS. Let $\mathbf{p} = (A_p, B_p)$ be any packet and $\mathbf{t} = (A_t, B_t)$ any time granule of $\mathfrak{C}$. Then $\mathbf{p}_{\mathbb{K}} := (A_p, (A_p)^{\mathbb{K}})$ and $\mathbf{t}_{\mathbb{K}} := (A_t, (A_t)^{\mathbb{K}})$ and their infimum $\mathbf{p}_{\mathbb{K}} \wedge \mathbf{t}_{\mathbb{K}}$ are formal concepts of the derived context $\mathbb{K}$ of $\mathfrak{K}$, and for any view Q in $\mathbb{K}$*

$$\alpha_Q(\mathbf{p}_{\mathbb{K}} \wedge \mathbf{t}_{\mathbb{K}}) = \{\gamma_Q(g) | g \in A_p \cap A_t\}.$$

Consequently we obtain for the special view $Q_L := \{(L, n) | n \in N_L\}$

$$\alpha_L(\mathbf{p}_{\mathbb{K}} \wedge \mathbf{t}_{\mathbb{K}}) = L(\mathbf{p}, \mathbf{t}).$$

The proof can be easily obtained from the well-known fact that the mapping $(A, B) \longmapsto (A, A^{\mathbb{K}})$ is an infimum-preserving order-embedding from the concept lattice of the Q-part into the concept lattice of $\mathbb{K}$ as described in [GW99], page 98, Prop. 31.

4   The Information Distribution of an Aspect of a Concept

For each Q-aspect of a concept we introduce its information distribution and its relative frequency distribution which can be understood as a very meaningful mathematical representation of the famous "probability distribution of a quantum mechanical system".

Definition 8. "The Information Distribution on a Q-aspect of a concept"
Let $\mathfrak{K} := (G, M, (\mathfrak{B}(\mathbb{S}_m))_{m \in M}, I)$ be a CSS with semantics $\mathbb{S}_m$ ($m \in M$) and $\mathbb{K} := (G, N, J)$ its derived context. Let $\mathbf{c} = (A_c, B_c)$ be a formal concept of $\mathbb{K}$ and $Q \subseteq N$. Then the mapping $\beta_{c,Q}$ is defined by

$$\beta_{c,Q} : \alpha_Q(\mathbf{c}) \longrightarrow \mathfrak{P}(A_c) \text{ where}$$
$$\beta_{c,Q}(\gamma_Q(g)) := \{h \in A_c | \gamma_Q(h) = \gamma_Q(g)\} \quad \text{for } \gamma_Q(g) \in \alpha_Q(\mathbf{c}).$$

$\beta_{c,Q}$ is called the information distribution of $\mathbf{c}$ on $\alpha_Q(\mathbf{c})$, $\beta_{c,Q}(\gamma_Q(g))$ the set of $\mathbf{c}$ − instances at $\gamma_Q(g)$, and $|\beta_{c,Q}(\gamma_Q(g))|$ the information frequency of $\mathbf{c}$ in $\gamma_Q(g)$. If $A_c \neq \emptyset$ and A_c is finite, then the relative information frequency $\varrho_{c,Q}$ is defined by

$$\varrho_{c,Q} : \alpha_Q(\mathbf{c}) \longrightarrow [0, 1]$$
$$\varrho_{c,Q}(\gamma_Q(g)) := |A_c|^{-1} |\beta_{c,Q}(\gamma_Q(g))|.$$

The relative information frequency $\varrho_{c,Q}$ generates the corresponding probability $\boldsymbol{P}_{c,Q}$ which has for each subset $S \subseteq \alpha_Q(\boldsymbol{c})$ the value

$$\boldsymbol{P}_{c,Q}(S) = \sum\{\varrho_{c,Q}(\gamma_Q(g))|\gamma_Q(g) \in S\}.$$

Remark: The relative information frequency should be distinguished from the proportion of **c**-instances in the contingent of $\gamma_Q(g)$:

$$\pi_{c,Q}(\gamma_Q(g)) := |\beta_{c,Q}(\gamma_Q(g))||\gamma_Q^{-1}(\gamma_Q(g))|^{-1} \quad (\gamma_Q(g) \in \alpha_Q(\boldsymbol{c})).$$

These definitions will be illustrated by examples in section 6.

5   States as Special Aspects

States of actual objects in a Conceptual Time System with Actual Objects and a Time Relation (CTSOT) have been introduced by the author [Wo02b] as object concepts of the derived context of the event part of a CTSOT. In the following definition of a state we generalize from single object concepts to sets of object concepts, and from the event part to arbitrary parts of the derived context. The main idea is to define in a given spatiotemporal CSS a Q-state of an "actual object" as a suitable aspect in $\mathbb{K}_Q$.

Definition 9. *"Q-state of an actual object of a spatiotemporal CSS"*
Let $\mathfrak{C} := (\mathfrak{K}, (P, T, L))$ be a spatiotemporal CSS where $\mathfrak{K} := (G, M, (\mathfrak{B}(\mathbb{S}_m))_{m \in M}, I)$ is a CSS with semantic scales $\mathbb{S}_m = (G_m, N_m, I_m)$ $(m \in M)$. Let $\mathbb{K} = (G, N, J)$ be the derived context of $\mathfrak{K}$, $\boldsymbol{p} := (A_{\boldsymbol{p}}, B_{\boldsymbol{p}}) \in \mathfrak{B}(\mathbb{S}_P)$ and $\boldsymbol{t} := (A_{\boldsymbol{t}}, B_{\boldsymbol{t}}) \in \mathfrak{B}(\mathbb{S}_T)$. Let $r_P(\boldsymbol{p})$ and $r_T(\boldsymbol{t})$ be the realizations of $\boldsymbol{p}$ and $\boldsymbol{t}$ in $\mathfrak{B}(\mathbb{K})$. For any view $Q \subseteq N$ we define the Q-state of the "actual object" $(r_P(\boldsymbol{p}) \wedge r_T(\boldsymbol{t}))$ in $\mathbb{K}_Q$ by

$$state_Q(\boldsymbol{p}, \boldsymbol{t}) := \alpha_Q(r_P(\boldsymbol{p}) \wedge r_T(\boldsymbol{t})).$$

The following Proposition 2 shows that the states of a CTSOT can be described as singleton states of a suitable CSS:

Proposition 2 *Let $\mathfrak{C} := (P_0, G_0, \Pi, \boldsymbol{T}_0, \boldsymbol{C}_0, R)$ be a CTSOT where P_0 is a set (of "persons", or "objects") and G_0 a set (of "time granules") and $\Pi \subseteq P_0 \times G_0$ a set (of "actual persons") and $\boldsymbol{T}_0 = ((\Pi, M, W, I_{T_0}), (\mathbb{S}_m|m \in M))$ is the time part and $\boldsymbol{C}_0 = ((\Pi, E, V, I_{C_0}), (\mathbb{S}_e|e \in E))$ is the event part and $R \subseteq \Pi \times \Pi$ is "the time relation" of $\mathfrak{C}$. Let $\mathbb{K}_E$ be the derived context of the event part of $\mathfrak{C}$.*
   Then there exists a spatiotemporal CSS $\hat{\mathfrak{C}} := (\mathfrak{K}, (P, T, L))$ with semantic scales $\mathbb{S}_P$, $\mathbb{S}_T$ and $\mathbb{S}_L$ with object concept mappings γ_P, γ_T and γ_L, and there exists a subset $\hat{E}$ of the attribute set of the semantically derived context of $\hat{\mathfrak{C}}$ and an isomorphism f from $\mathbb{K}_E$ onto the $\hat{E}$-part $\mathbb{K}_{\hat{E}}$ of the semantically derived context $\mathbb{K}_{\hat{\mathfrak{C}}}$ of $\hat{\mathfrak{C}}$ and an (f-induced) isomorphism $\hat{f}$ from $\mathfrak{B}(\mathbb{K}_E)$ onto the concept lattice $\underline{\mathfrak{B}}(\mathbb{K}_{\hat{E}})$ such that for all $p \in P_0$ and all $t \in G_0$ and $\boldsymbol{p} := \gamma_P(p)$ and $\boldsymbol{t} := \gamma_T(t)$

$$\{\hat{f}(\gamma_E(p,t))\} = \alpha_{\hat{E}}(r_P(\boldsymbol{p}) \wedge r_T(\boldsymbol{t})) = state_{\hat{E}}(\boldsymbol{p}, \boldsymbol{t}),$$

hence the singleton consisting of the image of a state of an actual object in the given CTSOT is the $\hat{E}$-state of the corresponding realized actual object.

Proof. From the CTSOT $\mathfrak{C}$ we construct a spatiotemporal CSS $\hat{\mathfrak{C}} := (\mathfrak{K}, (P, T, L))$ where $P := P_0$, $T := G_0$ and $L := E$ and $\mathfrak{K} := (\Pi, \{P, T, L\}, (\mathfrak{B}_P, \mathfrak{B}_T, \mathfrak{B}_L), I_{\mathfrak{K}})$. The semantic scales are chosen as the nominal scales $\mathbb{S}_P := (P, P, =)$, $\mathbb{S}_T := (T, T, =)$ and $\mathbb{S}_L := (G_L, N_L, I_L)$ where $G_L := \{(e(p, t)|e \in E)|(p, t) \in \Pi\}$, $N_L := \{(e, n)|e \in E, n \in N_e\}$ and N_e is the attribute set of the scale $\mathbb{S}_e := (G_e, N_e, I_e)$ in the given CTSOT $\mathfrak{C}$. We define the incidence relation I_L by

$$I_L := \{((\tilde{e}(p,t)|\tilde{e} \in E), (e, n)) \in G_L \times N_L|e(p, t)I_e n\}.$$

The ternary incidence relation $I_{\mathfrak{K}}$ is defined as $\{((p, t), m, m(p, t))|m \in \{P, T, L\}\}$ where the values of the instance (p, t) are defined as $P(p, t) := \gamma_P(p)$, $T(p, t) := \gamma_T(t)$, and $L(p, t) := \gamma_L((\tilde{e}(p, t)|\tilde{e} \in E))$. That is visualized in Table 1:

Table 1. A data table of a CSS representing the event part of a CTSOT

instances	P	T	L
(p,t)	$\gamma_P(p)$	$\gamma_T(t)$	$\gamma_L(\tilde{e}(p, t)|\tilde{e} \in E)$

To finish the proof we define $\hat{E} := \{L\} \times N_L$ which is a subset of the attribute set of the semantically derived context of $\hat{\mathfrak{C}}$. Let $\mathbb{K}_{\hat{E}} = (\Pi, \hat{E}, \hat{J})$ be the $\hat{E}$-part of the semantically derived context of $\hat{\mathfrak{C}}$. Then the following mapping f with $f(p, t) := (p, t)$ and $f(e, n) := (L, (e, n))$ is an isomorphisms from the formal context $\mathbb{K}_E = (\Pi, N_L, I_E)$ onto $\mathbb{K}_{\hat{E}}$ since for all $(p, t) \in \Pi$ and for all $(e, n) \in N_L$ $(p, t)I_E(e, n) \Leftrightarrow e(p, t)I_e n \Leftrightarrow ((\tilde{e}(p, t)|\tilde{e} \in E)I_L(e, n) \Leftrightarrow (e, n) \in int(\gamma_L((\tilde{e}(p, t)|\tilde{e} \in E))) \Leftrightarrow (e, n) \in int(L(p, t)) \Leftrightarrow (p, t)\hat{J}(L, (e, n)) \Leftrightarrow f(p, t)\hat{J}f(e, n)$.

Let $\hat{f} : \mathfrak{B}(\mathbb{K}_E) \to \mathfrak{B}(\mathbb{K}_{\hat{E}})$ where $\hat{f}(A, B) := (fA, fB)$ be the f-induced concept lattice isomorphism. Then for all $p \in P_0$ and all $t \in G_0$ let $\boldsymbol{p} := \gamma_P(p) = (\{p\}, \{p\})$ and $\boldsymbol{t} := \gamma_T(t) = (\{t\}, \{t\})$; from the definition of a Q-state we get that $state_{\hat{E}}(\boldsymbol{p}, \boldsymbol{t}) = \alpha_{\hat{E}}(r_P(\boldsymbol{p}) \wedge r_T(\boldsymbol{t}))$. Since the $ext(r_P(\boldsymbol{p}))$ is by definition of the realization the set $\{(P, p)^{\downarrow}\} = \{(q, s) \in \Pi|P(q, s) = p\} = \{p\} \times G_0$ and the $ext(r_T(\boldsymbol{t})) = \{(T, t)^{\downarrow}\} = \{(q, s) \in \Pi|T(q, s) = t\} = P_0 \times \{t\}$. Hence $(q, s) \in (ext(r_P(\boldsymbol{p})) \cap ext(r_T(\boldsymbol{t}))$ if and only if $(q, s) = (p, t)$. Hence $\alpha_{\hat{E}}(r_P(\boldsymbol{p}) \wedge r_T(\boldsymbol{t})) = \{\gamma_{\hat{E}}(p, t)\} = \{\hat{f}(\gamma_E(p, t))\}$ since $\hat{f}$ is the f-induced concept lattice isomorphism.

That demonstrates that the states of the possibly distributed actual objects of a spatiotemporal CSS cover the states of a CTSOT as special singleton cases.

6   An Example

The following example shows a small part of some industrial data measured at a distillation column during 20 days using certain "variables" (many-valued attributes) like "input" and "pressure" as indicated in Table 2. In this simple example there is only one object, namely the distillation column, and the time variable "day" is a key. Therefore this system can be described as a CTSOT, and hence as a CSS with a derived context $\mathbb{K}$ which contains all the attributes which are used in the following.

Table 2. A data table of a conceptual semantic system

instances	distillation column	day	input	pressure
1	1	1	616	119
.	.	.	.	.
20	1	20	664	120

For that example we now visualize the information distribution and the information frequency of a Q-aspect of the attribute concept $\mathbf{c} := \mu(pressure > 120)$ which means in this distillation column "high pressure". We take Q as the set of all attributes of the concept lattice of $\mathbb{K}_Q$ in Figure 2. The extent of $\mathbf{c}$ is $A_\mathbf{c} := \{2, 4, 6, 7, 10\}$ which is partitioned in the contingents $\beta_{\mathbf{c},Q}(\gamma_Q(3)) = \{2, 6\}$, $\beta_{\mathbf{c},Q}(\gamma_Q(1)) = \{4, 7\}$, and $\beta_{\mathbf{c},Q}(\gamma_Q(11)) = \{10\}$ as represented in Figure 1 which shows the concept lattice of $\mathbb{K}_{Q \cup B_\mathbf{c}}$ where $B_\mathbf{c}$ is the intent of $\mathbf{c}$. The corresponding relative information frequencies are $2/5$, $2/5$, $1/5$; the proportions $\pi_{\mathbf{c},Q}(\gamma_Q(g))$ of the $\mathbf{c}$-instances in these three contingents are $2/6$, $2/6$, $1/3$. In Figure 2 we visualize the aspect of "high pressure" by black circles in a line diagram of $\mathbb{K}_Q$. Clearly, by definition of an aspect, "high pressure" is reported at least once in each of the concepts of that aspect, and these concepts are all subconcepts of the attribute concept $\mu(input \leq 645)$. It is obvious that we can see that information much more clear in Figure 1 using the implication "$(pressure > 120) \Rightarrow (input \leq 645)$". But the concept lattice in Figure 1 is more complicated. That argument is not really convincing in that simple example, but in general the concept lattice of $\mathbb{K}_{Q \cup B_\mathbf{c}}$ may be much larger than that of $\mathbb{K}_Q$. Therefore a visualization of aspects in a line diagram of $\mathbb{K}_Q$ can be used very effectively in practice.

A well-known example of that kind is a (conceptually interpreted) meteorological chart where the nice and simple two-dimensional map is enriched by level lines for several pressure values. Clearly, each level line, say $\{(x, y)|f(x, y) = c_0\}$ can be represented as the aspect of a concept. The same holds for sets of the form $\{(x, y)|f(x, y) \geq c_0\}$ which are often represented on a meteorological chart by oval-like figures indicating "high pressure". When these oval-like figures move over the chart they are identified from one day to the next and get names - as if being a physical object.

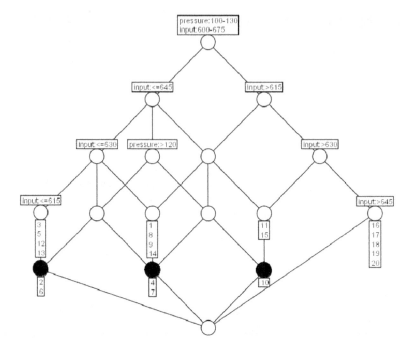

Fig. 1. Selection of the instances with high pressure

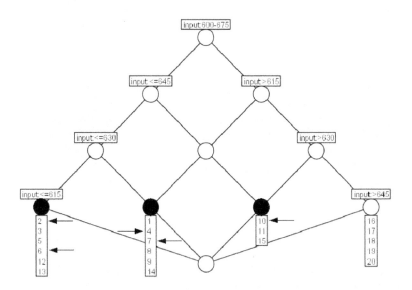

Fig. 2. In black: the aspect of "high pressure"; the arrows (→) show the five instances indicating "high pressure" and visualize the information frequencies 2/5, 2/5, and 1/5.

Clearly, such abstract objects like "high pressure" are represented in Conceptual Semantic Systems usually as concepts of semantic scales. While scale concepts in different scales cannot be compared, their corresponding realized concepts (see Def. 4) can be compared in the semantically derived context where they are connected by instances.

7 Connections to the Statistical Interpretation of Quantum Theory

Now we relate Conceptual Semantic Systems to some leading ideas in the statistical interpretation of Quantum Theory. For that purpose we quote some statements from the Nobel Lecture of Max Born on *The Statistical Interpretation of Quantum Mechanics* [Bor54]. On page 261 he wrote:

> Schrödinger thought that his wave theory made it possible to return to deterministic classical physics. He proposed (and he has recently emphasized his proposal anew's), to dispense with the particle representation entirely, and instead of speaking of electrons as particles, to consider them as a continuous density distribution $|\psi|^2$ (or electric density $e|\psi|^2$).

Following an idea of Einstein who interpreted the square of the wave amplitude of an optical wave as the probability density for the occurrence of photons Max Born [Bor54] (page 262) argued:

> This concept could at once be carried over to the ψ-function: $|\psi|^2$ ought to represent the probability density for electrons (or other particles). It was easy to assert this, but how could it be proved?

To my knowledge there is no generally accepted interpretation of the ψ-function in Quantum Theory. As a new step towards a meaningful interpretation of the foundations of Quantum Theory I plan to combine the experience of physicists as described for example in [Boh52, But99, Du01, Kuc99, Pa04] with the conceptual framework presented in this paper.

For the intended applications in Quantum Theory the definition of a Q-aspect of a concept $\mathbf{c}$ plays an important role since $\alpha_Q(\mathbf{c})$ represents nicely the idea of the set of "places" where a "distributed (quantum) object $\mathbf{c}$" has been observed. The concept lattice $\mathfrak{B}_Q$ plays the role of a general "space" into which an "object" is "embedded" by means of its Q-aspect.

Max Born [Bor54] (page 265-266) clearly distinguished between particles and objects in a similar sense as we do:

> The concept of a particle, e.g. a grain of sand, implicitly contains the idea that it is in a definite position and has definite motion. ... Every object that we perceive appears in innumerable aspects. The concept of the object is the invariant of all these aspects. From this point of view, the present universally used system of concepts in which particles and waves appear simultaneously, can be completely justified.

It was shown by the author in [Wo04c] that particles and waves can be understood as special cases of general objects, called "packets" which represent in the framework of Conceptual Semantic Systems the idea of a "wave packet" in physics. Born's "innumerable aspects" of an object are now mathematically well presented by the Q-aspects of a concept.

If we wish to "measure" the "probability that the general object $\mathbf{c}$ is at the location $\gamma_L(g)$" then we have to arrange our experiment, which yields the many-valued context of a spatiotemporal CSS, in such a way that the relative information frequency $\varrho_{\mathbf{c},L} := \varrho_{\mathbf{c},\{L\} \times N_L}$, which we call the *BORN-frequency*, can be interpreted as the intended probability, for example by choosing a "spatial uniform distribution" of the information units . Is there any other way to "measure that probability"?

Max Born finished his Nobel lecture [Bor54] (page 266) with the statement:

> The lesson to be learned from what I have told of the origin of quantum mechanics is that probable refinements of mathematically methods will not suffice to produce a satisfactory theory, but that somewhere in our doctrine is hidden a concept, unjustified by experience, which we must eliminate to open up the road.

Among the possible candidates for such hidden concepts is maybe the concept of "a continuously fine precision" which can not be justified by any experience in practice. It can be replaced in many applications by a pragmatically based usage of granularity in theory and practice.

8   Conclusion and Future Research

We have shown how the notion of a state of an actual object in a CTSOT can be generalized to the notion of a Q-state of a "distributed actual object" of a spatiotemporal CSS. That seems to be the suitable conceptual representations of the idea of "the state of a quantum mechanical system". Even more important is the definition of a Q-aspect of a concept $\mathbf{c}$ and its numerous applications in many fields where "distributed objects" are investigated. That seems to be a very powerful tool for the investigation of the foundations of Quantum Theory. First results are the formal definitions of particles and waves in Conceptual Semantic Systems and the proof that they cover the classical particles and waves in physics. The introduction of the information distribution and the (relative) frequency distribution of a concept with respect to a view Q yields a nice conceptual representation of "the probability distribution of a quantum mechanical system".

These first approaches have to be discussed in the near future with physicists and other scientist working with "distributed systems".

The application of Conceptual Semantic Systems to Conceptual Graphs and Power Context Families will be discussed in another paper.

References

[Boh52]   D. Bohm: A suggested interpretation of the quantum theory in terms of "hidden" variables. Phys.Rev. **85**, 1952.

[Bor54]   M. Born: Die statistische Deutung der Quantenmechanik. Nobelvortrag. In: [Bo63], II, 430-441. English translation: The statistical interpretation of quantum mechanics. Nobel Lecture, December 11, 1954. http://nobelprize.org/physics/laureates/1954/born-lecture.html

[Bor63]   M. Born: *Ausgewählte Abhandlungen.* Vandenhoeck and Ruprecht, Göttingen 1963.

[But99]   J. Butterfield (ed.): *The Arguments of Time.* Oxford University Press, 1999.

[Ca98]   E. Castellani (ed.): *Interpreting Bodies: Classical and Quantum Objects in Modern Physics.* Princeton University Press 1998.

[Du01]   D. Dürr: *Bohmsche Mechanik als Grundlage der Quantenmechanik.* Springer, Berlin 2001.

[GW99]   B. Ganter, R. Wille: *Formal Concept Analysis: mathematical foundations.* Springer, Heidelberg 1999; German version: Formale Begriffsanalyse: Mathematische Grundlagen. Springer, Heidelberg 1996.

[MLG03]   A. de Moor, W. Lex, B. Ganter (eds.): Conceptual Structures for Knowledge Creation and Communication. LNAI **2746**, Springer, Heidelberg 2003.

[Kuc99]   K. Kuchar: The Problem of Time in Quantum Geometrodynamics. In [But99], 169-195.

[Pa04]   O. Passon: *Bohmsche Mechanik. Eine elementare Einführung in die deterministische Interpretation der Quantenmechanik.* Verlag Harri Deutsch, Frankfurt 2004.

[PW99]   S. Prediger, R. Wille: The lattice of concept graphs of a relationally scaled context. In: W. Tepfenhart, W. Cyre (eds.): *Conceptual structures: standards and practices.* LNAI **1640**. Springer, Heidelberg 1999, 401–414.

[PCA99]   U. Priss, D. Corbett, G. Angelova (eds.): *Conceptual structures: integration and interfaces.* LNAI **2393**. Springer, Heidelberg 2002.

[Rei56]   H. Reichenbach: *The Direction of Time.* Edited by M. Reichenbach. Berkeley: University of California Press, 1991. (Originally published in 1956.)

[So84]   J. F. Sowa: *Conceptual structures: information processing in mind and machine.* Adison-Wesley, Reading 1984.

[SWi92]   S. Strahringer, R. Wille: Towards a Structure Theory for Ordinal Data. In: M. Schader (ed.): *Analysing and Modelling Data and Knowledge.* Springer, Heidelberg 1992; 129–139.

[TFr98]   G. Toraldo di Francia: A World of Individual Objects? In: [Ca98] 21–29.

[Wi97]   R. Wille: Conceptual Graphs and Formal Concept Analysis. In: D. Lucose, H. Delugach, M. Keeler, L. Searle, J. F. Sowa (eds.): *Conceptual structures: fulfilling Peirce's dream.* LNAI **1257**. Springer, Heidelberg 1997, 290–303.

[Wi00]   R. Wille: Contextual Logic summary. In: G. Stumme (ed.): *Working with conceptual structures: Contributions to ICCS 2000.* Shaker-Verlag, Aachen 2000, 265–276.

[Wi02]   R. Wille: Existential concept graphs of power context families. In: [PCA99], 382–395.

[Wi03]   R. Wille: Conceptual Contents as Information - Basics for Contextual Judgment Logic. In: [MLG03], 1–15.

[Wo00a]   K.E. Wolff: Concepts, States, and Systems. In: D.M. Dubois, (ed.): Computing Anticipatory Systems. CASYS99 - Third International Conference, Liège, Belgium, 1999, American Institute of Physics, Conference Proceedings 517, 2000, 83–97.

[Wo00b]   K.E. Wolff: Towards a Conceptual System Theory. In: B. Sanchez, N. Nada, A. Rashid, T. Arndt, M. Sanchez (eds.): Proceedings of the World Multiconference on Systemics, Cybernetics and Informatics, SCI 2000, Vol. II: Information Systems Development, International Institute of Informatics and Systemics, 2000, ISBN 980-07-6688-X, 124–132.

[Wo01]   K.E. Wolff: Temporal Concept Analysis. In: E. Mephu Nguifo et al. (eds.): ICCS-2001 International Workshop on Concept Lattices-Based Theory, Methods and Tools for Knowledge Discovery in Databases, Stanford University, Palo Alto (CA), 91–107.

[Wo02a]   K.E. Wolff: Transitions in Conceptual Time Systems. In: D.M. Dubois (ed.): International Journal of Computing Anticipatory Systems vol. 11, CHAOS 2002, 398–412.

[Wo02b]   K.E. Wolff: Interpretation of Automata in Temporal Concept Analysis. In: [PCA99], 341–353.

[Wo04a]   K.E. Wolff: Towards a Conceptual Theory of Indistinguishable Objects. In: Eklund, P. (ed.): Concept Lattices: Proceedings of the Second International Conference on Formal Concept Analysis. LNCS **2961**, Springer-Verlag, Heidelberg, 2004, 180–188.

[Wo04b]   K.E. Wolff: States, Transitions, and Life Tracks in Temporal Concept Analysis. Preprint Darmstadt University of Applied Sciences, Mathematics and Science Faculty, 2004.

[Wo04c]   K.E. Wolff: 'Particles' and 'Waves' as Understood by Temporal Concept Analysis. In: K.E. Wolff, H.D. Pfeiffer, H.S. Delugach (eds.): Conceptual Structures at Work. 12th International Conference on Conceptual Structures, ICCS 2004. Huntsville, AL, USA, July 2004. Proceedings. Springer Lecture Notes in Artificial Intelligence, LNAI **3127**, Springer-Verlag, Heidelberg 2004, 126–141.

[WY03]   K.E. Wolff, W. Yameogo: Time Dimension, Objects, and Life Tracks - A Conceptual Analysis. In: [MLG03], 188–200.

[WY05]   K.E. Wolff, W. Yameogo: Turing Machine Representation in Temporal Concept Analysis. In: B. Ganter, R. Godin (eds.): Formal Concept Analysis. Third International Conference ICFCA 2005. Springer Lecture Notes in Artificial Intelligence, LNAI **3403**, Springer-Verlag, Heidelberg 2005, 360–374.

Hierarchical Knowledge Integration Using Layered Conceptual Graphs

Madalina Croitoru, Ernesto Compatangelo, and Chris Mellish

Department of Computing Science, University of Aberdeen
King's College, Aberdeen, AB243UE

Abstract We describe the 'Hierarchical as View' approach to knowledge integration from heterogeneous sources. This is based on a novel representation called Layered Conceptual Graphs (LCGs), a hierarchical extension of conceptual graphs that address interoperability issues.
We introduce LCGs on the basis of a new graph transformation system, which could be an appropriate hierarchical graph model for applications that require consistent transformations. We highlight a new type of rendering based on the additional expansion of relation nodes.
Both the querying and integration capabilities of our approach are based on projection, an operation also defined for simple conceptual graphs. Integration is performed in a way that neither depends on the order in which the sources are combined nor on their physical availability.

1 Introduction

Combining implicit information, a *knowledge integration system* offers answers to queries that none of the isolated sources can answer on their own. Of course, the extent to which this is possible explicitly depends on the reasoning capabilities of the formalism used to represent the (combined) knowledge sources.

The typical architecture of global information systems consists of one well-designed global domain that integrates a number of wrapped sources. The integration process has been often performed describing the data sources as containing answers to views over the global schema. This first kind of approach, exemplified by the TSIMMIS system [1], is known as 'Global-As-View' (GAV). The integration process has also been performed by describing the mediated schema as containing answers to views over source relations. This second kind of approach, exemplified by the Information Manifold [2], is known as 'Local-As-View' (LAV).

While the LAV approach allows new sources to be added and removed in a modular manner, the GAV approach requires source descriptions to be modified when such changes occur. Consequently, query answering is straightforward in GAV-based systems, as the answers can be obtained by simply composing the query with the views. However, LAV-based systems require a more sophisticated form of query rewriting. An approach called 'Both As View', which is based on the use of reversible schema transformation sequences, was thus proposed in the *Automed* project [3] to combine the best of both the GAV and LAV worlds.

F. Dau, M.-L. Mugnier, G. Stumme (Eds.): ICCS 2005, LNAI 3596, pp. 267–280, 2005.
© Springer-Verlag Berlin Heidelberg 2005

In this paper, we present a formal approach to knowledge integration where the representation formalism is based on an extension of Conceptual Graphs (CGs for short). We denote such extension, which enables the creation of a hierarchical structure of CGs, a Layered Conceptual Graph (LCG for short). The hierarchical structure of LCGs is obtained by building a new layer every time new sources are integrated into the existing LCG. The alignment/articulation issues induced by overlapping knowledge are addressed by means of constraints encoded in the conceptual representation.

We introduce layered conceptual graphs in order to formalise a novel approach to knowledge integration called 'Hierarchical as View' (HAV for short). This allows for the easy addition of data sources as well as fast querying, exploiting the benefits of the Both as View approach in a structured way.

The presentation is structured as follows. Section 2 introduces the notion of layered conceptual graph and explain how LCGs are used to perform hierarchical integration. Section 3 formally describes LCGs. Section 3.2 analyses the projection operation for LCGs. Section 3.3 specifies the semantics of LCGs. Since this paper presents ongoing work, in section 4 we discuss the applications of LCGs from the interoperability perspective along with the further formalisation and practical issues to be addressed in the future.

2   Integration with Layered Conceptual Graphs

A simplified example of our knowledge integration approach based on hierarchical conceptual graphs is shown in Figure 1. At each level, several data sources

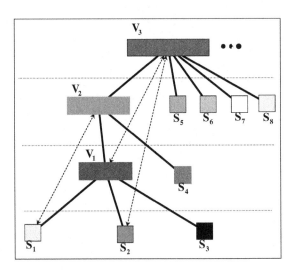

Fig. 1. Hierarchical architecture for integration

are depicted as small squares, while different articulation views are depicted as rectangles. The bidirectional links between sources and views represent pointers used for navigation. Due to the nature of the formalism employed, the final integration structure will not depend on the order of the sources to be integrated. Moreover, several sources can be integrated at the same time.

Whenever a new source is added to the existing structure, both the factual and ontological knowledge of the sources must be combined with those of the existing integrated structure. One way of solving such problem uses articulation [4]. Articulation is performed by linking two concepts that represent approximately the same notion and adding a node that denotes the two related concepts in the articulation. The new node and the link between the two concepts are collectively denoted as articulation rules. Together with the two data sources, these rule compose the articulation of the data sources.

For instance, let us consider Figure 2, originally introduced in the context of the ONION articulation system [5]. Here, the two graphs *Carrier* and *Factory*

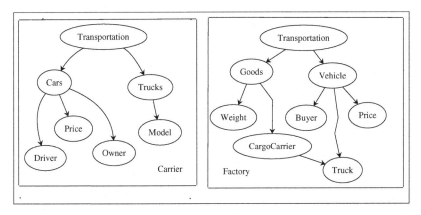

Fig. 2. The graphs describing two ontologies, Carrier and Factory

represent two simplified source ontologies, while the integration graph *Transportation* shown in Figure 3 represents their possible articulation.

The data structure discussed in this paper, namely Layered Conceptual Graphs (LCGs for short) is a rigorously defined representation formalism evolved from conceptual graphs [6]. We have introduced this new data structure to addresses the inherent ontological and factual issues of hierarchical integration. This allows us to highlight a new type of rendering based on the additional expansion of relation nodes. The idea of a detailed context of knowledge can be traced back to the definition of Simple Conceptual Graphs (SCGs) [6], to the work of [7] and to the definition of the more elaborate Nested Conceptual Graphs [8]. The querying capabilities associated with our approach are supported by the logically sound projection operation, which is defined between SCGs and

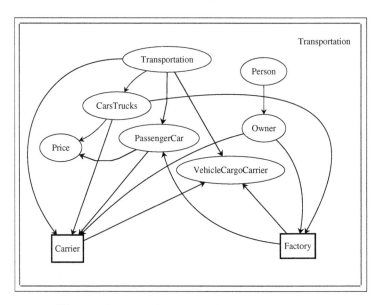

Fig. 3. The articulation ontology Transportation

LCGs. The difference between Layered Conceptual Graphs and Nested Graphs is due to the fact that nested graphs do not address the description of relations between the "zoomed" knowledge and its context. The nodes' context in nested graphs allows transition between conceptual graphs but the overall conceptual structure is no longer a properly defined bipartite graph.

3   Formalising Layered Conceptual Graphs

In this section we give the full mathematical definitions of the structures employed to perform integration, showing that our formalism meets the methodological requirements of such integration framework.

A *bipartite graph* is a graph $G = (V_G, E_G)$ with the nodes set $V_G = V_C \cup V_R$, where V_C and V_R are finite disjoint nonempty sets, and each edge $e \in E_G$ is a two element set $e = \{v_C, v_R\}$, where $v_C \in V_C$ and $v_R \in V_R$. Usually, a bipartite graph G is denoted as $G = (V_C, V_R; E_G)$. We call $G^{\emptyset}$ the empty bipartite graph without nodes and edges.

Let $G = (V_C, V_R; E_G)$ be a bipartite graph. The number of edges incident to a node $v \in V(G)$ is the degree, $d_G(v)$, of the node v. If, for each $v_R \in V_R$ there is a linear order $e_1 = \{v_R, v_1\}, \ldots, e_k = \{v_R, v_k\}$ on the set of edges incident to v_R (where $k = d_g(v)$), then G is called an *ordered bipartite graph*. A simple way to express that G is ordered is to provide a labelling $l : E_G \to \{1, \ldots, |V_C|\}$ with $l(\{v_R, w\}) = $ index of the edge $\{v_R, w\}$ in the above ordering of the edges incident in G to v_R. l is called a *order labelling* of the edges of G. We denote an ordered bipartite graph by $G = (V_C, V_R; E_G, l)$.

For a vertex $v \in V_C \cup V_R$, the symbol $N_G(v)$ denotes its neighbours set, i.e. $N_G(v) = \{w \in V_C \cup V_R | \{v, w\} \in E_G\}$. Similarly, if $A \subseteq V_R \cup V_C$, the set of its neighbours is $N_G(A) = \cup_{v \in A} N_G(v) - A$. If G is an ordered bipartite graph, then for each $r \in V_R$, the symbol $N_G^i(r)$ denotes the i-th neighbour of r, i.e. $v = N_G^i(r)$ if and only if $\{r, v\} \in E_G$ and $l(\{r, v\}) = i$.

Throughout this paper we use a particular type of subgraph of a bipartite graph: $G^1 = (V_C^1, V_R^1; E_G^1)$ is a subgraph of $G = (V_C, V_R; E_G)$ if $V_C^1 \subseteq V_C$, $V_R^1 \subseteq V_R$, $N_G(V_R^1) \subseteq V_C^1$ and $E_G^1 = \{ \{v, w\} \in E_G | v \in V_C^1, w \in V_R^1\}$. In other words, we require that the (ordered) set of all edges incident in G to a vertex from V_R^1 must appear in G^1. Therefore, a subgraph is completely specified by its vertex set. In particular, if $A \subseteq V_C$:

- The *subgraph spanned by A in G*, denoted as $\lceil A \rceil^G$, has $V_C(\lceil A \rceil^G) = A \cup N_G(N_G(A))$ and $V_R(\lceil A \rceil^G) = N_G(A)$.

- The *subgraph generated by A in G*, denoted as $\lfloor A \rfloor_G$, has $V_C(\lfloor A \rfloor_G) = A$ and $V_R(\lfloor A \rfloor_G = \{v \in N_G(A) | N_G(v) \subseteq A\}$.

- For $A \subseteq V_R$, the *subgraph induced by A in G*, denoted $[A]_G$, has $V_C([A]_G) = N_G(A)$ and $V_R([A]_G) = A$.

3.1   Layered Conceptual Graphs

The simple conceptual graphs are defined following the work of [9].

Definition 1. *A support is a 4-tuple* $S = (T_C, T_R, \mathcal{I}, *)$ *where:*

- T_C *is a finite partially ordered set (poset),* $(T_C, \leq)$, *of concept types, defining a type hierarchy which has a greatest element* $\top_C$, *namely the universal type. In this specialisation hierarchy,* $\forall x, y \in T_C$ *the symbolism* $x \leq y$ *is used to denote that* x *is a subtype of* y.

- T_R *is a finite set of relation types partitioned into* k *posets* $(T_R^i, \leq)_{i=1,k}$ *of relation types of arity* i *(*$1 \leq i \leq k$*), where* k *is the maximum arity of a relation type in* T_R. *Moreover, each relation type of arity* i, $r \in T_R^i$, *has an associated signature* $\sigma(r) \in \underbrace{T_C \times \ldots \times T_C}_{i \text{ times}}$, *which specifies the maximum concept type of each of its arguments. This means that if we use* $r(x_1, \ldots, x_i)$, *then* x_j *is a concept with type*$(x_j) \leq \sigma(r)_j$ *(*$1 \leq j \leq i$*). The partial orders on relation types of the same arity must be signature-compatible, i.e. it must be such that* $\forall r_1, r_2 \in T_R^i$ $r_1 \leq r_2 \Rightarrow \sigma(r_1) \leq \sigma(r_2)$.

- $\mathcal{I}$ *is a countable set of individual markers, used to refer specific concepts.*

- $*$ *is the generic marker to denote an unspecified concept (with a known type).*

- *The sets* T_C, T_R, $\mathcal{I}$ *and* $\{*\}$ *are mutually disjoint and* $\mathcal{I} \cup \{*\}$ *is partially ordered by* $x \leq y$ *if and only if* $x = y$ *or* $y = *$.

Definition 2. *A simple conceptual graph is a 3-tuple* $SG = [S, G, \lambda]$, *where:*

- $S = (T_C, T_R, \mathcal{I}, *)$ *is a support;*
- $G = (V_C, V_R; E_G, l)$ *is an ordered bipartite graph;*
- λ *is a labelling of the nodes of G with elements from the support S:*
  $\forall r \in V_R, \ \lambda(r) \in T_R^{d_G(r)}; \quad \forall c \in V_C, \ \lambda(c) \in T_C \times (\mathcal{I} \cup \{*\})$ *such that*
  if $c = N_G^i(r), \ \lambda(r) = t_r$ *and* $\lambda(c) = (t_c, ref_c)$, *then* $t_c \leq \sigma_i(r)$.

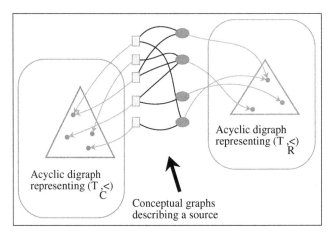

Fig. 4. A simple conceptual graph

An example of simple conceptual graph is shown in Figure 4. As suggested, a SCG provides a semantic set of pointers to two ontologies. This is the starting point of our research, in which we view an articulation as a semantic provider of pointers to sources like those portrayed in Figure 5.

In this paper the sources are represented as conceptual graphs. Therefore, in Figure 5 the squares and the ovals represent concept and relation nodes. These nodes can either be simple or complex as we will show in the next definitions. If the data sources are expressed by the means of other formalisms(e.g. RDF, OWL etc), then the sources "schema" will be represented using a CG like language, and the interaction between the sources and the integration system will be done by the means of a wrapper.

The pointers depicted in Figure 5 can either address the sources or a new hierarchy which is specific for this level of integration. In this new hierarchy, we added several new 'artificial' concepts that will be used for integration.

Note that we have associated a separate formal hierarchy, which belongs to the integrations structure as a whole, to those relations that have been specifically introduced for integration purposes. This relations were denoted as 'semantic bridges' in the ONION articulation system [5]. However, in our opinion

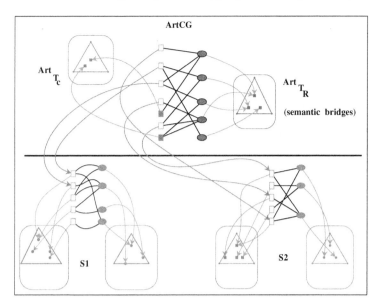

Fig. 5. Articulation principle

semantic bridges in ONION are not detailed enough. Conversely, we think that our hierarchy of integration relations makes semantic bridges easier to understand and to use from both navigational and reasoning viewpoints [10].

At each level, we highlight the specific support for that level; this support changes from level to level. In order to formalise this transition, we introduce a new concept, namely transitional description. This also facilitate the formal definition of Layered Conceptual Graphs. Note that transitional descriptions are not 'rules', as referred in the conceptual graphs literature.

Definition 3. *Let $G = (V_C, V_R; E_G)$ be a bipartite graph. A transitional description associated to G is a pair $\mathcal{TD} = (D, (G.d)_{d \in D \cup N_G(D)})$ where*

- $D \subseteq V_C$ *is a set of complex nodes.*
- *For each $d \in D \cup N_G(D)$ $G.d$ is a bipartite graph.*
- *If $d \in D$ then $G.d$ is the description of the complex node d. Distinct complex nodes $d, d' \in D$ have disjoint descriptions $G.d \cap G.d' = G^\emptyset$.*
- *If $d \in N_G(D)$ then either $G.d = G^\emptyset$ or $G.d \neq G^\emptyset$ and, in this case, $N_G(d) - D \subseteq V_C(G.d)$ and $V_C(G.d) \cap V_C(G.d') \neq \emptyset$ if and only if $d' \in N_G(d) \cap D$.*

Definition 4. *If $\mathcal{TD} = (D, (G.d)_{d \in D \cup N_G(D)})$ is a transitional description associated to the bipartite graph $G = (V_C, V_R; E_G)$, then the graph $\mathcal{TD}(G)$ obtained from G by applying $\mathcal{TD}$ is constructed as follows:*

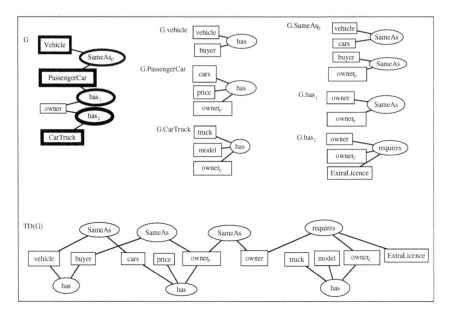

Fig. 6. LCG example

1. *Take a new copy of $\lfloor V_C - D \rfloor_G$.*

2. *For each $d \in D$, take a new copy of the graph $G.d$ and make the disjoint union of it with the current graph constructed.*

3. *For each $d \in N_G(D)$, identify the nodes of $G.d$ which are already added to the current graph (i.e. the atomic nodes of G that are neighbours of d and the nodes of $G.d'$ which appear in $G.d$). For each complex neighbour d' of d in G, add the remaining nodes of $G.d$ as new nodes in the current graph and link all these nodes by edges as described in $G.d$ (in order to have an isomorphic copy of $G.d$ as a subgraph in the current graph).*

Figure 6 suggests how transitional descriptions can be applied for the real world example introduced in Figure 3. The conceptual graph notations used are not rigorously depicted (e.g. generic markers, labelled edges, support details) the purpose of this picture being to highlight the benefits of a hierarchical approach. The graph G has $V_C = \{Vechicle, PassengerCar, owner, CarTruck\}$, $V_R = \{SameAs_G, has_1, has_2\}$ and the set of complex nodes from V_C (shown as bold rectangles) is $D = \{Vechicle, PassengerCar, CarTruck\}$. The complex relations are $SameAs_G$, has_1 and has_2. In the description of $G.has_2$, a new node appears besides already existing nodes.

Proposition 1. *If $G = (V_C, V_R; E_G)$ is a bipartite graph and $\mathcal{TD}$ is a transitional description associated to G, then the graph $\mathcal{TD}(G)$ obtained from G by applying $\mathcal{TD}$ is also a bipartite graph.*

Definition 5. *Let* $SG = [S, G, \lambda]$ *be a simple conceptual graph, where* $G = (V_C, V_R; E_G, l)$. *A transitional description associated to* SG *is a pair* $\mathcal{TD} = (D, (SG.d)_{d \in D \cup N_G(D)})$ *where:*

- $D \subseteq V_C$ *is a set of complex nodes.*
- *For each* $d \in D \cup N_G(D)$, $SG.d = [S.d, G.d, \lambda.d]$ *is a SCG.*
- *If* $d \in D$, *then* $SG.d$ *is the description of the complex node* d. *Distinct complex nodes* $d, d' \in D$ *have disjoint descriptions* $G.d \cap G.d' = G^\emptyset$.
- *If* $d \in N_G(D)$, *then either* $G.d = G^\emptyset$ *or* $G.d \neq G^\emptyset$. *In this case,* $N_G(d) - D \subseteq V_C(G.d)$ *and* $V_C(G.d) \cap V_C(G.d') \neq \emptyset$ *for each* $d' \in N_G(d) \cap D$. *Moreover,* $S_d \supseteq S \cup_{d' \in N_G(d)} S_{d'}$, $\lambda.d(v) = \lambda(v)$ *for* $v \in N_G(d) - D$ *and* $\lambda.d(v) = \lambda.d'(v)$ *for* $v \in V_C(G.d) \cap V_C(G.d')$.

Note that if $G.d = G^\emptyset$ for $d \in N_G(D)$, we have no description available for relation vertex d. This either depends on a lack of information or on an inappropriate expounding. The idea traces back to the notion of context in [6] or to the more elaborate notion of nested conceptual graph [11]. However, as our approach is not just a diagrammatic representation, the bipartite graph structure is taken into account.

Definition 6. *If* $\mathcal{TD} = (D, (SG.d)_{d \in D \cup N_G(D)})$ *is a transitional description associated to the SCG* $SG = [S, G, \lambda]$, *then the SCG* $\mathcal{TD}(SG)$ *obtained from* SG *by applying* $\mathcal{TD}$ *is* $\mathcal{TD}(SG) = [S', \mathcal{TD}(G), \lambda']$. *Here,* $\mathcal{TD}(G)$ *is the bipartite graph* $\mathcal{TD}(G)$ *obtained from* G *by applying* $\mathcal{TD}$, $S' = \cup_{d \in D \cup N_G(D)} S.d$ *and* λ' *is any legal labelling function defined on* $V(\mathcal{TD}(G))$ *which preserves the labels given to the vertices in* $V(G)$ *and* $V(G.d)$ *for all* $v \in D \cup N_G(D)$.

Using proposition 1, we can verify that $\mathcal{TD}(SG)$ is a simple conceptual graph. Note that last condition in definition 5 (concerning the support of a SCG that describes a relation vertex adjacent to at least one complex concept vertex) states that we have all the information which is needed to express the ontological and factual details which appear in this description.

The inclusion $S_d \supseteq S \cup_{d' \in N_G(d)} S_{d'}$ is conceived only as an availability of concept and relation types concerned. Moreover, the above union is a *semantic union* and problems related to compatibility (i.e. ontology alignment and/or matching [4]) must be solved by suitable constraints added in the SCG $SG.d$. The support of $\mathcal{TD}(SG)$, S' contains the supports of all the SCGs that describe the complex nodes from D. We denote these as *remote supports*, which are located in the sources of $\mathcal{TD}(SG)$. The remaining $S' - \cup_{d \in D} S.d$ contains S, the support of SG, and all the above mentioned ontological information needed to establish semantic bridges between sources. The latter kind of support is called the *articulation support* of $\mathcal{TD}(SG)$.

Definition 7. *Let* d *a non-negative integer. A layered conceptual graph (LCG) of depth* d *is a family* $\mathbf{LG} = \langle\ SG^0, \mathcal{TD}^0, \dots, \mathcal{TD}^{d-1}\ \rangle$ *where:*

- $SG^0 = [S^0, (V_C^0, V_R^0; E^0), \lambda^0)]$ *is a SCG,*
- $\mathcal{TD}^0$ *is a transitional description associated to* SG^0,

- *for each k, $1 \le k \le d - 1$, $\mathcal{TD}^k$ is a transitional description associated to*
  $SG^k = [S^k, (V_C^k, V_R^k; E^k), \lambda^k)] = \mathcal{TD}^{k-1}(SG^{k-1}).$

SG^0 is the base simple conceptual graph of the layered conceptual graph $\mathbf{LG}$ and $SG^k = \mathcal{TD}^{k-1}(SG^{k-1})$ $(k = 1, \ldots, d)$, are its layers.

In other words, if we have a interconnected world described by a SCG and if we can provide details about both some complex concepts and their relationships, then we can construct a second level of knowledge about this world, describing these new details as conceptual graphs and applying the corresponding substitutions. This process can be similarly performed with the last constructed level, thus obtaining a coherent set of layered representations of the initial world. The informal example in Figure 5 can be easily expressed as a LCG of depth 1.

We can define substructures of the LCG model which can be useful to devise a customisable and versatile functionality that deals with large graph structures.

Definition 8. *If $\mathbf{LG_1} = \{LG_1^0, \ldots, LG_1^{d_1}\}$ and $\mathbf{LG_2} = \{LG_2^0, \ldots, LG_2^{d_2}\}$ are two LCGs on a common support, then $\mathbf{LG_2}$ is a layered conceptual subgraph (LC subgraph) of $\mathbf{LG_1}$ if $d_2 \le d_1$ and $\exists k$, $0 \le k \le d_1 - d_2$, such that for all $i \in \{0, \ldots, d_2\}$: G_2^i is a subgraph of G_1^{k+i}, $D_2^i \subseteq D_1^{k+i}$ and $G_2.d$ is a subgraph of $G_1.d$ for each $d \in D_2^i \cup N_{G_2^i}(D_2^i)$.*

A simple way to obtain a layered conceptual subgraph of a given LCG, namely $\mathbf{LG} = \{LG^0, \ldots, LG^d\}$, is to consider an interval $[l..k] \subseteq [0..d]$, setting $D^k \leftarrow \emptyset$; the resulting layered conceptual subgraph is then denoted as $\mathbf{LG}_{[l..k]}$.

For integration applications, it is interesting to consider the following particular type of LC subgraph. If for each level k, $0 \le k \le d - 1$, of the LCG $\mathbf{LG} = \{LG^0, \ldots, LG^d\}$, a set FD^k of forbidden descriptions is provided, then $\mathbf{LG} - \cup_{k=0}^{d-1} FD^k$ is the LC subgraph of $\mathbf{LG}$ obtained by replacing $D^k \leftarrow D^k - FD^k$ in each $\mathcal{TD}^k$. The construction of the LC subgraph can be performed either in a top-down or in a bottom-up way, depending on how the forbidden descriptions (i.e. the not available sources) are provided.

3.2   Reasoning with Layered Conceptual Graphs

Projection [6] is the fundamental operation on simple conceptual graphs, since it can be used to define a pre-order on the set of SCGs based on the same support.

If $SG = (G, \lambda_G)$ and $SH = (H, \lambda_H)$ are two SCGs defined on the same support S, then a *projection from SG to SH* is a mapping

$$\Pi : V_C(G) \cup V_R(G) \to V_C(H) \cup V_R(H) \quad \text{such that:}$$

- $\Pi(V_C(G)) \subseteq V_C(H)$ and $\Pi(V_R(G)) \subseteq V_R(H)$;
- $\forall c \in V_C(G)$, $\forall r \in V_R(G)$ if $c = N_G^i(r)$ then $\Pi(c) = N_H^i(\Pi(r))$;
- $\forall v \in V_C(G) \cup V_R(G)$ $\lambda_G(v) \ge \lambda_H(\Pi(v))$.

If there is a projection $SG \rightarrow SH$, then SG subsumes SH (i.e. $SG \geq SH$). This *subsumption relation* is a pre-order on the set of all SCGs defined on the same support. Subsumption checking is an NP-complete problem [9].

Projection can be extended to LCGs; however, with knowledge integration in mind, we only consider the case where queries are SCGs.

Definition 9. *A descending path of length k in the layered conceptual graph* $LG = \langle SG^0, T\mathcal{D}^0, \ldots, T\mathcal{D}^{d-1} \rangle$ *is a sequence* $P = v_0, \ldots, v_k$ $(k \leq d)$,*where* $v_i \in V(SG^i)$ *and, for each i, $1 \leq i \leq k$, condition $v_i \in V(G.v_{i-1})$ holds. The last vertex of P is denoted as $end(P)$. Moreover k, i.e. the length of P, is denoted by $length(P)$. The set of all descending paths of LG is referred as $\mathcal{P}(LG)$.*

Definition 10. *Let $LG = \langle SG^0, T\mathcal{D}^0, \ldots, T\mathcal{D}^{d-1} \rangle$ be a layered conceptual graph of depth d and $SQ = (S_Q, Q, \lambda_Q)$ be a SCG such that $S_Q \subseteq S_{T\mathcal{D}^d}$. A projection from SQ to LG is a mapping $\Pi : V_C(Q) \cup V_R(Q) \rightarrow \mathcal{P}(LG)$ such that $\forall v \in V(Q)$, if $\Pi(v) = P_v$:*

- *if $v \in V_C(Q)$, then $end(P_v) \in V_C(SG^{length(P_v)}) - D^{length(P_v)}$ and if $v \in V_R(Q)$ then $end(P_v) \in V_R(SG^{length(P_v)}) - N_{SG^{length(P_v)}}(D^{length(P_v)})$;*
- *$\forall c \in V_C(Q)$, $\forall r \in V_R(Q)$ if $c = N_G^i(r)$, then $length(P_c) \leq length(P_r)$ and for each v on P_r at distance k from the start vertex of P_r such that $length(P_c) \leq k \leq length(P_r)$, we have $N_{SG^k}^i(v) = end(P_c)$.*
- *$\forall v \in V_C(Q) \cup V_R(Q)$, $\lambda_G(v) \geq \lambda_{SG^{length(P_v)}}(end(P_v))$.*

If there is projection from SQ to LG, then SQ subsumes LG ($SQ \geq LG$). With our integration objectives in mind, we preferred this (algorithmic, but somehow cumbersome) definition of projection, which mimics the GAV navigation through the mediated (i.e. articulated) schema.

3.3   Layered Conceptual Graph Semantics

The semantics associated with layered conceptual graphs are based on the semantics of conceptual graphs. A semantics Φ is provided which maps each SCG G based on a support S into a first order logic formula $\Phi(G)$ [6]. This is such that if $\Phi(S)$ is the set of formulas associated to the support S, then for any two simple conceptual graphs G and H defined on S, if $G \geq H$ then $\Phi(S), \Phi(H) \models \Phi(G)$ (soundness). Completeness, i.e. the fact that if $\Phi(S), \Phi(H) \models \Phi(G)$ then $G \geq H$, holds only if H is in *normal form*. In other words, it only holds if each individual marker appears at most once in concept node labels.

In our layered conceptual graphs formalism, we plan to use an extension of the semantics Ψ introduced in [8], for which projection is sound and complete with respect to first order logic without any restriction.

If $S = (T_C, T_R, \mathcal{I}, *)$ is a support, then a constant is assigned to each individual marker from $\mathcal{I}$, a binary predicate is assigned to each concept type from T_C, and an n-ary predicate is assigned to each relation type of arity n from T_R^n. For sake of simplicity, each constant or predicate has the same name as the support element it is associated to. Hence, if $t \in T_C$, then the binary predicate $t(y, x)$ holds. Intuitively, this means that the concept represented by node x is y and that its type is t. A set of formulae $\Psi(S)$, which corresponds to the interpretation of the partial orderings of T_C and T_R, is associated to S. For all $t, t' \in T_C$ such that $t \geq t'$, formula $\Psi(t, t') = \forall x \forall y (t'(x, y) \rightarrow t(x, y))$ is added to $\Psi(S)$. Moreover, for all $t, t' \in T_R^n$ such that $t \geq t'$, formula $\Psi(t, t') = \forall x_1 \ldots \forall x_n (t'(x_1, \ldots, x_n) \rightarrow t(x_1, \ldots, x_n))$ is added to $\Psi(S)$. This can be viewed as a single formula by taking the conjunction of all formulas introduced in it. Any simple conceptual graph G based on the support S is translated by Ψ into a formula $\Psi(G)$ constructed as follows.

- Firstly, to each concept node $c \in V_C$ a variable x_c is assigned.
- Afterwards, for each relation node $r \in V_R$ with $type(r) = t_r$, degree p and $N_G^i(r) = c_i$, $i = 1, \ldots, p$, the atom $\Psi(r) = t_r(x_{c_1}, \ldots, x_{c_p})$ is constructed.

The atom $\Psi(c) = t_c(y_c, x_c)$ is associated to each concept node $c \in V_C$ with $type(c) = t_c$. Here, x_c is the variable associated to the node c and y_c is a term built using the following rule:

If the marker of c is an individual marker, then y_c is the constant associated to this marker and if c has a generic marker in G, then y_c is a (distinct) new variable.

If $V_C = \{c_1, \ldots, c_n\}$ and m of these concept nodes have generic markers, then

$$\Psi(G) = \exists x_{c_1} \ldots \exists x_{c_n} \exists y_1 \ldots \exists y_m (\wedge_{i=1}^n \Psi(c_i) \wedge \wedge_{r \in V_R} \Psi(r))$$

The semantics Ψ^* of layered conceptual graphs is an inductive extension of the semantics Ψ of conceptual graphs. Let $\mathbf{LG} = \langle SG^0, T\mathcal{D}^0, \ldots, T\mathcal{D}^{d-1} \rangle$ be a LCG and $S_{LG} = (T_C, T_R, \mathcal{I}, *)$ be the union of the supports of its levels. A constant is assigned to each individual marker from $\mathcal{I}$, a ternary predicate is assigned to each concept type from T_C, and an $n + 1$-ary predicate is assigned to each relation type of arity n from T_R^n. As stated above, each constant or predicate has the same name as the element of the support it is associated to. If $t \in T_C$, then the ternary predicate $t(x, y, z)$ holds. Intuitively, this means that (i) at level x, y is a concept vertex, (ii) the concept represented by this vertex is z, and (iii) its type is t. Similarly, if $t \in T_R^n$, then predicate $t(x, z_1, \ldots, z_n)$ holds. This means that (i) a relation vertex on the level x exists and that (ii) the relation represented by this vertex is $t(z_1, \ldots, z_n)$.

The set of formulae $\Psi^*(S_{LG})$ is obtained as above, with the exception that the predicates associated to concept types became ternary ones in this case.

Formula $\Psi^*(LG)$ associated by Ψ^* to the LCG LG is constructed as follows. Let x_k, where $0 \leq k \leq d-1$, be the variables that represent the levels. Moreover, let $\Psi^*(SG^k)$ be the formula obtained from $\Psi(SG^k)$ by adding x_k as the first argument of each member predicate. Then,

$$\Psi^*(LG) = \exists x_0 \exists x_1 \ldots \exists x_{d-1} (\wedge_{k=0}^{d-1} \Psi^*(SG^k))$$

Using the corresponding result on conceptual graphs [8] and the definition of a transitional system, it can be proven that projection (as defined in the previous subsection) is sound and complete with respect to Ψ^*: $SQ \geq LG$ if and only if

$$\Psi^*(S_{LG}), \Psi^*(LG) \models \Psi^*(SQ)$$

Since the semantics of LCGs are similar to the semantics of nested graphs a few words to compare the two formalisms are necessary. Transitional descriptions for LCGs are a syntactical device which allows a successive construction of bipartite graphs. The knowledge detailed on a level of a hierarchy is put in context by using descriptions for relation nodes as well, while nested graphs only detail the concept nodes. Nested Graphs can be viewed as a particular instance of layered conceptual graphs when a complex relation description is the empty graph. From a semantic point of view, the logical variables that appear in the formula associated to LCGs are level related, and not context related like in nested graphs. We can say that while nested graphs provide a set of local contexts, layered conceptual graphs provide a global context for each level.

4   Discussion and Further Work

Data integration approaches must to find structural transformations and semantic mappings that lead to the correct merge of information and that allow users to query the so-called mediated schema. Articulation as a integration mechanism provides a valid solution for creating these semantic mappings.

We introduce a representation formalism that is suitable for the hierarchical integration of knowledge. The set of requirements for knowledge representation formalisms must include *(i)* the existence of a declarative semantics, *(ii)* a logical foundation, and *(iii)* the possibility of representing structured knowledge [12]. Our representation structures fulfil all these conditions in a formal way.

This paper presented ongoing work on the formalisation of layered conceptual graphs. These are defined on the basis of a new graph transformation system, which could also be an appropriate hierarchical model for real world applications that require consistent transformations such as hypermedia models.

The introduction of layered conceptual graphs is motivated by the development of an integration approach that performs combination in a structured way. However (as discussed in the previous section) several practical and algorithmic issues still need to be addressed before an implementation can be effectively finalised. A possible priority will be a methodology for the semantic integration of the different source supports.

The definition of layered conceptual subgraphs allows the run-time adaptation of the cropped mediated schema. This portion is only addressing the user needs and rights, thus reducing the search space in practical cases. Its construction relates to the Local-As-View integration methodology. Reasoning mimics the Global-As-View approach since is based on hierarchical projection.

References

[1] Garcia-Molina, H., Papakonstantinou, Y., Quass, D., Rajaraman, A., Sagiv, Y., Ullman, J.D., Vassalos, V., Widom, J.: The TSIMMIS approach to mediation: Data models and languages. Journal of Intelligent Information Systems **8** (1997) 117–132

[2] Levy, A.: The Information Manifold Approach to Data Integration. Intelligent Systems **13** (1998) 12–16

[3] Jasper, E., Tong, N., Brien, P.M., Poulovassilis, A.: Generating and optimising views from both as view data integration rules. Proc. of the 6th Baltic Conf. on Database and Information Systems (2004)

[4] Mitra, P., Wiederhold, G., Decker, S.: A scalable framework for the interoperation of information sources. In: Proc. of the 1st Semantic Web Working Symp. (SWWS'01). (2001) 317–329

[5] Mitra, P., Wiederhold, G., Kersten, M.L.: A Graph-Oriented Model for Articulation of Ontology Interdependencies. In: Proc. of the VII Conf. on Extending Database Technology (EDBT'2000). Lecture Notes in Computer Science, Springer-Verlag (2000) 86–100

[6] Sowa, J.: Conceptual Structures: Information Processing in Mind and Machine. Addison-Wesley (1984)

[7] Esch, J., Levinson, R.: An implementation model for contexts and negation in conceptual graphs. In Ellis, G., Levinson, R., Rich, W., Sowa, J.F., eds.: Proc. of the 3rd Int'l Conf. on Conceptual Structures (ICCS'95). Volume 954 of Lecture Notes in Computer Science., Springer (1995) 247–262

[8] Chein, M., Mugnier, M.L., Simonet, G.: Nested Graphs: A Graph-based Knowledge Representation Model with FOL Semantics. In Cohn, A.G., Schubert, L.K., Shapiro, S.C., eds.: in Proc. of the 8th Int'l Conf. on the principles and Knowledge Representation and Reasoning (KR'98), Morgan Kaufmann (1998) 524–535

[9] Chein, M., Mugnier, M.L.: Conceptual graphs: Fundamental notions. Revue d'Intelligence Artificielle **6-4** (1992) 365–406

[10] Compatangelo, E., Croitoru, M.: Domain knowledge articulation using integration graphs. In: in Proc. of the Eurolan W.shop on Ontologies and Information Extraction. (2003) 45–55

[11] Chein, M., Mugnier, M.L.: Positive nested conceptual graphs. In Lukose, D., Delugach, H.S., Keeler, M., Searle, L., Sowa, J.F., eds.: Proc. of the 5th Int'l Conf. on Conceptual Structures (ICCS'97). Volume 1257 of Lecture Notes in Computer Science., Springer (1997) 95–109

[12] Baader, F.: Logic-based knowledge representation. In: Artificial Intelligence Today, Recent Trends and Developments. Number 1600 in Lecture Notes in Computer Science (1999) 13–41

Evaluation of Concept Lattices in a Web-Based Mail Browser

Shaun Domingo and Peter Eklund

School of Economics and Information Systems
The University of Wollongong
Northfields Avenue, Wollongong, NSW 2522, Australia
sdom@advanced-stocktaking.com, peklund@uow.edu.au

Abstract Concept lattices assist human understanding in three ways: firstly, by collecting formal concepts that contain maximal sets of objects with shared attributes; secondly, the relatedness of concepts is revealed by providing a hierarchy of formal concepts in the information space. Finally, the concept lattice (drawn as a line diagram) reveals inferences that can automatically derive association rules. Therefore, a major hypothesis of the application of concept lattices is that they visually assist in understanding the structure of information contained within an information space. However, there has been little in the way of empirical tests to substantiate this hypothesis. This paper describes the process and results of a usability evaluation for a program called MAIL-STRAINER, a Web-based variant of the MAIL-SLEUTH program, which in turn is based on the Conceptual Email Manager (CEM).

1 Introduction

This paper presents software with a design methodology that places Formal Concept Analysis (FCA) as the core navigation and visualisation aid to manage Web-based email. The hypothesis is that if users are able to read and understand the MAIL-STRAINER line diagram, then the program serves as an appropriate tool to manage Webmail. In so far as MAIL-STRAINER is a authentic implementation of concept lattices for information visualisation, its evaluation can be used to draw conclusions about the ability of novice FCA users to interpret line diagrams.

The paper is presented in two parts. The first looks at the development of an open source Web-mail implementation based on SquirrelMail [1]: the resulting FCA Web-mail environment is called MAIL-STRAINER. To gauge how true an implementation MAIL-STRAINER is of FCA, we need to understand its pedigree. This description of the MAIL-STRAINER is intended to convince the reader it is an authenticate FCA implementation. The second part of the paper includes the presentation of results from a usability study with 16 University students. The conclusions drawn about the information visualisation aspects of MAIL-STRAINER are used to infer conclusions about the ease of use and readability of concept lattices by an audience untrained in Formal Concept Analysis (FCA).

F. Dau, M.-L. Mugnier, G. Stumme (Eds.): ICCS 2005, LNAI 3596, pp. 281–294, 2005.
© Springer-Verlag Berlin Heidelberg 2005

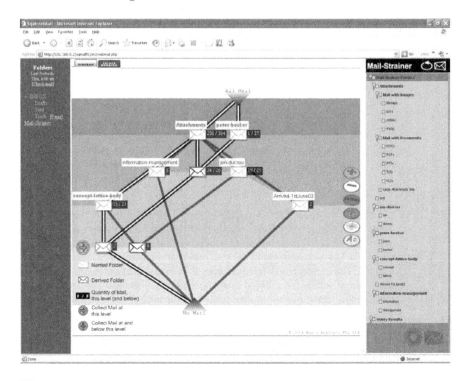

Fig. 1. A screenshot of the MAILSTRAINER program showing the re-used elements of the flash-lattice drawing program (center) and a multiple inheritance tree-widget on the right. On the left are typical Webmail controls such as those used in Mirapoint. MAILSTRAINER is built on top of the open-source SquirrelMail program and is a concept lattice-based Webmail environment.

2   Background and Survey: Email Management and FCA

Email management is a specific type of knowledge management problem. It is a sub-class of document management in that the typical methodologies for indexing, searching and modifying documents are employed. It differs from document management in that the task of processing email is based on a single identifiable process – namely, retrieving and browsing email in a single context. The research described in this paper looks at the issue of Email Management with this in mind.

Cole and Stumme [8] describe a technique for organizing email to be searched and browsed with a central visual concept lattice component. In a series of papers by Cole, Eklund and Stumme [8, 9, 6, 7], the authors show how a concept lattice can be used to navigate text document collections. More generally, information retrieval systems based on concept lattices have been experimentally evaluated in Godin et al [11] and Carpineto and Romano [2, 3] who have shown that the FCA-based technique is useful for document browsing and discovery.

2.1 MAILSTRAINER Objectives

Web-based mail (Webmail) is a popular client type for email. The hypothesis of this research is that if it is possible to build a Webmail interface using concept lattices: if users then understand how to examine concept lattices it follows that they will have a greater likelihood to search, browse and manage email in the Web-based context. One aim is to engineer an interactive environment for searching and browsing email in a Web-based context. The work therefore extends research carried out by Cole et al. [7], applying concept lattices as a lightweight visualisation package and applying a virtual file structure for managing email in an open-source, server-side environment. The functionality of a Conceptual Email Manager (CEM) [7] from a user's perspective is to:

1. *retrieve previously stored emails* – in a visual and logically clustered way;
2. *discover knowledge in the email collection* – to find collections of emails thematically linked and discover patterns of communication between different groups or to detect frequently recurring topics.

From an implementation perspective, the system functionality is to:

1. *engineer* a method for indexing mailboxes, via a protocol such as IMAP, and determine how to integrate the MAIL-STRAINER framework into an open-source Webmail package;
2. *adapt* the existing implementation of the concept lattice in the parent program – MAIL-SLEUTH – and apply it to a Web-based context, examining the index structures built in point 1.;
3. *create* the necessary navigational structures required to mimic the MAIL-SLEUTH (CEM) folder structure presented through a traditional tree widget.

2.2 FCA and Information Retrieval

In [4], Carpineto and Romano present an overview on Information Retrieval using Formal Concept Analysis (FCA). In applying FCA to the domain of email management (or document retrieval systems in general), we consider documents to be the objects, and a set of index-terms or keywords to be the attributes that describe a particular email or document. A concept within the domain of such a system would consist of a subset of the documents and a subset of the index-terms. Cole and Eklund [5] report that previous applications that have made use of FCA for Information Retrieval can be divided into two categories: (i) those that generate a large concept lattice, the number of concepts is roughly the square of the number of documents; (ii) those that use conceptual scaling and/or object zooming to reduce computational and display complexity.

Carpineto and Romano [4] explain that automatic generation of index terms can be done in several ways. While "full-text retrieval is easily handled by most statistical information retrieval systems, it is not practical for concept lattice-based applications, for which we need to generate a restricted set of terms". To overcome this problem they suggest that a five step approach involving *text segmentation, word stemming, stop wording, word weighting* (based on term frequency) and *word selection*, i.e. using some heuristic threshhold with which to select the index terms (attributes) with the highest weighting.

Cole and Stumme [8] describe the process of creating a *Hierarchy of Classifiers* for an email knowledge base. This Hierarchy of Classifiers is based on partial order theory, and allows a series of catchwords to be associated with an email. Further, the Hierarchy of Classifiers allows a specific ordering of these catchwords implied by the user's search query. The Hierarchy of Classifiers play an important role in internally mapping email concepts based on the email knowledge base.

2.3   Conceptual Email Manager (CEM)

There are several detailed papers that outline arguments for using a concept lattice within a Conceptual Email Manager (CEM). Cole and Stumme [8] tie the archetypical Conceptual Email Manager with Formal Concept Analysis, and hence, with concept lattices. They describe a Conceptual Email Manager as three high-level structures:

(i) a formal context which assigns to each email a set of catchwords;
(ii) a hierarchy on the set of catchwords in order to define an information ordering over the catchwords; and
(iii) a mechanism for creating conceptual scales which are used within a graphical interface for the retrieval of emails.

An important realisation about filing email is that it is usually multi-attributed. When we classify email under one attribute it detracts from the meaning of the email because we loose the ability to recall the email by its other attributes. This is a general problem with the traditional hierarchical filing methods. Creating folders for files for an email means it must fit into one directory or file otherwise copies need to be made. Meaning is lost as a result of decreasing the number of attributes an object is associated with. In a CEM, "the formal concepts replace the folders. In particular, this means that emails can appear in different concepts" [8]. A concept lattice is a proven visual tool for navigational aid and moreover, a good conceptual organiser.

2.4   Mail-Sleuth

In MAIL-SLEUTH, conceptual scales are determined by the user, and the realised scales are derived from them when a diagram is requested by the user. MAIL-SLEUTH is capable of restoring scales defined in previous sessions, via the storage of scales against unique names. There were four requirements in the design of MAIL-SLEUTH, namely:

1. to assist the user in editing and browsing a classifier hierarchy (the underlying folder structure): the classifier hierarchy is a partially ordered set $(M, \leq)$ where each element of M is a classifier. The structure of $(M, \leq)$ must be evident, and modifiable.

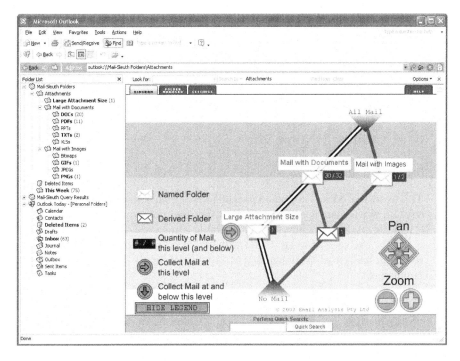

Fig. 2. The final "look" of MAIL-SLEUTH. The line diagram is highly stylized and inter-active. Folders "lift" from the view surface and visual clues (red and blue arrows) suggest the queries that can be performed on vertices. Layer colors and other visual features are configurable. Unrealized vertices are not drawn and "Derived" Virtual Folders are differ-entiated from Named Virtual Folders. A high level of integration with the Folder List to the left and the Folder Manager (see tab) is intended to promote a single-user Conceptual Information System task flow using small diagrams [13].

2. to help the client visualise and modify conceptual scales.
3. to allow the client to manage the assignment of classifiers to emails (query to structure attribution): MAIL-SLEUTH should (a) automatically associate emails with classifiers based on email content, and (b) view and modify email classifiers
4. to assist the client in searching the conceptual space of emails for both individual and conceptual groupings of emails: navigation through concept lattices derived from conceptual scales.

Eklund et al. [13] describe an independent test and survey that was carried out on MAIL-SLEUTH. In a discussion about the incorporation of the concept lattice in MAIL-SLEUTH they state that "participants were able to read the lattice diagrams without prompting" [13]. This gives partial evidence for the incorporation of line diagrams within an email management context but needs to be independently verified in the context of this study.

This research will therefore make use of concept lattices for two reasons. The first is that it is a authentic implementation of concept lattices for information visualization. The second is that, usability testing for MAIL-SLEUTH suggested user acceptance and understanding of concept lattices. This prompts speculation that if the interface of MAIL-STRAINER closely resembles that of MAIL-SLEUTH, then a similar usability rating would result.

3   Usability Evaluation

3.1   Test Methodology

This research adopts a quasi-naturalistic approach to HCI research methodology [12] where the experimenter attempts to set up an trial as close as possible to a real-life situation, while still maintaining some level of control over the experiment. In the case of MAIL-STRAINER, Webmail will look as similar as possible to the University of Wollongong's Mirapoint Webmail: a system the subjects are familiar with.

One of the driving forces behind this research is the opportunity to endorse awareness about Formal Concept Analysis and its use by way of developing a prototype that makes use of concept lattices. MAIL-STRAINER has two main user contexts. The first monitors the effect that visualisation – through the use of concept lattices – has on a user in a Web-based environment. This involves a usability analysis and exploration as to whether or not MAIL-STRAINER meets the requirements of a user configured email management tool. The second context examines how MAIL-STRAINER works within the environment in which it is embedded.

3.2   Test Subjects

The study seeks to create an understanding of users across a range of disciplines and computer usage capabilities and capacities. One of the necessary requirements in the design of MAIL-STRAINER is that it targets a wide range of users of diverse ability, culture and age. This study is aimed at sixteen test subjects sought to complete a usability test of MAIL-STRAINER based on a pre-conceived usability script. Twenty is a good sample for a study of this magnitude, and while there should have been 20 undertaking the usability sessions, 4 were unable to attend due to last minute difficulties. A commercial study might look at a sample of up to 300–400 people in a much broader demographic. Sixteen participants attended testing sessions for around 1 to 1.5 hours each. Attendance times differed substantially because users took different amounts of time to complete the script and answer the survey/feedback questions.

A reduced and specific target audience is chosen for pragmatic reasons and to eliminate several unnecessary variables. In an attempt to understand how MAIL-STRAINER affects its target audience, a sample of University students from a range of disciplines were chosen (see Fig. 5). These students were asked

	Description
1	Mail-Strainer makes it easier to locate emails - especially when the inbox is full and also makes it easier to locate specific attachments
2	The Concept Lattice is good because it shows links between folders and emails
3	A comprehensive and functional tool and fairly simple to operate, a powerful search facility with customisable queries for most useful email qualities
4	Appealing to visual senses
5	Great for searching for specific emails, when you are too lazy or have too much mail in your mailboxes
6	Easy to navigate - even for a computer dummy
7	Mail-Strainer would be an extremely useful tool, especially for someone who received > 200 emails per day (or just lots of email in general) - many people mentioned this
8	Extremely useful tool for email sorting and categorisation
9	Once over the initial learning curve, it is quite simple to use, it would be useful to someone who took the time to figure it out
10	Virtual Folders make viewing of a particular subset of emails much easier, and you don't have to remember where you put an email (like with traditional filing)
11	Allows the user to personalise their mailboxes to suit their requirements
12	Mail-Strainer is a much better way of organising email than simply placing separate emails into folders that have no connection/interaction, and provides a permanent way of searching and categorising mail without having to repeat the search query
13	Good idea to have the Mail-Strainer tree in conjunction with the lattice
14	Folder Manager is easy to use

Table 1. Mail-Strainer Positives as identified by Testing Participants

to use MAIL-STRAINER within a restricted test-bed. Within testing sessions, participants provide evaluative feedback in response to a series of survey questions (shown in Table 3), and in most cases, one-on-one interview questions as well as providing feedback through a test script (summarized in Tables 1 and 2). The test-bed includes an instance of MAIL-STRAINER embedded within SquirrelMail and an adequate set of test email. The test data is a single IMAP email account with an mailbox containing approximately 300 mail items. This mailbox is the same as that used in Access Testing's evaluation of the Mail-Sleuth product reported in [13].

To summarise previous ideas mentioned in this section, the second phase of this research focuses on hypothesis testing through qualitative cross-sectional study and assessment: encouraging users to evaluate MAIL-STRAINER as an email management tool through participation and observation. It observes group statistics that point out the participants abilities to complete the tasks in the test script. The second phase is a conversion of the results from this study into an argument in favour or against the use of Web-based email management using MAIL-STRAINER.

	Description
1	There is a large learning curve related to understanding the Line Diagram, and Mail-Strainer in general
2	The concept lattice is invaluable but needs more assistance and explanation
3	Should be able to delete emails though Mail-Strainer folders and deleting folders with a tick-box would also be a benefit
4	Creating substructures within the folder structure was an issue (programmer's fault)
5	One person said that "Mail-Strainer was annoying to navigate"
6	The "monster lattice" is a problem that occurs when there are too many structures within the folder structure. This is both a problem computationally and also in terms of readability. There should be a reduced Line Diagram (functionality wasn't turned on at testing time)
7	Would be less frustrating when the bugs are ironed out, e.g., Mail-Strainer became very slow when adding 2 extra 'queries' (structures)
8	There should be more limitations so that the items that are in trash are not indexed (this is available in Mail-Sleuth)
9	Labels on the lattice were hard to read (this was typically a user navigational issue - not understanding the components in the panel properly), but was a statement made on a number of other occasions also
10	Users couldn't understand what the bottom two buttons in the navigational component did
11	Some people didn't like the generate structure by keywords (especially having to supply 2 or more keywords)
12	Mail-Strainer may not be useful for a home user (who uses Client-Side Email Management Software)
13	Although the system is intended for people with many emails in their mailboxes, it might not be useful for this target audience since it gets slower as there are more emails

Table 2. Mail-Strainer negatives as identified by testing participants

3.3 Data Collection

This empirical study seeks to gather information in relation to the following questions. Firstly, from the test script (described in Section 3.4): 1. Can users define what each of the main components in Mail-Strainer actually does? 2. Are users able to perform email management using Formal Concept Analysis? 3. What are the positives and negatives about this tool?

Secondly, via self-assessed psychometric surveys (described in Table 3): (i) What are user's main goals in relation to using Webmail? (ii) Do Webmail users have a pre-established reliance on existing email solutions? (iii) What type of email manager are they (see Section 2.3.2)? (iv) Does Mail-Strainer actually perform the task it says it does? (v) What features presented by this tool assist in managing email? (vi) What features presented by this tool make it more difficult to manage email? (vii) Could Mail-Strainer or a related tool be used to improve Webmail in the future?

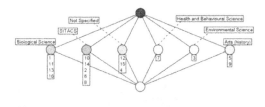
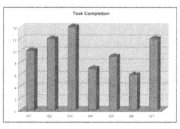

Fig. 3. (left) The concept lattice shows the distribution by discipline of the students involved in the study and (right) A bar graph displays the distribution of task completion. The number indicates those participants capable of completing any particular task in the test script. We take the word "complete" to mean both finished and accurate.

3.4   Test Script and Results

TASK1: Participants were asked to Find out how many emails there are in the 'This Week' structure. This task was relatively simple, and required the participants to be able to read straight off the line diagram, or if need be, from the contents of the Virtual Folder named "This Week". The question tests the participant's ability to determine two things: (i) if he/she understands the notion of intents and extents within the concept lattice; (ii) if he/she understands the meaning behind the numbers attached to nodes in the concept lattice.

It was worrying that only 10 out of the 16 participants could accomplish this task. On closer examination, two recipients didn't give a written answer to this question. Two participants gave the answer 9, which could very well have been accurate as a result of changing mailbox data due to real-time usability testing, and the other two gave the answer 2, which meant that they hadn't understood how contingents work within the concept lattice. Therefore, we conclude that as few as only 2 participants failed **TASK1**.

TASK2: Required the user to find all the emails that contained both .DOC and .PPT attachments, and to open the email and browse it. This question tested the participants ability to: (ii) drill down on the folder structure, and understand the nature of conceptual scales within the concept lattice; (ii) read the concept lattice and understand how to read intents and extents; (iii) Retrieve the folder contents of the folder containing this item and open the email within it.

The subject should ideally notice that they were able to retrieve all emails that contained .PPT and .DOC documents from two locations. They could have clicked on the concept "PPT" in the top-level lattice, or they could have scaled their diagram by clicking on the DOC drill-down node to reveal a reduced lattice. At this point the user should have seen that all emails in the PPT's concept also contained DOC attachments. For this task, 12 people were able to identify the email as having the subject "Secure Pay Information" sent by Collins. Of the other 4 participants, only 1 person was unable to complete this task and 3 others either completed the task and didn't respond.

TASK3: The purpose of this task was to encourage the participant to use the folder manager to Create a new virtual folder using the "Create Structures from Keywords" functionality in MAIL-STRAINER.

In 14 cases, users were able to test this functionality and create a structure relevant to the user's Inbox. The users that could perform this task also understood that they needed to refresh the left tree pane in order to reveal the new structure they had just added. In the other 2 cases there was no response from the user.

TASK4: This task required users to understand the nature of derived concepts within the concept lattice. In particular, they needed to be able to use the lattice to find "how many emails containing image attachments arrived in the last week".

Only 7 were able to come to the conclusion that there was one email containing an image attachment that arrived in the last week. There were 5 participants that said there were 2 emails that contained these two attributes because they were reading the contingent off the wrong lattice node.

TASK5: This task tested the participants ability to recall and act on what had been learned in TASK3. In this step subjects needed to count how many emails existed in their inbox, sent by "Jon Ducrou". The user was tested on their ability to: (i) create a structure from keywords, or create a folder structure manually; (ii) find this newly created structure within the folder structure within the top-level lattice, or as a conceptual scale; (iii) count how many emails in total were from the user "Jon Ducrou" using the lattice node labelling.

There were two emails embedded within this mailbox that were from "Jon", and 23 emails from the contact name "Jon Ducrou". However, we only wanted the emails that came from "Jon Ducrou" which meant that the user had to be able to understand that the bottom-most concept, or greatest subconcept contained in the lattice contained the number of emails from the contact "Jon Ducrou", namely 23.

There were a total of 6 people who said that 25 emails were from Jon Ducrou, which is correct because Jon Ducrou (the person) actually sent these extra two emails with an email-id "Jon" but incorrect in terms of the emails that came from the email identity "Jon Ducrou" read from the line diagram. The other person that got this question wrong answered with the answer 4. We cannot ascertain how the participant came to this conclusion. The 7 people who got this question wrong were still having problems reading the llne diagram at this point of the testing session.

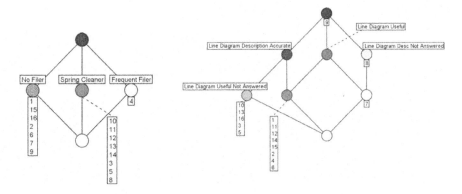

Fig. 4. (left) The concept lattice shows the distribution of email type by self-assessment of the subjects: 3 recognised email behavior types are frequent filer, spring cleaner and no filer (right) Question 9 in the self-assessment survey presented in Table 3 shows the distribution of subject responses on the key issue of the usefulness of the line diagram.

TASK6: this task was similar to that of **TASK5**. It introduces an extra concept to examine, appearing within the scale of "Mail with Documents". The user was asked to "find all emails that contain PDF and DOC attachments". This question clearly asks the user to find the emails that contain a conjunction of PDF and DOC attachments. In a similar way to **TASK5**, the user had to undertake the following tasks to retrieve the answer: seven. (i) use the top-level concept lattice (MAIL-STRAINER Folders drilldown node) to depict a count of all emails containing PDF and DOC attachments, or the easier way was to use the MAIL-STRAINER tree to find the drilldown folder "Mail with Documents"; (ii) If the scale "Mail with Documents" was used, then the user would find a much reduced lattice. The user had the simple task of then determining that there were 7 emails containing PDF and DOC attachments, not 10 or 9 or a mixture of responses as some participants recorded. All objects within the concept "PDFs" inherited the attributes "DOCs".

Two participants said that there were a total 9 documents containing PDFs and DOCs. This means one of two things. Either the 2 users were not aware of the conjunction in the question or they thought that the actual way to find a count of the emails was to add the number on the left of the label in each of the upper and lower-right concepts. 3 participants reported that there were 7 PDFs, 2 DOCs, and 1 PPTs. This answer was also completely wrong: it seemed these participants still did not understand the concept that every node in the concept lattice inherited the attributes of its intent. There were 5 participants that either could not complete the task or didn't give a response.

TASK7 This task helped the participants become familiar with manual creation of drilldown nodes and virtual folders, i.e. they were required to make

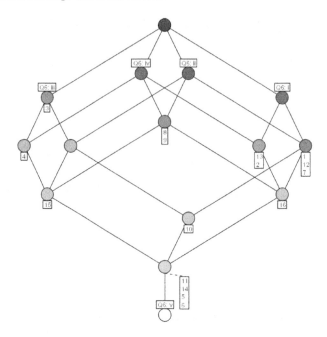

Fig. 5. This concept lattice shows the distribution by subject answer for the 10 questions under survey.

their own structures. This task was impeded severely by the restriction that participants were only able to create folders beneath the top level folder: Mail-Strainer Folders. Therefore Virtual Folders, and top-level drill-down folders were able to be created.

Of the 16 participants, 7 participants wrote down that they had achieved this task, and this can be verified. Of the other 9 participants who didn't respond 5 were verified to have played with the folder manager, attribute queries to folders, and so forth. In conclusion, a total of 12 people were able complete this task.

3.5   Outcomes

The results of the usability testing and self-assessment surveys lead to the following outcomes:

– users had definite problems reading the line diagram during the testing sessions. These problems were firstly associated with not understanding how to read them, and secondly to do with overcomplicated concept lattices which eventually overwhelmed them.
– further tutorial and help would be vital within any independent system. There is a definite adjustment phase incorporated with reading line diagrams for the novice user. MAIL-SLEUTH incorporates a good on-line help system,

1. What is your home Faculty?
2. Would you say you have a higher reliance on Web-based email (using a Web browser to access your email with things like Mirapoint) or client-based email (using a tool like Outlook Express, Outlook, Eudora) in order to send, receive and browse email?
3. For each of the following locations, please select the method (web-based, client-based or N/A) you would typically use to access your email. University? Work? Home? Friends? Interstate? Overseas? Other?
4. There are supposedly three methods of filing email:
   – Frequent Filer
   – No Filer
   – Spring Cleaner
   Which classification best describes your method for filing email?
5. On a scale of 1-10 (10 is highest), how correct is the following statement? "Mail-Strainer is a tool that gives you more power over searching and browsing your email on the Web".
6. What features of Mail-Strainer did you find helped you complete the tasks listed in the testing session script? (please circle all that apply).
   – Graphical representation of email through the Line Diagram
   – Ability to search email
   – Graphics and Design of Mail-Strainer
   – Virtual Folders
7. What features made it more difficult to complete the tasks listed in the testing session? (please circle all that apply).
   – Graphical representation of email through the Line Diagram
   – Ability to search email
   – Graphics and Design of Mail-Strainer
   – Virtual Folders
8. Which Mail-Strainer features could be used to improve Webmail for use in the future? Please select all that apply and write a couple of words suggesting why.
   – Graphical representation of email through the Line Diagram
   – Ability to search email
   – Graphics and Design of Mail-Strainer
   – Virtual Folders
9. In your own words, please describe what the Line Diagram allowed you to do, and comment on whether or not this tool helped you to complete the allocated tasks in your script. (short answer < 100 words).
10. Please rate your overall experience with MAIL-STRAINER from 1-10 (10 is best).

Table 3. MAIL-STRAINER self-assessment survey questions.

restructured using Formal Concept Analysis and line diagrams, which gives the user practice at reading Line Diagrams while they learn [10].

– there are a number of considerations to take on board in regards to the MAIL-STRAINER interface in relation to the concept lattice (especially the navigational pane).

– envisaging further tutorials, and debugged programming problems, MAIL-STRAINER would be a promising Email Management for Webmail.

4 Conclusion

This paper reports an applied software project to engineer an interactive environment for searching and browsing email in a Web-based architecture, employing MAIL-SLEUTH's component document visualisation technique using applied lattice and order theory. Therefore, the project includes elements of survey, design and software development for the purpose of building a prototype for evaluation:

MAIL-STRAINER. A resultant, yet core component of this research was to examine MAIL-STRAINER's acceptance as an Email Management tool through qualitative evaluation of users within the University of Wollongong domain, where Webmail is a core component of blended learning activities in that institution's undergraduate program. The prospects for novice Formal Concept Analysis users to read and interpret line diagrams remains promising but are not (as yet) considered overwhelming using the present tools.

References

[1] B. Bice. Building a web mail server with squirrelmail. *SysAdmin Magazine: The Journal for Uniz and Linux System Administrators*, 11(7): http://www.sysadminmag.com/articles/2002/0207/, 2002.

[2] C. Carpineto and G. Romano. Information retrieval through hybrid navigation of lattice representations. *International Journal of Human Computer Studies*, 45(5):553–578, 1996.

[3] C. Carpineto and G. Romano. A lattice conceptual clustering system and its application to browsing retrieval. *Machine Learning*, 24(2):95–122, 1996.

[4] C. Carpineto and G. Romano. *Concept Data Processing: Theory and Practice*. Wiley, 2004.

[5] R. Cole and P. Eklund. Scalability in formal concept analysis: A case study using medical texts. *Computational Intelligence*, 15(1):11–27, 1999.

[6] R. Cole, P. Eklund, and G. Stumme. CEM - a program for visualization and discovery in email. In *4th European conference on principles and practice of knowledge discovery in databases,*. Springer,, September 2000.

[7] R. Cole, P. Eklund, and G. Stumme. Document retrieval for email search and discovery using formal concept analysis. *Journal of Experimental and Theoretical Artificial Intelligence*, 17(3):257–280, 2003.

[8] R. Cole, P., and G. Stumme. CEM – a program for visualization and discovery in email. In D.A. Zighed, J. Kormorowski, and J. Zytkow, editors, *Proceedings of PKDD 2000*, LNAI 1910, pages 367–374, Berlin, 2000. Springer.

[9] R.J. Cole and P.W. Eklund. Analyzing an email collection using formal concept analysis. In *European Conference on Knowledge and Data Discovery, PKDD'99*, LNAI 1704, pages 309–315. Springer, 1999.

[10] P. Eklund and B. Wormuth. Restructuring help systems using formal concept analysis. In B. Ganter and R. Godin, editors, *Proceedings of the 3rd International Conference on Formal Concept Analysis - ICFCA'04*. Springer, February 2005.

[11] R. Godi, R. Missaoui, and A. April. Experimental comparison of navigation in a galois lattice with conventional information retrieval methods. *International Journal of Man-Machine Studies*, 38:747–767, 1993.

[12] K. Kirakowski and M. Corbett. *Effective Methodology for the Study of HCI*. Elsevier Science, 1990.

[13] J. Ducrou P. Eklund and P. Brawn. Concept lattices for information visualization: Can novices read line diagrams. In Peter Eklund, editor, *Proceedings of the 2nd International Conference on Formal Concept Analysis - ICFCA'04*. Springer, February 2004.

D-SIFT: A Dynamic Simple Intuitive FCA Tool

Jon Ducrou[1], Bastian Wormuth[2], and Peter Eklund[1]

[1] School of Economics and Information Systems
The University of Wollongong
Northfields Avenue, Wollongong, NSW 2522, Australia
jrd990@uow.edu.au, peklund@uow.edu.au
[2] Darmstadt University of Technology, Department of Mathematics,
Schloßgartenstr. 7, 64289 Darmstadt, Germany
bastian@wormuth.info

Abstract This paper introduces D-SIFT, a Web-based browser application that provides untrained users in Formal Concept Analysis with practical and intuitive access to core analysis functionality in Formal Concept Analysis. D-SIFT is an information systems architecture that supports natural search processes over a predefined database schema and its attribute values. This enables the user to build concept lattices interactively through the selection and refinement of dynamic definitions of search boundaries (via interaction with an object "zoom" feature), and dynamic selection of search scales (via interaction with an attribute "filter" feature), based on the attribute values contained within the database. The paper investigates the claim that D-SIFT systems are an advance on the search and analysis paradigm of the Toscana-system workflow. In detail, the paper presents the architecture of the D-SIFT browser and illustrates the resulting D-SIFT-systems on an example database. The two examples illustrate the generality of system integration outcomes from D-SIFT. The Conceptual Information Systems that result from applying the D-SIFT architecture present a new workflow for building and interacting with Formal Concept Analysis-based information systems. The workflow more closely aligns with dynamic schema interaction, a popular technique used in conceptual modeling.

Introduction

This paper presents a new application framework for Formal Concept Analysis (FCA). The initial idea behind the framework is the simplification of existing application development frameworks for FCA, in particular the way humans process standard searches in FCA. The software prototype – called D-SIFT– consolidates various features that have been introduced by other applications [1, 2, 3].

One of the new features of D-SIFT is highlighting the filter and ideal of any concept within the line diagram, and the dynamic creation and manipulation of concept lattices. This allows the user to add and remove attributes from the displayed concept lattice according to his current preferences. D-SIFT extends the

F. Dau, M.-L. Mugnier, G. Stumme (Eds.): ICCS 2005, LNAI 3596, pp. 295–306, 2005.
© Springer-Verlag Berlin Heidelberg 2005

classical features of FCA software with so-called *mandatory attributes*. Similar to zooming in TOSCANAJ, the user is able to restrict the displayed object set by the selection of mandatory attributes. The resulting concept lattice is limited to objects which share these attributes. The core difference to zooming in TOSCANAJ and its predecessors is the way the user selects the bounding attributes. This process in D-SIFT more closely resembles iterative search in information retrieval, whereby the user starts from one or two keywords and progressively refines the result set by the addition of further (or different) keywords

D-SIFT is also more easily accessible as a platform than existing FCA frameworks. The required plug-ins used are provided in standard configurations of most Web browsers, and the underlying database complies with the CSV file format (text files with comma-separated entries). These are easy to create and export from most applications. D-SIFT also follows the roles introduced in [4]; the separation of domain expert and the user of the Conceptual Information System is even stronger in D-SIFT, as the application does not provide the possibility for changes on the conceptual level of the data – such actions must be performed within the application that originated the CSV files on which D-SIFT operates.

Creating a Conceptual Information System from a Database

D-SIFT takes a user-supplied *comma separated values* database (CSV) and provides an interface to the database as a conceptual information system. For this reason the input format of D-SIFT closely aligns with a typical export format from a relational database management system (RDBMS) and common applications like Excel and OpenOffice.

The CSV format is simple, easy to read and edit. It is a common optional output format for most modern and legacy applications and database systems. The standard style for a CSV is to split text data into columns (or fields) separated by commas, and rows (or records) separated by newline characters. The first line (or record) is often used as a comma separated list of the column headings for the remaining records. CSV files are also forced to contain only data that can be expressed as text; this caters to the input requirements of D-SIFT.

To translate the CSV database into a Conceptual Information System, the user indicates how D-SIFT should treat each field. This requires the user to indicate a field which is each entry's identifier (entity in RDBMS modeling terms) and then group the remaining fields into nominal or numerical scale models.

In order to extract objects with meaningful names, the user identifies the field which provides an identifier for the database (e.g. a candidate key such as *name* in a database of people). Nominal data, in FCA terms, is usually text (e.g. *names* or *locations* in the people database), and sometimes represents boolean values (e.g. attributes such as *gender* or attributes with values such as *yes/no*). Numerical data is represented by numbers which, over the scope of the entire field, have some form of ordering (e.g. a schema attribute such *length in meters* with some entries longer than others).

There are instances where database attributes consisting of numeric data should not be scaled numerically; for example identifiers such as social security numbers, which may or may not be indicative of an order over the data values. D-SIFT also gives the option to drop fields that are not of interest to the user, by tagging those fields (e.g. *comment* or *ID* fields).

Interaction with the CSV file described to this point in the text allows D-SIFT to collect enough information to construct the context and scale information for the Conceptual Information System.

The user can now browse the Conceptual Information System via the D-SIFT interface. To build a query, query elements are added to either the 'Zoom' or 'Filter' lists. Query elements are made up of one or more *nominally-scaled* or *numerically-scaled* attributes. Nominally-scaled attributes comprise attribute groups and an attribute value. Numerically-scaled attributes comprise an attribute group, a *size* and an *order*. The size of a numerically scaled attribute can be thought of as the number of intervals which will be produced, while the order specifies the way in which the values should be compared. The orders are of three types, *Ordinal Up*, *Ordinal Down* and *Interordinal*. Ordinal Up and Down correspond to comparisons based on $\geq$ and $\leq$ respectively. Interordinal generates both Ordinal Up and Down simultaneously(See Fig. 1).

If the database contains columns which are not nominal and not numerical, they need to be scaled before the database is uploaded to D-SIFT. In this case the author of the CSV file needs to apply the process of conceptual scaling to the columns which cannot be handled with nominal, ordinal or interordinal scales. Conceptual scaling processes the values of a many-valued attribute and resolves the entries to several single-valued columns. After this process, D-SIFT can access all information hidden in the database. More on conceptual scaling can be found in [4] and [5].

A Typical Process of Searching and Exploring

While aiming to increase interface intuition and decrease the complexity of existing FCA application frameworks, the authors of D-SIFT intended to meet two of Wille's dimensions of Conceptual Knowledge Processing, namely **exploring** and **identifying** [6]. A typical process of exploring or identifying always starts with the aim to obtain an overview of a domain. In our example, a database concerning cellphones, we imagine a potential customer of a new phone. The user may have a rough idea of the technical features but no understanding which of these he really needs or wants. The case scenario follows the user's looking at all the features – or specifying known features. After obtaining an overview of the data, the user can sort the features into those that are essential and the remaining features as softer constraints on the search. The user may have already encountered dichotomous features, but not knowing which to eliminate may continue to use both. Step-by-step the user will make decisions and compromises before

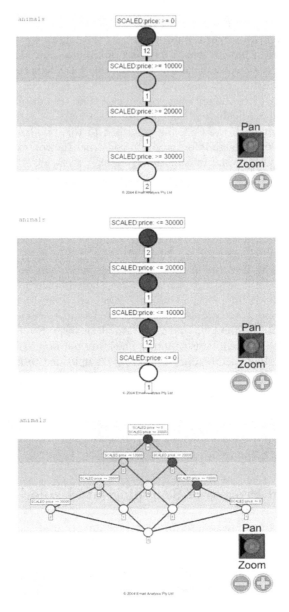

Fig. 1. Concept lattices showing the different types of numeric scaling available via the D-SIFT interface are (from top to bottom) Ordinal Up, Ordinal Down and Interordinal. All are shown with a size of 3.

selecting the phone with the features that he desires most and satisfy the search criteria. The last part of the search process will require many comparisons and iterations when exploring the information landscape with multiple dichotomous attributes. D-SIFT follows this basic process of searching and identifying. The user can start with 'all', an **exploration** of the data by selecting all features he is interested in. The resulting diagram presented by D-SIFT might be large, but the diagram provides instant access to an overview of the data; the user does not need to peruse tables or lists. The **identification** of features that are distinguishing is simple – once the user becomes familiar with interpreting lattices, he can easily select certain attributes from the top levels of the diagram, change them to mandatory attributes and then pursue his search on a simplified, smaller and less complex diagram. Here we also add the dimension **comparison** to Wille's tasks in Conceptual Knowledge Processing, because the diagram can always be used to compare the existence (or otherwise) of certain combinations of features.

Alternatively, the user can add and promote features from *filter* to *zoom* as a way of refining results. The user starts with two or three of the most desired features and uses the resulting diagram to decide whether to change attributes that are filter to zoom, or add additional filter attributes. This process of promoting filtered attributes to zoomed attributes reduces the size of the object set until the user reaches a point where the uncertainty of a decision can be minimized. By filtering only a few attributes at a time, this method produces smaller line diagrams which increases the ease of **identification** and makes **exploration** possible via incremental changes through interaction.

Case Scenario One: The Exploration Method

We now demonstrate the ideas described in the previous section with respect to a more concrete interaction scenario. In this scenario, the user knows every feature considered important (and desirable) in a new cellphone. In this case scenario the user wants:

> GPRS support
> Infrared capabilities
> Built-in MP3 player
> Built-in organiser
> Vibration alert
> Voice-dial
> WAP support

The user adds all the corresponding attributes as 'filter' attributes. The resulting line diagram from these filter attributes is too large for the user to make an instant decision, but the line diagram gives an overview of the search space and it is possible to conclude the following from it:

1. The top-most concept has a contingent of 60, therefore there are 60 phones with none of the desired features.
2. The bottom-most concept has an empty extent, therefore there is no phone with all desired features.

3. The attributes `vibe:yes`, `voicedial:yes` and `wap:yes` are most common in this diagram[3].

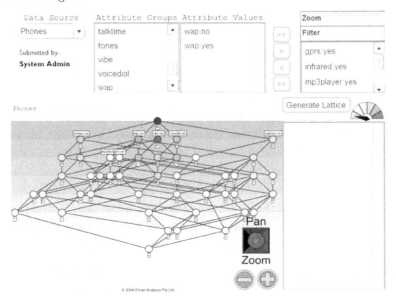

This knowledge leads the user to zoom on the three common attributes, which would seem a good way to reduce the complexity of the data while still maintain the majority of the phones.

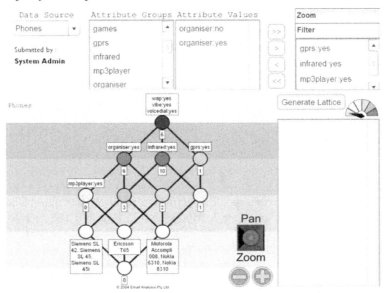

---

[3] Recognizable by the fact they are darker in colour - indicating a large extent compared to other concepts. This is the coloring style used in [7].

The result above shows that at least one desired feature can not be kept, and that the selection is from 7 phones in 3 groups – each group has one of the desired features missing.

- The *Siemens SL 42*, *Siemens SL 45* and *Siemens SL 45i* do not have GPRS Support.
- The *Ericsson T65* does not have infrared capabilities.
- The *Motorola Accompli 008*, *Nokia 6310* and *Nokia 8310* do not have a built-in MP3 player.

At this point the user could decide that infrared capability is the least desired feature and opt for the *Ericsson T65* as the phone to purchase.

Case Scenario Two: Attributes Addition Method

The user knows that two things he definitely wants in a cellphone are predictive text and infrared capabilities. He starts the search and adds `t9dict:yes` and `infrared:yes` as filter attributes.

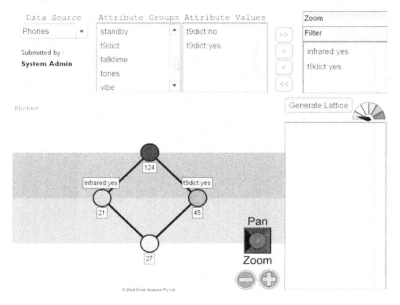

This produces the simple lattice above showing 27 phones on the bottom concept. This means there are 27 phones with both predictive text and infrared capabilities. The list of 27 phones is too large for the user to reach a decision straight away so he promotes `t9dict:yes` and `infrared:yes` to zoom attributes. The user decides that an organiser and a long stand-by time are also important features which would influence his purchase decision, so adds the corresponding attributes as filters on the data. When adding stand-by time – the aim being to emphasis phones with a greater stand-by time – he configures the `stand-by` attribute to be 'Ordinal Up' resulting in the following diagram.

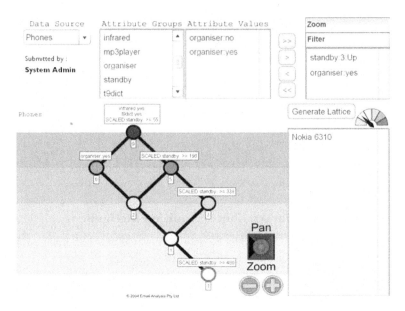

After looking at the generated lattice above, the user decides enough phones come with an organiser to warrant adding `organiser:yes` as a 'zoom' attribute. He realizes that a long stand-by time might come at the cost of increased phone weight. To ensure that the phone he gets is not too heavy for his needs, he adds the weight with the order 'Ordinal Down' so that lighter phones are emphasised resulting the in following diagram.

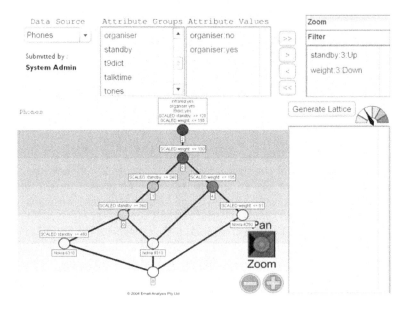

The diagram above allows the user to quickly choose an optimum weight/ stand-by time combination. It is easy to see in the above diagram that the phone most suitable to the requirements specified is the *Nokia 8310*.

Mathematical Characterization

In the following section we presume a basic knowledge of Formal Concept Analysis. For a introduction to FCA see [8], further mathematical foundations are given in [9].

The mathematical base for D-SIFT is already covered by the formal description of a Toscana System, the Conceptual Data System (CDS). CDSs are formalized in [2, 4], so we will give the most important definitions from Hereth Correira and Kaiser to provide D-SIFT with a formal base.

The creation of a Conceptual Information System always starts with the agreement on a formal context [4], in our case given by the uploaded CSV file. The formal context can consist of single-valued or many-valued attributes; therefore we recall the definition of a many-valued context as follows:

Definition 1 (many-valued context): A *many-valued context* is a structure $\mathbb{K} := (G, M, \bigcup_{m \in M} W_m, I)$ that consists of the set of *objects* G, the set of *attributes* M, a set of *values* $W := \bigcup_{m \in M} W_m$ and a ternary relation $I \subseteq (G \times M \times W)$ between G, M and W where

$$(g, m, w_1) \in I \text{ and } (g, m, w_2) \in I \Rightarrow w_1 = w_2 \in W_m.$$

The relation $(g, m, w) \in I$ is also written as $m(g) = w$ and stands for "the *attribute* m has the *value* w for the *object* g". By W_m we denote the potential set of values of an attribute m, while $m(G)$ is the actual set of values occurring in the context.

A *conceptual scale* for a subset of attributes of $\mathbb{K}$ is given by:

Definition 2 (conceptual scale): Let $\mathbb{K} = (G, M, W, I)$ be a many-valued context and $N \subseteq M$. Then we call a formal context $\mathbb{S}_N := (G_N, M_N, I_N)$ *conceptual scale* if $\{(m(g))_{m \in N} \mid g \in G\} \subseteq G_N \subseteq \bigtimes_{m \in N} W_m$. We say that $\mathbb{S}_N$ scales the attribute set N. A family of *conceptual scales* $(\mathbb{S}_{N_j})$ scales $\mathbb{K}$ if every *conceptual scale* $\mathbb{S}_{N_j}$ scales the attribute set $N_j \subseteq M$.

The conceptual scale and the corresponding line diagram appear as on entity to the user [2]. Formally we must differentiate between such a conceptual scale and its graphical representation, the line diagram. This difference is further explained in [2] using a mapping λ, that maps every conceptual scale to a corresponding line diagram. The diagram map λ is defined in [2] as well. We will use the diagram map λ to proceed with the conceptual schema:

Definition 3 (conceptual schema): Let $(\mathbb{S}_{N_j})_{j \in J}$ be a family of conceptual scales and let $(\lambda_j)_{j \in J}$ be a family of diagram maps where $dom(\lambda_j) =$

$\mathfrak{B}(\mathbb{S}_{N_j}) \cup \prec_j$. Then we call the vector $\mathcal{S} := (\mathbb{S}_{N_j}, \lambda_j)_{j \in J}$ *conceptual schema*. We say that a *conceptual schema* $\mathcal{S}$ and a many-valued context $\mathbb{K}$ are *consistent* if $(\mathbb{S}_{N_j})_{j \in J}$ scales $\mathbb{K}$.

The zooming realized in D-SIFT is slightly different from zooming implemented in Toscana Systems where the user bounds the object set through the selection of a concept as the point at which to zoom. In Toscana Systems, the selection of a concept chooses the contingent or the full extent of the concept. The current version of D-SIFT works differently: the zooming filter is defined by the selection of attributes (resp. attribute values in the many-valued case) from a list and these attributes are taken as mandatory for the set of displayed objects. The various diagrams that can be created by the combinations of constraints and selections of attributes can be captured by so called states introduced in [2]. The formal definitions from [2] can be used to develop D-SIFTs zooming process:

Definition 4 (state, initial state): Let $\mathcal{S} := (\mathbb{S}_{N_j}, \lambda_j)_{j \in J}$ be a conceptual schema consistent with a many-valued context $\mathbb{K} := (G, M, W, I)$. Then a *state* is a triple $s := (\sigma, F_1, F_2)$, where $F_1 \subseteq F_2 \subseteq G$ and $\sigma := (j_i)_{i=1}^n$ with $j_i \in J$ for $i \in \{1, ..., n\}$ and $i \neq k \Rightarrow j_i \neq j_k$. F_1 is called *exact zooming filter* and F_2 is called *full zooming filter*. An *initial state* is a state where $F_1 = F_2 = G$.

The core difference in the application of the structure of the Conceptual Data System to D-SIFT is in the understanding of *states*. States in Toscana Systems are based on conceptual scales, normally having several attributes (or values of many-valued attributes). The states in D-SIFT are based on scales normally containing a *single* attribute. The combination of several states is still made possible. However this represents a key difference from the processing of states in D-SIFT compared to Toscana Systems. In Toscana Systems the user works with existing predefined diagrams, a user of D-SIFT is changing the current diagram in order to suit expectations and constraints. The original conceptual schema of D-SIFT therefore consists of a set of scales which are all single-attribute scales. The displayed diagram is thus based on one or more scales, not just the graphical representation of a single conceptual scale. We therefore make use of conceptual scales and states as defined in [2] and define a display state for D-SIFT as follows:

Definition 5 (display state): Let $\mathcal{S} := (\mathbb{S}_{N_j}, \lambda_j)_{j \in J}$ be a conceptual schema consistent with a many-valued context $\mathbb{K} := (G, M, W, I)$. Then a *display state* is a tuple $d := (O, F)$ with $O \subseteq G, F \subseteq M$ and $Z' \cap F = \emptyset$. F are the attributes selected for filtering and $O = Z'$ where Z are the attributes selected as full zooming filter. These appear as label of the top element of the diagram representing the concept lattice $\mathfrak{B}(O, Z \cup F, W, I)$.

The display state integrates D-SIFT into the formalisation of Conceptual Information System as used and known by today.

Further Research

The final version of the user interface for the selection of mandatory attributes is planned to be similar to TOSCANA and TOSCANAJ where by clicking a concept thereby selects the concept's intent as a restriction on the objects. This represents a minor implementation extension to the existing D-SIFT.

Furthermore, we are investigating the possibility of using the human input coded in conceptual scales from already existing Toscana Systems to support the user search and exploration. After parsing the .CSX file of a Toscana System, D-SIFT could "offer" groups of attributes as in Toscana Systems. Then the interaction can start from a given diagram, extending and changing it using the existing dynamic creation features of D-SIFT.

The last issue of further research addresses the use of the multi-context as introduced in [10, 11]. On the basis of several contexts sharing sets of attributes or objects, the user would be enabled to "jump" from the perspective of one formal context to the corresponding perspective in another context by the use of coherence mappings.

Conclusion

This paper has presented the architecture of the D-SIFT browser and illustrates the resulting D-SIFT-systems on two case scenarios against a database of cellular phones. The two examples demonstrate the generality of system integration outcomes from D-SIFT. The Conceptual Information Systems which result from applying the D-SIFT architecture present a new workflow for building and interacting with Formal Concept Analysis-based information systems. The workflow more closely aligns with dynamic schema interaction used in conceptual modeling.

Acknowledgement This research results from a collaboration between the authors supported by an ARC International Linkage Grant and the DFG.

References

[1] Peter Becker, Richard J. Cole: Querying and analysing document collections with Formal Concept Analysis. In: Proceedings of the 8th Australasian Document Computing Symposium, Canberra (2003)

[2] Hereth Correira, J., Kaiser, T.B.: A Mathematical Model for TOSCANA-Systems: Conceptual Data Systems. In Eklund, P., ed.: Concept Lattices, Heidelberg, Berlin, New York, Springer (2004)

[3] Becker, P.: Multi-dimensional Representations of Conceptual Hierarchies. In Stumme, G., Mineau, G., eds.: Proceedings of the 9th International Conference on Conceptual Structures, Laval (2001)

[4] Kaiser, T.: Conceptual Data Systems - Providing a Mathematical Basis for TOSCANA-Systems (2003) Diploma thesis, University of Technology Darmstadt.

[5] Wormuth, B.: A Conceptual Information System for Topic Maps (2004) Diploma thesis, University of Technology Darmstadt.

[6] Wille, R.: Conceptual landscapes of knowledge: A pragmatic paradigm for knowledge processing. In Gaul, W., Locarek-Junge, H., eds.: Classification in the Information Age, Springer (1999) 344–356

[7] Becker, P., Hereth, J., Stumme, G.: ToscanaJ - an open source tool for qualitative data analysis. In: Advances in Formal Concept Analysis for Knowledge Discovery in Databases. Proc. Workshop FCAKDD of the 15th European Conference on Artificial Intelligence (ECAI 2002). Lyon, France. (2002)

[8] Wille, R.: Introduction to Formal Concept Analysis. In Negrini, G., ed.: Modelli e modellazione, Models and Modelling, Roma, Consiglio Nazionale delle Ricerche, Instituto di Studi sulle Ricerca e Documentazione Scientifica (1997)

[9] Ganter, B., Wille, R.: Formal Concept Analysis: Mathematical Foundations. Springer, Berlin (1999)

[10] Dörflein, S.K., Wille, R.: Coherence Networks of Concept Lattices: The Basic Theorem. Number 3403 in LNCS, Heidelberg - Berlin - New York, Springer (2005) to appear in the proceedings of the ICFCA05.

[11] Wille, R.: Conceptual Structures of Multicontexts. In Eklund, P.W., Ellis, G., Mann, G., eds.: Conceptual Structures: Knowledge Representation as Interlingua. Number 1115 in LNAI, Heidelberg - Berlin - New York, Springer (1996) 23–29

[12] Wille, R., Zickwolff, M.: Grundlagen einer Triadischen Begriffsanalyse. In Stumme, G., Wille, R., eds.: Begriffliche Wissensverarbeitung: Methoden und Anwendungen, Heidelberg - Berlin - New York, Springer (2000) 125–150

[13] Rock, T., Wille, R.: Ein TOSCANA—Erkundungsystem zur Literatursuche. In Stumme, G., Wille, R., eds.: Begriffliche Wissensverabeitung: Merthoden und Anwendungen, Berlin-Heidelberg, Springer (2000) 239–253

[14] Dieberger, A., Dourish, P., Höök, K., Resnick, P., Wexelblat, A.: Social navigation: Techniques for building more usable systems. ACM Transactions on Human-Computer Interaction 7 (2004) 26–58

[15] Horvotz, E., Kadie, C., Paek, T., Hovel, D.: Models of attention in computing and communication: From principles to applications. CACM 46 (2003) 52–59

[16] Carpineto, C., Romano, G.: Concept Data Processing: Theory and Practice. Wiley (2004)

[17] Carpineto, C., Romano, G.: Order-theoretical ranking. Journal of the American Society for Information Sciences (JASIS) 7 (2000) 587–601

[18] Cole, R., Eklund, P., Stumme, G.: Document retrieval for email search and discovery using formal concept analysis. Journal of Experimental and Theoretical Artificial Intelligence 17 (2003) 257–280

[19] Cole, R., Eklund, P.: Browsing semi-structured web texts using formal concept analysis. In: Proceedings 9th International Conference on Conceptual Structures. Volume 2120 of LNAI., Berlin, Springer (2001) 319–332

[20] Eklund, P., Ducrou, J., Brawn, P.: Concept lattices for information visualization: Can novices read line diagrams. In Eklund, P., ed.: Proceedings of the 2nd International Conference on Formal Concept Analysis - ICFCA'04, Springer (2004)

[21] Colomb, R.: Information Spaces: the Architecture of Cyberspace. Springer (2002)

Analyzing Conflicts with Concept-Based Learning

Boris A. Galitsky[1], Sergei O. Kuznetsov[2] and Mikhail V. Samokhin[2]

[1] School of Computer Science and Information Systems
Birkbeck College, University of London
Malet Street, London WC1E 7HX, UK
galitsky@dcs.bbk.ac.uk

[2] All-Russian Institute for Scientific and Technical Information (VINITI),
Usievicha 20, Moscow 125190, Russia
serge@viniti.ru, samohin_m@mtu-net.ru

Abstract. A machine learning technique for handling scenarios of interaction between conflicting agents is suggested. Scenarios are represented by directed graphs with labeled vertices (for mental actions) and arcs (for temporal and causal relationships between these actions and their parameters). The relation between mental actions and their descriptions gives rise to a concept lattice. Classification of an undetermined scenario is realized by comparing partial matchings of its graph with graphs of positive and negative examples. Developed scenario representation and comparative analysis techniques are applied to the classification of textual customer complaints.

1 Introduction: Reasoning with Conflict Scenarios

Scenarios of interaction between agents are an important subject of study in Artificial Intelligence. An extensive body of literature addresses the problem of logical simulation of agents' behavior, taking into account their beliefs, desires and intentions [1]. A substantial advancement has been achieved in building the scenarios of multiagent interaction, given properties of agent including their attitudes. Current approaches to the multiagent systems are either based on logical deduction [2,16] or simulation [4,14]; means of automated comparative analysis are still lacking [9].

In the former case, the sequence of mental states of agents is deduced from their initial mental states and initial attitudes. Deductive reasoning about actions and the logic with agents' attitudes as modalities are the most popular means to yield sequences of mental states of agents [20].

In the latter case, the system imitates the decision-making of agents, choosing the best action for each agent at each step, taking into account its current intentions, beliefs and desires, as well as those of others. Having the preference relation on the set of resultant states, each agent selects an action that is expected to lead to the most desired state [4].

However, a general framework to reuse the experience accumulated in previous scenarios of multiagent interaction has not been developed. For effective building and

F. Dau, M.-L. Mugnier, G. Stumme (Eds.): ICCS 2005, LNAI 3596, pp. 307-322, 2005.
© Springer-Verlag Berlin Heidelberg 2005

predicting of interaction between agents, it is helpful to augment reasoning and/or simulation with machine learning [9,15,18,19]. It would reduce the number of possible agents' actions at each step, taking into account how these agents acted in previous cases.

Recently, the issue of providing BDI (Belief-Desire-Intention) agents [1] with machine learning capabilities attracted interest; an application domain such as agents for intelligent information access was considered in [8]. Nevertheless, a BDI-based machine learning framework for operating with scenarios of inter-human interactions was not developed yet. A number of case-based reasoning approaches have been suggested to treat the scenarios of interaction between BDI agents [15,18,19]; however, description of agents' attitude is reduced to their beliefs, desires and intentions in these studies. Indeed, behavior of real-world conflicting agents is described in a richer language using a wide number of mental entities including *pretending, deceiving, offending, forgiving, trust*, etc.

In this paper we build the representation machinery for conflict scenarios and propose a simple machine learning technique for classifying scenarios of multiagent conflict. This technique can be implemented in a stand-alone mode or used in combination with deductive reasoning or simulation.

Multiagent conflict is a special case of scenarios where the agents have inconsistent goals and a negotiation procedure is required to achieve a compromise [14]. In this paper we discover that following the logical structure of how negotiations are represented in text, it is possible to judge about consistency of this scenario [6].

Scenarios suggest the usage of complex data structure. In this paper we employ labeled directed acyclic graphs with arcs for describing interaction of two parties in a conflict, thus being within the standard concept graph representation [17]. A learning model needs to be focused on a specific graph representation for these conflicts. The learning strategies used here are based on ideas similar to that of Nearest Neighbors (see, e.g., [13]), case-based [10,12] and concept-based learning [7,11] or JSM-method [3]. Having defined scenarios and the operation of finding common subscenarios, we use the Nearest Neighbors and concept-based learning approach to relate a scenario to the class of valid or invalid scenarios.

The paper is organized as follows. The introduction of the domain of conflict scenarios is followed by a formal treatment of mental actions and defining a conflict scenario as a graph with vertices labeled by mental actions. Having defined the similarity operation on graphs (finding maximal common subgraphs), we present the procedure of relating a scenario to a class. The paper concludes with the description of the application domain of understanding customer complaints and the preliminary evaluation of the complaint data set.

2 The Domain of Conflict Scenarios

In this section we present our model of a conflict scenario oriented to the use in a machine learning setting. Here we develop a knowledge representation methodology based on approximation of a natural language description of a conflict [5].

When modeling scenarios of inter-human conflict, it is worth distinguishing mental and non-mental states and actions. The former include *knowing, pretending* (states) and *informing* or *asking* (actions); the latter are related, for example, to *location, energy* and *account balance* (physical states), as well as *moving, heating* and *withdrawal* (physical actions). It has been shown that an adequate description of mental world can be performed using mental entities and merging all other physical action into a constant predicate for an arbitrary physical action and its resultant physical state [5]. Furthermore, we express a totality of sequential mental states for a scenario via a set of mental actions that would unambiguously lead to these mental states. Hence we approximate an inter-human interaction scenario as a sequence of mental actions, ordered in time, with a causal relation between certain mental actions. Our approximation follows the style of situation calculus, scenarios are simplified to allow for effective matching by means of graphs.

Only mental actions remain as a most important component to express similarities between scenarios. Each vertex corresponds to a mental action, which is performed by either *proponent*, or *opponent*, the latter are called *agents* (here we consider two-agent systems, however, the model is easily extended to involve multiple agents). An arc (oriented edge) denotes a sequence of two actions.

In our model mental actions have two parameters: *agent name* and *subject* (information transmitted, a cause addressed, a reason explained, an object described, etc.). Representing scenarios as graphs, we take into account both parameters. Arc types bear information whether the subject stays the same. Thick arcs link vertices that correspond to mental actions with the same subject, thin arcs link vertices that correspond to mental actions with different subject.

The curve arcs denote a causal link between the arguments of mental actions, e.g., *service is not as advertised* ⇒ *there are particular failures in a service contract, ask* ~> *confirm.*

Let us consider an example of a scenario and its graph (Figure 1).

---

*I **explained** that my cheque I wrote after I made a deposit bounced.*

*A customer service representative **accepted** that it usually takes some time to process the deposit.*

*I **reminded** that I was unfairly charged an overdraft fee a month ago in a similar situation.*

*They **denied** that it was unfair because the overdraft fee was disclosed in my account information.*

*I **disagreed** with their fee and wanted it deposited back to my account.*

*They **explained** that nothing can be done at this point and that I need to look into the account rules closer.*

Note that first two sentences (and the respective subgraph comprising two vertices) are about the current transaction, three sentences after (and the respective subgraph comprising three vertices) address the *unfair charge,* and the last sentence is probably related to both issues above. Hence the vertices of two respective subgraphs are linked with thick arcs (*explain-accept*) and (*remind-deny-disagree*).

In formal conflict scenarios extracted from text there can be multiple mental actions per step, for example *I disagreed ... and suggested...*. The former mental action describes how an agent receives a message (*accept, agree, reject*, etc.) from an opponent, and the latter one describes the attitude of this agent initiating a request (*suggest, explain*, etc.), or reaction to the opponent's action. This division into *passive* (response) mode and *active* (request) mode is represented in the second attribute of mental actions specified in the second column of Table 2. Sometimes, either of the above actions is omitted in textual description of conflicts. Frequently, a mental action, which is assumed but not mentioned explicitly, can be deduced. In this paper for the sake of simplicity we will consider single action per step, performing the comparative analysis of scenarios.

There is a commonsense causal link *between being charged an unfair fee* and *intention to have this amount of money back* which is expressed by the arc between *remind* and *disagree*. Semantically, arcs with causal labels between vertices for mental actions express the causal links between the arguments of mental actions rather than between the mental actions themselves.

In our further analysis we will show how to relate this scenario (denoted as U) to the class of negative (unjustified) complaints.

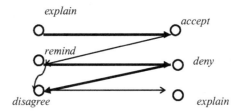

Fig. 1. The graph for approximated scenario

3 Semantics of Mental Actions

As to the choice of mental actions to adequately represent multiagent conflicts, we have selected the most frequently used from our structured database of complaints (Table 1).

To express the similarity between mental actions, we introduce five attributes each of which reflects a particular semantic parameter for mental activity (Table 2):

- *Positive/ negative attitude* expresses whether a mental action is a cooperative (friendly, helpful) move (1), uncooperative (unfriendly, unhelpful) move (-1), neither or both (hard to tell, 0).
- *Request / respond mode* specifies whether a mental action is expected to be followed by a reaction (1), constitutes a response (follows) a previous request, neither or both (hard to tell, 0).

- *Info supply / no info supply* tells if a mental action brings in an additional data about the conflict (1), does not bring any information (-1), 0; does not occur here.
- *High / low confidence* specifies the confidence of the preceding mental state so that a particular mental action is chosen, high knowledge/confidence (1), lack of knowledge/confidence (-1), neither or both is possible (0).

Intense / relaxed mode says about the potential emotional load: high (1), low (-1), neutral (0) emotional loads are possible.

Table 1. The set of mental actions from a typical complaint

Customer describes actions of **himself**	Customer describes actions of the **Company**
Agree, explain, suggest, bring company's attention, remind, allow, try, request, understand, inform, confirm ask, check, ignore, convince disagree, appeal, deny, threaten	Agree, explain, suggest, remind, allow, try, request, understand, inform, confirm, ask, check, ignore, convince, disagree, appeal, deny, threaten, bring to customer's attention, accept complaint, accept /deny responsibilities, encourage, cheat

Table 2. Many-valued context of mental actions

Mental action	Attributes				
	Positive/ negative attitude	Request / respond mode	Info supply / no info supply	High / low confidence	Intense / relaxed mode
agree	1	-1	-1	1	-1
accept	1	-1	-1	1	1
explain	0	-1	1	1	-1
suggest	1	0	1	-1	-1
bring_attent	1	1	1	1	1
remind	-1	0	1	1	1
allow	1	-1	-1	-1	-1
try	1	0	-1	-1	-1
request	0	1	-1	1	1
understand	0	-1	-1	1	-1
inform	0	0	1	1	-1
confirm	0	-1	1	1	1
ask	0	1	-1	-1	-1
check	-1	1	-1	-1	1
ignore	-1	-1	-1	-1	1
convince	0	1	1	1	-1
disagree	-1	-1	-1	1	-1
appeal	-1	1	1	1	1
deny	-1	-1	-1	1	1
threaten	-1	1	-1	1	1

Note that out of the set of meanings for each mental action (entity, speech act), we merge its subset into a single meaning, taking into account its relations to the meanings of other mental actions [5]. Our approach follows along the lines of the theory of speech acts [23, 25] in its ability to handle performatives [22]. The theory of performatives is proposed as a test case for the rationality-based theory of mental actions such as *threatening, warning*, or *promising* that are carried out simply by saying the appropriate words. It is shown how "I request you..." can be a request, and "I lie to you that ..." can be self-defeating. The analysis [22] supports and extends the systematic account [24] of the roles of the speaker's communicative intention and the hearer's inference in literal, nonliteral and indirect uses of sentences to perform speech acts. Table 2 is obtained as a reduction of speech act's attributes to the case of multiagent conflicts.

An alternative way to express the set of selected meanings for each mental action uses an expression in the *want-know-believe* basis (Figure 2), presented in [1,20], extending the BDI model [4,5]. Note that clauses may be embedded as arguments for mental actions as meta-predicates. We refer the reader to [5] for further details on defining mental actions and mental states in the above basis. For example, various meanings of mental action *inform* are expressed as follows:

inform(Who, Whom, What) :- want(Who, know(Whom, What)),
    believe(Who, not know(Whom, What)),
    believe(Who, want(Whom, know(Whom, What))).
    % The most general definition
inform(Who, Whom, What) :- believe(Who, know(Whom, What)),
    want(Who, believe(Whom, know(Who,What))).
    %to *inform Whom* that not only *Whom* but *Who knows What*
inform(Who, Whom, What) :- ask(Whom, Who, What),
        want(Who, know(Whom, What)).
    % *informing* as answering
inform(Who, Whom, What) :- ask(SomeOne, Who, believe(Whom, What)),
        want(Who, know(Whom, What).% following SomeOne's request for informing

---

disagree(A,B,W) :- inform(A,B,W), not believe(B,W), inform(B,A, not W).
agree(A,B, W) :- inform(A,B, W), believe(B, W), inform(B,A, W).
explain(A,B, W) :- believe(A, (W :- V)), not know(B, W), inform(A,B,V),
            inform(A,B,(W :- V)), believe(B,W).
confirm(A,B, W) :- inform(A,B,W), know(A, believe(B, W)).
bring_attention(A,B, W) :- want(A, believe(B, know(A, W))).
remind(A,B, W):- believe(A, believe(B, W)),
            inform(A,B,W), want(A, know(B, know(A, W))).
understand(A,W) :- inform(B,A,W), believe(B, not believe(A, (W :- V))),
want(B, believe(A, (W :- V))), inform(B, A,(W :- V)), believe(A,(W :- V),
            believe(A, W).
acceptResp(A, W) :- want(B, not W), believe(B, (W:-do(A,W1))),
        want(A, know(B, believe(A, (W:-do(A,W1)))))), inform(A,B, (W:-do(A,W1))).

Fig. 2. The clauses for the selected mental entities from Table 2

To represent the hierarchy of mental actions by a concept lattice, we scale nominally the first and second attributes (i.e., the attribute values -1, 0, and 1 are considered as completely dissimilar). The third, fourth, and fifth attributes are already two-valued. Thus, the scaled context has seven attributes and the resulting concept lattice is presented in Figure 3. *ConExp* [21] software was used to construct and visualize the concept lattice [8] of mental actions and their attributes.

The concept lattice illustrates the semantics of mental actions; it shows how the choice of natural language semantics for mental entities covers the totality of meanings in the knowledge domain of interaction between agents.

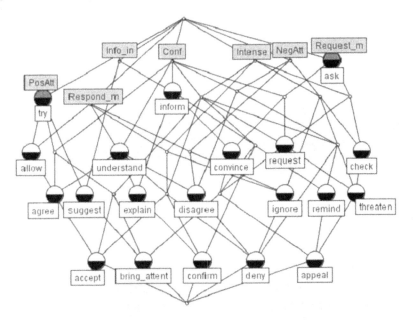

Fig. 3. The concept lattice for mental actions

4 Defining Scenarios as Graphs

We proceed with the description of our scenario dataset. This dataset contains two sets of complaint scenarios: showing a good attitude of a complainant (consistent plot with proper argumentation, a *valid complaint*) on the left, and a bad attitude of a complainant (inconsistent plot with certain flaws, implausible or irrational scenarios, an *invalid complaint*) on the right (Figure 4).

Each scenario includes 2-6 interaction *steps*, each consisting of mental actions with the alternating first attribute {*request – respond - additional request or other follow up*}. A step comprises one or more consequent actions with the same subject. Within a step, vertices for mental actions with common argument are linked with *thick* arcs.

For example, *suggest* from scenario V2 (Figure 4) is linked by a thin arc to mental action *ignore*, whose argument is not logically linked to the argument of *suggest* (the subject of suggestion). The first step of V2 includes *ignore-deny-ignore-threaten*; these mental actions have the same subject (it is not specified in the graph of conflict scenario). The vertices of these mental actions with the same argument are linked by the *thick* arcs. For example, it could be **ignored** *refund because of a wrong mailing address*, **deny** *the reason that the refund has been ignored* [*because of a wrong mailing address*], **ignore** *the denial* [*...concerning a wrong mailing address*], *and* **threatening** *for that ignorant behavior* [*...concerning a wrong mailing address*]. We have *wrong mailing address* as the common subject S of mental actions *ignore-deny-ignore-threaten* which we approximate as

$ignore(A1, S)$ & $deny(A2,S)$ & $ignore(A1,S)$ & $threaten(A2, S)$, *keeping in mind the scenario graph* . In such approximation we write $deny(A2, S)$ for the fact that $A2$ *denied the reason that the refund has been ignored because of S. Indeed,* $ignore(A1, S)$ & $deny(A2,S)$ & $ignore(A1,S)$ & $threaten(A2, S)$. Without a scenario graph, the best representation of the above in our language would be

$ignore(A1, S)$ & $deny(A2, ignore(A1, S))$ & $ignore(A1, deny(A2, ignore(A1, S)))$ & $threaten(A2, ignore(A1, deny(A2, ignore(A1, S))))$.

Let us enumerate the constraints for the scenario graph:

1) All vertices are fully ordered by the temporal sequence (earlier-later);
2) Each vertex has a special label relating it either to the proponent (drawn on the right side in Figure 4) or to the opponent (drawn on the left side);
3) Vertices denote actions either of the proponent or of the opponent;
4) The arcs of the graph are oriented from earlier vertices to later ones;
5) Thin and thick arcs point from a vertex to the subsequent one in the temporal sequence (from the proponent to the opponent or vice versa);
6) Curly arcs, staying for causal links, can jump over several vertices.

Similarity between scenarios is defined by means of maximal common subscenarios. Since we describe scenarios by means of labeled graphs, first we consider formal definitions of labeled graphs and domination relation on them (see, e.g., [7]).

Given ordered set G of graphs (V,E) with vertex- and edge-labels from the sets $(\mathcal{L}_V, \preceq)$ and $(\mathcal{L}_\mathcal{E}, \preceq)$. A labeled graph Γ from G is a quadruple of the form $((V,l),(E,b))$, where V is a set of vertices, E is a set of edges, $l: V \rightarrow \mathcal{L}_V$ is a function assigning labels to vertices, and $b: E \rightarrow \mathcal{L}_\mathcal{E}$ is a function assigning labels to edges. We do not distinguish isomorphic graphs with identical labelings.

The order is defined as follows: For two graphs $\Gamma_1 := ((V_1,l_1),(E_1,b_1))$ and $\Gamma_2 := ((V_2,l_2),(E_2,b_2))$ from G we say that Γ_1 **dominates** Γ_2 or $\Gamma_2 \leq \Gamma_1$ (or Γ_2 is a **subgraph** of Γ_1) if there exists a one-to-one mapping $\varphi: V_2 \rightarrow V_1$ such that it

- respects edges: $(v,w) \in E_2 \Rightarrow (\varphi(v), \varphi(w)) \in E_1$,
- fits under labels: $l_2(v) \preceq l_1(\varphi(v))$, $(v,w) \in E_2 \Rightarrow b_2(v,w) \preceq b_1(\varphi(v), \varphi(w))$.

Note that this definition allows generalization ("weakening") of labels of matched vertices when passing from the "larger" graph G_1 to "smaller" graph G_2.

Now, generalization Z of a pair of scenario graphs X and Y (or their similarity), denoted by $X \sqcap Y = Z$, is the set of all inclusion-maximal common subgraphs of X and Y, each of them satisfying the following additional conditions:

- To be matched, two vertices from graphs X and Y must denote mental actions of the same agent;
- Each common subgraph from Z contains at least one thick arc.

This definition is easily extended to finding generalizations of several graphs (e.g., see [7, 11]). The subsumption order $\sqsubseteq$ on pairs of graph sets X and Y is naturally defined as $X \sqsubseteq Y := X \sqcap Y = X$.

After scaling the many-valued context of mental actions, descriptions of mental action are given by 9-tuples of attributes, ordered in the usual way. Thus, vertex labels of generalizations of scenario graphs are given by intents of the scaled context of mental actions (see Figure 3).

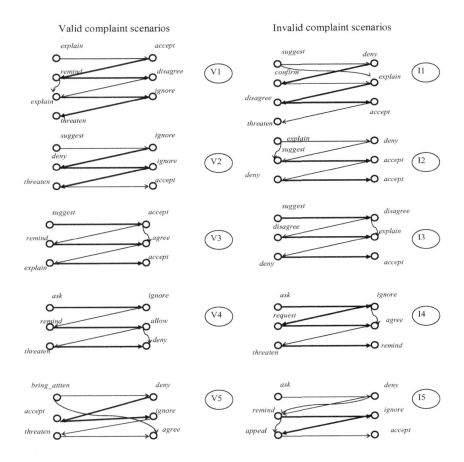

Fig. 4. The training set of scenarios

If the conditions above cannot be met then the common subgraph does not exist.

5 Relating a Scenario to a Class

Here we propose two schemes for classifying scenarios, given examples from positive and negative classes (see an example of a training sample in Figure 4).

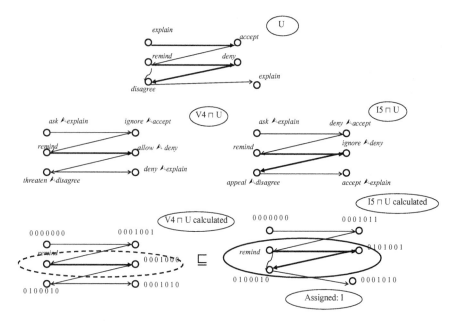

Fig. 5. A scenario with unassigned complaint status and the procedure of relating this scenario to a class

5.1 Nearest-Neighbor Classification

The following conditions hold when a scenario graph U is assigned to a class (we consider positive classification, i.e., to valid complaints, the classification to invalid complaints is made similarly):

1) U is similar to (has a nonempty common scenario subgraph of) a positive example R^+.

2) For any negative example R^-, if U is similar to R^- (i.e., $U \sqcap R^- \neq \varnothing$) then $U \sqcap R^- \sqsubseteq U \sqcap R^+$. This condition introduces the measure of similarity and says that to be assigned to a class, the similarity between the unknown graph U and the closest scenario from the positive class should be higher than the similarity between U and each negative example (i.e., representative of the class of invalid complaints).

Condition 2 implies that there is a positive example R^+ such that for no R^- one has $U \sqcap R^+ \sqsubseteq R^-$, i.e., there is no counterexample to this generalization of positive examples.

Let us now proceed to the example of a particular U in Figure 5 on the top. The task is to determine whether U belongs to the class of valid complaints (on the left of

Figure 4) or to the classes of invalid complaints (on the right); these classes are mutually exclusive.

We observe that V_4 is the graph of the highest similarity with U among all graphs from the set $\{V_1, \ldots V_5\}$ and find the common subscenario $U \sqcap V_4$. Its only thick arc is derived from the thick arc between vertices with labels *remind* and *deny* of U and the thick arc between vertices with labels *remind* and *allow* of V_4. The first vertex of this thick arc of $U \sqcap V_4$ is *remind* $\wedge$ *remind* = *remind*, the second is *allow* $\wedge$ *deny* = <0 0 0 1 0 0 0> ($U \sqcap V_4$ is calculated at the left bottom). Other arcs of $U \sqcap V_4$ are as follows: that from the vertex with the label *remind* to the vertex with the label <0 0 0 1 0 0 0>; the arc from the vertex with the label <0 0 0 1 0 0 1> to the vertex with the label *remind*; the arc from the vertex with the label <0 0 0 1 0 0 0> the vertex with the label <0 1 0 0 0 1 0>. These arcs are thin, unless both respective arcs of $U \sqcap V_4$ are thick (the latter is not the case here). Naturally, common subscenario may contain multiple steps, each of them may result in the satisfaction of conditions 1) - 2) for the class assignment above.

Similarly, we build the common subscenario $U \sqcap I_5$; I_5 delivers the largest subgraph (two thick arcs) in comparison with I_1, I_2, I_3, I_4. Moreover, $U \sqcap V_4 \sqsubseteq U \sqcap I_5$, this inclusion is highlighted by the ovals around the steps. Condition 2 is satisfied. Therefore, U is an invalid complaint as having the highest similarity to invalid complaint I_5.

In [7, 11] we considered a learning model from [3] formulated in FCA terms. As applied to scenarios, this model is described as follows. Given similarity (meet) operation $\sqcap$ on pairs of scenarios that defines a semilattice, sets of positive and negative examples, a (+)- hypothesis is defined as similarity of several positive examples which does not cover any negative example (for the lack of space we refer to [7, 11] for exact definitions). (-)-hypotheses are defined similarly. Now an undetermined scenario is classified positively if it contains (in terms of $\sqsubseteq$) a positive hypothesis and does not contain any negative hypothesis.

Finally, we discuss briefly the complexity of the approach. Generally, even testing $\sqsubseteq$ relation is an NP-complete problem. However, our scenario graphs are usually not large, the number of vertices not exceeding 20-30. In the simplest case where the vertex labels are incomparable, the large number of vertex labels reduces practical complexity, since a vertex can be matched only to a vertex with the same label. As for computing similarity operation $\sqcap$, the realization of which involves several testing of $\sqsubseteq$, with our modest computation resources (PC with 1.7 GHz and 1GB RAM) and training sample with 40 examples, it was feasible to compute classifications in 3-4 hours using a lower part of the concept lattice of vertex labels (labels with common generalization being too general were considered dissimilar).

6 Evaluations

In this section we present the results of preliminary evaluation of our classification model. Firstly, we evaluate the accuracy of our nearest-neighbors technique given the above dataset Figure 4. For each of ten scenarios, we set its class as unknown and verify if it can be related to its class properly, building common subscenarios with four representatives of its class and five foreign scenarios. Only scenarios I2 and V3 can be neither assigned to the proper class nor to a foreign class; the rest of scenarios were properly assigned.

Our further evaluation involved an improvement of existing software for processing customer complaints, called *ComplaintEngine*, available for download at http://www.dcs.bbk.ac.uk/~galitsky/ComplaintEngineSuite.html. Five attributes of mental actions, selected for the model presented in this paper, were indeed selected to improve the accuracy of scenario recognition, given the particular set of complaints from our database of formalized complaints. Our database primarily originates from the data on financial sector, obtained from the website of publicly available textual complaints PlanetFeedback.com.

Currently, *ComplaintEngine* uses anti-unification procedure to find a similarity between scenarios. Machine learning of ComplaintEngine uses the JSM-type plausible reasoning [3] augmented with situation calculus, reasoning about mental states and other reasoning domains. *ComplaintEngine* applies domain-independent anti-unification to formulas that include enumeration of mental actions in time.

As expected, graph representation of scenarios and employed nearest-neighbors technique allowed noticeable improvement of complaint recognition accuracy. Judging on the restricted dataset of 80 banking complaints (40 complaints make the training set and 40 complaints have to be classified), the performance of *ComplaintEngine* was improved by 6% to achieve the resultant recognition accuracy of 89%. Relating a scenario to a class, *ComplaintEngine* is capable of explaining its decision by enumeration of similar and dissimilar scenarios, as well as particular mental actions which led to its decision.

Since such accuracy was achieved by manual adjustment of the model of multi-agent scenario, we expect it to be much lower for other complaint domains. However, we believe that the role of improved machine learning technique for functioning in a new complaint domain will be substantial.

We would like to briefly familiarize the reader with the functionality of *Complaint Engine* [6]. The user interface to specify a complaint scenario is shown at Figure 6a. Figures 6b and 6c depict the fragments of this form where complainant selects his mental actions and mental actions of his opponent (a company) respectively. Mental actions are selected from the list of twenty or more, depending on the industry sector of a complaint. The parameters of mental actions are specified as text in the Interactive Form; however they are not present in the formal graph-based scenario representation.

Having performed the justification of complaint validity, *ComplaintEngine* sets the list box for complaint status at "unjustified". *ComplaintEngine* provides the explanation of its decision, highlighting the cases which are similar to U (unjustified), and which are different from U (justified). Moreover, *ComplaintEngine* indicates the mental actions (steps) that are common for U and other unjustified complaints to further back up its decision.

A similar form to Figure 6 is used for a complainant to file a complaint, and for a company to store complaints, analyze them, determine validity, explain how the decision has been made, and finally advise on a strategy for complaint resolution (see the demo at www.dcs.bbk.ac.uk/~galitsky/CLAIMS/ComplaintEngineSuite.zip).

input your complaint

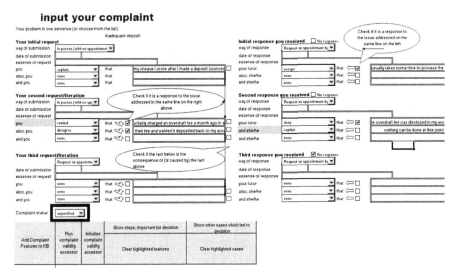

Fig.6a. The screen-shot of the Interactive Complaint Form where the complaint scenario U from Figure 4 is specified.

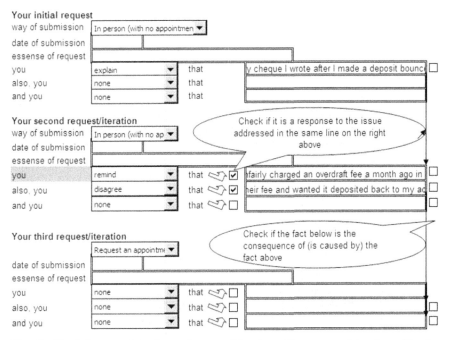

Fig. 6b: The left pane of the interactive complaint form. Here a complainant specifies her mental actions and their parameters.

A complainant has a choice to use the above form or to input complaint as a text so that the linguistic processor processes the complaint automatically and fills the form for her. Using the form encourages complainants to enforce a logical structure on a complaint and to provide a sound argumentation. After a complaint is partially or fully specified, the user evaluates its consistency. *ComplaintEngine* indicates whether the current complaint (its mental component) is consistent or not, it may issue a warning and an advice concerning improvement of the logical structure of this complaint.

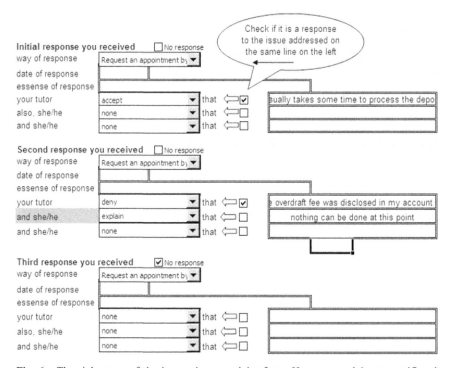

Fig. 6c: The right pane of the interactive complaint form. Here a complainant specifies the mental actions and their parameters of his opponent (a company).

When the complainant is satisfied with the response of *ComplaintEngine*, he submits the completed form. The other option is if a user observes that it is not possible to file a reasonable complaint, it may be dismissed at this early stage by the complainant.

7 Conclusions

In this paper we proposed a Nearest Neighbors-based approach to relate a formalized conflict scenario to the class of valid and the class of invalid complaint scenarios. The representation language is that of labeled directed acyclic graphs with generalization operator on them. We considered the concept lattice of mental actions and showed how the procedure of relating a complaint to a class can be implemented.

This is an initial attempt to build a machine learning technique for formal scenarios as graphs. The further steps of the research along the line of machine learning for multiagent scenarios will be as follows:

- Developing a more precise representation languages for scenarios of multi-agent interactions; adding more features to scenario representation in addition to temporal and causal links;
- In terms of applications, proceeding beyond the domain of customer complaints;
- Performing comparison with other classification techniques.

Building a framework for comparative analysis of formal scenarios, one or another way to express the similarity between the main entities has to be employed. In our earlier studies we approximated the meanings of mental entities using their definitions via the basis of *want-know-believe*. However, building the concept lattice for mental actions was found to be more suitable on the way to define a concept lattice for scenarios themselves.

Also, here we suggested a novel approach to building a semantic network between linguistic entities in the basis of selected attributes. The choice of attributes in this study is motivated by the task of scenario comparison; these attributes may vary from domain to domain. Twenty selected mental entities are roughly at the same level of generality – there are "horizontal" semantic relations between them.

We believe the current work is one of the first targeting machine learning in such domain as multiagent interactions described in natural language. A number of studies have shown how to enable BDI-agents with learning in a particular domain (e.g. information retrieval). In BDI settings the description of agents' attitudes is quite limited: only their beliefs, desires and intentions are involved. Moreover, just the automated (software) agents are addressed. In this paper we significantly extended the expressiveness of representation language for agents' attitudes, using twenty mental actions linked by a concept lattice. The suggested machinery can be applied to an arbitrary domain including inter-human conflicts, obviously characterized in natural language.

The preliminary evaluation of our model shows it is an adequate technique to handle such complex objects (both in terms of knowledge representation and reasoning) as mental actions of scenarios of multiagent interactions. Nearest Neighbors approach was found suitable to relate an inter-human conflict scenario to a class. Evaluation using our limited dataset, as well as the dataset of formalized real-world complaints showed a satisfactory performance.

Acknowledgements The second and third authors were supported by the Russian Foundation for Humanities, project no. 05-03-03019a.

References

1. Bratman, M.E.: Intention, plans and practical reason. Harvard University Press (1987)
2. Fagin,R., Halpern, J.Y., Moses,Y., Vardi,M.Y. : Reasoning about knowledge *MIT Press*, Cambridge, MA, London, England (1995)
3. Finn, V.K. : Plausible Reasoning in Systems of JSM-type, Itogi Nauki I Techniki, Seriya Iformatika, 15, 54-101, 1991 [in Russian].
4. Galitsky, B. : A Library of Behaviors: Implementing Commonsense Reasoning about Mental World. 8th Intl Conf on Knowledge-Based Intelligent Info Syst. (2004)
5. Galitsky, B.: Natural Language Question Answering System: Technique of Semantic Headers. *Advanced Knowledge International*, Adelaide, Australia (2003)
6. Galitsky, B. and Tumarkina, I.: Justification of Customer Complaints using Emotional States and Mental Actions. *FLAIRS* Miami, FL. (2004)
7. Ganter, B. and Kuznetsov, S.O.: Pattern Structures and Their Projections,   Proc. 9th Int. Conf. on Conceptual Structures, ICCS'01, G. Stumme and H. Delugach, Eds., Lecture Notes in Artificial Intelligence, vol. 2120 (2001) 129-142
8. Ganter, B. and Wille R.: Formal Concept Analysis. Mathematical Foundations, Springer (1999)
9. Guerra-Hernandez, A.1, Fallah-Seghrouchni,A. E. and Soldano, H.: Learning in BDI Multi-agent Systems CLIMA IV - Computational Logic in Multi-Agent Systems Fort Lauderdale, FL, USA (2004)
10. Kolodner, J.: Case-Based Reasoning. Morgan Kaufmann (1993)
11. Kuznetsov S.O.: Learning of Simple Conceptual Graphs from Positive and Negative Examples. In: J. Zytkow, J. Rauch (eds.), Proc. Principles of Data Mining and Knowledge Discovery, Third European Conference, PKDD'99, Lecture Notes in Artificial Intelligence, vol. 1704 (1999) 384-392
12. Laza, R. and Corchado, J. M.: CBR-BDI Agents in Planning. *Symposium on Informatics and Telecommunications* (SIT'02), Sevilla, Spain, September 25-27, pp. 181-192 (2002).
13. Mitchell, T.: Machine Learning, McGraw-Hill (1997)
14. Muller, H.J. Dieng, R. (eds.): Computational Conflicts: Conflict Modeling for Distributed Intelligent Systems. Springer-Verlag New York (2000)
15. Olivia,C., Chang, C.F., Enguix, C.F. and Ghose A.K.: Case-Based BDI Agents: an Effective Approach for Intelligent Search on the World Wide Web. Intelligent Agents in Cyberspace. AAAI Spring Symposium (1999)
16. Shanahan, M.: Solving the frame problem. MIT Press (1997)
17. Sowa, J.: Conceptual Graphs, Conceptual Structures: Information Processing in Mind and Machine, Addison-Wesley, Reading, MA (1984)
18. Stone, P., Veloso, M.: Multiagent Systems: A Survey from a Machine Learning Perspective. Autonomous Robotics, 8(3) (2000) 345-383
19. Weiss, G., Sen, S.: Adaptation and Learning in Multiagent Systems. Lecture Notes in Artificial Intelligence, Vol. 1042. Springer-Verlag, Berlin Heidelberg New York (1996)
20. Wooldridge, M.: Reasoning about Rational Agents. The MIT Press Cambridge MA (2000)
21. Yevtushenko, S.A.: http://www.sf.net/projects/conexp. Last accessed April 7 2005.
22. Cohen, P.R, Levesque, H.J.: Performatives in a Rationally Based Speech Act Theory. Proceedings of the 28th conference on Association for Computational Linguistics (1990) 79-88
23. Searle, J.: Speech Acts: An Essay in the Philosophy of Language, Cambridge, Eng.- Cambridge University Press (1969)
24. Bach, K. and Harnish, R.M. Linguistic Commuication and Speech Acts, Cambridge, Mass.: MIT Press (1979)
25. Searle, J.: Expression and Meaning: Studies in the Theory of Speech Acts.Cambridge University Press. (1979)

Querying a Bioinformatic Data Sources Registry with Concept Lattices

Nizar Messai, Marie-Dominique Devignes, Amedeo Napoli, and
Malika Smaïl-Tabbone

UMR 7503 LORIA, BP 239, 54506 Vandœuvre-lès-Nancy, FRANCE
{messai,devignes,napoli,smail}@loria.fr
http://www.loria.fr/equipes/orpailleur

Abstract Bioinformatic data sources available on the web are multiple and heterogenous. The lack of documentation and the difficulty of interaction with these data banks require users competence in both informatics and biological fields for an optimal use of sources contents that remain rather under exploited. In this paper we present an approach based on formal concept analysis to classify and search relevant bioinformatic data sources for a given user query. It consists in building the concept lattice from the binary relation between bioinformatic data sources and their associated metadata. The concept built from a given user query is then merged into the concept lattice. The result is given by the extraction of the set of sources belonging to the extents of the query concept subsumers in the resulting concept lattice. The sources ranking is given by the concept specificity order in the concept lattice. An improvement of the approach consists in automatic refinement of the query thanks to domain ontologies. Two forms of refinement are possible by generalisation and by specialisation.

1 Introduction

Bioinformatics is facing the great challenge of enabling biologists to effectively and efficiently access to data stored in distributed data sources. The large number of sources, their heterogeneity and the complexity of the biological objects they refer to, make it difficult to adequately relate the sources with a user query. The query itself often needs to be processed and distributed over several data sources. Different approaches are being experienced through data warehouses (e.g. GUS [6]), federated databases (e.g. SEMEDA [14]) or mediators (e.g. TAMBIS [11]), all aiming at organizing access to several data sources in order to satisfy user queries. Systems such as TAMBIS or SEMEDA are capable of taking into account semantic processing of user query. However, most available systems only deal with a limited number of data sources that do not satisfy a large proportion of user queries. The work presented here aims at modeling knowledge about bioinformatic data sources in order to propose to users, given a query, the best-suited available bioinformatic data sources. The problem here is not the querying of the data sources themselves but rather the identification and selection among

F. Dau, M.-L. Mugnier, G. Stumme (Eds.): ICCS 2005, LNAI 3596, pp. 323–336, 2005.
© Springer-Verlag Berlin Heidelberg 2005

all existing data sources of the most appropriate ones given the query. In this paper we propose a solution to a particular information retrieval (IR) problem where data sources instead of documents are searched and indexation is based on metadata reflecting information about sources rather than on data extracted from documents. Formal concept analysis (FCA) is used here for improving the retrieval of relevant data sources thanks to a dynamic and flexible classification of existing sources. In addition domain ontologies have been taken into account for processing the query in a semantic manner.

We will first review in section 2 related works that combine FCA and IR, as well as ontology usage in similar problems. Section 3 presents the BioRegistry project as a new repository for metadata about bioinformatic data sources. The formalisation of our problem using FCA is detailed in section 4 and the querying aspects are developed in section 5 including an original query refinement method. Finally some perspectives of this research work are discussed in section 6.

2   Related Work

2.1   Concept Lattices for Information Retrieval (IR)

The application of concept lattices in information retrieval was originally present at the beginning of FCA [24]. Indeed, an obvious analogy exists between object-attribute and document-term tables. Information retrieval was then mentioned as one application field for concept lattices usage [12]. The formal concepts in the lattice are seen as classes of relevant documents that match a given user query with the subsumption relation (i.e. the partial ordering relation within the concept lattice) between concepts allowing moving from one query to another more general or more specific. A lattice-based information retrieval approach is proposed in [4]. In both propositions [12] and [4], lattice-based information retrieval shows performances that are better than boolean information retrieval. One limitation is the complexity of the lattice (regarding the size and the needed computation) for large contexts. But in the real applications it is estimated that this maximum complexity is not reached [12]. However some works such as multi-level strategies developed in ZooM [19] and iceberg lattices [22] propose solutions for such complex applications either by expanding or refining a subpart of a lattice (ZooM) or by decreasing the overall size of the lattice by limiting the exploration depth of the set of concepts (icebergs).

2.2   Concept Lattice Construction

Several works deal with the problem of generating the set of concepts and the concept lattice of a given formal context. A detailed comparison between performances of algorithms for generating concept lattices and their diagram graphs can be found in [15]. Some of the proposed algorithms allow an incremental construction of concept lattice for a given formal context such as proposed in [13, 5, 23]. This aspect is particulary beneficial for the information retrieval applications in general and for our bioinformatic problem in particular for two reasons. First, user queries need to be merged into the set of concepts in order to retrieve relevant documents (or bioinformatic data sources) included in these concepts as in [12, 4]. Second, incremental lattice construction allows the insertion of new concepts, that in our case takes into account the availability of new bioinformatic data sources on the web. This kind of insertion is essential for keeping the BioRegistry repository in accordance with the web content as explained in the following.

2.3   Improvement of FCA-based IR Performance Using Ontologies

Query refinement is an IR mechanism aiming at improving retrieval performance by adding to user's query new terms related to the query terms [2]. Propositions combining ontologies and FCA in the purpose of improving retrieval performance are found in [3, 20, 21]. In the two first works, a thesaurus is included to enhance the retrieval process by enriching the indexation in the lattice. In the last work, domain ontologies are used to build refined lattices according to user preferences thus avoiding complete lattice construction. Both approaches work directly on the lattice either by adding terms or by considering only parts of it.

In our work, domain ontologies are taken into account at early stage of the information retrieval process, i.e. during the BioRegistry construction (see below). This leads us to propose a mechanism for IR improvement based on query rather than lattice modification.

3   The BioRegistry Project

3.1   Bioinformatic Data Sources

Hundreds of biological data sources are known today [8]. Most efforts so far have been devoted to unifying the access to these sources, facilitating query processing and distribution over relevant sources, integrating answers, etc. These tasks involve designing appropriated workflows and require seamless interoperation of resources. Integrated systems are available that rely on data warehouses or mediation architectures. Today solutions are also envisaged in the context of semantic web, involving composition of web services [1, 26, 17].

The maximal efficiency of these solutions is reached when the whole knowledge available about all existing data sources can be exploited. For example the apparently simple query: *"What are the genes from human chromosome X that are preferentially expressed in brain?"* deals with both so-called *mapping data* and *expression data* which may or may not be contained in a single source at a given time. Probably more than one data source can be found for each part of the query. The user may select one of these sources because of given quality criteria (e.g. manual revision of the data or update frequency) or availability information (e.g. access constraints).

The largest existing catalog for bioinformatic data sources is certainly DBCAT[1] [7]. However, this flat file repository contains a rather small metadata set and offers limited query capabilities because most fields domains are open (free text). Registries are being developed for bioinformatic web services such as in the BioMoby[2] and MyGrid[3] projects [16]. Today the proportion of biological information accessible through web services is far too limited and does not properly answer users needs. However this situation may change and the need for modeling and organizing knowledge about web services in order to give access to relevant services for a given query will become as pressing as today for biological data sources. In order to build a specific environment for bioinformatic data source classification and searching and to test our propositions, we have decided to build our own registry called BioRegistry, in which the various metadata attached to biological data sources are organized in a dynamic, flexible and structured manner.

3.2 The BioRegistry Model

A hierarchical model has been designed to organize four categories of metadata attached to a data source: source identification, topics covered by the source, data and data source quality, availability. At present, all these metadata are manually extracted from the documentation associated to the data sources [18]. Topic information is divided in two parts: the subjects covered by the data sources and the organisms concerned. In both domains, existing controlled vocabularies, ontologies are used to valuate metadata fields by choosing the most specific terms and therefore minimizes the redundancy. The BioRegistry model thus includes a sub-hierarchy for describing and referencing the ontologies. To illustrate this point, figure 1 shows the ontology used to represent the phylogeny of model organisms in the purpose of indexing bioinformatic data sources. This ontology has been extracted from the NCBI taxonomy[4] that is used to index the Genbank sequences entries. Model organisms are lying at the leaves of the ontology and only the structuring nodes have been retained. Assuming that each node represents a concept defined by common properties shared by the corresponding group of organisms, the relation between nodes can be considered as a specialization (a partially ordering) relation. The MeSH thesaurus[5] has been used to valuate the subjects metadata field. The BioRegistry has been implemented as an XML schema compatible with semantic web languages such as OWL. Instances of the BioRegistry model organizing metadata relative to certain bioinformatic data sources can be visualized at the BioRegistry home page[6].

---

[1] http://www.infobiogen.fr/services/dbcat/
[2] http://mobycentral.cbr.nrc.ca/cgi-bin/gbrowse_moby
[3] http://mobycentral.cbr.nrc.ca/cgi-bin/gbrowse_moby
[4] http://www.ncbi.nlm.nih.gov/entrez/query.fcgi?db=Taxonomy
[5] http://www.nlm.nih.gov/mesh/meshhome.html
[6] http://bioinfo.loria.fr/Members/devignes/Bioregistry/presentationBioregistry/view

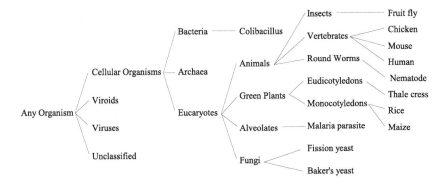

Fig. 1. Ontology of living organisms (defined for the BioRegistry)

3.3   BioRegistry Exploitation and FCA

First exploitation of the BioRegistry is form-based querying, allowing structured information retrieval of the metadata. This should allow the biologist to formulate a multi-criteria query combining various metadata categories and to recover a sorted list of data sources matching the query. For example the query cited in section 3.1 would be composed of the following criteria: subjects concerned = mapping or expression, organism concerned = human, manual revision = yes, update frequency = monthly, and access constraints = free.

However, this approach requires the user to formulate a query which may reveal inefficient without an overall knowledge on the data sources described in the BioRegistry. To overcome this limit, we decided to apply FCA to the BioRegistry content. Indeed this type of formalisation could enable flexible classification of data sources on the basis of metadata sharing as well as querying of the registry. The resulting classification of the sources present in the BioRegistry in the form of a concept lattice enables the user to discover relevant sources simply by browsing the lattice itself. Given a query, it is also possible to recover from the concept lattice data sources sharing all or a subset of metadata with the query. One advantage over classical information retrieval is that in the BioRegistry, the set of data sources is far smaller (about one thousand) than most sets of documents, thus constraining search space and query processing under certain limits. This can be considered as a condition for scalability.

Domain ontologies, used to valuate the BioRegistry metadata, are also intended to help the user to select query terms. In addition, they will be exploited as a mean for query refinement in order to improve the recall (see sections 5.2, 5.3, 5.4 and 5.5).

4   Concept Lattices for Classifying BioRegistry Data Sources

4.1   Construction of BioRegistry Concept Lattice

In this section, we show how the FCA framework applies to the formalisation of BioRegistry content. More detailed FCA related definitions can be found in [10].

In the following, the formalisation of the BioRegistry is given by a formal context $\mathcal{K}_{bio} = (G, M, I)$ where G is a set of bioinformatic data sources (e.g. Swissprot, RefSeq,...), M is a set of metadata (e.g. manual revision, human organism,...) and I is a binary relation between G and M called the incidence of $\mathcal{K}_{bio}$ and verifying: $I \subseteq G \times M$ and $(g, m) \in I$ (or gIm) where g, m are such that $g \in G$ and $m \in M$ means that the data source g has the metadata m. An example of formal context is given in table 1 with bioinformatic data sources and metadata full names in table 2 (symbols and abbreviations are used for a better visibility in the lattice). Consider $A \in G$ a set of data sources, then the set of

Table 1. Example of BioRegistry formal context $\mathcal{K}_{bio}$.

Sources\Metadata	NS	PS	AS	AO	An	Ve	Hu	Mo	MR
S1	0	1	0	1	0	0	0	0	1
S2	1	1	1	1	0	0	0	0	1
S3	1	0	0	0	0	0	1	0	0
S4	0	1	0	1	0	0	0	0	1
S5	1	1	1	0	0	0	1	0	0
S6	1	0	0	0	1	0	0	0	0
S7	0	1	0	0	0	0	0	1	0
S8	0	1	0	0	0	1	0	0	0

Table 2. Complete names of bioinformatic data sources and their metadata.

Source name	Symbol
Swissprot	S1
RefSeq	S2
TIGR-HGI	S3
GPCRDB	S4
HUGE	S5
ENSEMBL	S6
Mouse Genome DB	S7
Vega Genome Browser	S8

Metadata (attributes)	Abbreviation	Category
Nucleic Sequences	NS	Subject
Proteic Sequences	PS	Subject
Any Sequence	AS	Subject
Any Organism	AO	Organism
Animals	An	Organism
Vertebrate	Ve	Organism
Human	Hu	Organism
Mouse	Mo	Organism
Manual Revision	MR	Quality

metadata common to all the sources in A is $A' = \{m \in M \mid \forall g \in A, \, gIm\}$. Dually for a set $B \in M$ of metadata, the set of data sources sharing all the metadata in B is $B' = \{g \in G \mid \forall m \in B, \, gIm\}$.

A formal concept in the BioRegistry formalisation $\mathcal{K}_{bio}$ is a data source set sharing a metadata set. It is formally presented by a pair (A, B) where $A \subseteq G$, $B \subseteq M$, $A' = B$, and $B' = A$; A and B are called the *extent* and the *intent* of the concept, respectively. We denote by $\mathcal{C}$ the set of all formal concepts of $\mathcal{K}_{bio}$. Consider $C_1 = (A_1, B_1)$ and $C_2 = (A_2, B_2)$ in $\mathcal{C}$. C_1 is subsumed by C_2 if $A_1 \subseteq A_2$ or dually $B_2 \subseteq B_1$ (denoted by $C_1 \sqsubseteq C_2$). $(\mathcal{C}, \sqsubseteq)$ is a complete lattice [25] called the concept lattice corresponding to the context $\mathcal{K}_{bio}$. In the following, $(\mathcal{C}, \sqsubseteq)$ will be denoted by $\mathcal{L}(\mathcal{C})$. Figure 2 shows the concept lattice $\mathcal{L}(\mathcal{C})$ corresponding to the BioRegistry formal context example $\mathcal{K}_{bio}$ given in table 1.

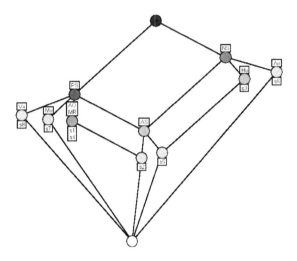

Fig. 2. The concept lattice $\mathcal{L}(\mathcal{C})$ corresponding to $\mathcal{K}_{bio}$

One important characteristic of the formal context $\mathcal{K}_{bio}$ is that the set M of metadata is carefully delineated during the BioRegistry construction so that its cardinality remains small. This particularity led us to choose the Godin algorithm [13] to generate the corresponding concept lattice since the context is small and sparse [15]. In addition, as mentioned in section 2.2 this algorithm allows the addition of new concepts to an existing lattice. This aspect is useful for the querying method described in section 5.

4.2 Flexible BioRegistry Data Sources Classification

Because various sets of data sources and/or various sets of metadata can easily be extracted from the BioRegistry (as a structured document), numerous possibilities can be offered to customize the views on the overall organization of bioinformatic data sources. For example, a user interested in the sharing of *subjects* across the data sources (see section 3.2 for a definition of *subject* in the

BioRegistry) may define a modified formal context where the attribute set is only composed of subject metadata. The object set of this modified formal context would be constituted by all the data sources being indexed with these metadata. Alternatively, a user may wish to visualize the classification of a subset of data sources dealing for instance with human data. A modified formal context may be constructed where the object set is a subset of the data sources retrieved from the BioRegistry on the basis of the metadata *organism* (see section 3.2) valued as *human*. The attribute set of this modified formal context would then be composed by the set of all metadata associated to the selected subset of data sources.

This flexibility in customizing the views over the BioRegistry content is for the moment very different from the solutions [19, 22] discussed above (section 2.1). It simply relies on a new automatic lattice construction every time that a new formal context can be created as an answer to a user need.

5   Querying BioRegistry Concept Lattices

5.1   Relevant Bioinformatic Data Sources Retrieval

Once the concept lattice $\mathcal{L}(\mathcal{C})$ is generated begins the retrieval of relevant sources. In the same way as in [12] and [4], we define a query as a formal concept $Q = (Q_A, Q_B)$ where $Q_A = \{Query\}$, i.e. a name for the extent to be formed (it can also be seen as a name for denoting an empty extent or a virtual class to be instantiated) and Q_B is the set of metadata to be used during the search. Actually, using the name *Query* is an artifact for allowing the extent of the lattice by classifying the query $Q = (Q_A, Q_B)$. As an example consider a query that searches for data sources with the metadata *Nucleic Sequences*, *Human* and *Manual Revision*. Using the abbreviations given in table 2, the query is given by $Q = (\{Query\}, \{NS, Hu, MR\})$.

Once Q is given, it has to be classified in the concept lattice $\mathcal{L}(\mathcal{C})$ using the incremental classification algorithm of Godin et al. [13]. The resulting concept lattice is noted $(\mathcal{C} \oplus Q, \sqsubseteq)$ where $\mathcal{C} \oplus Q$ denotes the new set of concepts once the query has been added. In the following the concept lattice $(\mathcal{C} \oplus Q, \sqsubseteq)$ will be denoted by $\mathcal{L}(\mathcal{C} \oplus Q)$. For the given example the modified concept lattice $\mathcal{L}(\mathcal{C} \oplus Q)$ is shown on figure 3. Dashed circles point out new or modified concepts due to the insertion of the query. Only these concepts share properties with the query and could thus be interesting for the user.

The query concept is denoted by Q either in $\mathcal{L}(\mathcal{C})$ or in $\mathcal{L}(\mathcal{C} \oplus Q)$. If there exists in the lattice $\mathcal{L}(\mathcal{C})$ a concept of the form $(A, Q_B \cup B)$, then the classification of Q in $\mathcal{L}(\mathcal{C})$ will produce a subsumer concept of the form $(\{Query, A\}, Q_B)$ that will be the new query concept to be considered. For the sake of simplicity, we continue to denote by Q the query concept in $\mathcal{L}(\mathcal{C} \oplus Q)$ whatever the case.

Definition 1. *A data source is relevant for a given query if and only if it shares at least one metadata mentioned in the query. The degree of relevance is given by the number of metadata shared with the query and by the stage during which the data source is added to the result.*

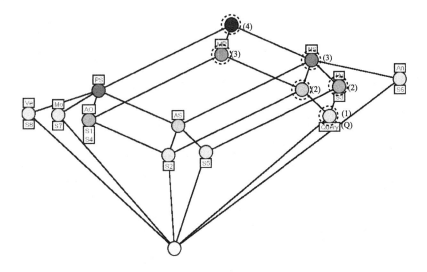

Fig. 3. The concept lattice $\mathcal{L}(\mathcal{C} \oplus Q)$

This definition of relevance is the basis of the retrieval process in the lattice and differs from the neighborhood notion used in [4]. The latter can lead to retrieved documents lacking any query term which is acceptable in document retrieval but not suitable to our needs. The above definition of relevance is sufficient to explain the retrieval algorithm detailed hereafter.

Considering the above definition, all the relevant sources are in the extents of Q and its subsumers in the concept lattice (indicated by dashed circles other than the top concept in figure 3) since the intent of each one of these concepts is a subset of Q_B (the intent of the query concept). In the following we will denote by $\mathcal{R}_{sources}$ the set of relevant data sources for the considered query. It is important to mention here that all the sources in $\mathcal{R}_{sources}$ do not have the same relevance. In fact, they are ranked according to the number of shared metadata with the query and according to the stage during which they have been added to $\mathcal{R}_{sources}$.

Intuitively, the relevant data source retrieval algorithm consists first in classifying the query concept in the lattice, operation that instantiates the extension $\{Query\}$ (actually, $Query$ could be considered as a variable to be instantiated). Then, the set of data sources that are inherited from the subsumers of the query concept Q in the lattice are gathered in the result $\mathcal{R}_{sources}$. The rank of the returned data sources may be memorized according to the distance of the sources to the query concept. Consider $\mathcal{C}_1$ the direct (most specific) subsumers of Q in the concept lattice. The set of data sources in the extents of the concepts in $\mathcal{C}_1$ and not already in the result are added to the result. The next step consists in considering the direct subsumers of concepts in $\mathcal{C}_1$ (subsumers of distance two of Q) and adding new emerging data sources to the result set $\mathcal{R}_{sources}$. Then we continue in the same way for C_2, C_3 etc until we reach an empty set C_n. In each step, the data sources in the extent of concept with an empty intent are ignored since they may not share metadata with the query.

In figure 3, the numbers near the concepts show the iterations of the explained algorithm. In the first iteration, Q is considered. In this case there is no data source in the extent of the query Q. In the second iteration the data sources S_3, S_5 and S_2 are added to the result. In the third iteration the data sources S_6, S_1 and S_4 are added to the result.

Finally the set $\mathcal{R}_{sources}$ includes the data sources ranked as follows:

1. S3 (TIGR-HGI) and S5 (HUGE) share *Nucleic Sequences* and *Human* with Q

1. S2 (RefSeq) shares *Nucleic Sequences* and *Manual Revision* with Q
2. S6 (ENSEMBL) shares *Nucleic Sequences* with Q
2. S1 (Swissprot) and S4 (GPCRDB) share *Manual Revision* with Q

Additional ranking criteria can be defined according to given preferences (e.g. user preferences).

5.2 Ontology-Based Query Refinement

The query concept may not be filled with any result. For example in the BioRegistry formal context presented in table 1 a user searching data sources containing data relative to the organism *Chicken* will not get any answer. However, there may be data sources relevant to the query described by metadata that do not directly map the query metadata. To help further the user we propose a query refinement procedure based on domain ontologies.

Contrasting with the propositions [3, 20, 21] mentioned in section 2.3, we modify the query instead of the lattice. In fact, we preserve the whole lattice structure and we modify the query by inserting metadata related to metadata of the query in a given ontology. This strategy, which can be automated, avoids introducing redundancy in the lattice.

Added metadata are either more specific or more general than those initially in the query. This leads to two types of query refinement: refinement by generalisation and refinement by specialisation. It is important to recall here that we are not facing any synonymy problem between metadata in the query and terms in the ontology since metadata valuation in the BioRegistry involves terms extracted from domain ontologies.

The generalisation refinement w.r.t. a metadata adds more general metadata represented by its ancestors in the ontology. In the example cited above (metadata *Chicken*), considering the ontology shown in figure 1, the metadata that can be added to the query are *Vertebrates, Animals, Eucaryotes, Cellular Organisms,* and *Any Organism*. However some of these metadata (*Eucaryotes* and *Cellular Organisms*) are not in the formal context $\mathcal{K}_{bio}$ given in the table 1. This means that these metadata are not shared by any source in this context so adding them to the query will not lead to any result enrichment. Only new metadata already present in $\mathcal{K}_{bio}$ are considered during the generalisation refinement process.

In a dual way the specialisation refinement w.r.t. a metadata adds semantically more specific metadata represented by its descendants in the ontology. In the given example the metadata *Chicken* has no descendant and thus could not be specialised. A better example would be a query composed by the metadata *Eucaryotes* which does not retrieve any answer since this metadata is absent from the formal context $\mathcal{K}_{bio}$. Specialisation refinement leads to inspect all descendants of *Eucaryotes* in the ontology and select only those that appear in the formal context (*Animals, Vertebrate, Human*, and *Mouse*) to add them to the query.

It is possible to combine both types of query refinement. This means, for a given query metadata, adding both its ancestors and its descendants in the corresponding domain ontology. In all cases of query refinement the number of added metadata can be controlled by considering only the nearest ancestors to the considered metadata in the ontology (generalisation refinement) or its nearest descendants (specialisation refinement).

Once the ontology-based query refinement done, the refined query has to be inserted into the original lattice $\mathcal{L}(\mathcal{C})$ and the algorithm detailed above can be applied to the new resulting lattice $\mathcal{L}(\mathcal{C} \oplus Q)$. The next section presents the ontology-based query refinement.

5.3 The Generalisation Query Refinement

Consider the query with the metadata *Chicken* represented by the formal concept $Q = (\{Query\}, \{Ch\})$. The result for this query is empty since the metadata it contains is not in the context. Applying the generalisation query refinement, we obtain the following result as response to the refined query:

1. S6 (ENSEMBL) shares *Animals* with the refined query
1. S8 (Vega Genome Browser) shares *Vertebrate* with the refined query
1. S1 (Swissprot), S2 (RefSeq) and S4 (GPCRDB) share *Any Organism* with the refined query

Each source of the result has a part satisfying the query and a part that does not (e.g. S8 is concerned with *Chicken* but with *Mouse* and *Human* as well). Furthermore the shorter the distance between the query metadata and the added metadata the more relevant the resulting sources (S8 is preferable to S6). This aspect motivates the possibility of controlling the added metadata during the generalisation refinement process mentioned above. Hence to avoid introducing less relevant (or irrelevant) sources in the result we have to consider only the nearest ancestors of the considered metadata in the domain ontology.

5.4 The Specialisation Query Refinement

Consider the query with the metadata *Eucaryotes* represented by the formal concept $Q = (\{Query\}, \{Eu\})$. The result for this query is empty since the metadata it contains is not in the context. Applying the specialisation query refinement, we obtain the following result for the refined query:

1. S6 (ENSEMBL) shares *Animals* with the refined query
1. S8 (Vega Genome Browser) shares *Vertebrate* with the refined query
1. S5 (ENSEMBL) shares *Human* with the refined query
1. S7 (Mouse Genome DB) shares *Mouse* with the refined query

In this case each source of the result gives a partial answer to the query and a composition of these data sources could provide a complete answer to the query if each descendant (of the query metadata) indexes one data source. Similarly as in the generalisation refinement the distance between the original metadata and the added ones explains the difference of sources relevance. In fact sources dealing with a far descendant of the query metadata give precise information that is not always needed by the user. The level of specialisation can be controlled by considering the nearest descendants of the metadata in the domain ontology that constitute the best coverage of the query.

5.5   Choice Between Generalisation and Specialisation Query Refinement

When the considered metadata is a leaf or is the root of the domain ontology there is no problem of choice since in both case only one type of refinement is possible (generalisation in the first case and specialisation in the second). But when the metadata is neither a leaf nor the root the two types of refinement are possible. The choice can be done with relation to the user preferences. In fact if the user accepts to get data sources a part of which corresponds to his need then the generalisation refinement is adopted. If he accepts to get data sources that correspond to a part of his need the specialisation refinement is used. In both cases it is useful to have a post ranking of the new selected data sources reflecting the similarity between their indexing metadata and the query one [9].

6   Conclusion and Future Work

In this paper, we have presented an approach combining formal concept analysis and domain ontologies for an information retrieval problem in bioinformatics. The BioRegistry as a structured repository of metadata relative to bioinformatic data sources (including data quality information) constitutes a well-suited application domain for the FCA theory allowing scalability and flexibility. The approach is intended for the problem of relevant bioinformatic data sources selection for a further interrogation. Indeed concept lattices appear as a mean to provide customised views about bioinformatic data sources and to organize knowledge about these sources. This in turn can help the user in the process of data sources retrieval to answer his query. Furthermore ontology-based query refinement mechanisms have been proposed to improve this retrieval process.

An implementation of our proposition is currently underway. Preliminary testing have shown the usefulness of a post-processing mechanism to improve the ranking of retrieved data sources.

Acknowledgments

We would like to thank Shazia Osman for her contribution to the conception of the BioRegistry model, its construction and indexation of the data sources it contains. This work was supported by the "PRST Intelligence Logicielle" from the Région Lorraine.

References

[1] D. Buttler, M. Coleman, T. Critchlow, R. Fileto, W. Han, C. Pu, D. Rocco, and L. Xiong. Querying Multiple Bioinformatics Information Sources: Can Semantic Web Research Help? *SIGMOD Record*, 31(4):59–64, December 2002.

[2] D. Carmel, E. Farchi, Y. Petruschka, and A. Soffer. Automatic query refinement using lexical affinities with maximal information gain. In *SIGIR '02: Proceedings of the 25th annual international ACM SIGIR conference on Research and development in information retrieval*, pages 283–290. ACM Press, August 2002.

[3] C. Carpineto and G. Romano. A lattice conceptual clustering system and its application to browsing retrieval. *Machine Learning*, 24(2):95–122, August 1996.

[4] C. Carpineto and G. Romano. Order-theoretical ranking. *Journal of the American Society for Information Science*, 51(7):587–601, May 2000.

[5] C. Carpineto and G. Romano. *Concept Data Analysis: Theory and Applications*. John Wiley & Sons, 2004.

[6] S. B. Davidson, J. Crabtree, B. P. Brunk, J. Schug, V. Tannen, G. C. Overton, and C. J. Stoeckert. K2/Kleisli and GUS : experiments in integrated access to genomic data sources. *IBM systems journal*, 40(2):512–531, 2001.

[7] C. Discala, X. Benigni, E. Barillot, and G. Vaysseix. DBCAT: a catalog of 500 biological databases. *Nucleic Acids Research*, 28(1):8–9, January 2000.

[8] M. Y. Galperin. The Molecular Biology Database Collection: 2004 update. *Nucleic Acids Research*, 32:D4–D22, 2004.

[9] P. Ganesan, H. Garcia-Molina, and J. Widom. Exploiting hierarchical domain structure to compute similarity. *ACM Transactions on Information Systems (TOIS)*, 21(1):64–93, January 2003.

[10] B. Ganter and R. Wille. *Formal Concept Analysis*. Mathematical Foundations, Springer-Verlag, 1999.

[11] C. A. Goble, R. Stevens, G. Ng, S. Bechhofer, N. W. Paton, P. G. Baker, M. Peim, and A. Brass. Transparent Access to Multiple Bioinformatics Information Sources. *IBM Systems Journal*, 40(2):532–551, 2001.

[12] R. Godin, G. W. Mineau, and R. Missaoui. Méthodes de classification conceptuelle basées sur les treillis de Galois et applications. *Revue d'intelligence artificielle*, 9(2):105–137, 1995.

[13] R. Godin, R. Missaoui, and H. Alaoui. Incremental Concept Formation Algorithms Based on Galois (Concept) Lattices. *Computational Intelligence*, 11:246–267, 1995.

[14] J. Kohler, S. Philippi, and M. Lange. SEMEDA : ontology based semantic integration of biological databases. *Bioinformatics*, 19(18):2420–2427, December 2003.

[15] S. O. Kuznetsov and S. A. Obiedkov. Comparing Performance of Algorithms for Generating Concept Lattices. *Journal of Experimental & Theoretical Artificial Intelligence*, 14:189–216, 2002.

[16] P. Lord, S. Bechhofer, M. D. Wilkinson, G. Schiltz, D. Gessler, D. Hull, C. Goble, and L. Stein. Applying semantic web services to Bioinformatics: Experiences gained, lessons learnt. In F. v. H. Sheila A. McIlraith, Dimitris Plexousakis, editor, *The Semantic Web ISWC 2004: Third International Semantic Web Conference, Hiroshima, Japan, November 7-11, 2004. Proceedings*, volume 3298, pages 350–364. Springer-Verlag GmbH, 2004.

[17] T. Oinn, M. Addis, J. Ferris, D. Marvin, M. Greenwood, T. Carver, Matthew, Pocock, A. Wipat, and P. Li. Taverna : a tool for the composition and enactment of bioinformatics workflows. *Bioinformatics*, 20:3045–3054, 2004.

[18] S. Osman. Réalisation d'un annuaire de sources de données génomiques en vue de la collecte et de l'intégration de données sur le web. Rapport de master professionnel sciences et techniques mention informatique, spécialité bio-informatique, Université Bordeaux I, Université Victor Segalen, Bordeaux II, Septembre 2004.

[19] N. Pernelle, M.-C. Rousset, H. Soldano, and V. Ventos. ZooM: a nested Galois lattices-based system for conceptual clustering. *Journal of Experimental and Theoretical Artifial Intelligence (JETAI)*, 14(2):157–187, September 2002.

[20] U. Priss. Lattice-based Information Retrieval. *Knowledge Organization*, 27(3):132–142, 2000.

[21] B. Safar, H. Kefi, and C. Reynaud. OntoRefiner, a user query refinement interface usable for Semantic Web Portals. In *Proceedings of Application of Semantic Web technologies to Web Communities, Workshop ECAI'04*, pages 65–79, Valencia, Spain, August 2004.

[22] G. Stumme, R. Taouil, Y. Bastide, and L. Lakhal. Conceptual Clustering with Iceberg Concept Lattices. In *Proceeding GI-Fachgruppentreffen Maschinelles Lernen (FGML'01)*, Universitat Dortmund 763, Oktober 2001.

[23] D. van der Merwe, S. A. Obiedkov, and D. G. Kourie. AddIntent: A New Incremental Algorithm for Constructing Concept Lattices. In P. W. Eklund, editor, *ICFCA Concept Lattices, Second International Conference on Formal Concept Analysis, ICFCA 2004, Sydney, Australia, February 23-26, 2004, Proceedings*, volume 2961, pages 372–385. Springer, 2004.

[24] R. Wille. Restructuring lattice theory: an approach based on hierarchies of concepts. *Ordered sets*, pages 445–470, 1982.

[25] R. Wille. Line diagrams of hierarchical concept systems. *International Classification*, 2:77–86, 1984.

[26] C. Wroe, R. Stevens, C. Goble, A. Roberts, and M. Greenwood. A suite of DAML+OIL Ontologies to Describe Bioinformatics Web Services and Data. *International Journal of Cooperative Information Systems*, 12(2):197–224, March 2003.

How Formal Concept Lattices Solve a Problem of Ancient Linguistics⋆

Wiebke Petersen

Institute of Language and Information
Department of Computational Linguistics
University of Düsseldorf
petersew@uni-duesseldorf.de

Abstract. In his grammar of ancient Sanskrit, Pāṇini represents the phonological classes as intervals of a list. This representation method and especially the actual list constructed by Pāṇini, which is called the *Śivasūtras*, earns universal admiration. The legend says that god *Śiva* revealed the *Śivasūtras* to Pāṇini in order to let him start developing his grammar of Sanskrit. A question still discussed is whether it is possible to shorten the *Śivasūtras*. In the course of this paper, I am going to prove that this question can be reduced to a question about the graph-theoretical form of a particular formal concept lattice. Furthermore, I show how the *Śivasūtras* can be reconstructed from Pāṇini's grammar.

1 Introduction

1.1 Pāṇini's Grammar of Sanskrit

Pāṇini's grammar of Sanskrit (see [1], commented edition) is not only one of the oldest recorded grammars (according to [2], it dates from around 350 BC), but also one of the most complete grammars of any language ever written. It earns universal admiration among linguists: "The descriptive grammar of Sanskrit, which Pāṇini brought to its perfection, is one of the greatest monuments of human intelligence and an indispensable model for the description of languages", ([3]). Pāṇini developed a number of ingenious techniques to represent his grammar system in a very compact and concise way, including the introduction of a semi-formalized meta-language and an intricate system of conventions governing rule applications (e.g. [4], [5]). Since the grammar was designed for oral tradition, its compactness was particularly desirable, and the linear form of the whole grammar was a prerequisite.

The science of Sanskrit developed as a tool used for the preservation and propagation of the Vedas, the religious scriptures of ancient Hindus. The various Vedic texts were produced in different regions of India, beginning about 1500 BC (see [6]). In the course of time, a gap developed between the language of

---

⋆ Thanks to James Kilbury for providing me with the subject of this paper as a nice riddle.

the ancient scriptures and the colloquial use of Sanskrit. This gap affected both the phonetic form of the orally preserved texts and the comprehension of them. The late Vedic texts tell an anecdote to illustrate the importance of correct pronunciation and stressing (Śatapatha-Brāhmaṇa 1.6.3.8): The demon Tvaṣṭṛ longed for a son who would kill the war god Indra. But instead of begging for an *indra-śatrú* ('Indra-killer'), he asked for an *índra-śatru* and got a son who was killed by Indra. Hence, a guidance for the correct recitation of the religious texts became necessary. Pāṇini's grammar comprises both the Vedic Sanskrit and bhāṣā, the language spoken by the priestly class of his time, to which we refer as *classical* Sanskrit nowadays. Pāṇini's grammar and its canonization laid the foundations for the development of Sanskrit into a *lingua franca* of adminsitration and science.

Sanskrit is an inflected language with a rich morphology and sandhi.[1] The mastery of the sandhi rules is particularly important since two methods of reciting are used in rituals: the standard method of reciting whole continuous sūtras and the *padapāṭha* – 'word for word'-recitation – of the Veda. For the latter, the sūtras must be analyzed and decomposed into single words and all sandhi-processes must be canceled. This explains why the phonology of Sanskrit stands in the center of interest.

Pāṇini's grammar consists of four components: *Aṣṭādhyāyī*, *Śivasūtras*, *Dhātupāṭha*, and *Gaṇapāṭha*. The *Aṣṭādhyāyī* is the central component consisting of about 4.000 rules, which make references to classes defined on the elements of the other three components. The *Śivasūtras* (see Fig. 1) are the smallest component and consist of only 14 sūtras, which comprise a list of the phonological segments of Sanskrit and meta-linguistically used stop markers. According to this list, the natural phonological classes of Sanskrit are defined by a representation method specified in the *Aṣṭādhyāyī*. The *Aṣṭādhyāyī* refer to the phonological classes defined by the *Śivasūtras* in 100s of rules.

अइउण् ॥ऋऌक् ॥एओङ् ॥ऐऔच् ॥हयवरट् ॥लण् ॥ञमङणनम् ॥झभञ् ।
घढधष् ॥जबगडदश् ॥खफछठथचटतव् ॥कपय् ॥शषसर् ॥हल् ॥

$a \cdot i \cdot uṇ \,||\, r \cdot ḷk \,||\, e \cdot oṅ \,||\, ai \cdot auc \,||\, hayavaraṭ \,||\, laṇ \,||\, ñamaṅaṇanam \,||\, jhabhañ \,||$
$ghaḍhadhaṣ \,||\, jabagaḍadaś \,||\, khaphachaṭhathacaṭatav \,||\, kapay \,||\, śaṣasar \,||\, hal \,||$

Fig. 1. Pāṇini's *Śivasūtras*

Figure 1 shows the *Śivasūtras* in a linear sequence of sūtras, as constructed by Pāṇini for oral tradition. Nowadays, it is common to present them in the tabular form of Tab. 1 to support readability. Each sūtra consists of a sequence

---

[1] Sandhi refers to the systematic phonological modifications morphemes and words undergo if they are combined. An example of a sandhi phenomenon in English is the variation of the indefinite determiner (a/an), which depends on the first sound of the following word.

Table 1. Pāṇini's *Śivasūtras* in tabular form

1.	a	i	u				Ṇ
2.				ṛ	ḷ		K
3.		e	o				Ṅ
4.		ai	au				C
5.	h	y	v	r			Ṭ
6.						l	Ṇ
7.	ñ	m	ṅ	ṇ	n		M
8.	jh	bh					Ñ
9.			gh	ḍh	dh		Ṣ
10.	j	b	g	ḍ	d		Ś
11.	kh	ph	ch	ṭh	th		
			c	ṭ	t		V
12.	k	p					Y
13.		ś	ṣ	s			R
14.	h						L

of phonological segments, denoted in the table by lower case letters, followed by one stop marker (called *anubandha*), identified by using capitals. The *anubandhas* are taken from the set of consonants of Sanskrit. As a result, some consonants occur twice in the list: once as an *anubandha* and once as a phonological segment. Furthermore, there is one phonological segment, namely *h*, and one *anubandha*, Ṇ, occurring twice in the same role.

Pāṇini represents the phonological classes of Sanskrit as intervals of the list given by the *Śivasūtras*. Thereby, each class is encoded as a continuous sequence by giving its start segment and the marker element immediately following the last segment of the sequence. Two questions, which are discussed to this day, are whether it is possible to optimize the *Śivasūtras* with respect to the length and how Pāṇini was able to construct the *Śivasūtras*. About the latter, the legend says that god *Śiva* revealed the *Śivasūtras* to Pāṇini in order to let him start developing his grammar of Sanskrit. In the course of this paper, I am going to prove that the question of optimality can be reduced to a question about the graph-theoretical form of a special formal concept lattice. Furthermore, I prove that the *Śivasūtras* can be reconstructed from the *Aṣṭādhyāyī* without making a claim on additional aids. Hence, the hypothesis that the *Śivasūtras* must be necessarily older than the *Aṣṭādhyāyī* proves untenable (e.g. [7]).

We will use a simple phonological rule of Sanskrit as an example to show how the *Śivasūtras* interact with the *Aṣṭādhyāyī*. Phonological rules are operational rules of the form, "*A* is replaced by *B* if preceded by *C* and succeeded by *D*," or in modern notation:[2]

$$A \rightarrow B/C\_D \ . \tag{1}$$

As mentioned before, Pāṇini's grammar is designed for oral tradition and hence, non-linguistically signs like '$\rightarrow$' cannot be used in rule representations. Pāṇini

---

[2] Alternatively, such a rule can be denoted as a context-sensitive rule: $CAD \rightarrow CBD$.

denotes operational rules by using grammatical case markers to prescribe the role an expression plays in a rule. The functions of such meta-linguistically used case suffixes are laid down in individual sūtras of the *Aṣṭādhyāyī*; e.g., the interpretation of the genitive suffix is laid down in sūtra 1.1.49:

1.1.49 *ṣaṣṭhī stāneyogā* (**षष्ठी स्तानेयोगा**)[3] "The function of the genitive case in a sūtra is that of the phrase 'in the place of' when no special rules qualify the sense of the genitive" ([8]).

The four components of (1) are marked by case markers as follows: A is marked by the *genitive*, B by the *nominative*, C by the *ablative*, and D by the *locative* suffix.

Sūtra 6.1.77. of the *Aṣṭādhyāyī* serves us as an example for the interaction of the *Śivasūtras* with the *Aṣṭādhyāyī*; it encodes the phonological rule of Sanskrit that the vowels of the class $\langle i, u, ṛ, ḷ \rangle$ are replaced by their non-syllabic (consonantal) counterparts $\langle y, v, r, l \rangle$ if they are followed by a vowel:

6.1.77. *iko yaṇ aci* (**इको यण् अचि**)

Cancelling all sandhi processes results in the *padapāṭha* or 'word-for-word'-form *ikaḥ yaṇ aci*, which is morphologically analyzed as:

$$[ik]_{genitive} \quad [yaṇ]_{nominative} \quad [ac]_{locative} \; .$$

Like the grammatical case markers, the technical expressions **ik**, **yaṇ** and **ac** belong to Pāṇini's meta-language, too; they are called *pratyāhāras* and denote phonological classes. A *pratyāhāra* consists of a phonological segment followed by an *anubandha*. The vowel 'a' in the expression **yaṇ** fulfills two tasks: first, it serves as a linking vowel which turns the *pratyāhāra* into a pronounceable syllable, and second, it prevents the consonant 'y' from being mistaken for the *anubandha* 'Y'. Using the convention of distinguishing the *anubandhas* by capitals, we can write the *pratyāhāras* of sūtra 6.1.77 as **iK**, **yN** and **aC**. *Pratyāhāras* denote intervals of the *Śivasūtras*, their interpretation is defined by sūtra 1.1.71 of the *Aṣṭādhyāyī*: A *pratyāhāra* consisting of a phonological marker a and an *anubandha* M denotes the continuous sequence of phonological segments in the *Śivasūtras* which starts with a and ends with the phonological segment which is the direct predecessor of the *anubandha* M. Table 2 shows the interpretation of the *pratyāhāras* which are involved in sūtra 6.1.77. Note that although **N** denotes two distinct *anubandhas*, the meaning of the *pratyāhāra* **yN** is unambiguous since only one of the *anubandhas* is a successor of **y**.

Now we are able to state the phonological rule encoded in sūtra 6.1.77 in a modern form:

$$iK \rightarrow [yN]/\_\,[aC] \; .$$

It states that the elements of the class iK= {i,u,ṛ,ḷ} are replaced by their counterparts of the class yN= {y, v, r, l} if they occur right in front of a member of the

---

[3] The sūtras are given in Latin transliteration and *Devanāgarī*.

Table 2. Interpretation of the *pratyāhāras* of sūtra 6.1.77. (**iK**= $\{i,u,ṛ,l̥\}$, **yN̄**= $\{y,v,r,l\}$, **aC**= $\{a,i,u,ṛ,l̥,e,o,ai,au\}$)

class aC= {a,i,u,ṛ,l̥,e,o,ai,au}. Other sūtras ensure that the correct 'counterpart' is selected.

There are 100s of sūtras in the *Aṣṭādhyāyī* using *pratyāhāras* for the denotation of phonological classes, but altogether not more than 41 different *pratyāhāras* are actually used for this purpose. We will refer to those 41 phonological classes of Sanskrit identified by Pāṇini in the *Aṣṭādhyāyī* as *Pāṇini's phonological classes*.

Phonological segments form a natural phonological class if they behave similarly in similar phonological contexts. This analogous behavior can be expressed in generalized rules as we have seen in sūtra 6.1.77. The phonological classes of a grammar are mutually related: classes can be subclasses of other classes, two or more classes can have common elements, etc. These connections are naturally represented in a hierarchy. The *pratyāhāra* representation encodes such connections in a linear form. An aim of this paper is to determine the conditions under which a set of sets in fact has a *Śivasūtra*-style linear representation.

1.2   The Economy Problem of the *Śivasūtras*

Pāṇini does not discuss the criteria on which he constructed the *Śivasūtras*; but, by looking at the intricate methods he used in the *Aṣṭādhyāyī* to make it as compact as possible, it becomes clear that he aimed at an economical representation.

Since the *Śivasūtras* denote a list of phonological segments and *anubandhas*, two interesting sublists can be regarded: the list of sounds and the *anubandha* list. Hence, the *Śivasūtras* can be optimized in three respects, concerning the length of the lists:

1. The length of the whole list is minimal.
2. The length of the sublist of the *anubandhas* is minimal and the length of the whole list is as short as possible.
3. The length of the sublist of the sounds is minimal and the length of the whole list is as short as possible.

It should be noted that none of these minimality criteria implies one of the others. Looking at the *Śivasūtras*, the double occurrence of the sound h is especially astonishing. This is why the third minimality criterion is in the focus of attention in research on the economy of Pāṇini's *Śivasūtras*. For example, [9] and [10] argue that Pāṇini's *Śivasūtras* are optimal with respect to the third criterion, using linguistic principles and arguments referring to the construction principles of the other parts of the grammar.

By looking at the phonological segments as attributes and at the phonological classes as objects, a formal context can be associated with the *Śivasūtras*.[4] I am going to prove that Pāṇini's *Śivasūtras* respect the third minimality criterion and that this property depends solely on two facts: first, the corresponding formal concept lattice is planar and second, there exists a plane drawing of the Hasse diagram of the lattice in which each attribute concept lies at the boundary of the infinite face if one removes the top node of the lattice and all edges connecting it with co-atoms of the lattice (see Fig. 8).

The rest of this paper is organized as follows: This introductory section will be completed by some preliminary definitions formalizing the third minimality criterion and two short subsections about Formal Concept Analysis and the theory of planar graphs. Section 2 shows that Pāṇini was forced to duplicate at least one of the phonological segments and presents a sufficient condition for the existence of a *Śivasūtra*-style representation. Finally, Sect. 3 explains, how an optimal *Śivasūtra*-style representation can be constructed if it exists. Furthermore, it is proven that solely the duplication of the h enabled Pāṇini to construct the *Śivasūtras* such that they fulfill the third minimality criterion.

1.3   Preliminary Definitions

The following definitions formalize the main concepts of the preceding sections: Definition 1 derives the notion of an *S-alphabet* from the linear form of the *Śivasūtras*. Definition 2 and Def. 3 generalize Pāṇini's method of using *pratyāhāras* to represent phonological classes. The phenomenon of the duplicated phonological segment h in Pāṇini's *Śivasūtras* is covered by the notion of an *enlarged S-alphabet*. Finally, Def. 4 formalizes the third minimality criterion.

Definition 1. *A* well-formed Śivasūtra-alphabet *(short* S-alphabet*) is a triple* $(\mathcal{A}, \Sigma, <)$ *consisting of two disjoint finite sets* $\mathcal{A}$ *and* Σ, *and a total order* $<$ *on* $\mathcal{A} \cup \Sigma$. $\mathcal{A}$ *is called the* alphabet *and* Σ *the* marker set.

Definition 2. *A subset* T *of the alphabet* $\mathcal{A}$ *is* S-encodable *in* $(\mathcal{A}, \Sigma, <)$ *if and only if there exists* $a \in \mathcal{A}$ *and* $M \in \Sigma$, *such that* $T = \{b \in \mathcal{A} | a \leq b < M\}$. aM *is called the* pratyāhāra *or* S-encoding *of* T *in* $(\mathcal{A}, \Sigma, <)$.

In the following, we call a pair $(\mathcal{A}, \Phi)$ consisting of a finite set $\mathcal{A}$ and a set Φ of subsets of $\mathcal{A}$ (i.e., $\Phi \subseteq \mathbf{P}(\mathcal{A})$) a *system of sets*.

---

[4] Formal contexts and formal concept lattices are defined in Sec. 1.4.

Definition 3. *An S-alphabet* $(\mathcal{A}', \Sigma, <)$ *corresponds to a system of sets* $(\mathcal{A}, \Phi)$ *if and only if* $\mathcal{A} = \mathcal{A}'$ *and each element of* Φ *is S-encodable in* $(\mathcal{A}', \Sigma, <)$. *An S-alphabet which corresponds to* $(\mathcal{A}, \Phi)$ *is called an* S-alphabet *of* $(\mathcal{A}, \Phi)$. *A system of sets for which a corresponding S-alphabet exists is said to be* S-encodable.

For example, take the set of subsets

$$\Phi = \{\{d, e\}, \{b, c, d, f, g, h, i\}, \{a, b\}, \{f, i\}, \{c, d, e, f, g, h, i\}, \{g, h\}\} \quad (2)$$

of the alphabet $\mathcal{A} = \{a, b, c, d, e, f, g, h, i\}$; it is S-encodable and

$$a\, b\, M_1\, c\, g\, h\, M_2\, f\, i\, M_3\, d\, M_4\, e\, M_5 \quad (3)$$

is one of the corresponding S-alphabets. The S-encodings of Φ are: dM_5, bM_4, aM_1, fM_3, cM_5, and gM_2.

In order to formalize the third minimality criterion and to deal with the double occurrence of h in the *Śivasūtras*, we need the concept of *enlarging* an S-alphabet: $\hat{\mathcal{A}}$ is said to be an *enlarged alphabet* of $\mathcal{A}$ if there exists a surjective map $\vartheta : \hat{\mathcal{A}} \to \mathcal{A}$. ϑ extends naturally to sets: $\vartheta : \mathbf{P}(\hat{\mathcal{A}}) \to \mathbf{P}(\mathcal{A})$. It is clear that for every system of sets $(\mathcal{A}, \Phi)$ we can find an enlarged alphabet $\hat{\mathcal{A}}$ and a set of subsets $\hat{\Phi}$ with $\Phi = \{\vartheta(\varphi') : \varphi' \in \hat{\Phi}\}$ such that $(\hat{\mathcal{A}}, \hat{\Phi})$ is S-encodable. To achieve such an S-encodable system of sets $(\hat{\mathcal{A}}, \hat{\Phi})$ we enlarge $\mathcal{A}$ so that the sets of $\hat{\Phi}$ are pairwise disjoint. Then we arrange the sets of $\hat{\Phi}$ in an arbitrary sequence and separate them by markers. The induced S-alphabet $(\hat{\mathcal{A}}, \hat{\Sigma}, \hat{<})$ corresponds obviously to $(\hat{\mathcal{A}}, \hat{\Phi})$.

An S-alphabet of $(\hat{\mathcal{A}}, \hat{\Phi})$ will sometimes be called an *enlarged S-alphabet* of $(\mathcal{A}, \Phi)$. Since we always find a finite, enlarged S-alphabet of $(\mathcal{A}, \Phi)$, a minimally enlarged S-alphabet exists.

Definition 4. *An enlarged S-alphabet* $(\hat{\mathcal{A}}, \hat{\Sigma}, \hat{<})$ *of* $(\mathcal{A}, \Phi)$ *is said to be* optimal *if and only if it fulfills the following conditions: First, there exists no other enlarged S-alphabet* $(\tilde{\mathcal{A}}, \tilde{\Sigma}, \tilde{<})$ *of* $(\mathcal{A}, \Phi)$, *the alphabet* $\tilde{\mathcal{A}}$ *of which has fewer elements than* $\hat{\mathcal{A}}$ *and furthermore, as a secondary condition, no other enlarged S-alphabet* $(\tilde{\mathcal{A}}, \tilde{\Sigma}, \tilde{<})$ *of* $(\mathcal{A}, \Phi)$ *exists with* $|\tilde{\mathcal{A}}| = |\hat{\mathcal{A}}|$ *and* $|\tilde{\Sigma}| < |\hat{\Sigma}|$.

1.4 Formal Concept Analysis

Formal Concept Analysis (see [11]) starts with the definition of a *formal context* K as a triple (G, M, I) consisting of a set of objects G, a set of attributes M, and a binary *incidence relation* $I \subseteq G \times M$. For any subset of objects $A \subseteq G$, their set of common attributes is defined as $A' := \{m \in M | \forall g \in A : (g, m) \in I\}$. Analogously, the set of common objects for $B \subseteq M$ is $B' := \{g \in G | \forall m \in B : (g, m) \in I\}$. A *formal concept* is a pair (A, B) with the properties $A = B'$ and $B = A'$, where A is called the *extent* and B the *intent* of the concept. The set of all formal concepts of a context is partially ordered by the subconcept-superconcept-relation: $(A_1, B_1) \leq (A_2, B_2) \Leftrightarrow A_1 \subseteq A_2 \Leftrightarrow B_1 \supseteq B_2$. The set of formal concepts together with this partial order forms a complete lattice, called the *formal concept lattice*. As usual, we denote the set of all formal concepts of a formal context (G, M, I) by $\mathcal{B}(G, M, I)$ and the corresponding concept lattice by $\underline{\mathcal{B}}(G, M, I)$. The *attribute concept* $\mu(m)$ associated with an attribute m is the greatest concept whose intent contains m, and analogously, the *object concept* $\gamma(g)$ of an object g is the smallest concept whose extent contains g.

Formal Concept Analysis has been applied to a number of linguistic problems before. A survey of these linguistic applications can be found in [12]. What is remarkable about the application discussed in the present paper is that the attention is focused on the graph-theoretical form rather than the order-theoretical form of the formal concept lattices as it is generally the case.

1.5   Criterion of Kuratowski on Planar Graphs

If a graph can be drawn in the Euclidean plane, such that neither a vertex nor a point of an edge lies in the inner part of another edge (i.e., no crossing edges exist), then it is said to be *planar*.[5] A lattice is said to be a *planar lattice* if its Hasse-diagram augmented by an extra edge from the top to the bottom node is a planar graph. One of the most important criteria for the planarity of graphs is the criterion of Kuratowski, which is based on the notion of minors of a graph. A graph M is said to be a *minor* of a graph G if it can be arrived from G by first removing a number of vertices and edges from G and then contracting some of the remaining edges.

Proposition 1 (Criterion of Kuratowski). *A graph G is planar if and only if G contains neither a K^5 nor a $K_{3,3}$ as a minor (see Fig. 2).*

Fig. 2. The complete graph K^5 with 5 vertices (left) and the complete bipartite graph $K_{3,3}$ with $2 \cdot 3$ vertices (right)

2   Planar Formal Concept Lattices and S-encodability

2.1   Are Pāṇini's Phonological Classes of Sanskrit S-encodable?

To each system of sets $(\mathcal{A}, \Phi)$ we define the *corresponding context* $(\Phi, \mathcal{A}, \ni)$. The Hasse-diagram of $\underline{\mathcal{B}}(\Phi, \mathcal{A}, \ni)$ gives us a first hint on whether $(\mathcal{A}, \Phi)$ is S-encodable:

---

[5] Formal definitions of planar graphs and planarity can be found in most textbooks on graph theory (e.g. [13]).

Proposition 2. *If $(\mathcal{A}, \Phi)$ is S-encodable, then the corresponding formal concept lattice $\underline{\mathcal{B}}(\Phi, \mathcal{A}, \ni)$ is planar.*

Proof. Let $(\mathcal{A}, \Sigma, <)$ be an S-alphabet of $(\mathcal{A}, \Phi)$. The proof is based on the construction of a plane drawing of the Hasse-diagram of $\underline{\mathcal{B}}(\mathcal{A}, \Phi, \in)$ with stair-shaped edges.[6] For each formal concept (A, B) of $(\mathcal{A}, \Phi, \in)$, its coordinates in $\mathbf{R}^2$ are given as follows: The smallest object of A w.r.t. $(\mathcal{A}, \Sigma, <)$ determines the x-coordinate of the vertex;[7] its y-coordinate is given by the length of the longest descending chain between (A, B) and $(\emptyset'', \emptyset')$ in $\underline{\mathcal{B}}(\mathcal{A}, \Phi, \in)$.

The edges of the constructed Hasse-diagram are stair-shaped polygonal arcs: Let (A, B) and $(\bar{A}, \bar{B})$ be two formal concepts of $(\mathcal{A}, \Phi, \in)$ with $(A, B) \prec (\bar{A}, \bar{B})$. If $\min(A) = \min(\bar{A})$, then the edge between (A, B) and $(\bar{A}, \bar{B})$ is a straight line; else the vertices (A, B) and $(\bar{A}, \bar{B})$ are connected by the polygonal arc (see Fig. 3)

$$((\bar{A}, \bar{B})_x, (\bar{A}, \bar{B})_y), ((\bar{A}, \bar{B})_x, (\bar{A}, \bar{B})_y - \frac{1}{2}),$$

$$((A, B)_x - \frac{1}{2}, (\bar{A}, \bar{B})_y - \frac{1}{2}), ((A, B)_x - \frac{1}{2}, (A, B)_y), ((A, B)_x, (A, B)_y) \ .$$

The only exception to this edge-construction rule is that every edge between a concept (A, B) and $(\emptyset'', \emptyset')$ is just a straight line. The construction of the edges guarantees that no vertex of the Hasse-diagram lies in the inner part of an edge.

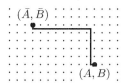

Fig. 3. Stair-shaped edge of the Hasse-diagram

With a simple but detailed case distinction it can be proven that every conflict which occurs between two edges (i.e. every crossing of two edges) can be solved by slightly transforming one of the edges in such a way that the distance between the transformed and the original edge does not exceed $\frac{1}{4}$.   □

Figure 4 shows the resulting plane Hasse-diagram of $\underline{\mathcal{B}}(\mathcal{A}, \Phi, \in)$ with Φ and $\mathcal{A}$ taken from example (2); the vertices are placed w.r.t. the S-alphabet given in (3).

It follows as a corollary that a system of sets is not S-encodable whenever the corresponding formal concept lattice is not planar.

Together with Kuratowski's criterion this proves that Pāṇini's phonological classes are not S-encodable since Fig. 5 shows a section of the concept lattice

---

[6] Rotating the constructed Hasse-diagram by $180°$ yields in a plane Hasse-diagram of $\underline{\mathcal{B}}(\Phi, \mathcal{A}, \ni)$.

[7] If the smallest element of A is the n-th smallest element of $\mathcal{A}$ in $(\mathcal{A}, \Sigma, <)$, then the x-coordinate is n. If $A = \emptyset$, then the x-coordinate is 0.

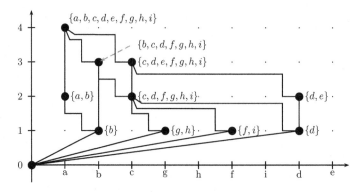

Fig. 4. Stair-shaped plane Hasse-diagram of the concept lattice of $(\mathcal{A}, \Phi, \in)$ with Φ and $\mathcal{A}$ taken from (2); the vertices are placed w.r.t. the S-alphabet in (3).

corresponding to the phonological classes, which has K^5 as a minor. Hence, Pāṇini was forced to duplicate at least one of the phonological segments. But it remains to prove that h is the best candidate for the duplication; this discussion will be postponed.

Proposition 3. *Pāṇini's phonological classes of Sanskrit are not S-encodable.*

2.2   Excursus: A Sufficient Condition of S-encodability

The condition for S-encodable systems of sets given in Prop. 2 is necessary but not sufficient, however. Figure 6 (left) shows an example of a system of sets which is not S-encodable, although its corresponding concept lattice is planar. We need a stronger condition to fully identify those systems of sets which are S-encodable.

Proposition 4. *Let $(\mathcal{A}, \Phi)$ be a system of sets and $\bar{\Phi} = \Phi \cup \{\{a\} : a \in \mathcal{A}\}$. The following statements are equivalent:*

1. $(\mathcal{A}, \Phi)$ is S-encodable.
2. $\underline{\mathcal{B}}(\bar{\Phi}, \mathcal{A}, \ni)$ is planar.

Proof. By adding a new singleton $\{a\}$, $a \in \mathcal{A}$, to Φ, the S-encodability is preserved (at most one new marker immediately following a has to be inserted in an S-alphabet of $(\mathcal{A}, \Phi)$). Hence, $(\mathcal{A}, \Phi)$ is S-encodable if and only if $(\mathcal{A}, \bar{\Phi})$ is S-encodable. Together with Prop. 2, this proves that statement 1. implies statement 2.

Given a plane drawing of $\underline{\mathcal{B}}(\bar{\Phi}, \mathcal{A}, \ni)$, all singletons of $\bar{\Phi}$ are co-atoms of the lattice and the x-coordinates of their corresponding attribute concepts induce a total order on $\mathcal{A}$. Inserting a marker behind each element of $\mathcal{A}$ yields an S-alphabet encoding of $(\mathcal{A}, \bar{\Phi})$.   □

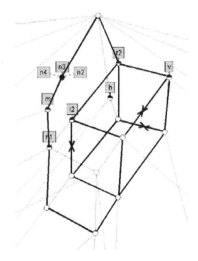

Fig. 5. The figure shows a section of the concept lattice corresponding to Pāṇini's phonological classes which has K^5 as a minor. The minor K^5 can be derived by deleting all but the highlighted edges and contracting the edges marked by crosses. The figure shows that the class memberships of the phonological segments h, v and l (denoted by $l2$) are independent of each other.

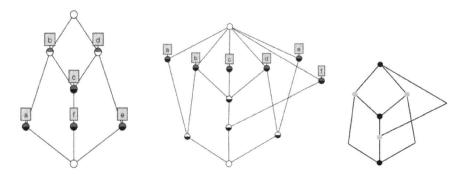

Fig. 6. Left: formal concept lattice of $(\Phi, \mathcal{A}, \ni)$ with $\Phi = \{\{d, e\}, \{a, b\}, \{b, c, d\}, \{b, c, d, f\}, \{a, b, c, d, e, f\}\}$; middle: formal concept lattice of $(\bar{\Phi}, \mathcal{A}, \ni)$ which has $K_{3,3}$ as a minor (see right figure)

It follows as a corollary, that a system of sets $(\mathcal{A}, \Phi)$ is S-encodable whenever a plane Hasse-diagram of $\underline{\mathcal{B}}(\Phi, \mathcal{A}, \ni)$ exists in which each attribute concept lies at the boundary of the infinite face if one removes the vertex $(\emptyset', \emptyset'')$ from the Hasse-diagram. This boundary graph is called the *S-graph of* $(\mathcal{A}, \Phi)$ and it is fixed up to isomorphism. The left part of Fig. 7 shows a plane drawing of the formal concept lattice corresponding to (2) in which the S-graph is highlighted.

Looking back at the example given in Fig. 6, it is clear that by moving from Φ to $\bar{\Phi}$ the concept lattice loses the quality of being planar; $\underline{\mathcal{B}}(\bar{\Phi}, \mathcal{A}, \ni)$ has the bipartite graph $K_{3,3}$ as a minor (figure on the right side). Hence, $(\mathcal{A}, \Phi)$ cannot be S-encodable; this can be also derived from the fact that the attribute concept $\mu(f)$ does not lay in the S-graph of the Hasse diagram of $\underline{\mathcal{B}}(\Phi, \mathcal{A}, \ni)$ (left figure).

3   Constructing *Śivasūtra*-Style Representations

3.1   The S-graph Determines the S-alphabet

If $(\mathcal{A}, \Phi)$ is a system of sets which is S-encodable, then an S-alphabet $(\mathcal{A}, \Sigma, <)$ of $(\mathcal{A}, \Phi)$ can be found as follows: Take the labeled S-graph of $(\mathcal{A}, \Phi)$ and a path in it, that starts and ends at the vertex corresponding to $(\mathcal{A}', \mathcal{A}'')$. The path must meet the following conditions: First, the path passes each attribute concept at least once; second, none of the edges occurring more than once in the path is part of a cycle in the S-graph. By looking at the S-graph as a subgraph of the directed Hasse-diagram, the edges of the path can be directed.

The S-alphabet, seen as a sequence of markers and elements of $\mathcal{A}$, can be constructed from the empty sequence by traversing the path from the beginning to the end: If an attribute concept $\mu(a)$ is reached, then add a to the sequence. If an edge is passed whose direction contradicts the traversal direction, a new, previously unused, marker element is added to the sequence, unless the last added element is already a marker. Finally, after the end of the path is reached, revise the sequence as follows: If an element of $\mathcal{A}$ appears more than once in the sequence, delete all but the first occurrences. The definition of the S-graph guarantees that, if the path passes an attribute concept $\mu(a)$ more than once, the path goes upwards immediately after it reaches $\mu(a)$ for the first time. Hence, eliminating all but the first occurrence of a reduces the number of markers in the resulting S-alphabet.

Applied to our small example (2), we may choose the path illustrated in Fig. 7, which fulfills the required conditions. Traversing the path, we pass first $\mu(a)$ and $\mu(b)$ without using an edge against its destined direction. Now we move downwards and violate the direction of the edge, and therefore we have to add a marker to our sequence, so that it starts with $a\,b\,M_1$. Now moving upwards we collect the c and the g, but since $\mu(g) = \mu(h)$ we also have to collect the h. After this we move downwards again, and that is why we add a new marker. We again reach $\mu(c)$ and add c a second time to our sequence. So far our sequence is $a\,b\,M_1\,c\,g\,h\,M_2\,c$, and if we continue we end up with the S-alphabet depicted in (3).

Note that this procedure does not yield a unique S-alphabet since we have several decisions to make: (a) If two attribute concepts are identical the order of the attributes in the S-alphabet is arbitrary; (b) from $\mu(c)$ we can either go to $\mu(g)$ or $\mu(i)$; (c) the path can be traversed clockwise or anti-clockwise.

Whenever a run violates the direction of an edge immediately after passing an attribute concept, a new marker has to be added to the S-alphabet. Hence,

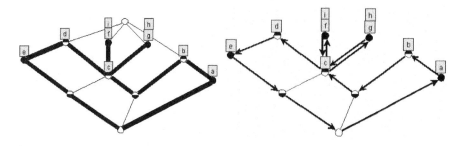

Fig. 7. Left: plane Hasse-diagram of the formal concept lattice corresponding to (2) with highlighted S-graph; right: possible path in the S-graph from which the S-alphabet $a\,b\,M_1\,c\,g\,h\,M_2\,f\,i\,M_3\,d\,M_4\,e\,M_5$ can be achieved

every optimal S-alphabet of $(\mathcal{A}, \varPhi)$ can be constructed by finding a run through the S-graph which minimizes the number of such marker-insertion situations.

3.2 Pāṇini's *Śivasūtras* Are Optimal

Figure 8 shows a plane Hasse-diagram of the concept lattice of $(\hat{\varPhi}, \hat{\mathcal{A}}, \ni)$, where $\mathcal{A}$ is the alphabet of phonological segments of Sanskrit, $\varPhi$ is the set of Pāṇini's phonological classes, and $\mathcal{A}$ and $\varPhi$ are enlarged by duplicating the segment h according to Pāṇini (the duplication of h is denoted by h_-).[8] The black and the striped rectangles next to some of the vertices mark the places where markers have to be added, depending on the traversal direction (black: anti-clockwise [14 markers], striped: clockwise [17 markers]). It is obvious that no S-encoding can have less than 14 markers and the optimal S-alphabets are the various combinatorial variants of

$$\langle a,i,u,M_1,\underline{r},\underline{l},M_2,\{\langle\{e,o\},M_3\rangle,\langle\{ai,au\},M_4\rangle\},h,y,v,r,M_5,l,M_6,$$
$$\tilde{n},m,\{\bar{n},\underline{n},n\},M_7,jh,bh,M_8,\{gh,\underline{d}h,dh\},M_9,j,\{b,g,\underline{d},d\},M_{10},$$
$$\{kh,ph\},\{ch,\underline{t}h,th\},\{c,\underline{t},t\},M_{11},\{k,p\},M_{12},\{\acute{s},\underline{s},s\},M_{13},h,M_{14}\rangle.$$

among which Pāṇini's *Śivasūtras* can be found. [9] argues in detail that the order chosen by Pāṇini out of the set of possibilities is unique if one requires a subsidiary principle of restrictiveness.

So far we have argued that Pāṇini was forced to enlarge the alphabet of phonological segments, but it remains to show why duplicating the h is the best choice. If h is entirely removed from the 41 phonological classes, then the optimal S-alphabet has only one marker less, namely 13.

In Pāṇini's phonological classes the phonological segments h, v, and l occur independently of each other. The Hasse-diagram of a concept lattice of a formal context which contains three independent attributes has K^5 as a minor and is therefore not planar (see Fig. 5). Triples of three independent attributes are called K^5-*triples*. Hence, to get a planar concept lattice it is necessary to duplicate at least one element of each K^5-triple.

---

[8] Drawings done by 'Concept Explorer' (http://www.sourceforge.net/projects/conexp).

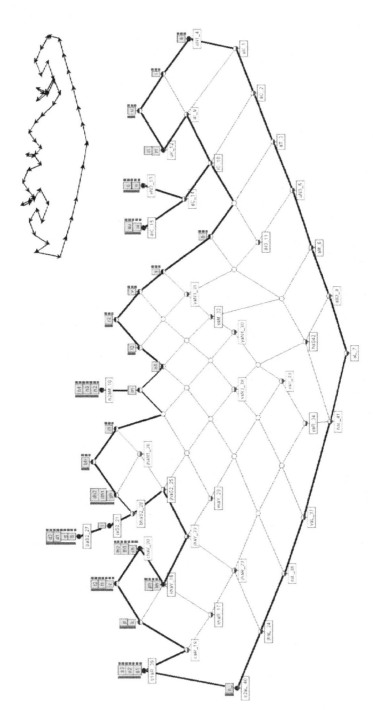

Fig. 8. Plane Hasse diagram of the concept lattice of Pāṇini's phonological classes of Sanskrit. The denotations in the figure are as follows: *h*⌣ is the duplicate of *h* and r1:r, l1:l, r2:r, l2:l, n1:ñ, n2:n, n3:ṇ, n4:n, dh1:ḍh, dh2:dh, dh2:ḍh, d1:ḍ, d2:d, th1:ṭh, th2:th, t1:ṭ, t2:t, s1:ś, s2:ṣ, s3:s.

The white boxes mark the phonological classes. The small figure on top shows the path in the S-graph we have to choose in order to construct Pāṇini's *Śivasūtras*.

Looking at Pāṇini's phonological classes, we find 249 K^5-triples; each of them contains h, and no other element is contained in each of them. Hence, to avoid the duplication of h it would be necessary to duplicate more than one element. For this reason, there is no other choice then duplicating h in order to get an optimal S-alphabet corresponding to Pāṇini's phonological classes.

This answers the question whether Pāṇini's Śivasūtras are optimal in the sense that there exists no other sequence of the phonological segments interrupted by less stop markers.

Proposition 5. *Pāṇini's* Śivasūtras *form an optimal S-alphabet.*

4 Outlook

Pāṇini's method of linearly encoding a subset of a power set could also be interesting from the viewpoint of other coding and sorting problems. However, it should be noted that the property of S-encodability with no or only moderate enlargement of the alphabet set is rare, at least among the phonological systems of natural languages (see [14]). Hence, linguists should investigate how Pāṇini's phonological analysis of Sanskrit differs from phonological classifications of other languages. Furthermore, it would be interesting to give an alternative mathematical description of S-encodable systems of sets and enlarged S-alphabets for testing and using the property.

References

1. Katre, S.M.: Aṣṭādhyāyī of Pāṇini. University of Texas Press, Austin (1987)
2. Kiparsky, P.: Paninian linguistics. In Asher, R.E., ed.: The Encyclopedia of Language and Linguistics. Volume 6. Pergamon Press, Oxford (1994)
3. Bloomfield, L.: Review of Liebich, Konkordanz Panini-Candra. Language **5** (1929) 267–276
4. Kiparsky, P.: On the architecture of Pāṇini's grammar. Lecture notes (2002)
5. Ostler, N.: Sanskrit studies as a foundation for computational linguistics. In: Proceedings of the LESAL Workshop, Mumbai (2001)
6. Deshpande, M.M.: Ancient Indian phonetics. In Koerner, E.F.K., Asher, R.E., eds.: Concise history of the language sciences: From the Sumerians to the cognitivists. Elsevier, Oxford, New York, Tokyo (1995) 72–77
7. Faddegon, B.: The mnemotechnics of Pāṇini's grammar I: The Śiva-Sūtra (1929). In Staal, F., ed.: A Reader on the Sanskrit Grammarians. Motilal Banarsidass (1985) 275–285
8. Vasu, S.C., ed.: The Aṣṭādhyāyī of Pāṇini, Allahabad (1891) (2 volumes). Reprint: Delhi (1962)
9. Kiparsky, P.: Economy and the construction of the Sivasutras. In Deshpande, M.M., Bhate, S., eds.: Paninian Studies. Ann Arbor, Michigan (1991)
10. Staal, F.J.: A method of linguistic description. Language **38** (1962) 1–10
11. Ganter, B., Wille, R.: Formal Concept Analysis. Mathematical Foundations. Springer, Berlin (1999)

12. Priss, U.: Linguistic applications of Formal Concept Analysis. (In: Proceedings of ICFCA 2003)
13. Diestel, R.: Graph Theory. Springer, New York (1997)
14. Petersen, W.: A mathematical analysis of Pāṇini's Śivasūtras. Journal of Logic, Language and Information **13** (2004) 471–489

A New Method to Interrogate and Check UML Class Diagrams

Thomas Raimbault, David Genest, and Stéphane Loiseau

Université d'Angers, Laboratoire d'Etude et de Recherche d'Angers (LERIA),
2 boulevard Lavoisier 49000 ANGERS Cedex 01 - France
{thomas.raimbault, david.genest, stephane.loiseau}@info.univ-angers.fr

Abstract. We present a new method for graphically interrogating and checking *UML class diagrams*. We employ the model of *conceptual graphs* (CGs) as representation, calculation and visualisation model. The key idea of our work is to translate UML class diagrams into the formalism of CGs. First, UML notations are encoded into *UML Ontology* that is a *support* of CG. Second, using the UML Ontology, a UML class diagram can be translated into a CG, called *CG class diagram*. Third, CG class diagrams can be interrogated via the elementary operation of CG, named *projection*. Fourth, constraints and rules provides a way to model specifications for checking CG class diagrams. We use two approaches to check a class diagram: *object-oriented specifications* and *field specifications*.

1 Introduction

Unified Modeling Language (UML) [BJR98] is the graphic modeling language, which becomes the industrial de-facto reference notation to express object-oriented systems. UML defines different diagram types representing different aspects and views of a system, especially the *class diagram*. The class diagram is considered in this paper, because it is the core of object modeling. However, UML is just a language: it does not specify any means to interrogate or check diagrams. To obtain a reliable modeling of a system, the designer needs on the one hand to look at his design objectively. So, he has to interrogate the diagrams. On the other hand, the diagrams must be checked: the designer has to specify his own checking specifications and understand the results provided by checks.

In spite of there are commercial tools today like Rational Software Rose [IBM04] or Borland Together [Bor04] for designing UML diagrams, they work as follows. First, the process of asking query on class diagrams are limited to pre-formatted queries. Second, the suggested checks are only standard: they implicitly check object-oriented concepts. The goal of our work is to offer to the designer a more complete method for interrogating and checking UML class diagrams. We employ the model of *conceptual graphs* (CGs) as representation, calculation and visualisation model.

The CG model is introduced by [Sow84]. A formulation of simple CGs is proposed by [CM92, MC96]. Thereafter, many extensions of this basic model were proposed. We use in this article the model of *typed nested CGs* [CM97, CMS98],

F. Dau, M.-L. Mugnier, G. Stumme (Eds.): ICCS 2005, LNAI 3596, pp. 353–366, 2005.

with *coreference links* [CM04], *rules* [Sal98, BM02] and *constraints* [BGM99a] [BM02]: the Appendix recalls the fundamental definitions. The CG model presents many interests. The CG is a formal and visual knowledge representation model. It provides reasoning operations, which are sound and complete with respect to deduction in first order logic (see [Sow84, CM92] for simple graphs, [CMS98] for nested graphs, and [Sow84, Wer95, KS97] for more general graphs equivalent to first order logic). Throughout this article, we will employ the term "conceptual graph" for "typed nested conceptual graph".

To answer the requirements of quality and of human-system interaction in modeling, our contribution is a graphic method: first a translation of UML class diagrams into CGs is provided. Second, a new method to interrogate and to check class diagrams is presented. Figure 1 presents a global schema of our approach. The key idea of our work is to translate UML class diagrams into the formalism of CGs, where UML notations are encoded into *UML Ontology* that is a *support* of CG. Thus, a given class diagram is automatically translated into a CG, called *CG class diagram*, that is defined on UML Ontology. As a result we can reuse the reasoning power of the CG model, with graphic operators, rules and constraints. Our first fundamental contribution is a graphic method to check the validity of class diagrams. We use two approaches to check a class diagram. We propose on the one hand *object-oriented specifications*. These specifications are used to carry out checks on class diagrams, which have to be in keeping with object-oriented concepts. For example, "there should not be any inheritance cycle in a class diagram". Compared to the existing UML tools, our method works similarly but with this time an explicit language to represent the specifications. The checking specifications are not in "black boxes". One the other hand, we offer the opportunity for the user to create or to adapt, by the intuitive and drawing aspect of our method, particular specifications in the area of corporate knowledge. In addition to the object-oriented specifications, these particular specifications, that we call *field specifications*, are applied to make class diagrams in keeping with the firm's needs. For instance, "each class of a class diagram have to be associed with another class". Both categories of specifications are given as *positive or negative constraints* in the CG model. Our second fundamental contribution is the possibility to visually interrogate the syntactic contents of class diagrams. For example, "Does the *FourWheelDrive* class inherit from the *Car* class?". The user can draw a query as he wants it: a query is simply expressed as a class diagram in the formalism of CGs, called *CG query*.

This paper is organized as follows. Section 2 presents the *UML Ontology* that defines and organizes UML notations. Section 3 presents any propositions to translate a UML class diagram into a *CG class diagram*. Section 4 indicates how to interrogate a CG class diagram with *CG queries*. Section 5 presents two kind of specifications: *object-oriented specifications* and *field specifications*, which must be satisfy when a CG class diagram is checked.

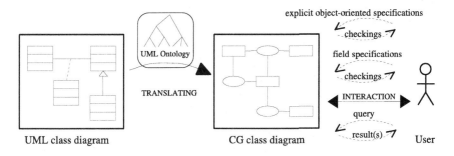

Fig. 1. Global presentation of our approach

2 UML Notations Modeled into *UML Ontology*

A short description of UML class diagram and its components is recalled in Section 2.1. Section 2.2 defines and organizes the UML notations into the set of concepts types, relation types and nested types of our *UML Ontology*.

2.1 A Short UML Class Diagram Description

Class diagram shows statical structure of a system, i.e. the system elements, their internal features and their relationships to other system elements. The system elements are modeled as classes in the class diagram. Two kinds of statical relations exist between classes: generalization and associations.

We present with an example, in Figure 2, the main elements of a UML class diagram. The *Car* class is drawn as a solid-outline rectangle. It contains the name of the class in the top compartment, the attributes *band* and *makingDate* in the middle compartment, and the *moveOff* operation in the bottom compartment. This operation has two parameters, one is an instance of *Driver* class and the other is a data type *int*, and returns a *bool*. The generalization relation is represented by an arrowed line drawn from the specialised class to the general class. Then, the *2CV* class has for generalization the *FrontWheelDrive* class, and the *Car* class has for subclasses the classes *FourWheelDrive*, *FrontWheelDrive* and *2CV*. An association between the classes *Driver* and *Car* is defined by the *Rental* association class and by the properties present at the ends of the association, like the multiplicity '0..N'. The association class is shown as a class symbol linked by a dashed line to a line symbol of association. More details on UML class diagrams are available in [BJR98, IBM04].

2.2 UML Ontology

We define and organize the ontology of object-oriented models into a structure that we call *UML Ontology*. The UML Ontology is based on [BJR98].

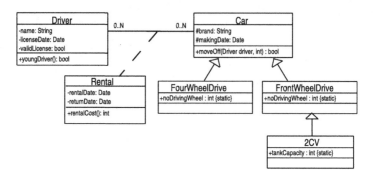

Fig. 2. UML class diagram

Definition 1 (UML Ontology). *UML Ontology is a support of CG, where UML notations are defined and organized into the set of concepts types (Figure 3), the set of relation types (Figure 4) and the set of nested types (Figure 5).*

Figure 3 presents the UML concepts that are ordered on a hierarchical set. Figure 4 shows the different associations that are possible between classes or between data. Figure 5 gives the different localizations where UML notations are placed to describe more specifically a concept.

3 CG Class Diagrams

We indicate now how to translate a UML class diagram into an equivalent CG class diagram.

Definition 2 (CG class diagram). *A CG class diagram is a CG, which is for one thing correctly formed on the UML Ontology, respecting the propositions in Sections 3.1 and 3.2; for another is in local normal form.*

Each class, attribute, operation and property (Section 3.1) is modeled by an own concept node with an appropriate type. The various elements, which describe the same concept node, are nested within this last. A nesting of a concept node constitutes an internal level of reading of the node. Note that a concept node can have defferent typed nestings.

Each relationship (Section 3.2) between classes is modeled by a binary-arity (resp. ternary-arity) relation node of 'generalization' (resp. 'association') type. This relation node links two concept nodes of 'class' type (resp. two concept nodes of 'class' type and one concept node of 'associationClass' type).

3.1 Internal Feature of Classes

The following propositions indicate how to translate a class, their properties and their members [1] in the formalism of CGs.

---

[1] Attribute and operation of a class are called members of this class.

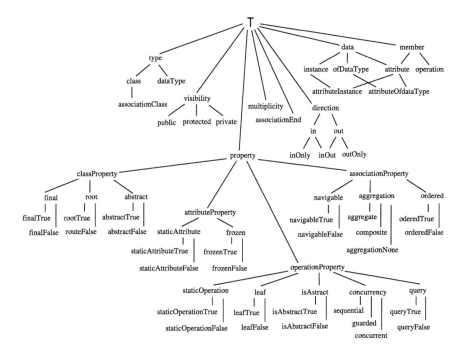

Fig. 3. Hierarchy of concept types

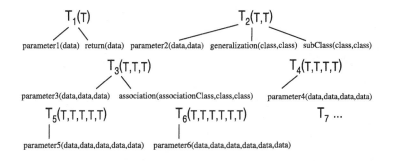

Fig. 4. Hierarchy of relation types

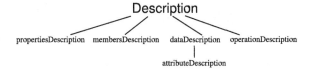

Fig. 5. Hierarchy of nesting types

Proposition 1 (Class in a CG class diagram). *A class is an individual concept node* c *of* class *type whose individual marker is the name of the class.*

A member of a class is represented by a concept node. The members of a class are nesteded into a typed nesting membersDescription *of* c.

A property of a class (resp. member) corresponds to a concept node from the general concept type visibility *and* classProperty *(resp.* attributeProperty *or* operationProperty*). The properties are nested within* c *into a typed nesting* propertiesDescription.

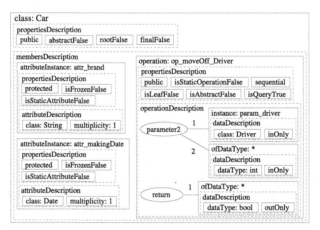

Fig. 6. Concept node of *class* type and its nestings

Proposition 2 (Attribute in a CG class diagram). *An attribute of a class is an individual concept node* a *whose the associed type can be of two kinds: either* attributeInstance *if the attribute is the instance of a class or* attributeOfDataType *if the attribute comes from a data-type. The individual marker is the name of the attribute, which is prefixed by 'attr_'.*

The type of an attribute is an individual concept node in the form of [class: className] *if the attribute is modeled by a concept node of a kind-of* instance *type. If not its form is* [dataType: typeName].

The multiplicity of an attribute is an individual concept node whose the associed type is multiplicity *and the individual marker is the multiplicity value.*

Both attribute's type and its multiplicity are nested within a *into a typed nesting* attributeDescription.

Proposition 3 (Operation and parameter in a CG class diagram). *An operation of a class is an individual concept node whose the associed type is* operation. *The individual marker is formed with the signature[2] of the operation. It is built in the following way: a prefix 'op_', followed name of the operation, the character '_', and the types of each parameter separated by '_'.*

---

[2] The signature of an operation is a set that gathers the name and the parameter types of the operation. The return parameter type is not use.

Just as an attribute, a parameter is a concept node in the form of [class: className] *or* [dataType: typeName], *which is nested within the concept node representing the operation into a typed nesting of* operationDescription.

The direction of a parameter is a concept node d, *which is nested within the concept node representing the parameter into a typed nesting* dataDescription. *The type of* d *comes from the general concept type* direction.

The parameters of an operation are associated to the same relation node (parameter$X$), *where X indicates the number of parameters, respecting the order between parameters. The return parameter is associated with the unary relation node* (return).

Figure 6 shows the *Car* class, described in Figure 2. It is represented by a concept node of 'class' type with an individual marker equals to 'Car'. This node has two nestings: one of 'propertiesDescription' type and the other of 'membersDescription' type. The types of concept nodes that are nested into the first nesting correspond to the properties of the class. Thus, the *Car* class is public and non abstract for example. The concept nodes into the second nesting represent the members of the class. In the same way, the *brand* attribute is represented by a concept node of 'attributeInstance' type with an individual marker equals to 'attr_brand'. This node has two nestings. The first one of 'descriptionProperties' type includes the attribute's properties. The second one of 'descriptionAttribute' type includes the type of data and the multiplicity of the attribute. Then, the *brand* attribute is an instance of 'String' class with a multiplicity equals to '1'. The *moveOff* operation has two parameters: first, *driver* as an instance of *Driver* class and second, a data without name of *int* data-type. Indeed, the individual concept node 'param_driver' of 'instance' type is linked on the first edge to a relation node of 'parameter2' type, and the generic concept node of *int* data-type on the second edge to this binary relation node. The operation returns a data of *bool* data-type.

The choice to prefix the names of attributes by 'attr_' and those of operations by 'op_' makes it possible to model for example an attribute and an operation, of the same class, which have the same name. The choice to use the signature of an operation and not only the name of the operation makes it possible to model the overload of operations.

3.2   Relationships Between Classes

There are two kinds of relationships between classes: generalization and association. The following definitions indicate how to translate generalization and association in the formalism of CGs.

Proposition 4 (Generalization in a CG class diagram). *A relation of generalization between two classes is modeled by a relation node of* generalization *type. The first neighbor if this relation node is the more specific class, and the second neighbor is the more general class.*

An association specifies a relation between several classes. Association is an abstraction which represents connections that exist between the classes of a given system.

We made the choice to always represent an association in its most complete form, i.e. by using an association class and informing how each class of an association takes part in it. Thus, an association consists of a central element: an *association class*. The classes that take part in association are called *class ends*.

Proposition 5 (Association in a CG class diagram). *An association class is modeled by a concept node of* associationClass *type.*

A relation node of association *type is first of all linked to the concept node of* associationClass *type, then at the concept node that correspond to a class end, and finally at a concept node of* associationEnd *type.*

The properties that characterize the way in which the class end takes part in an association are represented by a concept node of multiplicity *type and the concept nodes from the general concept type* associationProperty.

The Figure 7 shows the relationships between classes of the UML class diagram in Figure 2, which is translated in a CG class diagram. For readability purpose in this figure, properties' classes and their members are suggested by '...' within concept nodes. Generalization between the classes *Car* and *FourWheelDrive* for exemple is represented by a relation node of 'generalization' type. It is read in the following way: "the *FourWheelDrive* class has for generalization the *Car* class". Association between the classes *Driver* and *Car* is centralized by a concept node of 'associationClass' type. Concept nodes of 'association' type links the association class, the class ends and some details about these latest.

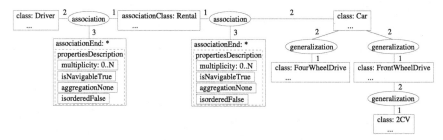

Fig. 7. Generalization and association links

4   Interrogate CG Class Diagrams

It is possible to make reasoning on a CG class diagram with *projection*. A form of reasoning is to interrogate a class diagram. For example, it can be interesting to know "the public operations of a given class", or to search "the subclasses of a given class".

Before interrogating a class diagram, we have to express the transitivity of generalization between classes. The application of rules makes possible this action. The closure of a CG class diagram has to be obtained before interrogating it.

Section 4.1 states what is a rule in the CG model, and presents two rules that represent transitivity of generalization. Section 4.2 presents how to interrogate CG class diagrams.

4.1   Rules and CGs

A rule [Sal98, BM02] allows one to add new knowledge. It is in the form of "if G_1 then G_2", where G_1 and G_2 are CGs. Thus, it is composed of a hypothesis and a conclusion, and is used in the following way: given a graph, if the hypothesis of the rule matches the graph, then the information contained in the conclusion is added to the graph. We follow here the notations of [BM02]: the representation of a rule is a bi-colored graph, which is extended to typed nested CGs.

Definition 3 (Rule and closure of CG [Sal98, BM02]). *A rule is a bi-colored CG, located by* $\boxed{\Rightarrow}$,*where the hypothesis is formed of the set of the nodes on white bottom, and the conclusion of the set of the nodes on black bottom.*

A rule R is said to be applicable *to a CG G if there is a projection, named Π, from the hypothesis of R into G. In this case, the result of the application of R to G following Π is the simple graph G_R obtained from G and the conclusion of R by restricting the label of each frontier node c in the conclusion to the label of its image $\Pi(c)$ in G, then joining c to $\Pi(c)$.*

A CG G is said to be closed *such a set of rules $\mathcal{R}$ if all information that can be added by a rule is already present in G.*

The Figure 8 shows two rules R_1 and R_2, which explicitly express the transitivity of generalization link. The hypothesis of R_1, on white bottom, is a class that has for generalization another class. The conclusion of R_1, on black bottom, is the second class has for subclass the second class. In other words, R_1 expresses that "If X class has for generalization Y, then Y has for subclass X". The rule R_2 expresses that "If X class has for subclass Y and Y has for subclass Z, then X has for subclass Z".

The application of the two rules in Figure 8 to the class diagram Figure 2 in the formalism of CGs has for result the CG in Figure 9.

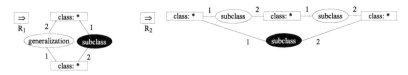

Fig. 8. Two rules to express the transitivity of generalization link

4.2   Interrogate CG Class Diagrams

With the UML Ontology, it is possible to interrogate a class diagram. A query is simply expressed as a class diagram in the formalism of CGs, called *CG query*. The idea is to find if the query could be matched on a subpart of the class diagram. In that case, the answer(s) of the query is (are) the generic concept node(s) of the query that correspond to its (their) matching(s) on the CG class diagram.

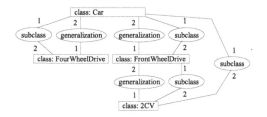

Fig. 9. Closure of CG with the rule of transitivity

Definition 4 (CG query). *Having the UML Ontology and a CG class diagram, a query on this class diagram is a CG that is defined on this same ontology, called CG query.*

Definition 5 (Result of a CG query). *The result of a CG query on a CG class diagram is the set of projections from the CG query to the CG class diagram.*

Suppose that we would like to interrogate the class diagram on Figure 2 with the following terms: "Does the *Car* class have any subclasses that are public?". The query is represented by the CG on the left on Figure 10. Indeed, this CG shows the *Car* class that represented by a concept node of 'class' type with an individual marker equals to 'Car'. This class has as a subclass a still unknown class. The latest is thus represented by a concept node of 'class' type without individual marker (i.e. a generic concept node). This query admits three results. Therefore results are: "Yes, the *Car* class has tree subclasses; and these subclasses are *FourWheelDrive*, *FrontWheelDrive* and *2CV*".

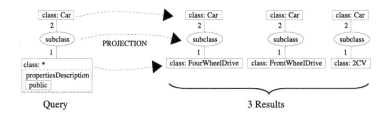

Fig. 10. Query and its results

5   Check CG Class Diagrams

There are two categories of checkings, which are expressed by *object-oriented specifications* and *field specifications*. First specifications are used to carry out checks on class diagrams that have to be in keeping with object-oriented concepts. The second specifications express needs of a specific project according to corporate objectives. Both object-oriented specifications and field specifications are represented by *constraints* of CGs.

Definition 6 (Validity of a CG class diagram). *A CG class diagram is valid iff it satisfies both* object-oriented specifications *and* field specifications.

Definition 7 (Object-oriented and field specifications). *Object-oriented specification and field specification are negative or positive constraints of CGs.*

Section 5.1 states what is a constraint. Section 5.2 and 5.3 present respectively any samples of object-oriented specifications and field specifications.

5.1   Constraints and CGs

We use two types of constraints: *positive constraint* and *negative constraint*, with the notations of [BGM99b, BM02] that are extended to typed nested CGs.

Definition 8 (Constraints of typed nested CGs). *A positive or negative constraint is a bi-colored typed nested CG. The condition of a constraint is drawn on white bottom, and the obligation for a positive constraint or the prohibition for a negative constraint on black bottom. Positive and negative constraints are respectively located by $\boxed{+}$ and $\boxed{-}$.*

Definition 9 (Constraint satisfaction [BGM99b, BM02]). *A concepual graph G satisfies a positive constraint C, if any projection of the condition of C to G can be extended to a projection of C to G. A concepual graph G satisfies a negative constraint C, if any projection of the condition of C to G can not be extended to a projection of the prohibition of C to G.*

5.2   Object-Oriented Specifications

To be in keeping with object-oriented concepts, a class diagram has to satisfy any appropriate specifications. These specifications, called *object-oriented specifications*, are given in the formalism of CG as positive and negative constraints.

Figure 11 shows one negative constraint C_1 and one positive constraint C_2. C_1 expresses the following prohibition: "an A class cannot be a subclass of a B class, which is a subclass of A". Then, this constraint checks that there is no inheritance cycle. The condition of C_2 is a class with an abstract operation, and this class has for subclass another class that is non-abstract. The obligation of C_2 is if the subclass has an operation o that as the same identifier of the first class' operation, then o must be not abstract. Thus, this positive constraint specifies that if a X class has abstract operations, a Y subclass of X can be not abstract only if the abstract oprerations of X are redefined as not abstract in Y. The coreference link between the two generic concept nodes of 'operation' type indicates that they are the same individual.

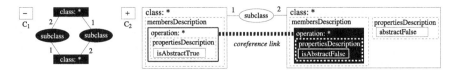

Fig. 11. Object-oriented specifications

5.3 Field Specifications

The *field specifications* can model any requirement in addition to object-oriented specifications that the user need according to corporate objectives on a specific project. These specifications are given in the formalism of CG as positive and negative constraints.

For instance, the positive constraint C_3 in Figure 12 specifies that any class has to be associed at another class. Note that all association relationships use an association class in this paper. The positive constraint C_4 represents the fact that a class cannot be abstract.

Fig. 12. Field specifications

6 Conclusion

A prototype was developed to translate a UML class diagram in XMI format [OMG02, Gen04b] into a CG class diagram in BCGCT format [Hae95]. In addition, rules and constraints were written with BCGCT format to check class diagrams and to satisfy the object-oriented concepts. Query and check on CG class diagrams were tested whith a tool that we have developped, basing on the CoGITaNT platform [Gen04a]. The results of interrogating and checking, with examples of class diagrams, are satisfactory. They provide the desired requirements in a quasi-instantaneous way.

Currently, an interface is under development to formulate queries and checks in the UML formalism, to which one adds a minimum of notations. Thus, although the CG model is the external format of our method of interrogating and checking UML diagrams, it can be transparent for the user if he wants.

We wish to extend our graphic method of translating, of interrogating and of checking to all type of UML diagrams. This extension will firstly offer a unique and visual formalism to express at the same time various UML diagrams, queries and checks. Secondly, queries and checks will be used between several UML diagrams.

References

[BGM99a]   J.F. Baget, D. Genest, and M.L. Mugnier. Knowledge acquisition with a pure graph-based knowledge representation model. In *Proc. of KAW'99*, volume 2, pages 7.1.1–7.1.20, 1999.

[BGM99b]   J.F. Baget, D. Genest, and M.L. Mugnier. A pure graph-based solution to the SCG-1 initiative. In *Proc. of ICCS'99*, volume 1640 of *LNAI*, pages 355–376. Springer, 1999.

[BJR98]   G. Booch, C. Jacobson, and J. Rumbaugh. *The Unified Modeling Language - a reference manual*. Addison Wesley, 1998.

[BM02]   J.F. Baget and M.L. Mugnier. Extensions of Simple Conceptual Graphs: the Complexity of Rules and Constraints. *JAIR*, 16(12):425–465, 2002.

[Bor04]   Borland. Together software, 2004. http://www.borland.com/together/.

[CM92]   M. Chein and M.L. Mugnier. Conceptual Graphs: Fundamental Notions. *Revue d'intelligence artificielle*, 6(4):365–406, 1992.

[CM97]   M. Chein and M.L. Mugnier. Positive nested conceptual graphs. In ICCS'97 [ICC97], pages 95–109.

[CM04]   M. Chein and M.L. Mugnier. Concept types and coreference in simple conceptual graphs. In *Proc. of ICCS'04*, volume 3127 of *LNAI*, pages 303–318. Springer, 2004.

[CMS98]   M. Chein, M.L. Mugnier, and G. Simonet. Nested graphs: A graph-based knowledge representation model with FOL semantics. In *Proc. of KR'98*, pages 524–534. Morgan Kaufmann Publishers, 1998.

[Gen04a]   D. Genest. CoGITaNT 5.1.5, 2004. http://cogitant.sourceforge.net.

[Gen04b]   Gentleware. Poseidon SE, 2004. http://www.gentleware.com/.

[Hae95]   O. Haemmerlé. La plate-forme CoGITo : manuel d'utilisation. Technical Report 95012, LIRMM, 1995.

[IBM04]   IBM. Rational rose, 2004. http://www-306.ibm.com/software/rational/.

[ICC97]   *ICCS'97*, volume 1257 of *LNAI*. Springer, 1997.

[KS97]   G. Kerdiles and É. Salvat. A sound and complete CG proof procedure combining projection with analytic tableaux. In ICCS'97 [ICC97], pages 371–385.

[MC96]   M.L. Mugnier and M. Chein. Représenter des connaissances et raisonner avec des graphes. *Revue d'intelligence artificielle*, 10(1):7–56, 1996.

[OMG02]   OMG. XML Metadata Interchange (XMI) Specification, 2002. http://www.omg.org/technology/documents/formal/xmi.htm.

[Sal98]   É. Salvat. Theorem proving using graph operations in the conceptual graph formalism. In *Proc. of ECAI'98*, 1998.

[Sow84]   J.F. Sowa. *Conceptual Structures: Information Processing in Mind and Machine*. Addison Wesley, 1984.

[Wer95]   M. Wermelinger. Conceptual graphs and first-order logic. In *Proc. of ICCS'95*, volume 964 of *LNAI*, pages 323–337. Springer, 1995.

7   Appendix

Definition 10 (Support of TNCGs [CMS98]). *A support of typed nested conceptual graph (TNCGs) is a 6-tuple $S = (T_C, T_R, T_U, \sigma, I, \tau)$. T_C is a partial ordered set of concept types, whose greatest element is $\top$ (the universal type). T_R is a partial ordered set of relation types, with a partition: $T_R = T_{Ri_1} \cup \ldots \cup T_{Ri_p}$*

where T_{Ri_j} is a set of i_j-ary relations, $i_j > 0$; thus, two comparable relation types must have the same arity n, and their greatest element is $\top_n$. T_U is a partial ordered set of nested types, whose greatest element is Description. In addition there is an empty description, noted $**$. The partial orders on T_C, T_R and T_U correspond to a kind-of *relation between types. σ is a mapping, which associates a signature with each relation type; the signature of a relation specifies the arity and greatest possible concept type for each argument. For any relation type $r \in T_{Ri_j}$, $\sigma(r) \in (T_C)^{i_j}$. I is a set of individual markers. In addition there is a generic marker, noted $*$, as non-specified individual. $I \in \{*\}$ is provided with an order, such that $*$ is the greatest marker, and individual markers are pairwise non-comparable. τ is a mapping from I to T_C.*

Definition 11 (Typed Nested Conceptual Graph [CM97, CMS98]). *A TNCG related to a support S, is a labelled bipartite graph $G = (R, C, U, lab)$. R and C are the node sets, respectively relation and concept node set. U is the set of edges. Edges incident on a relation node are totally ordered, they are numbered from 1 to the degree of the relation node. The i-th neighbor of a relation r is denoted by $G_i(r)$. Each node has a label given by the mapping lab such that $\forall r \in R, lab(r) \in T_R$ and $\forall c \in C, lab(c) = (type(c), ref(c), Desc(c))$ with $type(c) \in T_C$, $ref(c) = *$ or $ref(c) \in I$, $Desc(c) = **$ or $Desc(c)$ is sets of (nested type,a TNG). A concept node c such as $ref(c) \in I$ is said to be an individual concept node, if not (i.e. $ref(c) = *$) it is said to be a generic concept node. Moreover, lab respects the constraints fixed by the applications σ and τ: $\forall r \in R, type(G_i(r)) \leq \sigma_i(type(r))$, and $\forall c \in C$ if $ref(c) \in I$ (i.e. $ref(c) \neq *$) then $\tau(ref(c)) \leq type(c)$.*

Definition 12 (Projection of TNCGs [CM97, CMS98]). *A projection form a TNCG $H = (R_H, C_H, U_H, lab_H)$ to a TNCG $G = (R_G, C_G, U_G, lab_G)$ is a couple of applications $\Pi = (f, g)$, with $f : R_H \to R_G, g : C_H \to C_G$, which:*

1. *preserves the edges and the classification of the edges: for all edge rc of U_H, $f(r)g(c)$ is an edge of U_G. Plus, if $c = H_i(r)$ then $g(c) = G_i(f(r))$.*
2. *may decrease labels: $\forall r \in R_H, lab_G(f(r)) \leq lab_H(r)$; and $\forall c \in C_H$, given $lab_H(c) = (t, m, d)$ and $lab_G(g(c)) = (t', m', d')$, $lab_G(g(c))$ satisfies:*
   - *$(t', m') \leq (t, m)$*
   - *If d is a graph, then d' is a graph such as a projection from d to d' exists. If $d = **$, there is no constraint on d' that is a graph or equal to $**$.*

Definition 13 (Coreference link [CM04]). *A coreference link between two generic concept nodes represents the same individual in different nestings of a nested CG. A coreference link is symbolized by a dotted link connecting the two concept nodes representing the same individual.*

Definition 14 (Local normal form [CM04]). *A TNCG is in local normal form if there is no two concept nodes that have the same individual marker within a same nesting.*

Language Technologies Meet Ontology Acquisition

Galia Angelova

Institute for Parallel Processing, Bulgarian Academy of Sciences
25A, Acad. G. Bonchev Str., 1113 Sofia, Bulgaria
galia@lml.bas.bg

Abstract. This paper overviews and analyses the on-going research attempts to apply language technologies to automatic ontology acquisition. At first glance there are many successful approaches in this very hot field. However, most of them aim at the extraction of named entities as well as draft taxonomies and partonomies. Only few attempts exist for enriching ontologies by applying word-sense disambiguation. There are principle obstacles to extract automatically coherent conceptualisations from raw texts: it is impossible to identify exactly the types and their instances as well as the word meanings which denote types. It is also impossible to validate a text-based conceptual model against the real world. Thus we can expect only partial success in the semi-automatic acquisition in specific (limited) domains, by workbenches supporting the human knowledge engineer in the final ontological choices.

Keywords: natural language processing, information extraction, automatic knowledge acquisition from text

1 Introduction

Recent developments in artificial intelligence, knowledge representation, WWW, and information-processing applications resulted in the advent of the Semantic Web [1]. Its ultimate aim is to make the web resources more meaningful to computers by augmenting the presentation markup with semantic markup, i.e. meta-data annotations that describe the content. It is widely expected that the innovation will be provided by agents and applications dealing with ontology acquisition, merging and alignment, annotation of www-pages towards the underlying ontologies as well as intelligent, semantic-based text search and intuitive visualisation. However, the current progress in all these directions is not very encouraging despite the impressive number of running activities. Isolated results and tools are available for e.g. automatic and semi-automatic annotation of web pages, for knowledge-based information retrieval, for partial ontology learning and so on but it is still difficult to grasp a coherent picture of how the Semantic Web will drastically change the information age by offering quite new kinds of services. Another discouraging obstacle is that the ontologies for the Semantic Web are not clearly seen at the horizon. And, after five years of investments and active investigation, the vision of the "universal" Semantic Web becomes a "futuristic" target. Instead, the goals shift to realistic developments and notions like *"intelligent semantic-based Web applications"*, *"ontology-based tools"*, *"application ontologies"*, *"light-weight ontologies"*, *"bridging corporative views via negotiation"*,

F. Dau, M.-L. Mugnier, G. Stumme (Eds.): ICCS 2005, LNAI 3596, pp. 367-380, 2005.
© Springer-Verlag Berlin Heidelberg 2005

"mediation between entities in order to integrate them", *"use cases which demonstrate the viability of the approach"*[1] and so on.

In this way the development of the Semantic Web and the ontology-driven applications is currently slowed down due to principal problems, among them the knowledge acquisition bottleneck. Therefore, language technologies for free text processing are extensively applied to create or grow ontologies *"in a period as limited as possible with a quality as high as possible"* (citation from [2], a nice summary of the dominating project-oriented perspective). We can certainly remind that the idea of automatic knowledge acquisition from text is an older dream. However, the present achievements rely on *(i)* advanced language processing tools and *(ii)* very large linguistics resources – two artifacts that were not available twenty or thirty years ago. That is why the activities in this sphere deserve careful observation and analysis, as they are attempts to process all kinds of raw texts and really massive amounts of data.

This article analyses the on-going work in automatic knowledge acquisition by choosing several representative papers (instead of listing hundreds of titles), which contain research and application innovation from a language technology perspective. In addition, they try to map the results (extracted language units) to ontological elements and structures. We consider the latter mapping a decisive step in the automatic knowledge acquisition. Due to this reason we skip here many papers which deal with natural language understanding and word-sense disambiguation from the computational linguistics perspective, as they are focused on extraction of text semantics rather than on world knowledge. Section 2 overviews the progress in Named Entity Recognition (NER) – which is in fact a collection of NE instances in texts with annotations of their types. Section 3 presents approaches for learning of concept types from texts, by clustering of natural language terms into semantic sets standing for domain concepts. Section 4 summarises the achievements in detection of conceptual relationships and automatic constructions of taxonomies and enlargement of ontologies. Section 5 discusses kinds of ontological constructions which are *not* targets of automatic knowledge acquisition today.

2 Acquisition of Instances (Named Entities)

The tasks of identification, extraction and classification of named entities from texts have attracted special attention in the 80-ies (last century) in the emerging arear of Information Extraction (IE) and Message Understanding [3]. Indeed, all named entities – proper names, names of organizations, companies, locations, even dates – bear much information about the text content, genre, sometimes author etc. NER is usually approached by grammar-based text analysis. Often the extracted text items are mapped to very large lexicons of names which are created in advance. The collection of such multilingual lexicons is an important IE activity. The IE systems are compared at special competitions according to several parameters, among them the NER success rate. NER is evaluated via the percentage of correctly recognized named entities which is a measure for the grammar coverage as well as an indirect hint for

---

[1] Citations from the Semantic Web session at the event "Information Society Technologies 2004" (IST-04), organised by the European Commission, The Hague, 15-17 November 2004.

Entity	**Algorithm**
Abstract	**Application**
Happening	**Application_Domain**
Object	**Event**
Agent	**Conference**
Organisation	**Tutorial**
Person	**Workshop**
BusinessObject	**Formal_Language**
InformationResource	**Method**
Location	**Organisation**
AstronomicalObject	**Association**
Facility	**Department**
GlobalRegion	**Enterprise**
LandRegion	**Institute**
NonGeographicalLocation	**Research_Funding_Institution**
PoliticalRegion	**University**
PopulatedPlace	**Person**
StreetAddress	**Project**
WaterRegion	**Publication**
Product	**Book**
Statement	**InProceedings**
Vehicle	**Report**
	Tool
	Topic

Fig.1. *Left:* KIM Ontology. *Right*: a sample ontology in OntoMat Annotizer

the amount of the available lexical resources of names. The most advanced IE systems today recognize some kinds of named entities in unknown texts with more than 90% accuracy. For instance, KIM [4] recognizes English *person names* and *locations* in inter-domain web content with 89.09% and 91.23% accuracy correspondingly[2]. KIM's current lexicon of names (resource of instances) contains more than 200,000 items. Each location has several aliases (English, French, Spanish and sometimes the local transcription). The IE systems need such data, because locations are difficult to recognize otherwise. KIM tries to extract more information than the isolated instances; it searches for patterns defining attributes and relations of the featured entity, like: ***subRegionOf*** *property for* **Location**-*s*, ***hasPosition*** *for* ***Persons***, ***locatedIn*** *for* ***Organizations***, *etc*. The automatic extraction of such features is much less successful than the recognition of single, isolated named entities.

The essence of named entities extraction is to recognize them as instances according to some ontology of concepts. Figure 1 presents the KIM ontology, which is used for automatic classification of names. This ontology consists of 250 classes and has about 100 relations. Ontologies are also applied for manual annotation – i.e. classification - of named entities. A sample ontology in OntoMat Annotizer [5] is given at Figure 1 too. The two ontologies clearly illustrate that there are many kinds

---

[2] More precisely, IE applies the classical *recall R* (the ratio of correctly extracted entities against all the available entities present in the texts) and *precision P* (the ratio of correctly extracted entities against all the extracted entities). What we called accuracy above is the harmonic mean of *R* and *P* – known as *F-measure*, where $F = (2*P*R)/(P+R)$.

of named entities in the texts and obviously, their automatic recognition and classification require substantial efforts. The NER result - extracted "drafts of named entities" - needs further manual validation and human correction, in case we want to build correct lists of concept instances with some associated attributes.

In the context of NER, let us comment on several issues which are relevant to our discussion regarding automatic ontology acquisition.

First, we emphasize that the extracted named units are symbol strings that are "mechanistically" collected from different texts. For instance, the NER module might extract *Sofia* as a person name (most probably female) and as a city name (hopefully a city located in Bulgaria). But it is unlikely to extract information that there was a Byzantine princess *Sofia* who lived in the city of *Sofia*, at present in Bulgaria. This fact is too complex to be extracted at the NER stage. Usually only titles and professional positions are recognized successfully "around" the person names (e.g. *President Bush*). In addition, the task of NER cannot infer that many women are named *Sofia* or *Sophia* and often the two names are language-dependent variants. So the result of NER is a list of isolated items tagged via ontology types.

Second, the NER task does not distinguish well between different variants of the same name. For instance, *"The Bulgarian Academy of Sciences"* is organization name which can appear in the text as *"The Academy of Sciences"* or simply *"The Academy"* in some unambiguous contexts (while *the academy* often refers to the main building of the organization). In addition NER does not resolve the references in the text as this requires deep natural language understanding. However, the recognition of referential citations is important prerequisite for successful acquisitions of facts. Consider the following discourse, which consists of sentence 1 and sentence 2:

Sofia was a Byzantine princess. She lived in the city of Serdica, known today as Sofia.

The resolution of the pronominal anaphor *"she"* in sentence 2 to *"Sofia"* in sentence 1 is a necessary condition for extraction of the historical fact we want to encode. Now we make the following important observation: even the simplest facts regarding instances might be communicated by complex language structures in free texts. One may assume that the recognition of named entities is simpler – which is true in a sense – but the extraction of related facts may require implementation of the full potential of natural language processing and natural language understanding, which looks impossible today. The variety of natural language expressions is so high that important facts cannot be encoded without ambiguity even in controlled languages (at least there are still no experimental evidences for optimism beyond the naturally restricted languages of some very specific technical domains, e.g. aircraft industry).

So at this point of our discussion we already realise why we lack successful applications which acquire knowledge about arbitrary entities from free texts. Today propositions are acquired only from suitably formulated text statements. Due to the "natural understanding bottleneck", the automatic ontology construction is (still) not an alternative which may replace the precise manual definition and construction. The present approaches are focused on the acquisition of domain concepts as well as relevant statements that are easy to identify and extract. Another important activity is the integration of concepts into ontologies.

3 Acquisition of Concepts from Texts

The main machine learning approach, applied to extract concepts (semi-) automatically from texts, is clustering of co-occurrences of words. The idea is that similar words appear as collocations with the same verbs. Moreover, words are similar to the extent they appear in similar contexts (this is the so called distributional hypothesis). No annotation of the input text is needed beforehand. Usually the first step is to perform lexical and morphological analysis of the input. Then complex terms of several tokens are grouped as single units and afterwards (partial) syntactic analysis is performed to extract main sentence phrases or to fully parse each sentence. Below we briefly summarize some relevant activities and their results, keeping our focus to the language technologies involved in the process.

The paper [6] presents techniques applied in the ontology learning environment *Text-To-Onto*, such as ontology learning from free text, from dictionaries, or from legacy ontologies. A convincing example illustrates the benefit of processing a domain thesaurus, where the concepts are described together with their definitions and hierarchy. The environment *Text-To-Onto* was applied to a machine-readable dictionary of an insurance company which contained entries like the following one:

> ***Automatic Debit Transfer:*** *Electronic service arising from a debit authorization of the Yellow Account holder for a recipient to debit bills that fall due direct from the account.*

Several heuristics are applied to this morphologically analysed definition. One simple heuristic relates the definition term *automatic debit transfer* to the first noun phrase occurring in the definition - *electronic service*. Their corresponding concepts are linked in a draft hierarchy:

AUTOMATIC DEBIT TRANSFER *IS-A* ELECTRONIC SERVICE.

Applying this heuristic iteratively, one may propose large parts of the target ontology. The process of ontology learning is semi-automatic with human intervention, as the obtained hierarchy is further refined by a human engineer using the ontology engineering workbench *OntoEdit*. In this way, specific language resources like thesauri are very useful for the concept acquisition phase. However, often dictionaries with definitions are not available and then the acquisition starts from free texts.

The papers [2,7] present systematic attempts to build an ontology from scratch. The research is done within the project *OntoBasis*, which deals with elaboration and adaptation of text analysis tools for the construction of specific domain ontologies. Here we use a very simple example to illustrate the approach. The tools extract automatically binary relations (lexons) by mining the Verb-Object dependency, applying selectional restrictions and functional relations. In fact they extract pairs

[*Main Verb - Nominal String*],

where the Nominal String is a string of adjectives and nouns. Then the nominal strings are clustered according to the cooccuring verbs. A sample cluster from a Hepatitis corpus is:

{ *liver transplantation, transplantation, orthotopic liver transplantation* }.

The suggested concept is described in the corresponding medical text by three nominal strings.

In addition to the mining of the verb-object pairs, pattern matching is done for triples

[*Nominal string - Preposition - Nominal string*].

The resulting phrases look as follows:

```
blood_vessel_growth on ribonucleolytic_activity,
amino_acid_residue in polymerase,
primer from amino_acid_sequence.
```

These triples are organised in classes of prepositional structures and compared to the clusters obtained after mining the verb-object pairs. In case of similarity, clusters are augmented to the following extracts:

[*Nominal String   Preposition   AugmentedCluster*].

The obtained "correct" structures look as follows:

```
[dose, injection, vaccination] of
                   [hepatitis B vaccine, HBV vaccine, vaccine]
[use] of [face mask, mask, glove, protective eyewear]
[vaccination, vaccine] against [disease, virus, virus type]
```

Other resulting constructs are:

```
[level, expression] of …
[effect] on …
[increase] in …
```

Obviously, these lists need further human intervention for refinement and organization into ontology.

This unsupervised approach for ontology initiation looks promising, as it requires no preliminary tagging of the training data and relies on relatively simple language technologies and tools which are more or less available for many natural languages. However, it is not very successful. It was evaluated against UMLS (the Unified Medical Language System) by comparison of the extracted nominal strings to the UMLS labels. The *recall* (percentage of correctly recognized nominal strings against all relevant nominal strings) and the *precision* (percentage of correctly recognized nominal strings against all recognized nominal strings) were computed according to the quantity of UMLS pairs found in the clusters. The reported results for the Hepatitis corpus are less than 33% recall and less than 17% precision, for clustering of 100-500 words [7]. Adding prepositional information to the clusters increases the recall but the precision remains very low. The authors consider the approach useful for a preliminary step but only for a limited amount of words. The resulting structures obviously need human intervention for further refinement and structuring. It also becomes evident - because of the relatively low recall - that there are many UMLS concepts whose names are not formed by nominal strings or prepositional structures from the kind which is treated in [2, 7]. Please note that this well-documented work deals with ontology initiation using a minimum of preliminary available tools and resources; that is why we consider it very important and present it here in more details.

ASIUM, a more sophisticated system for concept acquisition, is presented in [8]. ASIUM learns knowledge from syntactically parsed text, so the input for the learning algorithm – conceptual clustering method - is fully analysed text. No concepts are given in advance. From the parsed input, ASIUM learns verb frames like:

```
<verb> <preposition | syntactic role: concept>.
```
A sample is given below:
```
<to drop> <object: Explosive> <in: Public_Place>
```

(the pairs `object:Explosive` and `in:Public_Place` are *subcategories*, `object` is a *syntactic role* and `in` is a *preposition* but `Explosive` and `Public_Place` are concepts used as *restrictions of selection*). The method learns by grouping similar subcategories into clusters. The resulting concepts are labeled by a human expert.

ASIUM relies on a language technology which is relatively sophisticated – a full syntactical analyser of French. The parser outputs also roles in the verb frames which look like the subcategorization frames but with concepts replaced by nouns:

```
<verb> <preposition | role: head noun>.
```
By grouping head nouns and semantic roles, ASIUM generalises the initial syntactic frames and covers by induction examples that did not occur as such in text. As shown in [8], ASIUM is a rather powerful concept acquisition system. Starting with the syntactic frames:

```
<to travel> <subject:[father,neighbour,friend]> <by:[car,train]>
<to drive> <subject:[friend,colleague]><object:[car,motorcycle]>
```

ASIUM will learn two concepts:
```
        Human: father; neighbor; friend; colleague.
        Motorized Vehicle: car; train; motorcycle.
```
and two subcategorization frames:
```
    <to travel><subject: Human> <by:Motorized Vehicle>
    <to drive><subject: Human> <object: Motorized Vehicle>
```
As the authors explain in [8], human experts control the link between the new concepts and the verbs because the threshold, fixed preliminary by the expert, does not avoid over-generalisation.

ASIUM inspired further research work for learning concepts from parsed texts. For instance [9] expands the ontology of the CIRCM-TUTOR system using a similar approach. It is difficult to compare directly all such systems for concepts acquisition, as they employ different modules for syntactic analysis and start from different inputs. But it is evident that *(i)* the deeper linguistic analysis has a positive effect to concept acquisition and *(ii)* the resulting concepts (clusters) always need to be revised by a human expert who also assigns them labels.

4 Learning Taxonomies and Enriching Ontologies

Here we briefly overview three approaches. One of them is based on deep linguistic information and very complex word sense processing. Its result is enriching WordNet with new domain concepts. The second one combines patterns from different sources to build taxonomic relations. The third one processes the output of a parser and organises the words/concepts in a hierarchy, following the Formal Concept Analysis. In this section it becomes evident that the task of taxonomic structuring always exploits information from deeper text analysis, which provides additional evidence concerning the meaning of the words and concepts met in the texts.

The tool OntoLearn [10] extracts domain ontologies from documents shared among the members of virtual organizations. Its first step is to extract and filter the domain terminology from the available documents. Because of their low ambiguity and high specificity, these words or phrases are very good candidates to label domain concepts, as they denote important domain concepts and relations. There are well-known methods for extracting stable collocations from text. OntoLearn uses rule-based tools developed earlier by the team and extracts domain terms, which are further evaluated by two specific measures: *domain relevance* (how often the term appears in the domain corpus, compared to a larger collection of corpora) and *domain consensus* (which measures the distributed use of a term in a domain corpus). Combining both measures, OntoLearn computes the *weight* of each term. Accepted terms have weight over a threshold which is set experimentally. In this way, accepted terms for *tourism* are *travel information, shopping street, airline ticket, booking form*, etc. while for *finance* accepted terms are *vice president, net income, executive officer, composite trading* and so on.

After selecting the domain terminology, OntoLearn builds subtrees of terms (concept labels) according to simple string inclusion. For instance, an initial hierarchy of terms in the travel domain may look like the one shown at Fig. 2. Without semantic

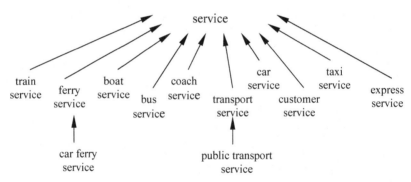

Fig. 2. A lexicalized tree in the domain of tourism [10]

interpretation, Fig.2 contains erroneous classifications, e.g. *bus service* is not classified as *public transport service*. The main innovation of OntoLearn is that it performs semantic interpretation of the complex terms by constructing it compositionally. OntoLearn uses the general WordNet senses and appropriate conceptual relations that hold among the concept components. The semantic resources are processed by the so-called *structural semantic interconnection* algorithm, a novel knowledge-based iterative approach to word sense disambiguation. Very roughly, as an illustration only, we remind that in WordNet *bus* is a kind of (*public*) *transport*, so it is possible to compute that *bus service* has to be classified as a *public transport service*. In this way the complex domain terms can be semantically interpreted and arranged in a hierarchical manner (please note that the terms – single words have to be defined somehow, either in WordNet or in thesauri, otherwise OntoLearn has no way to calculate their meaning). Figure 3 shows a domain concept tree, obtained from the lexicalized tree after the semantic interpretation. We see there that an *express* can be a *bus* or a *train*, and both interpretations are valid because they are obtained from relations between terms within the domain. Conceptual relations play important role in the semantic interpretation. The chosen kernel of conceptual relations is built using the definition of basic relations given in [11].

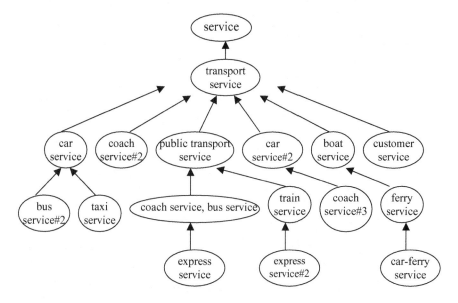

Fig. 3. Domain concept tree [10]

OntoLearn evaluation is difficult as the ontology assessment is still an open problem. There may be experimental validation and evaluation, by users who apply the ontology, as well as assessment against methodological evaluation guidelines. OntoLearn was evaluated as a tool supporting ontology engineers in the Harmonise project. During the first year of the project, a core ontology with 300 concepts was manually defined. Simultaneously OntoLearn extracted an initial list of 14,383 candidate terms and organised them in a domain concept forest of 3,840 concepts. The latter resource was submitted to the domain experts for ontology updating and integration. The obtained precision ranged from 72.9% to about 80% and the recall was 52.74%. The precision shift is due to the well-known fact that experts may have different intuitions about the relevance of a concept. The authors conclude: "…In any case, OntoLearn favoured a considerable speed up in ontology development, since shortly the Harmonise ontology reached about 3,000 concepts. Clearly, the definition of an initial set of basic domain concepts is crucial, so as to justify long lasting and even harsh discussions. But once an agreement is reached, filling the lower levels of the ontology can still take a long time simply because it is a tedious and time-consuming task." Therefore the authors consider OntoLearn as a very useful tool within the Harmonise project [10]. A drawback of OntoLearn is that it is unable to analyze totally unknown terms as it exploits much linguistic information which is available for the words known to the system only.

In the computational linguistics, OntoLearn is a state-of-the-art result for ontology construction using semantic interpretation. The aim is to enrich a general purpose ontology – WordNet – with the detected domain concepts (as WordNet does not cover complex domain terms). Today only few papers propose to enrich existing ontologies with new concepts. Another approach addressing documents in Internet is reported in [12], where the authors collect for each concept in WordNet the words that appear most distinctively in texts related to it (*topic signatures*) and apply them for clustering the concepts that lexicalize the word senses of a given word.

The results reported in [13] illustrate the spirit of the information retrieval approaches to taxonomy learning (in contrast to the rule-based solution of OntoLearn presented above). The system described in [13] builds taxonomic links by solving a classification tasks for every concept pair. Let us consider two terms, say *conference* and *event*. They could be either unrelated, or taxonomically related in three different ways: *is-a(conference,event)*, *is-a(event,conference)*, *siblings(conference,event)*. Therefore, intuitively, the decision is to gather as many different sources of evidence as possible and choose the relation with maximal evidence according to all of them. This approach combines information from:

(i) Linguistic patterns explicating *is-a* relations matched to a large text corpus [14];
(ii) Linguistic patterns explicating *is-a* relations matched to the Web [14];
(iii) WordNet and its hyponyms, and
(iv) the lexicalized *is-a* relation, as exemplified at Fig. 2.

For instance, a pattern for searching hyponyms for NounPhrases (NP) in corpus is
NP0 such as NP1, NP2, …, NPn-1 and/or NPn (see [14] for other patterns).

Similar patterns for searching in Internet may reveal descriptions of hyponymic relations. Table 1 summarises some results for *is-a* and the related probabilities:

conference is-a	*conference* is-a
event 0.44	*service* 0.27,
meeting 0.11	*meeting* 0.11,
activity 0.11	*activity* 0.11
A. Extraction from corpus	**B.** Extraction from Internet

Table 1. Evidence for taxonomic relations, extracted from different sources

The four sources of evidence are normalized in order to be comparable and combined by two very simple techniques (to take the *mean* of all possible values and to use the *max*imum). The result shows that combining diverse and heterogeneous information indeed leads to better results than classification according to a single source (and *conference is-a event*). The approach is evaluated against a handcrafted ontology for the touristic domain. The *mean* strategy with all WordNet senses yielded precision 17.16% and recall 29.84% with threshold t=0.01; and the one with the first WordNet senses – precision 17.38% and recall R=29.24% at t=0.01. The *max* strategy with all senses yielded a precision 16.03% and recall 29.87% at t=0.04. These results are comparable and show that there is a lot of potential in the combination of different approaches for text extraction.

The last result which we overview is presented in the paper [15]. It builds a taxonomy starting from the results of a parser. We explain the construction by example. Let us consider the sentences:

People book hotels. The man drove the bike along the beach.

After parsing them, and after identifying the basic word forms (which turns *hotels* to *hotel* and *drove* to *drive*), the following semantic relations can be extracted from the parsing results:

book_subj(people)   drive_obj(bike)   drive_subj(man)
book_obj(hotel)   drive_along(beach)

Not all dependencies from this kind are interesting, but the repeating ones reveal semantic links between the verb, the thematic role and the filler (word or concept). Thus *hotel*s are *book*able, *bike*s are *driv*able, *men* are *driv*ing etc. And in contrast to the similarity-based clustering, which is popular in taxonomy construction, one can consider the Formal Concept Analysis [16] as an alternative set-theoretic classification. The authors show a taxonomy built for the tourism domain, using the attributes as presented in the example above (Table 2 and Figure 4). The claim is that Figure 4 is much easier to read and trace for the human engineer, as in keeps the labels of the classification features. The conceptual hierarchy is built on the basis of the inclusion relations between the selectional restrictions of all the verbs. Thus FCA supports tracing of reasons why a taxonomy looks the way it is, according to linguistic knowledge acquired from text. Figure 4 is in fact an interesting idea of linking natural language processing and FCA in order to trace the properties of the extracted knowledge chunks. If the input corpora are updated regularly, there is a natural way to justify the ontology evolution accordingly. A more detailed evaluation of this approach is presented in [17].

	bookable	rentable	driveable	rideable	joinable
apartment	X	X			
car	X	X	X		
motor-bike	X	X	X	X	
excursion	X				X
trip	X				X

Table 2. Concepts and attributes as a formal context in the domain of tourism

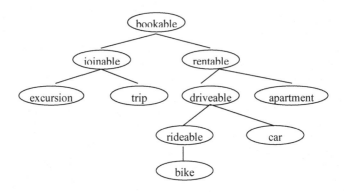

Fig. 4. A taxonomy built for Table 2, which reflects features extracted by parsing.

5 Discussion and Conclusion

In this paper we summarise typical approaches of how to employ language technologies in order to acquire knowledge from texts. We prefer to focus on prototypes which clearly organize the extracted language units into conceptual structures. For clarity and brevity, we grouped the approaches into three main categories: acquisition of instances, concepts, hierarchies. We emphasize that the extracted structures are often domain-dependent (because they are based on the input corpus) and need further refinement and formalization.

Despite the variety of the particular applications, there are some established strategies of how to apply language technologies in knowledge acquisition. For instance, it is common to extract phrasal descriptions after parsing and to encode the corresponding knowledge chunks. The assumption is that verbs pose more or less strong selectional restrictions on their arguments and this linguistic phenomenon reflects semantic relationships. The better the parser is, the more sentence phrases it extracts and the final result is more sophisticated and more accurate. However, the relatively low recall of this task is an evidence that many concepts are not communicated via compact phrasal descriptions and therefore are difficult to grasp by automatic text processing.

Together with the main language technologies in use (lexical and morphological analysers, POS taggers, tools for recognition of named entities, parsers, extractors of collocations, semantic interpreters of word senses) we exemplified the main classification technology applied for building a taxonomy – the unsupervised clustering. Mostly the *is-a* relation is considered. The resulting ontological structures are not perfect but they are viewed as drafts that need further justification by human experts. The concepts in the taxonomies mirror the words (strings) occurring in the source texts and the human expert may choose to delete some of them. If a word has several meanings and the input corpora are really large, then mining by patterns or counting the verb-object collocations will hopefully reveal all senses as shown at

Table 1. It is still unclear how to evaluate the usefulness and the correctness of a particular ontology acquired for application in certain domain.

However, word senses are *not* ontological concepts although there are many similarities between lexicons and ontologies [18]. An ontology is a set of categories and relationships among them while a lexicon depends on the word senses in particular natural language. In the same way the text semantics, extracted in formal propositions, is not conceptual model of the information encoded in the text. Moreover, gathering facts from Internet, we have no methods to evaluate whether the extracted knowledge is relevant to the real world. For instance, the NER module might collect named entities from science fiction sites. So even the results of the successful NER task, which are extracted with more than 90% linguistic accuracy, need conceptual verification before asserting them into a formal conceptual model. From this perspective, extracting knowledge from dedicated domain sites (e.g. ePortfolio portals) is a more promising approach for automatic ontology construction.

We have already said that attempts for automatic extraction of logical propositions are rare as the complexity of the task is well-known. Extraction of one fact from two sentences is not trivial, especially using the methods applied today, as the current parsers work over single sentences only. Another ontological construct that is not tackled automatically is the description of the classification perspective, since it is rarely encoded as a compact statement in free texts.

One may argue that the results presented above are discouraging because of their low accuracy. However the author prefers the optimistic vision. All the prototypes are at their initial implementation stage and the cited papers report work in progress. Moreover, the quality of the language technology tools is growing incrementally and this contributes to the increasing quality of the very large linguistic resources. In addition, there are many publicly available tools (e.g. POS-taggers and parsers for English) and open resources like WordNet. The interest in the multilingual resources and tools is growing too and much bilingual terminology is already extracted automatically. This facilitates the work of research groups in smaller countries who deal with languages other than English. The author believes that soon there will be large open archives of electronic text resources in a number of languages and many publicly available tools for language processing (similarly to the paper archives which were collected also incrementally). All these stimuli motivate further research attempts in automatic ontology acquisition. It seems unlikely that the quality of the automatically extracted ontologies will be satisfactory enough but at least drafting ontologies will be much easier. The author also believes that methods for ontology assessment against domain corpora will be developed in the near future.

References

[1] T. Berners-Lee. *Weaving the Web*. Harper, 1999.
[2] Reinberger, M.-L., Spyns, P., Pretorius, A.J. and Daelemans, W. *Automatic Initiation of an Ontology*. In R. Meersman, Z. Tari (Editors), Proc. CoopIS/DOA/OBDASE 2004, Springer, Lecture Notes in Computer Science 3290, 2004, pp. 600-617.
[3] L. Hirschman. *The Evolution of Evaluation: Lessons from the Message Understanding Conferences*. Computer Speech and Language 12, 1998, pp. 281-305.

[4] Popov, B., Kiryakov, A., Ognyanoff, D., Manov, D. and Kirilov, A. *KIM – a Semantic Platform for Information Extraction and Retrieval.* Journal of Natural Language Engineering 10 (3/4), 2004, pp. 375-392, see also URL: http://62.213.161.156/KIM/screen/KWUIMain.jsp

[5] Handschuh, S. and S. Staab. *CREAM: CREAting Metadata for the Semantic Web.* Computer Networks: The International Journal of Computer and Telecommunications Networking, Volume 42 , Issue 5, 2003, pp. 579 – 598.

[6] Maedche, A. and S. Staab. *Ontology Learning for the Semantic Web.* IEEE Intelligent Systems 16 (2), Special Issue on Semantic Web, 2001, pp. 72-79.

[7] Reinberger, M.-L. and P. Spyns: *Discovering Knowledge in Texts for the Learning of DOGMA-inspired Ontologies.* In the Proc. ECAI-2004 Workshop on Ontology Learning and Population: *Towards Evaluation of Text-based Methods in the Semantic Web and Knowledge Discovery Life Cycle,* August 2004, pp. 19-24.

[8] Faure, D. and T. Poibeau. *First Experiments of Using Semantic Knowledge Learned by ASIUM for Information Extraction Task Using INTEX.* In the Proc. of the Workshop on Ontology Learning, ECAI 2000, pp. 7-12.

[9] Lee, C.H., Seu, J. H. and M. Evens. *Building an Ontology for CIRCSIM-Tutor.* Proc. 13th Midwest AI and Cognitive Science Society Conference, MAICS-2002, Chicago, pp. 161-168.

[10] Navigli, R. and P. Velardi. *Learning Domain Ontologies from Document Warehouses and Dedicated Web Sites.* Journal of Computational Linguistics, Vol. 30, Issue 2, June 2004, pp. 151 - 179.

[11] Sowa, J. *Conceptual Structures: Information Processing in Mind and Machine,* 1984, Addison-Wesley, Reading, MA.

[12] Agirre E., Ansa1, O., Hovy E. and D. Martínez. *Enriching very large ontologies using the WWW.* In the Proc. of the Workshop on Ontology Learning, ECAI 2000, pp. 37-42.

[13] Cimiano P., Pivk A., Schmidt-Thieme L. and S. Staab. *Learning Taxonomic Relations from Heterogeneous Evidence.* In the Proc. ECAI-2004 Workshop on Ontology Learning and Population, 2004.

[14] Hearst, M.A., *Automatic Acquisition of Hyponyms from Large Text Corpora,* in Proc. COLING-1992, pp. 539-545.

[15] Cimiano P., Hotho, A. and S. Staab. *Comparing Conceptual, Divisive and Agglomerative Clustering for Learning Taxonomies from Text.* In the Proc. ECAI 04, IOS Press, 2004, pp. 435-439.

[16] Ganter, B. and R. Wille, *Formal Concept Analysis – Mathematical Foundations,* Springer Verlag, 1999.

[17] Cimiano, P., Staab, S. and J. Tane. *Automatic Acquisition of Taxonomies from Text: FCA meets NLP.* In the Proc. of the ECML/PKDD Workshop on Adaptive Text Extraction and Mining, Cavtat--Dubrovnik, Croatia, 2003, pp. 10-17.

[18] Hirst, G. *Ontology and the lexicon.* In: Staab, S. and R. Studer (Eds.), *Handbook on Ontologies,* Berlin: Springer, 2004, pp. 209-229.

Weighted Pseudo-distances for Categorization in Semantic Hierarchies

Cliff A. Joslyn[1] and William J. Bruno[2]

[1] Computer and Computational Sciences, Los Alamos National Laboratory, Los Alamos, NM, 87545, USA, joslyn@lanl.gov
[2] Theoretical Division, Los Alamos National Laboratory, Los Alamos, NM, 87545, USA, billb@lanl.gov

Abstract. Ontologies, taxonomies, and other semantic hierarchies are increasingly necessary for organizing large quantities of data. We continue our development of knowledge discovery techniques based on combinatorial algorithms rooted in order theory by aiming to supplement the pseudo-distances previously developed as structural measures of vertical height in poset-based ontologies with quantitative measures of vertical distance based on additional statistical information. In this way, we seek to accommodate weighting of different portions of the underlying ontology according to this external information source. We also wish to improve on the deficiencies of existing such measures, in particular Resnik's measure of semantic similarity in lexical databases such as Wordnet. We begin by recalling and developing some basic concepts for ordered data objects, including our pseudo-distances and the operation of probability distributions as weights on posets. We then discuss and critique Resnik's measure before introducing our own sense of links weights and weighted normalized pseudo-distances among comparable nodes.

1 Introduction

We are pursuing approaches to knowledge discovery based on combinatorial algorithms rooted in order theory [8], casting databases as ordered combinatorial data objects equipped both with inherent semantics and appropriate quantitative measures to support user-guided discovery tasks such as search, retrieval, discovery, anomaly detection, linkage, and alignment, with applications in intelligence analysis, homeland defense, computational biology, and law enforcement.

Ontologies, taxonomies, and other semantic hierarchies are increasingly necessary for organizing large quantities of data. Recent years have seen the emergence of a prominent example in the Gene Ontology (GO)[3] [5], a large, ($> 16K$ node) multi-taxonomy of biological functions and processes annotated to thousands of genes. Other cases include the UMLS Meta-Thesaurus [2], object-oriented typing hierarchies [12], and verb typing hierarchies in computational linguistics

---

[3] http://www.geneontology.org

F. Dau, M.-L. Mugnier, G. Stumme (Eds.): ICCS 2005, LNAI 3596, pp. 381–395, 2005.

[3]. Cast as Directed Acyclic Graphs (DAGs), these all entail partially ordered sets (posets) [16]. And other order-theoretic techniques such as formal concept analysis [4] and reconstructibility analysis, which uses lattices of reconstruction hypotheses of relational databases cast as irredundant covers of a space of variables [10], are available to provide order-theoretical representations of relational data objects. Research in fundamental properties and techniques for ordered data objects is sorely lacking; and their increasing size and the need for such tasks as linkage and federation of multiple ontologies constructed on similar domains by different organizations, creates unmet challenges in computer science and combinatorial algorithms.

In prior work we have explored some order-theoretical properties of semantic hierarchies such as measures of distance, level, and size in posets [6]. We have also developed the POSet Ontology Categorizer (POSOC)[4] for the task of "categorization" of gene lists in the GO, represented as a multi-labeled partially ordered set (poset), which we call a POSet Ontology (POSO) [7,9,17].

POSOC takes as a query a list of genes of interest, and then calculates a score for each node in the GO to represent how well that node best captures the overall distribution and location of the query in the poset. This score, whose ranks are as illustrated in an example in Fig. 1, depends on the vertical "separation" from the scored node to those nodes below it where query terms sit. We thus developed the concept of a pseudo-distance $\delta(a, b)$ between comparable nodes $a \leq b$ to measure this vertical distances as a property of the collection of lengths of the chains in the chain decomposition of the poset interval $[a, b]$.

It has been noted that our approach suffers from a deficiency, in that the structure of the GO is not uniform, but may be "denser" in certain regions and "sparser" in others, depending, for example, on the attention paid by the authors, or even the vagaries of funding of the research which supports construction of the GO. As an illustration, consider the taxonomy on the left of Fig. 2. While it appears that "grey wolf" should be as "close" to "animal" as is "ungulate", in fact we know that this isn't the case.

So our structural measure based on chains lengths should be supplemented by an approach to "stretch" or "shrink" links in the structure based on other available information. We propose to capture this additional information by a probability distribution p cast onto the POSO, and then use p to modify POSOC's current pseudo-distances $\delta(a, b)$ to become a weighted pseudo-distance $\delta^w(a, b)$ reflecting this degree of "stretch".

Here we do not presume any particular source for these probabilities. In practice, we're inspired by similar motivations as Lord *et al.* [13], who constructed p as the frequency with which GO node terms appeared in SWISS-PROT-Human protein database. They then used Resnik's measure of semantic similarity [15] developed for Wordnet[5] to measure the semantic distance between GO nodes.

---

[4] http://www.c3.lanl.gov/~joslyn/posoc.html
[5] http://www.cogsci.princeton.edu/~wn

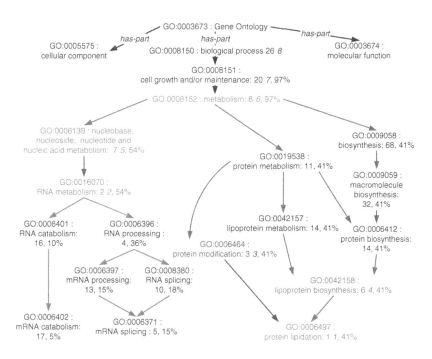

Fig. 1. Partial output from POSOC for a sample query [7]. Nodes in the GO are annotated by the rank of their score and the percentage of the query they cover.

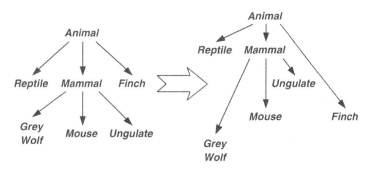

Fig. 2. A semantic hierarchy (left); with stretched links (right).

In this paper, we begin by recalling and developing some basic concepts for ordered data objects, including our current pseudo-distances δ and the operation of probability distributions as weights on posets. We then discuss and critique Resnik's measure of "semantic similarity" in weighted POSOS before introducing our own sense of link weights and weighted normalized pseudo-distances among comparable nodes.

2   Preliminaries on DAGs, Posets, and Covers

Assume a finite set of nodes P with $|P| \geq 2$. Then, assume a **directed acyclic graph (DAG)** $\Gamma \subseteq P^2$ where $\gamma = \langle a, b \rangle \in \Gamma$ is a directed edge from a node $b \in P$ to another $a \in P$. We also sometimes use $\gamma(a, b)$ to indicate a particular edge $\langle a, b \rangle \in \Gamma$. Then a **chain** (specifically, a **DAG chain**) of length $h \geq 2$ is a collection of edges

$$C = \langle \gamma_1, \gamma_2, \ldots, \gamma_i, \ldots, \gamma_{h-1} \rangle \subseteq \Gamma \tag{1}$$

where, letting $\gamma_i = \langle a_i, b_i \rangle$, for $h > 2$ and $1 \leq i \leq h - 1$, we have $b_i = a_{i+1}$.

We also represent Γ as a relational structure $\Gamma = \langle P, \Leftarrow \rangle$, where $\Leftarrow \subseteq P^2$ is a relation on P such that $\forall a, b \in P, a \Leftarrow b \leftrightarrow \langle a, b \rangle \in \Gamma$, so that there's a direct link in the DAG from b "down" to a.

The DAG Γ uniquely generates two mathematical structures:

- Let $\mathcal{V}(\Gamma) := \langle P, \prec \rangle$, be the **cover relation**, or just **cover**, where $\prec$ is the transitive reduction of $\Leftarrow$ [1]. So $a \prec b$ when $\gamma(a, b)$ is an edge in Γ which is non-transitive, that is, $\langle \gamma(a, b) \rangle$ is the only chain C with $a_1 = a, b_{h-1} = b$.
- Let $\mathcal{P}(\Gamma) := \langle P, \leq \rangle$ be the **partially ordered set (poset)** defined by Γ by transitive and reflexive closure of $\Leftarrow$. Below we sometimes use just $\mathcal{P}$ when clear from context. So $a \leq b$ when there's any chain C with $a_1 = a, b_{h-1} = b$.

Fig. 3 shows an example DAG Γ on $P = \{0, A, B, \ldots, K, 1\}$, with two transitive edges $D \Leftarrow B, I \Leftarrow 1$ shown as dashed lines, removal of which yields the cover relation $\mathcal{V}(\Gamma)$. $\mathcal{P}(\Gamma)$ would result from both retaining $D \Leftarrow B$ and $I \Leftarrow 1$ and adding all other transitive links, e.g. $H \Leftarrow B, E \Leftarrow C$. Note the inclusion of the special nodes $0, 1 \in P$ as global bounds, which we will always assume are present.

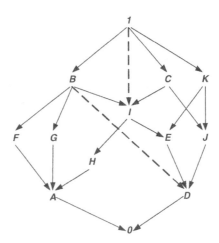

Fig. 3. A bounded DAG.

Given a DAG Γ, we generally focus on the corresponding poset and cover relations. In particular, below we will use $a \leq b$ to mean that $a \leq b$ in the poset $\mathcal{P}(\Gamma)$, and $a \Leftarrow b$, or $\gamma(a, b)$, to mean that $a \prec b$ in the cover $\mathcal{V}(\Gamma)$. $\forall a \in P$, let

$$\downarrow a := \{b \in P : b \leq a\}, \quad \uparrow a := \{b \in P : a \leq b\},$$

$$\dot{\downarrow} a := \{b \in P : b \prec a\}, \quad \dot{\uparrow} a := \{b \in P : a \prec b\}$$

be its **ideal, filter, children**, and **parents** respectively.

Two nodes $a, b \in P$ are called **comparable**, denoted $a \sim b$, when either $a \leq b$ or $b \leq a$. When $a, b \in P, a \neq b$ and $a \leq b$, we simply say $a \leq b \in P$. Two nodes $a, b \in P$ are **non-comparable** if $a \not\sim b$. A collection of nodes $A \subseteq P$ is also called a **chain** (specifically, a **poset chain**) if $\forall a, b \in A, a \sim b$, and an **antichain** if $\forall a, b \in A, a \not\sim b$. The **height** $\mathcal{H}(\mathcal{P})$ of a poset, or just $\mathcal{H}$, is the size of the largest chain, and the **width** $\mathcal{W}(\mathcal{P})$ is the size of the largest anti-chain.

Note that unlike in lattices, when $\mathcal{P}$ is a general poset then for a pair of nodes $a, b \in P$ the concepts of least upper bound $a \vee b$ and greatest lower bound $a \wedge b$, when they exist, are not singular. Rather, $a \vee b, a \wedge b \subseteq P$ are the sets of the (possibly multiple) least upper (greatest lower) bounds of a and b. For example, in Fig. 3 we have $E \vee J = \{C, K\} \subseteq P$.

Below assume two comparable nodes $a \leq b \in P$. Then let

$$\mathcal{C}(a, b) := \{C_1, C_2, \ldots, C_j, \ldots, C_M\} \subseteq 2^{2^P}$$

be the set of all DAG chains from a to b. $[a, b] := \{c \in P : a \leq c \leq b\}$ is the **interval** from a to b and is always a bounded sub-poset of $\mathcal{P}$ with $[a, b] = \bigcup_{j=1}^{M} C_j$. From Dilworth's theorem [16], we know that $M \geq \mathcal{W}([a, b])$. But otherwise, while bounded, the number M of chains is generally arbitrary with respect to a and b.

Let $h_j := |C_j| - 1$ be the length of a particular such chain $C_j \in \mathcal{C}(a, b)$. While in principle $0 \leq h_j \leq \mathcal{H} - 1$, note that $h_j = 0 \leftrightarrow a = b$, so in practice $h_j \geq 1$. Also define $\bar{h}_j := h_j/(\mathcal{H} - 1)$ as a chain length normalized to the height of $\mathcal{P}$. Let

$$\boldsymbol{h}(a, b) := \langle h_1, h_2, \ldots, h_j, \ldots, h_M \rangle, \quad \bar{\boldsymbol{h}}(a, b) := \boldsymbol{h}/(\mathcal{H} - 1)$$

be the **vectors of chain lengths** connecting a to b. We also denote

$$h_*(a, b) = \min_{h_j \in \boldsymbol{h}(a,b)} h_j, \quad h^*(a, b) = \max_{h_j \in \boldsymbol{h}(a,b)} h_j,$$

$$\bar{h}_*(a, b) = \min_{\bar{h}_j \in \bar{\boldsymbol{h}}(a,b)} \bar{h}_j, \quad \bar{h}^*(a, b) = \max_{\bar{h}_j \in \bar{\boldsymbol{h}}(a,b)} \bar{h}_j.$$

Finally, note that for $a \neq b$ and DAG chain C_j, we have $C_j = \{a, b\} \cup C$ for some, possibly empty, chain $C = \{c_2, c_3, \ldots, c_{h_j-1}\} \subseteq P$, so that for $2 \leq i \leq h_j - 1$,

$$a \prec c_2 \prec c_3 \prec \ldots \prec c_{h_j-2} \prec c_{h_j-1} \prec b. \tag{2}$$

3   Pseudo-distances

Our approach begins with measures between comparable nodes $a \leq b$, which indicate how "high" b is above a. A **pseudo-distance** is a function $\delta\colon P^2 \to \mathbb{R}$ where $\forall a \leq b \in P, h_*(a,b) \leq \delta(a,b) \leq h^*(a,b)$, provideing some aggregate measure of the number of "hops" between two comparable nodes. There is also a normalized form $\bar{\delta} := \delta/(\mathcal{H}-1)$ which measures what proportion of the height of the whole poset $\mathcal{P}$ is taken up between a and b, so that $\bar{\delta}(a,b) \in [0,1]$.

Current pseudo-distances implemented in POSOC include:

- **Minimum chain length:** $\delta_m(a,b) := h_*(a,b), \bar{\delta}_m(a,b) := \bar{h}_*(a,b)$
- **Maximum chain length:** $\delta_x(a,b) := h^*(a,b), \bar{\delta}_x(a,b) := \bar{h}^*(a,b)$
- **Average of extreme chain lengths:**

$$\delta_{ax}(a,b) := \frac{h_*(a,b)+h^*(a,b)}{2}, \qquad \bar{\delta}_{ax}(a,b) := \frac{\bar{h}_*(a,b)+\bar{h}^*(a,b)}{2}.$$

- **Average of all chain lengths:**

$$\delta_{ap}(a,b) := \frac{\sum_{h_j \in \boldsymbol{h}(a,b)} h_j}{M}, \qquad \bar{\delta}_{ap}(a,b) := \frac{\sum_{\bar{h}_j \in \bar{\boldsymbol{h}}(a,b)} \bar{h}_j}{M}.$$

In our example in Fig. 3 (recalling that transitive edges are removed from the cover relation $\prec$), we have $\mathcal{H}(\mathcal{P}) = 6$. Considering $D \leq 1$, then $\mathcal{W}([D,1]) = 3$, while $M = 5$, and

$$\mathcal{C}(D,1) = \{ \, D \prec E \prec I \prec B \prec 1, D \prec E \prec I \prec C \prec 1,$$
$$D \prec E \prec K \prec 1, D \prec J \prec C \prec 1, D \prec J \prec K \prec 1\},$$
$$\boldsymbol{h}(D,1) = \langle 4,4,3,3,3 \rangle, \quad \bar{\boldsymbol{h}}(D,1) = \langle 4/5, 4/5, 3/5, 3/5, 3/5 \rangle,$$
$$\delta_m(D,1) = 3, \delta_x(D,1) = 4, \delta_{ax} = 3.5, \delta_{ap}(D,1) = 3.4,$$
$$\bar{\delta}_m(D,1) = 0.60, \bar{\delta}_x(D,1) = 0.80, \bar{\delta}_{ax} = 0.70, \bar{\delta}_{ap}(D,1) = 0.68.$$

4   Weighted Posets

Our overall motivation is to improve on the quantitative approach which Lord et al. [13] and Resnik [15] brought to measures of distance (in their language, "semantic similarity") in taxonomies equipped with probability distributions. To do so, we need to understand the basic operations of probabilities on posets.

Definition 1 (Weighted Poset). *Define a weighted poset as a structure* $\mathcal{O} := \langle \mathcal{P}(\Gamma), p \rangle$, *where* $p\colon P \to [0,1]$ *is a probability distribution on the nodes P of the poset $\mathcal{P}(\Gamma)$, so that* $\sum_{a \in P} p(a) = 1$.

For any node $b \in P$, what we will call a "beta" function $\beta\colon P \to [0,1]$ is a kind of probability measure over Γ, defined as

$$\beta(b) := \sum_{a \leq b} p(a) = \sum_{a \in \downarrow b} p(a). \tag{3}$$

Fig. 4 continues our example with $\beta(a)$ to the right of each node, and $p(a)$ below it. Weights are also shown on links, which will be discussed in Sec. 7 below.

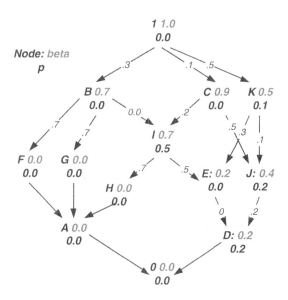

Fig. 4. An example of a weighted poset $\mathcal{O}$.

β is what's known as an isotone, or order-preserving, map on $\mathcal{P}$, which is crucial in Monjardet's general theory of metrics in posets [14]:

Proposition 1. $a \leq b \to \beta(a) \leq \beta(b)$.

Proof. Follows from $a \leq b \to \downarrow a \subseteq \downarrow b$, Eq. (3), and $p(a) \geq 0$.   □

5   A Mathematical Aside

We are regretfully ignorant of literature concerning discrete probability distributions like p on finite ordered sets, and yet we can note some intriguing connections to some powerful formalisms. In particular, let $\mathbf{B}(\mathcal{O}) := \{a \in P : p(a) > 0\} \subseteq P$

be the **base** of $\mathcal{O}$. Then consider the case when $\mathcal{P}$ is a Boolean lattice, in particular the power set 2^{Ω} on some underlying finite set Ω. β then is the **belief function** Bel on Ω from Dempster-Shafer evidence theory [11], and Eq. (3) becomes $\forall A \subseteq \Omega, \text{Bel}(A) = \sum_{B \subseteq A} p(A)$. Bel is super-additive, with the modular property

$$\forall A, B \subseteq \Omega, \quad \text{Bel}(A \cup B) \geq \text{Bel}(A) + \text{Bel}(B) - \text{Bel}(A \cap B). \tag{4}$$

Expressed back in the lattice $\mathcal{P}$, Eq. (4) becomes

$$\forall a, b \in P, \quad \beta(a \vee b) + \beta(a \wedge b) \geq \beta(a) + \beta(b), \tag{5}$$

recalling that the single point $a \vee b \in P$ now always exists. There are some important special cases, for example:

- If $\mathbf{B}(\mathcal{O})$ is the antichain of the atoms of $\mathcal{P}$, then Eq. (5)

$$\forall a, b \in P, \quad \beta(a \vee b) + \beta(a \wedge b) = \beta(a) + \beta(b),$$

  so that Bel becomes a classical probability measure Pr with p its discrete distribution, and

$$\forall A, B \subseteq \Omega, \quad \Pr(A \cup B) = \Pr(A) + \Pr(B) - \Pr(A \cap B).$$

- If $\mathbf{B}(\mathcal{O})$ is a maximal chain $C \subseteq \mathcal{P}$ with $|C| = \mathcal{H}(\mathcal{P})$, then

$$\forall a, b \in P, \quad \beta(a \wedge b) = \min(\beta(a), \beta(b))$$

  so that Bel becomes a so-called "necessity measure" η with

$$\forall A, B \subseteq \Omega, \quad \eta(A \cap B) = \min(\eta(A), \eta(B)).$$

Open questions remain about these properties when $\mathcal{P}$ is a general lattice, a complemented lattice, or a general poset (all finite). However, it's interesting to consider a potential form of Eq. (5) when $\mathcal{P}$ is a general poset:

$$\forall a, b \in P, \quad \sum_{c \in a \vee b} \beta(c) + \sum_{c \in a \wedge b} \beta(c) \geq \beta(a) + \beta(b),$$

and consider connections to general families of semimodular maps on posets [14].

6   Resnik's Semantic Similarity

A currently attractive way to approach semantic distance in semantic hierarchies is to use Resnik's measure of "semantic similarity" [15], originally developed for application to Wordnet, but then applied successfully by Lord *et al.* to the GO [13]. Not only do these concepts have a natural interpretation in our language, they also serve as a point of departure for our development.

Definition 2 (Resnik Semantic Similarity). *Given a weighted poset $\mathcal{O}$, then $\forall a, b \in P$, define*

$$\delta_{Resnik}(a, b) = \max_{c \in a \vee b} \left[-\log_2(\beta(c)) \right]. \tag{6}$$

Some issues are immediately evident when comparing the Resnik measure to our pseudo-distances δ. First, unlike δ_{Resnik}, our measure δ is definitely *not* a distance, most significantly because it is not defined for pairs of general nodes $a, b \in P$, but rather only for comparable nodes $a \leq b \in P$.

There is also some ambiguity as to the semantics of δ_{Resnik} as a measure of information content, since as discussed in Sec. 5, while the p's are definitely values of a discrete distribution on P, β is almost never actually a probability measure on P.

Also, if a portion of $\mathcal{P}$, and in particular the ideal $\downarrow b$ of a node $b \in P$, is a lattice, then the similarities between all the children of b are identical, no matter their β values.

Theorem 1. *Let $b \in P$ with $\downarrow b \subseteq P$ a lattice. Then $\forall a_1, a_2, a_3, a_4 \in \downarrow b$, $\delta_{Resnik}(a_1, a_2) = \delta_{Resnik}(a_3, a_4)$.*

Proof. Since $\downarrow b$ is a lattice, therefore $\forall a, a' \in \downarrow b, a \vee a'$ exists uniquely, and in particular $a \vee a' = \{b\}$. Thus $\forall a_1, a_2, a_3, a_4 \in \downarrow b, \delta_{\text{Resnik}}(a_1, a_2) = -\log(\beta(b)) = \delta_{\text{Resnik}}(a_3, a_4)$. □

But most significantly, δ_{Resnik} cannot distinguish among links in a chain.

Theorem 2. *If $a \leq b \leq c$, then $\delta_{Resnik}(a, c) = \delta_{Resnik}(b, c) = \delta_{Resnik}(c, c)$.*

Proof. Since $a \leq b \rightarrow a \vee b = \{b\}$ uniquely in P, then $\delta_{\text{Resnik}}(a, c) = \delta_{\text{Resnik}}(b, c) = \delta_{\text{Resnik}}(c, c) = -\log(\beta(c))$. □

This is precisely contrasted with our pseudo-distances δ and $\bar{\delta}$, which satisfy

$$a \leq b \leq c \quad \rightarrow \quad \delta(b, c) \leq \delta(a, c), \quad \bar{\delta}(b, c) \leq \bar{\delta}(a, c).$$

7   Link Weights in Weighted Posets

Despite the weaknesses of δ_{Resnik}, it is valuable in pointing the way towards information theoretical distance measures in probability weighted posets, and in being defined on all $a, b \in P$. While we are also actively researching such measures for multiple purposes [6], categorization in general and POSOC in particular depends only on proper measures among comparable nodes $a \leq b \in P$.

We are therefore looking for a different mechanism to introduce link weights into our categorization algorithm. Our aim is to "lengthen" chains in the weighted poset proportional to the amount of probability concentrated along them, while leaving them at "original length" where there is none. To do so, we also take an information theoretical approach, although somewhat distinct from Resnik's.

Definition 3 (Information Gain). *For all pairs of comparable nodes $a \leq b \in P$, let*

$$\iota(a, b) := \beta(b) - \beta(a) \tag{7}$$

be the amount of information gained when moving from b down to a. For each edge $\gamma(a, b)$, let $\iota(\gamma) := \iota(a, b)$ be the edge weight.

As an example, in Fig. 4 we have

$$\iota(D, K) = \beta(K) - \beta(D) = 0.5 - 0.2 = 0.3.$$

Also in Fig. 4, we have labeled each of the arcs γ with the information gain $\iota(\gamma)$.

It is obvious that $\iota(a, b) \in [0, 1]$. It is also comforting that for all pairs of comparable nodes $a \leq b \in P$, no matter which chain is traversed from b down to a, the sum of the edge weights is always equal to the information gain from b to a.

Proposition 2.

$$\forall a \leq b \in P, \quad \forall C_j \in \mathcal{C}(a, b), \quad \sum_{\gamma_i \in C_j} \iota(\gamma_i) = \iota(a, b).$$

Proof. Fix $a \leq b \in P$, and a chain $C_j \in \mathcal{C}(a, b)$ of length h_j, and use the notation from Eq. (1) and Eq. (2). Then we have

$$\sum_{\gamma_i \in C_j} \iota(\gamma_i) = (\beta(c_2) - \beta(a)) + (\beta(c_3) - \beta(c_2)) + \ldots +$$

$$(\beta(c_{h_j - 1}) - \beta(c_{h_j - 2})) + (\beta(b) - \beta(c_{h_j - 1}))$$
$$= \beta(b) - \beta(a) = \iota(a, b).$$

$\square$

This development follows Monjardet almost precisely [14], with β an isotone map on $\mathcal{P}$, $\iota(\gamma)$ an edge weight, and $\iota(a, b)$ a path weight.

Continuing our example, for $D \leq 1$, where we still have the $M = 5$ chains, we also have

$$\iota(D, 1) = .8$$
$$= \iota(\langle D, E \rangle) + \iota(\langle E, I \rangle) + \iota(\langle I, B \rangle) + \iota(\langle B, 1 \rangle) = 0.0 + 0.5 + 0.0 + 0.3$$
$$= \iota(\langle D, J \rangle) + \iota(\langle J, K \rangle) + \iota(\langle K, 1 \rangle) = 0.2 + 0.1 + 0.5$$

and similarly for the other chains. This is very convenient, because then we can deal with the information gain between comparable nodes and edge weights indiscriminately: information gain can always be calculated by edge weights, but we can also work with information gains whenever we want.

8   Weighted Normalized Pseudo-distances

For comparable nodes $a \leq b \in P$, we now want to take the chain lengths h_j (or the normalized chain lengths $\bar{h}_j$) and adjust them by the information gain $\iota(a, b)$. Thus for a particular chain $C_j \in \mathcal{C}(a, b)$ of length h_j $(\bar{h}_j)$, we wish to construct a weighted chain length $v_j(a, b)$ (or a normalized weighted chain length $\bar{v}_j(a, b)$) as a function of h_j $(\bar{h}_j)$ and $\iota(a, b)$, which will be equal to h_j $(\bar{h}_j)$ scaled up with $\iota(a, b)$. In this paper, we have only developed the normalized case for $\bar{v}_j$.

Considering a generic pair of comparable nodes $a \leq b \in P$ with information gain $\iota = \iota(a, b) = \beta(b) - \beta(a)$, and a particular chain $C_j \in \mathcal{C}(a, b)$ with length $\bar{h} = \bar{h}_j$, then how should the weighted, normalized chain length $\bar{v}$ be determined? Our motivations are as follows.

First, h and ι are in some sense independent quantities. As illustrated in cartoon form in Fig. 2, the size of ι indicates the amount of "stretch" in the chain C, no matter the number of "hops" along its length. So, "grey wolf" is farther from "animal" than "ungulate" is, despite the fact that they're both chains of length two. Similarly, "finch" and "mouse" are stretched to be a similar distance from "animal", despite the differences in those chain lengths. So despite the fact that by Thm. 1, as any particular chain C_j grows, ι increases with the chain length $\bar{h}_j$, nonetheless in any particular case $\bar{h}_j$ could be small while ι is large, or *vice versa*.

We thus wish to identify a function $f: [0, 1]^2 \to [0, 1]$ such that $\bar{v}_j := f\left(\bar{h}_j, \iota(a, b)\right)$, and list the properties desirable for $f(h, \iota)$ with $h, \iota \in [0, 1]$:

1. In one limit, ι could be zero, indicating that no probability mass was encountered when traversing from b down to a. In this case, we wish $\bar{v}$ to be recovered simply as $\bar{h}$, requiring that $f(h, 0) = h$.
2. Otherwise, we want ι to add in to increase the total length, requiring $\bar{v} \geq \bar{h}$, or in other words $f(h, \iota) \geq h$.
3. Then, considering some limit cases, if the entire structure is traversed by a maximal chain, then $\bar{v}$ should be maximum, so that $f(1, \iota) = 1$.
4. Similary, in the degenerate case of $a = b$ so that there are no chains and no ι gain, then $\bar{v}$ should be minimal, so that $f(0, 0) = 0$.
5. Finally, if $\iota = 1$ so that *all* the probability mass is encountered when traversing from b down to a, then $\bar{v}$ is maximal, no matter than length h: $f(h, 1) = 1$.

Thus we arrive at the following.

Definition 4 (Weighted Normalized Chain Lengths). *For $a \leq b \in P$, and for each chain $C_j \in \mathcal{C}(a, b), 1 \leq j \leq M$, define $\bar{v}_j := f\left(\bar{h}_j, \iota(a, b)\right)$, where*

$$f(h, \iota) := h^{1-\iota}, \quad h, \iota \in [0, 1], \tag{8}$$

*as the **weighted chain length**. Construct $\bar{\boldsymbol{v}}(a, b) := \langle \bar{v}_1, \bar{v}_2, \ldots, \bar{v}_j, \ldots, \bar{v}_M \rangle$ as the **vector of weighted chain lengths**, with $\bar{v}_j \in \bar{\boldsymbol{v}}(a, b)$, and let*

$$\bar{v}_*(a, b) = \min_{\bar{v}_j \in \bar{\boldsymbol{v}}(a,b)} \bar{v}_j, \qquad \bar{v}^*(a, b) = \max_{\bar{v}_j \in \bar{\boldsymbol{v}}(a,b)} \bar{v}_j.$$

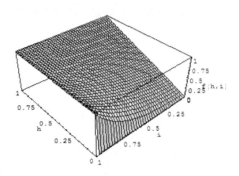

Fig. 5. $\bar{v}_j$ as a function of $\bar{h}_j$ and $\iota(a,b)$.

Theorem 3. $\bar{v}_j$ as defined using f in Eq. (8) satisfies the conditions 1–5 above.

Proof. Condition 2 follows from the fact that both $h, 1 - \iota \in [0,1]$, so that $f(h, \iota) = h^{1-\iota} \geq h$. All the other conditions follow directly from the form of f.

Fig. 5 shows f as a surface in $[0, 1]^2$, and Fig. 6 shows level curves of f for various values of h (left) and ι (right). Inspection of the figures reveals some interesting behaviors. For example, a large ι will bring up even a small h to a high value of f; for high h, f degrades gradually to be bounded below by h as ι drops, whereas for low h, this dropoff is more sudden.

We are now ready to introduce a modification to the prior pseudo-distances.

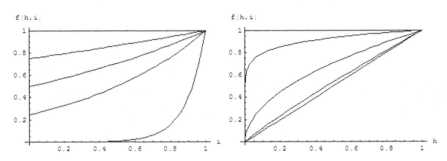

Fig. 6. (Left) Level curves of $f(h, \iota)$ for $h = 1(f \equiv 1), .75, .5, .25, .0001$; (Right) Level curves for $\iota = 1(f \equiv 1), .9, .5, .1, 0$.

Definition 5 (Weighted Normalized Pseudo-Distance). *Given a weighted poset $\mathcal{O}$, then for all $a \leq b \in P$, let a weighted normalized pseudo-distance $\bar{\delta}^w(a, b)$ be any function such that $\bar{v}_*(a, b) \leq \bar{\delta}^w(a, b) \leq \bar{v}^*(a, b)$. In particular, define the following weighted normalized pseudo-distances:*

- **Minimum Normalized Weighted Chain Length:** $\bar{\delta}_m^w(a,b) := \bar{v}_*(a,b)$.
- **Maximum Normalized Weighted Chain Length:** $\bar{\delta}_x^w(a,b) := \bar{v}^*(a,b)$.
- **Average of Extreme Normalized Weighted Chain Lengths:**

$$\bar{\delta}_{ax}^w(a,b) := \frac{\bar{v}_*(a,b) + \bar{v}^*(a,b)}{2}.$$

- **Average of All Normalized Weighted Chain Lengths:**

$$\bar{\delta}_{ap}^w(a,b) := \frac{\sum_{\bar{v}_j \in \bar{v}(a,b)} \bar{v}_j}{M}.$$

Given a weighted pseudo-distance, then the POSOC methodology [7] is simply modified to substitute $\bar{\delta}^w$ for $\bar{\delta}$. The results of our example are shown in Tab. 1.

j	h_j	$\bar{h}_j$	$\bar{v}_j$
1	3.000	0.600	0.903
2	3.000	0.600	0.903
3	3.000	0.600	0.903
4	4.000	0.800	0.956
5	4.000	0.800	0.956

	δ_*	$\bar{\delta}_*$	$\bar{\delta}_*^w$
m	3.000	0.600	0.903
x	4.000	0.800	0.956
ax	3.500	0.700	0.930
ap	3.400	0.680	0.924

Table 1. Results for $D \leq 1$. (Top) Chain lengths h_j, normalized chain lengths $\bar{h}_j$, and weighted normalized chain lengths $\bar{h}_j$ for chains $C_j \in \mathcal{C}(D,1), 1 \leq j \leq M = 5$; (Bottom) Pseudo-distances δ_*, normalized pseudo-distances $\bar{\delta}_*$, and weighted normalized pseudo-distances $\bar{\delta}_*^w$ for $* \in \{x, m, ax, ap\}$.

9   Conclusion and Discussion

We have demonstrated a method by which probabilities can be used to weighte pseudo-distances in ordered data objects. Space allows only an explication of this basic step, and not its application within the overall POSOC ontology categorization algorithm. While straightforward, this requires the determination of a real probability distribution p on the GO, perhaps in a manner similar to Lord *et al* [13]. The details and results of this application, along with development for non-normalized weighted chain lengths v_j and pseudo-distances δ^w, await further work.

We should also mention an additional step we wish to take in the future. As we've seen, $\bar{v}$ is not especially sensitive to even moderate information gains ι even when the chain lengths $\bar{h}_j$ are reasonable. But moreover, the example in Fig. 4 is perhaps somewhat unfortunately chosen, in that the probability mass is concentrated on the right-hand side of the structure. The pair $D \leq 1$ we considered has a relatively large gain $\iota(D, 1) = 0.8$. In practice, in a real ontology, mass can be expected to be widely distributed over a very wide and shallow structure (the width of the GO in some places exceeds 6000, while the height is only 16), meaning that it can be expected that for any typical pair of nodes $a \leq b$, $\iota(a, b)$ would actually be expected to be very low, and thus $\bar{v}$ quite close to $\bar{h}$. Thus we are also considering the viability of a non-linear function for information gain instead of Eq. (7) to heighten or tune the affect of small information gains. In particular, we are considering various logarithmic forms similar in spirit to Resnik's Eq. (6).

10   Acknowledgments

This work was sponsored by the Department of Energy under contract W-7405-ENG-36 to the University of California. We would like to thank the Los Alamos National Laboratory Protein Function Inference Group for their contributions to this work, and in particular Michael Wall (LANL Computer and Computational Science) for his consultation. We also appreciate the dialog currently ongoing with Alex Sanchez, Phillip Lord, and Robert Stevens of U. Manchester. Finally, we'd like to thank Petko Valtchev of the University of Montreal for the reference to Monjardet.

References

1. Aho, AV; Garey, MR; and Ullman, JD: (1972) "The Transitive Reduction of a Directed Graph", *SIAM Journal of Computing*, v. **1**:2, pp. 131-137
2. Bodenreider, Olivier; Mitchell, Joyce A; and McCray, Alexa T: (2002) "Evaluation of the UMLS As a Terminology and Knowledge Resource for Biomedical Informatics", in: *AMIA 2002 Annual Symposium*, pp. 61-65
3. Davis, Anthony R: (2000) *Types and Constraints for Lexical Semantics and Linking*, Cambridge UP
4. Ganter, Bernhard and Wille, Rudolf: (1999) *Formal Concept Analysis*, Springer-Verlag
5. Gene Ontology Consortium: (2000) "Gene Ontology: Tool For the Unification of Biology", *Nature Genetics*, v. **25**:1, pp. 25-29
6. Joslyn, Cliff: (2004) "Poset Ontologies and Concept Lattices as Semantic Hierarchies", in: *Conceptual Structures at Work, Lecture Notes in Artificial Intelligence*, v. **3127**, ed. Wolff, Pfeiffer and Delugach, pp. 287-302, Springer-Verlag, Berlin
7. Joslyn, Cliff; Mniszewski, Susan; and Fulmer, Andy; and GG Heaton: (2004) "The Gene Ontology Categorizer", *Bioinformatics*, v. **20**:s1, pp. 169-177
8. Joslyn, Cliff; Oliverira, Joseph; and Scherrer, Chad: (2004) "Order Theoretical Knowledge Discovery: A White Paper", *LAUR = 04-5812*, ftp://ftp.c3.lanl.gov/pub/users/joslyn/white.pdf

9. Joslyn, Cliff; Cohn, JD; Verspoor, KM; and Mniszewski, SM: (2005) "Automating Ontological Function Annotation: Towards a Common Methodological Framework", submitted to *2005 Bio-Ontologies Meeting, ISMB 05*

10. Klir, George and Elias, Doug: (2003) *Architecture of Systems Problem Solving*, Plenum, New York, 2nd edition

11. Klir, George and Yuan, Bo: (1995) *Fuzzy Sets and Fuzzy Logic*, Prentice-Hall, New York

12. Knoblock, Todd B and Rehof, Jakob: (2000) "Type Elaboration and Subtype Completion for Java Bytecode", in: *Proc. 27th ACM SIGPLAN-SIGACT Symposium on Principles of Programming Languages*

13. Lord, PW; Stevens, Robert; Brass, A; and C Goble: (2003) "Investigating Semantic Similarity Measures Across the Gene Ontology: the Relationship Between Sequence and Annotation", *Bioinformatics*, v. **10**, pp. 1275-1283

14. Monjardet, B: (1981) "Metrics on Partially Ordered Sets - A Survey", *Discrete Mathematics*, v. **35**, pp. 173-184

15. Resnik, Philip: (1995) "Using Information Content to Evaluate Semantic Similarity in a Taxonomy", in: *Int. Joint Conf. on Artificial Intelligence*, pp. 448-452, Morgan Kaufmann

16. Schröder, Bernd SW: (2003) *Ordered Sets*, Birkhauser, Boston

17. Verspoor, Karin; Cohn, J; Joslyn, C; SM Mniszewski, A Rechtsteiner, LM Rocha, and T Simas: (2004) "Protein Annotation as Term Categorization in the Gene Ontology Using Word Proximity Networks", *BMC Bioinformatics*, v. 6, suppl 1

Games of Inquiry
for Collaborative Concept Structuring

Mary A. Keeler[1] and Heather D. Pfeiffer[2]

[1]Center for Advanced Research Technology in the Arts and Humanities
University of Washington, Seattle, Washington 98117, USA
mkeeler@u.washington.edu
[2]Department of Computer Science
New Mexico State University, Las Cruces, New Mexico 88003-8001, USA
hdp@cs.nmsu.edu

Abstract. Google's project to digitize five of the world's greatest libraries will dramatically extend their search engine reach in the future. Current search-engine philosophy, which asserts that "any search starts with a question to be answered," will need to be advanced in terms of Peirce's philosophy: "Any inquiry begins by creating an hypothesis to be tested, or with abduction." As conceptual structures researchers prepare to meet access challenges in the world of large Internet knowledge stores, they have a solid foundation in Peirce's theorized stages of inquiry: abduction, deduction, and induction. To indicate how conceptual structures tools must augment collaborative, Internet-based inquiry, we imagine a future scenario in the context of a user-centered testbed, where Peirce scholars apply Peirce's pragmatic theory in their complex manuscript reconstruction work. We suggest that games of inquiry can be developed to formalize user collaboration and technology needs, for improved specification of tool requirements in the testbed context.

1 Scenario Context and Characters

Imagine some particular day in the future, when the technology to support collaborative inquiry has advanced through rapid-prototyping in user-centered testbeds, so that researchers can work together efficiently, no matter where they are located and no matter where their primary data is safely stored in library archives. In this particular scenario, the researchers are Peirce scholars and the data of their research are Peirce's manuscripts archived at Harvard's Houghton Library. These scholars are remotely located from one another, so must use the Internet to collaborate in their work to transcribe the digitized images of Peirce's handwritten manuscripts into digital text and to catalog a coherent sequence of Peirce's manuscript pages. Transcription involves deciphering what Peirce wrote, which becomes more difficult as his handwriting deteriorated in his declining years when he suffered from cancer and other ailments.

During the last seventeen years of his life (1897-1914), Peirce produced his most intensive theoretical work, on semiotic, pragmatism, and his system of Existential Graphs (EG) as a notation for the study of logic [see 1 and 2]. Amounting to some 40,000 pages, very few of these manuscripts have ever been published. And very few scholars have ever studied these pages, which their Houghton Library curator

F. Dau, M.-L. Mugnier, G. Stumme (Eds.): ICCS 2005, LNAI 3596, pp. 396-410, 2005.

estimates will survive no more than a few dozen more years (due to their high-acid paper). By 2005, scholars and technologists realized that Internet-based technology had advanced to make possible the reconstruction and representation of Peirce's polymathic corpus (ranging from religion, philosophy, and cosmology, to computer science, mathematics, logic, geophysics, and many others fields). Five, especially qualified scholars joined to pursue some of that reconstruction, collaboratively, in the Peirce Online Resource Testbed (PORT) project [3].

Jay Zeman, Professor Emeritus of Philosophy and specialist in Peirce's EG, lives in Florida; Bob Burch, Professor of Philosophy at Texas A&M is also an expert in Peirce's EG; Peter Øhrstrøm, Professor of Communication at the University of Aalborg in Denmark is an specialist in modal logic (including Peirce's "gamma graphs"); Robert Marty is Professor Emeritus at the University of Perpignon in France, a specialist in Peirce's "semiotic logic"; and Frithjof Dau, in Germany, completed his doctoral degree studying concept graphs in 2002, and a treatise on Peirce's EG in 2005. They joined PORT's collaboratory testbed, in partnership with technologists, to help develop imaging and transcribing methods and tools that would improve the quality and efficiency of their work. They are particularly motivated to create better access to the manuscript data because of inadequacies and misrepresentations in previous editions of Peirce's writings. The *Collected Papers of Charles Sanders Peirce* (referred to as *"CP"*), an eight-volume edition produced more than fifty years ago[1], remains (in spite of its many deficiencies) the most complete print resource of his work[2].

On this particular imaginary day in the future, the scholars are engaged in perhaps the most difficult detective problem during the initial stage of reconstructing Peirce's corpus. As in the initial stage of scientific investigations, the scholars must prepare the raw data (in the form of digitized manuscript page images) to serve as reliable data for interpretation, by *calibrating* (ordering the pages in some sequence and grouping the sequences accurately). For scholarly interpretation of his ideas as they

---

[1] The misnamed *Collected Papers of Charles Sanders Peirce* is the only large-scale and purportedly comprehensive edition of Peirce's writings. Published in six volumes, from 1931 to 1935, and enlarged by two volumes in 1958, its topical selections omit Peirce's writings on science and mathematics almost entirely. Although, it contains nearly 150 selections from his unpublished manuscripts, only one-fifth of those selections consist of complete manuscripts, and many of them are inaccurately dated; parts of some manuscripts appear in up to three of the eight volumes, and at least one series of papers is scattered throughout seven of the eight. Most manuscripts appear in excerpts (with only rare indication of how much has been left out), and different parts of the same manuscript appear in different volumes (because of the topical order), and in some cases parts of different manuscripts are grafted together (without mention), sometimes consisting of writings composed more than three decades apart. As a scholarly tool, this edition is unreliable and often frustrating [See 4 for detailed discussion].

[2] A textually and chronologically reliable print edition of Peirce's writings was begun in the late 1980s (*Writings of Charles S. Peirce: A Chronological Edition*) [5]. As of 2005, only six of the originally projected thirty volumes have been published, and even if completed, that edition would present less than one-third of Peirce's entire work. More significantly, in its paper-print format, cost of production may prohibit representation of Peirce's late manuscripts, which are characterized by text enclosed in graphical figures, graphics embedded in text, text contoured around graphics, whole pages of graphics with no text at all, and graphical figures with as many as four colors.

evolved, the page order should be chronological, as they were written by Peirce. Unfortunately, achieving that ideal may never be entirely possible, since only about one third of the corpus was originally dated by Peirce. At least, the order of each manuscript should be as close to Peirce's own apparent compositional order as possible. Even that goal is challenging, since some 10,000 pages in the Harvard collection[3] are considered "lost." These pages are present in the collection, but not properly placed in any manuscript sequence. Many more pages have been tentatively placed in supplementary folders, and labeled, for example, "145(s)," to indicate pages thought to belong to MS 145. In other cases, if scholars have not yet considered some group of "lost pages," which seem to belong together, they are simply labeled with a manuscript number. Such is the case with manuscript "507," a folder of nine pages, listed in the "Robin catalogue" of the Houghton collection[4] as undated and unnumbered.

2   A Manuscript Mystery

We enter the scenario with the five scholars attempting to solve a "lost page" mystery, as an example of the complexity of work in the initial stage of reconstruction. The 32 pages of the folder labeled "145(s)" have been previously identified as possibly belonging with the folder labeled "145," which the Robin catalogue lists as a manuscript of 12 pages (undated and unnumbered). Quoting from the manuscript title page, the catalogue entry reads: "An Attempt to state systematically the Doctrine of the Census in Geometrical Topics or Topical Geometry, more commonly called 'Topologie' in German books; Being a Mathematical-Logical Recreation of C.S. Peirce following the lead of J.B. Listing's paper in the 'Gottinger Abhandlungen'" [6:17]. The scholars each verify this catalogue entry with what they find in the "145" folder (using Hoffmann's online version of the Robin catalogue and PORT's online manuscript data archive). They find that the association of "145" pages with the manuscript pages of "145(s)" is supported by empirical evidence they can see in the manuscript page images displayed on each of their screens simultaneously. The pages of "145" are in a notebook with lined pages that appear to match the pages of "145(s)." But they question whether the folder labeled "145(s)" (of 32 pages) should be designated "145," while the folder labeled "145" (of 12 pages) should be "145(s)," indicating that its fewer pages are supplementary material for what should be the primary material in the folder currently marked "145(s)"?

---

[3] When Peirce's papers (amounting to between eighty and one-hundred thousand pages) were given to Harvard, shortly after his death (1915), they were not properly stored for several years (apparently, even distributed as scratch paper in war-time paper shortages), during which they fell into disarray. The microfilm (high-contrast, black and white, produced in 1967) and photo copies from it cannot be used in place of the original pages, which must be carefully examined for any discriminating features that might be clues as to where they belong (such as color and width of ruled lines and handwritten script, type of paper, watermark, shape of a torn or cut edges, weathering effects, and so on).

[4] Richard Robin's *Annotated Catalogue of the Papers of Charles S. Peirce* [6] provides the most valuable and comprehensive view of the disordered manuscripts (and their incomplete representation in the microfilm edition). See Michael Hoffmann's digital version of the Robin catalogue: http://www.iupui.edu/~peirce/robin/robin.htm.

Meanwhile, at least one page in the "507" folder (of unplaced pages) has been recognized as very similar in appearance and content to page number 22 in "145" (so close as to be considered possibly an earlier or later-written version of page 22). However, even though the "507" page has obviously been cut from a notebook, its cut edge does not match any cut fragment in the "145" or "145(s)" notebooks. And though its textual content appears to be nearly identical to that on page 22, the EGs across the top of the page are not identical. The "507" catalogue entry gives almost no clue: "Beta and gamma graphs, with algebraic translations. Rules of transformation" [6: 65].

Previously, the scholars had encountered the comparable case of a notebook catalogued and labeled as "464" and a folder of 15 pages labeled "464(s)" on notebook paper that looks exactly like that of the 35 pages of "464." In this case too, "464," but not "464(s)," has an entry in the Robin catalogue designating it as "Part 1 of the 3rd draft of 3rd Lecture" of his 1903 "Lowell Lecture" series.

As with "145," the content of one page of "464(s)" has obviously been cut from a notebook, and it appears very similar to that of a page in the "464" notebook; but in this case the cut edge of the "464(s)" page matches exactly with a cut fragment in "464," and even the marginal text that has been cut through matches. So there is strong empirical evidence that "464" and "464(s)" belong together, while the proper designation of "145" and "145(s)," together with the mystery of the "507" page (which is extremely similar in content but does not empirically match page 22 in "145"), remains in doubt. Where does this page belong, and how should "145" and "145(s)" be properly designated in the catalogue of manuscript material?

3   Essential Collaboration Capabilities

For the critical identification work in our future scenario of scholars at work, the digital images of the manuscript pages must be extremely accurate representations of the original pages. They must provide evidence not only of what Peirce rendered in graphical forms and in handwritten text but also of other bibliographic features (such as color, kind, and condition of the paper). The manuscript images must exhibit to all collaborating scholars the same critical features, no matter what may be the variations among the qualities of their computer display systems. Any visible feature on a page may be needed in their judgments as to where the page belongs, what was its significance to Peirce, and what insights it might convey. The mediating tools of their virtual collaboration context make it possible for the scholars to confer effectively in such minute judgments, in spite of the differences in display capabilities among their computer systems.

3.1   Consistent Image Quality

An image-display mediator (originally developed as "AI-Trader," by Puder and Römer [7]), monitors the differences among the scholars' five different computer displays and matches them across all system factors (such as file-compression ratio, screen size and resolution, and color calibration). This automatically standardized image is the reference quality for all collaborators, and enables the mediator to notify the scholars when their views do not correspond. On request, the mediator negotiates among the collaborator's individual requests to establish the optimum perceived attribute values for any selected purpose. Matching views is especially important for collaboration on detailed features such as tone of paper (for page matching). Subtle qualities that are not immediately recognized as significant might be identified as crucial evidence by one scholar and then confirmed in collaborative interpretation. All scholars have access to reliably comparable views of the same artifact, while individually each is able to explore the primary data for potentially relevant features identified from the perspective of his particular background knowledge of Peirce's work.

The mediation service avoids the problem of simply serving the lowest display system capabilities, and allows the scholars to intervene in the process at any point needed to resolve discrepancies because of differences in image quality among their displayed views of manuscript pages. The lowest quality display system of any scholar does not prevent those with higher quality systems from seeking and finding crucial evidence, and bringing it to the collaboration context as the object for consideration and discussion. The image-display mediator negotiates among complexly interrelated system variance factors (including resolution and color accuracy, compressed file quality, and speed of transmission) to serve all participants efficiently. Concurrently, the scholars can readily monitor this mediation process, as it reconciles "competing factors," to remain aware in viewing the images as the evidence of what Peirce rendered on paper. Any relative bibliographic invariance they identify in the evidence (such as a particular type of notebook paper or color of ink used in particular graphic forms) might then serve to establish meaningful segmentation, to be indexed for catalogue identifications that can be used in automatically searching for possible matching pages.

3.2 Concurrent Catalogue Documentation

PORT's concept-based catalogue for the Peirce archive, which has evolved from the data of print finding aids, has been digitally linked to both transcribed text and digital images of the original manuscript pages. By means of the catalogue, each manuscript page's digital image and transcription is coded to all existent corresponding print edition entries of its content, and each image appears on screen with its actual or estimated date of composition, if known. The scholars can propose corrections to the catalogue entries reflecting any additions and alterations accumulated during their intensive transcription and calibration work. Variations among their transcriptions and calibrations are monitored automatically, by digital matching, and these points of disagreement are highlighted in the digital transcriptions.

An ontology of the collection (or a comprehensive framework capable of relating the digitized archive data in a conceptual structure, for logical representation of inference relations) interrelates the catalogue entries on many conceptual levels. Special ontologies function as specialized database indexes (or views), representing the conceptual perspectives of particular disciplinary conventions with respect to the archive and any related data. These specialized indexes continue to evolve as each discipline evolves, keeping all views conceptually coordinated. Formal Concept Analysis [see 8] provides a graphical interface of lattices displaying the related views of catalogue entries in a relational database [see 9]. Graphical views (or visualizations) of this data-under-evolution enable disciplinary specialists to remain critically aware of the complex implications of each researcher's contribution (including their own) with respect to others in the scheme of inquiry as a whole. Other methods of knowledge representation and processing have made possible the essential functions of this collaborative resource evolution (see section 5 for scenario examples of the user-driven development process).

3.3   Essential Testbed Functions

Scholars or researchers in any specific discipline can *report* their reconstructions of Peirce's complex arguments, as hypotheses, based on their interpretations of the imaged and transcribed manuscript content. Editors and other scholars or researchers are then able to *track* their own hypotheses as well as those of others, from the image evidence supporting them, through the systematic deduction of their implications for manuscript order and content, to the ongoing inductive testing by scholars or researchers who interpret and employ content of the reconstructed arguments. Not only is the order of pages in a manuscript frequently hypothetical because of lost or misplaced pages, but Peirce's complex style of composition increases the uncertainty of accurate reconstruction. He sometimes uses the same page in several versions of elaborated discussion, the course of which sometimes even doubles-back on itself to pick up an unexplored path.

In our future scenario, each scholar in the testbed creates a *map* of the page-by-page course of a particular hypothetical Peircean "exploration." These computer-generated diagrams record each (possibly unique) reconstruction of Peirce's multi-path arguments[5]. These "S-diagrams" can then be matched, by either human perception or computer-generated graph methods. Detected relative invariance among maps reveals agreements in sequence across all scholars' hypothetical reconstructions, and pin-points where controversies in reconstructions lie and further investigation should be pursued. As interpretations proceed and scholars' hypothetical reconstructions evolve, a "dynamic map" is generated, which continues to display the evolution of the collaborative effort (see section 5). Many tools (for database, document, and knowledge-based management, search and retrieval, knowledge acquisition, interlingua for both natural language translation and system integration, knowledge-based communication services support and discourse management) have been integrated and customized to serve the scholars' reporting, tracking, and mapping needs (de Moor suggests an architecture [10:269-70]).

4   Stages of Inquiry

As our scenario proceeds, the scholars are engaged in hypothetical inference, the first stage of inquiry as Peirce conceived its three stages of abduction, deduction, and induction. In the first stage, *abduction*:

---

[5] As early as 1986, one Peirce scholar (Shea Zellweger) mapped on paper the pattern of several Peirce manuscripts, creating an elaborate, page-by-page, "organic" figure of the course of his writings (see 1 for examples).

1. they experience a mysterious or disturbing phenomenon (P): that page 1 of manuscript "507" looks strikingly similar in content to but does not empirically match page 22 of manuscript "145(s)";
2. they suppose that P would be explicable as not surprising, if hypothetical condition (H) were true, and then;
3. they accept in these conditional terms that H might be true [see *CP* 5.189].

Using the facts observable in the manuscript evidence, together with previous known facts of Peirce's work, each scholar attempts to formulate an H to explain P. Abduction's role in inquiry is to begin argumentation by providing claims that suggest how further inquiry might proceed and what might be its aim. Peirce stresses its significance in relation to the other two stages of inquiry: "Its only justification is that from its suggestion deduction can draw a prediction which can be tested by induction, and that, if we are ever to learn anything or to understand phenomena at all, it must be by abduction that this is to be brought about" [*CP*: 5.171].

4.1   Collaborative Abduction

In the scenario, Jay Zeman studies his S-diagram for manuscript "145" and notices that the content of the questionable "507" page might indicate a "branch" in Peirce's argument toward a discussion of gamma graphs, which could connect with the content of "145(s)" at page 22. He uses mediated standard-quality images to refer to both pages, together, indicating to the other scholars on an overlaid commentary image file what passages and points on the pages of text and images lead him to make his conjecture. Meanwhile, Bob Burch has been comparing the image of the "507" page with images of "145(s)" and "145" pages, arranging them side-by-side on his display screen. Since all the manuscript material was imaged at a standard focal length, the pages of each folder are represented in actual size relationships. Bob finds that the "507" page matches the "145(s)" paper (in terms of size and ruled line spacing), and appears to match that of "145." He double-checks the match of paper tones with a color-calibrated detector and finds that they apparently measure the same color; but to be certain, he runs a test on the full-size image files (with more color-depth data than can be displayed on his screen). That test indicates that the "507" paper is the same as the "145," but not quite the same as that of "145(s)."

Peter Øhrstrøm is working on the content alone, and finds that the folder labeled "145(s)" picks up the topic of "145" soon after the first page, and that the second page has a verso with only gamma graphs, possibly indicating where Peirce could have made a branch in his trend of thought. Peter reports that hypothesis, and Frithjof Dau confirms it as plausible from his knowledge of Peirce's EG notation (alpha, beta, and gamma). Frithjof's analysis of why Peirce created the "507" page indicates that he apparently redrew the EGs of "507" on the "145(s)" page 22, leaving out one figure, but Frithjof is not entirely sure that "507" is a bridge between "145" and "145(s)." Meanwhile, Robert Marty wonders (based on his knowledge of Peirce's semeotic logic) if the "507" page may have been cut from some as yet undiscovered version of "145," because Peirce decided it simply headed too quickly into the gamma graph part of EG, without explaining significant sign theory context. Although Marty suspects, he is not yet ready to assert that conjecture formally as an hypothesis, and must consult some of Peirce's other discussions of his logic and its graphical notation.

Jay Zeman mentions some significant bibliographic evidence, at this point: that "145(s)" has a title page very similar to that of "145," although with fewer colors in its title lettering and a shorter title, "Geometrical Topics;" and it is dated "April - May 1905." If Peirce wrote this manuscript over a month's time, that might explain the slight difference found between its tone of paper and that of "145"? Could the fact that "145(s)" has fewer colors throughout its 32 pages (only green, blue, and black) indicate it was Peirce's first attempt, while the five colors (green, blue, red, brown, and black) of "145" indicate a more advanced, succeeding attempt? As Jay closely compares the images of pages from both on his screen, he finds that "145" is definitely more elaborate in content, with both colorful text and graphics. The mystery remains, where does the "507" page then belong?

4.2 Directing Inquiry

Abduction refers to any operations by which theories and concepts are engendered [see *CP* 5.590 (1903)]. As inquiry proceeds, the consequences (or implications) of each scholar's hypothesis must be traced out by deduction, and compared with the results of experimentation by induction (in further efforts to represent his corpus accurately and, beyond, in the interpretation of Peirce's ideas in the many fields where it is relevant, including knowledge representation). As soon as one hypothesis is refuted, another one is subjected to the same stages of argumentation: Abduction merely suggests what may be; Deduction proves what must be; Induction shows what actually is operative [see *CP* 5.590]. Peirce identifies other modes of reasoning that mingle these three forms of argument [see *CP* 2.103 (1902)], for example, analogy [see *CP* 7.97 (1902); also see Sowa's discussion of analogy as a prerequisite for logical reasoning: 11], but he considers these three to be logically elementary, with distinct roles and limitations.

> Abduction furnishes all our ideas concerning real things, beyond what are given in perception, but is mere conjecture, without probative force. Deduction is certain but relates only to ideal objects. Induction gives us the only approach to certainty concerning the real that we can have. In forty years diligent study of arguments, I have never found one which did not consist of those elements. [CP 8.209 (1905)]

Peirce criticizes the common mistake of mixing the three as a simple argument, and even more commonly of confusing abduction with induction. They are "opposite poles of reason": one the most ineffective argument, the other the most effective.

> The method of either is the very reverse of the other's. Abduction makes its start from the facts, without, at the outset, having any particular theory in view, though it is motivated by the feeling that a theory is needed to explain the surprising facts. Induction makes its start from an hypothesis, which seems to recommend itself, without at the outset having any particular facts in view, though it feels the need of facts to support the theory. Abduction seeks a theory. Induction seeks for facts. In abduction the consideration of the facts suggests the hypothesis. In induction the study of the hypothesis suggests the experiments which bring to light the very facts to which the hypothesis had pointed. [CP 7.218 (1901)]

Induction (the experimental testing of a theory) is the most effective because, "although the conclusion at any stage of the investigation may be more or less erroneous, yet the further application of the same method must correct the error." Induction measures the degree of concordance of theory with fact [*CP* 5.145 (1903)].

Peirce conceived his theory of inquiry to explain the "marvelous self-correcting property of reason," and how "making experiments" upon his logical graphs can come into play [*CP* 5.579 (1898); and see Dau's analysis of Peirce's EG: 12]. After abduction allows a mass of facts to suggest a theory, deduction constructs an *image* or *diagram* of that suggested ideal interpretation by which relations of parts that are implicit in the proposed theory become explicit and convincingly predictable. Induction then tests to verify that predictability under real conditions. One of his latest accounts stresses the directive nature of inquiry.

> Abduction having suggested a theory, we employ deduction to deduce from that ideal theory a promiscuous variety of consequences to the effect that if we perform certain acts, we shall find ourselves confronted with certain experiences. We then proceed to try these experiments, and if the predictions of the theory are verified, we have a proportionate confidence that the experiments that remain to be tried will confirm the theory. [CP 8.209 (1905)]

In inquiry, we ask the pragmatic question: If what we conjecture were true, as the cause of something unexpected, what would be the consequences? We conjecture what might have been the antecedents from the consequences we observe, provisionally accept that explanation as an hypothesis worth testing, then proceed to the deductive stage of observing what we have asserted as the state of things, in diagrams (or graphs) of its premises. We can then perceive in the parts of the diagram relations not explicitly mentioned in the premises, as implications that should subsist, at least in a certain proportion of cases, and conclude their necessary or probable truth [see *CP* 1.66 (c. 1896)]. Abduction and deduction establish the rational directions for inductive investigation [see 13, for a detailed analysis verifying the order of these stages].

5   Games of Inquiry

We can correlate Peirce's theory and logical stages of inquiry with Robert Brandom's pragmatic *inferentialist* theoretical framework and his *model of discursive practice* [see 14], for a *functional* analysis of the scenario. Under that model, when a scholar asserts a claim (an hypothesis) about the manuscript pages, he expresses his commitment to a belief, which gives its conceptual content his authority and licenses others (including himself) to undertake a corresponding commitment to use that assertion as a premise in further inferences. In Brandom's theory, inferences involve both intercontent and interpersonal relations in the discursive practice of "giving and asking for reasons." When one scholar expresses a claim that entails a claim expressed by another, anyone who is thereby committed to the first is committed to the second. In Brandom's model, the direction and strength of any scholar's assertion (of its conceptual content) are defined by two *normative* dimensions: a speaker's *commitment* to that content, and what serves as *entitlement* to (as reasons for) that commitment. *Communication* becomes essential to rational practice, as "the interpersonal, intracontent inheritance of entitlement to commitments" [15:165].

Pragmatically, Brandom and Peirce agree that logic is "the linguistic organ of semantic self-consciousness and self-control," making it possible "to criticize, control and improve our concepts" [15:19, 149]. Logic's instrumental task is to make explicit what is implicit in the use of ordinary vocabulary, to reveal patterns of inference that are formally valid because they are invariant. But Brandom emphasizes that these formal properties of inference derive from *material* inferences that must be explained in terms of ordinary language; so the *primary* task of logic is to help us express something about the conceptual contents of material inferences. Rather than a standard of right reasoning, logic is "a 'distinctive set of tools' which we can use to make explicit (and hence available for criticism and transformation) the inferential commitments that govern the use of all our vocabulary, and hence the contents of concepts" [15:30; 14:246-6]. Consistent with Peirce's pragmatic theory, Brandom warns that when we display relevant grounds and consequences in logic to assert their inferential relations, we must never expect to achieve complete transparency of commitment and entitlement. That ideal serves as only a conceptual limit to hope for in pursuing inquiry [see 15:149, 76].

Brandom's functional theory of concepts focuses on their role in reasoning as prior to their supposed origin in experience [see 15:25; 14:243]. Conceptual content is determined by *how* concepts are structured by inferences, prior to their use in effectively referring to anything. He stresses: "the representational aspect of the propositional contents that play the inferential roles of premise and conclusion should be understood in terms of the social or dialogical dimension of communicating reasons, of assessing the significance of reasons offered by others," rather than as arrived at noninferentially from sense perception [see 15:166]. His model of discursive practice treats concepts as norms, not as single signs but as nodes in an inferential network of related concepts. Those norms determine "what counts as a *reason* for particular beliefs, claims, and intentions," as rules determine the correctness of moves in a game [15:25; 14:243][6]. In fact, Brandom's model identifies conceptual content in an expression by whether it can *play a role in the inferential game of making claims and giving and asking for reasons*[7].

Brandom contends that we give beliefs, desires, and intentions conceptual content when we ask such pragmatic questions as "under *what* circumstances would what is believed, desired, or intended be *true?*" [15:158]. His theory and model enforce the distinction between what is *said* or *thought* (the propositional dimension) and what it is said or thought *about* (the representational dimension), and he stresses that even our reports of observable properties, such as color, must be inferentially articulated to have conceptual content. Otherwise, we could not distinguish these noninferential reports from the automatic responses of machinery, such as thermostats and photocells. To understand a concept is to be able (or know how) to distinguish what follows from its application and what it follows from in the practical sense [see 15:108]. Any expressed report that functions conceptually must at least serve as a premise from which to draw inferential consequences: there can be no autonomous language game of entirely noninferential reports. The *representational* use of

---

[6] Peirce stresses a similar semeotic understanding in indicating how his Existential Graphs should be applied dialogically in inquiry (see *CP* 4.429-431).

[7] Brandom's model is based on Wilfred Sellars's suggested game of "giving and asking for reasons" [see 15:35; and 14:251-54].

propositional concepts relies on the *social* structure of their inferential articulations in the game of giving and asking for reasons [see 15:183]. Conceptual content is *collaborative, inferential* content.

5.1   The Manuscript Reconstruction Game

The scholars in the scenario report their claims in conditional form, to make explicit what is not otherwise made explicit in their ordinary discourse involving the informal material inferences that reflect what they consider significant during their empirical observations of the manuscript material. Using these formal expressions (or asserted hypotheses), they engage in collaborative discourse. In the testbed context [see 16], the game mode of discourse helps the scholars formulate and select valid hypotheses to explain their surprising experience of the evidence (the mystery of the "507" page). Their game, called the Manuscript Reconstruction Game (MRG), is played as a formal deductive and inductive test of each scholar's proposed hypothesis (of its strength in terms of commitments and entitlements as rationally and empirically valid for accurately reconstructing the manuscript pages).

Those playing the game must *report*: by selecting propositionally contentful expressions capable of serving in inferences as both premise and conclusion (what can be offered as, and itself stand in need of, reasons). Learning to play the game involves being able to tell what is a reason for what, distinguishing good reasons from bad, by keeping score on what other players are committed and entitled to, as two dimensions of *normative status*. Understanding the conceptual content of a claim is being able to accord it proper significance, or knowing how a commitment would change the score in various inferential contexts with other claims [see 15:165-66]. So the players must also *track* the relations of commitments and entitlements in the game's progression.

If Bob Burch makes a commitment by expressing the declarative sentence: "The paper of page 1 of manuscript '507' does not match that of page 22 in '145(s)' (with the color match test results as entitlement), indicating it was not originally written in that notebook," then a commitment by Jay Zeman, such as, "Manuscript '507' was cut from '145(s)'," is not compatible. And other possible sentences uttered by the other scholars may be incompatible. For each scholar's explicitly expressed observation, there will be a set of sentences that are logically incompatible with it; and inclusion relations among the sets of sentences asserted in a dialog among the five scholars will then indicate inferential relations among the sentences. That is, for example, the content of the claim expressed by asserting, "Page 1 of '507' could mark a branch of '145'," entails the content of the claim expressed by asserting, "'507' is the same paper tone as '145'," because everything incompatible with being the same paper tone is incompatible with marking a branch of "145"—perhaps. The two sorts of normative status (commitments and entitlements) and their interactions among the sentences asserted in the play of the game, are tracked in the three sorts of inferential relations:

committive (that is, commitment-preserving) inferences,
         a category that generalizes deductive inference;
permissive (that is, entitlement-preserving) inferences,
         a category that generalizes inductive inference; and
incompatibility entailments, a category that generalizes
         modal (counterfactual-supporting) inference [see 15:194].

The incompatibilities among commitments and their entitlements are automatically displayed to the scholars for their examination and consideration.

The three sorts of inferential consequence relations are ranked (all incompatibility entailments are commitment-preserving, though not vice versa, and all commitment-preserving inferences are entitlement preserving, though not vice versa) [see 15:195]), and can be used to *map* the discussion's evolution as the scholars produce and consume reasons.   Inheritance of commitment, inheritance of entitlement, and entailments according to the incompatibilities defined by the interactions of commitments and entitlements compose three axes that "inferentially articulate." The map reveals what Brandom calls the *normative fine structure of rationality* (NFS), describing the progress of the game in making implicit content explicit [see 15:195; also see Wille's rendering of Brandom's "rich inference structures" in Contextual Judgment Logic: 17]. The scholars rely on the mechanism of the game, as a logical editor or "logical lens," to help them focus on and clarify the complexities of inference and conceptual content in their collaborative view of the manuscript evidence.   Incompatibilities that emerge mark possibly missing hypothetical inferences that should direct further inquiry. Meanwhile commitment-preserving inferences trace the implications of validly related claims, and entitlement-preserving inferences record the tested reliability of those claims with respect to the evidence.

Under Brandom's framework, each scholar's commitments are his beliefs and the entitlements are his reasons for believing, based on the evidence. They express these commitments and entitlements to make explicit their implicit material inferences in an "inferentially articulated network," as the game requires. The game offers a method of "harmonizing" these rational and empirical elements of their collaborative conceptual content, by directing attention to incompatibilities to be resolved. In attempting to maximize their own scores, players must minimize the incompatibility entailments to improve their collaborative hypotheses, as the ultimate objective of the game. Brandom compares that process to judges formulating principles of common law: "intended both to codify prior practice, as represented by precedent, expressing explicitly as a rule what was implicit therein, and to have regulative authority for subsequent practice" [15: 75-76; and see 11:19]. Peirce's theory of inquiry has the same normative aim: "to articulate the conditions for veracity, under which thinking can reasonably be considered to increase order, harmony, and connectedness in the world of thought" [18:35; see 14:256].

5.2   The Tools Improvement Game

In the scenario, Peter Øhrstrøm, Robert Marty, and Frithjof Dau need to find more material evidence in the manuscript collection from which to determine their asserted commitments that are based on valid entitlements.   They begin a game with conceptual structures tools builders to specify requirements for an automated method of finding all instances of Peirce's gamma graph work in the archive. Selecting Martin's WebKB search engine [19], for testing, they first attempt to locate all manuscripts where Peirce discusses both his sign theory and his EG together. The same harmonizing framework helps the scholars jointly clarify their collaboration tool requirements in their testbed partnerships with tool developers, by making explicit their implicit augmentation-tool needs[8]. Emerging requirements can be tracked to eliminate incompatibilities, and an NSF concept map is generated as a virtual prototype (or hypothetical tool) for evaluation, before any actual system is built. Testing the actual prototype is carried out in the same recursive process of inquiry, with the scholars now playing the game as they assess their needs in using it; and each prototype generation is tracked to eliminate (or make compliant) incompatibilities with other tools.   The NFS map represents each tool system's evolving documentation, and reveals incompatibilities among tool specifications to aid in developing standardization that is directed *de jure* by tool usage, rather than *de facto*, by tool developers.   The "tools improvement game" (TIG; to be described in future work) can be played concurrently with the MRG.

6   Conclusions

In 1868, Peirce began his study of inquiry with an essay intended for a series called "Search for a Method." He began by contending that in some or all respects our "modern minds" still operate under *Cartesian* influence when we assume that argumentation is represented by "a single thread of inference depending often upon inconspicuous premisses." Peirce urged that we should "stand upon a very different platform from this." We must begin inquiry with all the prejudices we actually have that are based on valid entitlements: "no one who follows the Cartesian method will ever be satisfied until he has formally recovered all those beliefs which in form he has given up" [*CP* 2.265]. And because of those beliefs: "all higher animals have some insight into what is passing in the minds of their fellows. Man shows a remarkable faculty for guessing at that. Its full powers are only brought out under critical circumstances" [CP 7.40]. In the course of inquiry, we may find reason to doubt what we began by believing; but in that case we doubt because we have a positive reason for it, not because we began with complete doubt.

   Brandom's inferentialist model gives Peirce's pragmatic theory of inquiry the functional context of a game. In asserting sentences, we explicitly undertake commitments to the correctness of our material inferences (our implicit beliefs) from the circumstances to the consequences of their application. In making these commitments, we understand propositional content, "not as the turning on of a Cartesian light, but in practical mastery of inferentially articulated responses." Clear thinking and expression are matters of knowing what we are committed to in making claims, and what entitles us to those commitments. If we fail in either of these components, we fail to understand conceptual content, because we fail to grasp the inferential commitments that our use of concepts involves [see 15:63-64; 14:252-53]. Although Brandom's model obscures the initial significance of abduction that Peirce emphasizes in ordering the three stages of inquiry, Brandom clearly incorporates it as

---

[8] See http://port.semanticweb.org/ICCS2005/gametools.pdf for a diagram indicating how the tools operations can be connected and communicate with other components of the testbed. John Sowa's Flexible Modular Framework (FMF) [20] and Heather Pfeiffer's Conceptual Programming Environment (CPE) [21] integrate the tools and data needed in the testbed.

a directing influence in the *recursive process* of conducting inquiry, which is functionally consistent with Peirce's theory [see 14:255-57].

Whether we hope to reconstruct manuscripts or to improve conceptual structures tools, Brandom's model and Peirce's theory of inquiry promise a solid foundation for meeting the challenges of Google's "digital library of the 21st century." This paper attempts to take its own pragmatic advice: Before entering inquiry, we should mark out the proposed course of it, even if circumstances subsequently require the plan to be modified, as they usually will [see *CP* 5.34]. Pragmatically, our scenario offers an abduction, from which we may deduce what *would be the necessary specific tool requirements,* for that hypothetical concept of collaborative scholarship to become true.

References

General Notes: "MS" references are to Peirce's manuscripts archived at the Houghton Library, Harvard; for *CP* references, *Collected Papers of Charles Sanders Peirce,* 8 vols., ed. Arthur W. Burks, Charles Hartshorne, and Paul Weiss (Cambridge: Harvard University Press, 1931-58).

1. Keeler, M. [2003]. "Hegel in a Strange Costume: Reconsidering Normative Science in Conceptual Structures Research." In: de Moor, A., Lex, W., and Ganter, B. (Eds.): *Lecture Notes in Artificial Intelligence,* Vol. 2746, Springer-Verlag, pp. 37-53.
2. Keeler, M. [forthcoming, 2005]. "The Philosophical Context of Peirce's Existential Graphs," *Cognito, Centro de Estudos do Pragmatismo Filosofia.*
3. Keeler, M. and Kloesel, C. [1997]. "PORT: A Testbed Paradigm for Knowledge Processing in the Humanities." In: Lukose, D., Delugach, H., Keeler, M., and Searle, L. (Eds.): *Lecture Notes in Artificial Intelligence,* Vol. 1257, Springer-Verlag, pp. 505-520.
4. Keeler, M. and Kloesel, C. [1997]. "Communication, Semiotic Continuity, and the Margins of the Peircean Text." In David Greetham (ed.), *Margins of the Text.* Ann Arbor: University of Michigan Press.
5. Peirce, C. S. [1982-97]. *Writings of Charles S. Peirce: A Chronological Edition.* Six volumes. Edited by Christian J.W. Kloesel et al. Bloomington: Indiana University Press (see the Peirce Edition Project <http://www.iupui.edu/~peirce>).
6. Robin, R. S. [1967]. *Annotated Catalogue of the Papers of Charles S. Peirce.* Amherst: University of Massachusetts Press.
7. Puder, A. and Römer, K. [1997]. "Generic Trading Service in Telecommunication Platforms." In: Lukose, D., Delugach, H., Keeler, M., and Searle, L. (Eds.): *Lecture Notes in Artificial Intelligence,* Vol. 1257, Springer-Verlag, pp. 551-565.
8. Ganter, B. and Wille, R. [1999]. *Formal Concept Analysis: Mathematical Foundations.* Berlin Heildelberg New York : Springer-Verlag.
9. Priss, U. and Old, L. J. [2004]. "Modelling Lexical Databases with Formal Concept Analysis." *Journal of Universal Computer Science,* Vol 10: 8, pp. 967-984.
10. de Moor, A. [2004]. "Improving the Testbed Development Process in Collaboratories." In: Wolff, K.E., Pfeiffer, H.D. and Delugach, H.S. (Eds.): *Lecture Notes in Artificial Intelligence,* Vol. 3127, Springer-Verlag, pp. 261-274.
11. Sowa, J.F. and Majumdar, A.K. [2003]. "Analogic Reasoning." In: de Moor, A., Lex, W., and Ganter, B. (Eds.): *Lecture Notes in Artificial Intelligence,* Vol. 2746, Springer-Verlag, pp.16-36.
12. Dau, F. [2004]. "Types and Tokens for Logic with Diagrams." In: Wolff, K.E., Pfeiffer, H.D. and Delugach, H.S. (Eds.): *Lecture Notes in Artificial Intelligence,* Vol. 3127, Springer-Verlag, pp. 62-93.
13. Staat, W. [1993]. "On Abduction, Deduction, Induction, and the Categories," *Transactions of the Charles S. Peirce Society,* Vol. XXIX, No. 2, pp. 225-237.

14. Keeler, M. [2004]. "Using Brandom's Framework to Do Peirce's Normative Science." In: Wolff, K.E., Pfeiffer, H.D. and Delugach, H.S. (Eds.): *Lecture Notes in Artificial Intelligence*, Vol. 3127, Springer-Verlag, pp. 37-53.
15. Brandom, R. [2000].   *Articulating Reasons: An Introduction to Inferentialism.* Cambridge, MA: Harvard University Press.
16. de Moor, A., Keeler, M. and Richmond, G. [2002].   "Towards a Pragmatic Web." In: Priss, U., Corbett, D. and Angelova, G. (Eds.): *Lecture Notes in Artificial Intelligence*, Vol. 2393, Springer-Verlag, pp. 235-249.
17. Wille, R. [2003]. "Conceptual Contents as Information—Basics for Contextual Judgment Logic. In: de Moor, A., Lex, W., and Ganter, B. (Eds.): *Lecture Notes in Artificial Intelligence*, Vol. 2746, Springer-Verlag, pp. 1-15.
18. Parker, Kelly A. [2003]. "Reconstructing the Normative Sciences." *Cognito, Centro de Estudos do Pragmatismo Filosofia*, Vol. 4, no. 1, pp. 27-45.
19. Martin, P. [1997].   "The webKB set of tools: A common scheme for shared www, annotations, shared knowledge bases and information retrieval."   In: Lukose, D., Delugach, H., Keeler, M., and Searle, L. (Eds.): *Lecture Notes in Artificial Intelligence*, Vol. 1257, Springer-Verlag, pp. 585-588.
20. Sowa, J.F. [2002]. "Architectures for Intelligent Systems." In: *Special Issue on Artificial Intelligence of IBM Systems Journal,* Vol. 41, pp. 331-349.
21. Pfeiffer, H.D. [2004].   "An Exportable CGIF Module from the CP Environment: A Pragmatic Approach." In: Wolff, K.E., Pfeiffer, H.D. and Delugach, H.S. (Eds.): *Lecture Notes in Artificial Intelligence*, Vol. 3127, Springer-Verlag, pp. 319-332.

Toward Cooperatively-Built
Knowledge Repositories

Philippe Martin, Michael Blumenstein, and Peter Deer

Griffith University - School of I.T. - PMB 50 Gold Coast MC - QLD 9726 Australia*
pm@phmartin.info

Abstract. After noting that informal documents and formal knowledge bases are far from ideal for discussing or retrieving technical knowledge, we propose mechanisms to support the sharing, re-use and cooperative update of semi-formal semantic networks, assign values to contributions and credits to the contributors. We then propose ontological elements to guide and normalize the construction of such knowledge repositories, and an approach to permit the comparison of tools or techniques.

1 Introduction

The majority of technical information is currently published in mostly *unstructured forms* within documents such as articles, e-mails and user manuals. Thence, finding and comparing tools or techniques to solve a problem is a lengthy process (with often sub-optimal results) that involves reading many documents partly redundant with each other. This process heavily relies on memory and manual cross-checking, and its outcomes, even if published, are lost to many people with similar goals. Writing documents is also a lengthy process that involves summarizing what has been described elsewhere and making choices on which ideas or techniques to describe and how: level of detail, order, etc.

To sum up, whatever the field of study, there is *currently no well structured semantic network of techniques or ideas* that a Web user could (i) navigate to get a synthetic view of a subject or, as in a decision tree, quickly find its path to relevant information, and (ii) easily update to publish a new idea (or a new way to explain an idea) and link it to other ideas via semantic relations. Such a system is indeed difficult to build and initialize. However, it is part of a vision for a semi-formal "standardized online archive of computer science knowledge" [8] and dwarfed by the much more ambitious visions of a "Digital Aristotle" which would be capable of teaching much of the world's scientific knowledge by (i) adapting to its students' knowledge and preferences [3], and (ii) preparing and answering (with explanations) test questions for them [9]. The current approaches that are related to the above-cited problems can be divided into three groups.

First, the approaches *indexing (parts of) documents by metadata* (generated or manually inputed) such as Dublin Core metadata, DocBook metadata, topics,

---

* The 1st author began this article at the LOA, Italy

F. Dau, M.-L. Mugnier, G. Stumme (Eds.): ICCS 2005, LNAI 3596, pp. 411–424, 2005.

categories from ontologies, and formal summaries. These approaches are useful for retrieving or exploiting a large base of documents but do not lead to any browsable/updatable semantic network organizing facts or ideas. The same can be said about most document-related query answering systems.

Second, the approaches aiming to represent elements of a domain into formal or semi-formal *knowledge bases* (KBs). Some KBs essentially contain *term definitions*, formal ones as in Open GALEN (an ontology of medical knowledge) and semi-formal ones as in Fact Guru [7]; they are interesting to re-use for representing or indexing parts of documents but are *insufficient* to learn about a domain or compare techniques to solve a problem. Other KBs are mainly intended to *support problem solving*, e.g. the KBs of the QED Project (which aims to build a formal KB of all important established mathematical knowledge) and those of the Halo project [9] (which has for long term goal the design of a "Digital Aristotle"), and hence are *not meant to be directly read and browsed*.

Third, the *hypertext-based Web sites* describing and organizing resources (researchers, discussion lists, journals, concepts, theories, tools, etc.) of a domain, e.g. MathWorld. Some sites permit their users to collaborate or discuss by adding or updating documents, e.g. via wiki systems or annotation systems. Because these systems *do not use semantic relations*, the resulting information is often as poorly structured as in mailing lists and hence includes many redundancies, and arguments and counter-arguments for an idea are hard to find and compare. However, Wikipedia, an on-line hypertext encyclopedia which is also a wiki, albeit a very controlled one, has good quality articles on a wide variety of domains. These articles are well connected and permit their readers to get an overview of a subject and explore it to find information. Yet, Wikipedia's content structure and support for collaboration and IR could be improved. An easy-to-use and easy-to-implement feature would be a support for some semantic relations (e.g. subtypeOf, instanceOf, and partOf) and especially argumentation relations (e.g. proof, example, hypothesis, argument, correction; pre-Web hypertext systems like AAA [6] supported predefined sets of such relations). A semi-formal English-like syntax such as ClearTalk (the notation used in Fact Guru and CODE4 [7]) would support more knowledge processing while still being user-friendly.

It would be utopic to think that even motivated knowledge engineers would be (in the near future) able/willing to represent their research ideas completely into a formal, shared, well-structured readable semantic network that can be explored like a decision tree: there are too many things to enter, too many ways to describe or represent a same thing, and too many ways to group and compare these things. On the other hand, representing the most important structures into such a semantic network and interconnecting them with informal representations seems achievable and extremely interesting for education and IR purposes. Section 2 proposes some mechanisms to support the sharing, re-use and cooperative update of such semantic networks, including some mechanisms to assign values to the contributions and credits to the contributors. Section 3 proposes some ontological elements to guide and normalize the construction of these knowledge repositories. Section 4 shows an approach to permit the comparison of tools or techniques.

2 Support of Cooperation Between Knowledge Providers

2.1 Making Knowledge Explicit and Sharing It

We only consider asynchronous cooperation since it both underlies and is more scalable than exchanges of information between co-temporal users of a system.

The most decentralized knowledge sharing strategy is the one the W3C envisages for the "Semantic Web": many small ontologies, more or less independently developed, thus partially redundant, competing and very loosely interconnected. Hence, these ontologies have problems similar to those we listed for documents: (i) finding the relevant ontologies, choosing between them and combining them is difficult and sub-optimal even for a knowledge engineer (let alone for a machine), (ii) a knowledge provider cannot simply add one concept or statement "at the right place", and is not guided by a large ontology (and a system exploiting it) into providing precise concepts and statements that complement existing ones and are more easily re-used, and (iii) the result is not only more or less lost to others but increases the amount of "data" to search.

A more knowledge-oriented strategy is to have a knowledge server permitting registered users to access and update a single large ontology on a domain and upload files mixing natural language sentences with knowledge representations (e.g. in a controlled language). WebKB-1, WebKB-2, OntoWeb/Ontobroker and Fact Guru are examples of servers allowing this. This was also the strategy used in the well publicized KA2 project [1] which re-used Ontobroker and aimed to let Knowledge Acquisition researchers index their resources, but (i) the provided ontology was extremely small and could not be directly updated by users, and (ii) the formal statements had to be stored within an invented attribute (named "onto") of HTML hyperlink tags via a poorly expressive language. Thus, this approach was limiting and frustrating, and this project was not followed.

We know of only two knowledge servers having special protocols to support users' cooperation[1]: Co4 [2] and WebKB-2 [4]. The approach of Co4 is based on peer-reviewing; the result is a hierarchy of KBs, the uppermost ones containing the most consensual knowledge while the lowermost ones are the private KBs of contributing users. We believe the approach of WebKB-2 which has a KB shared by all its users leads to more relations between categories (types or individuals) or statements from the different users and is easier to handle (by the system and the users) for a large amount of knowledge and large number of users. Details can be found in [4] but here is a summary of its principles.

To avoid lexical problems, each category identifier is prefixed by a short identifier of its creator (who is also represented by a category). Each statement also has an associated creator and hence, if it is not a definition, may be considered as a belief. Any object (category or statement) may be re-used by any user within his/her statements. The removal of an object can only be done by

---

[1] Most servers, including WebKB-2, support concurrency control (e.g. via KB locking) and several, like Ontolingua, support users' permissions on files/KBs. Cooperation support is not so basic: it is about helping knowledge re-use, conflict prevention and the solving of each conflict once it has been detected by the system or a user.

its creator but a user may "correct" a belief by connecting it to another belief by a "corrective relation" (e.g. `pm#corrective_specialization`). (Definitions cannot be corrected since they cannot be false). If entering a new belief introduces a redundancy or an inconsistency that is detected by the system, it is rejected. The user may then either correct his/her belief or re-enter it again but connected by a "corrective relation" to each belief it is redundant or inconsistent with: this allows and makes explicit the disagreement of one user with (his/her interpretation of) the belief of another user. This also technically removes the cause of the problem: a proposition A may be inconsistent with a proposition B but a belief that "A is a correction of B" is not technically inconsistent with a belief in B. (Definitions from different users cannot be inconsistent with each other, they simply define different categories/meanings). Choices between beliefs may have to be made by people re-using the KB for an application, but then they can exploit the explicit relations between beliefs, e.g. by always selecting the most specialized ones. WebKB-2 displays a statement *with* its meta-statements, hence with the associated corrective relations. Finally, to avoid seeing the objects of certain creators during browsing or within query results, a user may set filters on these creators, based on their identifiers, types or descriptions.

For the construction of knowledge repositories, an interesting aspect of this approach to encourage re-use, precision and object connectivity is that it also works for semi-formal KBs. Here, regarding a statement, semi-formal means that if it is written in a natural language (whether it uses formal terms or not) it must *at least* be related to another statement by a formal relation, e.g. a generalization relation (`pm#corrective_generalization`, `pm#summary`, etc.) or an argumentation relation. Thus, to minimize redundancies and to help information retrieval within information repositories, this minimal semantic structure (which in many situations is the only one bearable by people) could be used to organize ideas that are otherwise repeated in many documents. For instance, for a Web site that centralizes and organizes/represents in a formal, semi-formal and informal way resources (tools, techniques, publications, mailing list, teams, etc.) related to a domain (e.g. CGs), it would be very interesting to have some space where discussions could be conducted in this minimal semi-formal way, and hence index or partly replace the mailing list: this would avoid recurring discussions or presentations of arguments, show the tree of arguments and counter-arguments for an idea, permit incremental additions, encourage deeper or more systematic explorations of each idea, and record the various reached status-quos.

Below is an example of what the top-levels of a semi-formal discussion could look like when displayed in an indented linear form (we do not expect this organization to be the direct result of a discussion but it may be the result of a semi-automatic re-organization of this discussion and then be refined by further semi-formal discussions). To save space, we have replaced (counter-)argument relations by '+' and '-', used comments to give an idea of lower levels, and not represented the authors of the statements. The (counter-)arguments for a statement are valid for its specializations and the (counter-)arguments of the specializations are (counter-)examples for their generalizations.

```
A KRL should (also) have an XML-based notation to ease knowledge sharing
 +: this permits to use URIs and Unicode
      -: most syntaxes can be adapted for that  //+: as noted by Berners-Lee
 +: this permits to re-use XML tools (parsers, XSLT, ...)
      -: useless additional step since KBSs do not use XML internally
      > classic XML tools re-usable even if a graph-based model is used
          > classic XML tools work on RDF/XML
              -: some XML tools expect a classic XML tree to be followed
              -: with difficulties since RDF/XML has multiple serialisations
 -: XML-based KRLs are hard to read without XML tools
      +: this is acknowledged by about everyone  //including by the W3C
      -: this does not matter since XML tools exist
          -: this is impossible or inconvenient in many tools or situations
          -: a good notation and a good text editor is often more convenient
 +: other notations may still be used  //-: but they are not standards ...
 +: Not using XML would require a separate plug-in for each syntax
      -: installing a plug-in takes less time than always loading XML files
 > The Semantic Web (SW) KRL should have an XML notation
      > The SW KRL should have an XML syntax but a graph-based model
          +: for flexibility and normalization reasons  //+: TBL's arguments
          +: classic XML tools re-usable even if a graph-based model is used
          -: then there are partially redundant standards (e.g. with RDF/XML)
          -: a subset of XML can be used  //examples: EARL, MPV, RSS, etc.
          > RDF should have the current RDF/XML syntax
              -: it is particularly arbitrary and hard to understand
              -: it leads to errors //+:cannot use schema validation languages
              -: it is inefficient //+:parsers 5-20 times slower than with XML
```

2.2   Valuating Contributions and Contributors

The above described knowledge sharing mechanism of WebKB-2 records and exploits annotations by individual users on statements but does not record and exploit any measure of the "usefulness" of each statement, a value representing its "global interest", popularity, originality, etc. Yet, this is interesting for a knowledge repository and especially for semi-formal discussions: statements that are obvious, un-argued, or for which each argument has been counter-argued, should be marked as such (e.g. via darker colors or smaller fonts) in order to make them less visible (or invisible, depending on the selected display options) and discourage the entering of such statements. More generally, the presentation of the combined efforts from the various contributors may then take into account the usefulness of each statement. Furthermore, given that the creator of each statement is recorded, (i) a value of usefulness may also be calculated for each creator (and displayed), and (ii) in return, this value may be taken into account to calculate the usefulness of the creator's contributions; these are two additional refinements to both detect and encourage argued and interesting contributions, and hence regulate them.

Ideally, the system would allow user-defined measures of usefulness for a statement or a creator, and adapt its presentation of the repository accordingly. Below we present the *default* measures that we shall implement in WebKB-2 (or more exactly, its successor and open-source version, AnyKB). We may try to support *user-defined* measures but since each step of the user's browsing would imply re-calculating the usefulness of all statements (except those from WordNet) and all creators, the result is likely to be very slow. For now, we only consider beliefs: we have not yet defined the usefulness of a definition.

To calculate a belief usefulness, we first associate two more basic attributes to the belief: 1) its "state of confirmation" and 2) its "global interest".

1) The first is equal to 0 if the belief has no (counter-)argument linked to it (examples of counter-argument relation names: "counter-example", "counter-fact", "corrective-specialization"). It is equal to 1 (i.e. the belief is "confirmed") if (i) each argument has a state of confirmation of 0 or 1, and (ii) there exists no confirmed counter-argument. It is equal to -1 if the belief has at least one confirmed counter-argument. It is also equal to 0 in the remaining case: no confirmed counter-argument but each of the arguments has a state of confirmation of -1. All this is independent of whom authored the (counter-)arguments.

2) Each user may give a value to the interest of a belief, say between -5 and 5 (the maximum value that the creator of the belief may give is, say, 2). Multiplied by the usefulness of the valuating user, this gives an "individual interest" (thus, this may be seen as a particular multi-valued vote). The *global interest* of a belief is defined as the average of its individual interests (thus, this is a voting system where more competent people in the domain of interest are given more weight). A belief that does not deserve to be visible, e.g. because it is clearly a particular case of a more general belief, is likely to receive a negative global interest. We prefer letting each user explicitly give an interest value rather than taking into account the way the belief is generalized by or connected to (or included in) other beliefs because interpreting an interest from such relations is difficult. For example, a belief that is used as a counter-example may be a particular case of another belief but is nevertheless very interesting as a counter-example.

Finally, the *usefulness of a belief* is equal to the value of the global interest if the state of confirmation is equal to 1, and otherwise is equal to the value of the state of confirmation (i.e. -1 or 0: a belief without argument has no usefulness, whether it is itself an argument or not).

In argumentation systems, it is traditional to give a type to each statement, e.g. fact, hypothesis, question, affirmation, argument, proof. This could be used in our repositories too (even though the connected relations often already give that information) and we could have used it as a factor to calculate the usefulness (e.g. by considering that an affirmation is worth more than an argument) but we prefer a simpler measure only based on explicit valuations by the users.

Our formula for a user's usefulness is: `sum of the usefulness of each belief from the user + square root (number of times he/she voted on the interest of beliefs)` . The second part of this equation acknowledges the participation of the user in votes while decreasing its weight as the number of votes increases. (Functions decreasing more rapidly than `square root` may perhaps better balance originality and participation effort).

These measures are simple but should incite the users to be careful and precise in their contributions (affirmation, arguments, counter-arguments, etc.) and give arguments for them: unlike in anonymous reviews, careless statements here penalise their authors. Thus, this should lead users not to make statements outside their domain of expertise or without verifying their facts. (Using a different pseudo when providing low quality statements does not seem to be an helpful strategy to escape the above approach since this reduces the number of authored statements for the first pseudo). On the other hand, the above measures should hopefully not lead "correct but outside-the-main-stream contributions" to be under-rated since counter-arguments must be justified. Finally, when a belief is counter-argued, the usefulness of its author decreases and hence he/she is incited to remove it or deepen the discussion.

In his description of a "Digital Aristotle" [3], Hillis describes a "Knowledge Web" to which researchers could add ideas or explanations of ideas "at the right place", and suggests that it should "include the mechanisms for credit assignment, usage tracking, and annotation that the [current] Web lacks", thus supporting a much better re-use and evaluation of the work of a researcher than via the current system of article publishing and reviewing. However, Hillis does not give any indication on such mechanisms. Although the mechanisms we proposed in this sub-section and the previous one were intended for one knowledge repository/server, they seem usable for the Knowledge Web too. To complement the approach with respect to the Knowledge Web, the next sub-section proposes a strategy to achieve knowledge sharing between knowledge servers.

2.3   Combining the Advantages of Centralization and Distribution

One knowledge server cannot support the knowledge sharing of all researchers. It has to be specialized and/or act as a broker for more specialized servers. If competing servers had an equivalent content (today, Web search engines already have "similar" content), a Web user could query or update any server and, if necessary, be redirected to use a more specialized server, and so on recursively (at each level, only one of the competing servers has to be tried since they mirror each other). If a Web user directly tried a specialized server, it could redirect him/her to use a more appropriate server or directly forward him/her query to other servers.

To permit this, our idea is that each server periodically checks *related servers* (more general servers, competing servers and slightly more specialized servers) and 1) integrates (and hence mirrors) all the objects (categories and statements) generalizing the objects in a *reference set* that it uses to define its "domain" (if this is a general server, this set is reduced to pm#thing, the uppermost concept type), 2) integrates either all the objects that are more specialized than the objects in the reference set, or if a certain depth of specialization is fixed, associates to its most specialized objects the URLs of the servers that can provide specializations for these objects (note: classifying servers according to domains is far too coarse to index/retrieve knowledge from distributed knowledge servers, e.g. knowledge about "neurons" or "hands" can be relevant to many domains; thus, a classification by objects is necessary), and 3) also associates the URLs of more general servers to the direct specializations of the generalizations of the objects in the reference set (because the specializations of some of these specializations do not generalize nor specialize the objects in the reference set).

Integrating knowledge from other servers is certainly not obvious but this is a more scalable and exploitable approach than letting people and machines select and re-use or integrate dozens or hundreds of (semi-)independently designed small ontologies. A more fundamental obstacle to the widespread use of this approach is that many industry-related servers are likely to make it difficult or illegal to mirror their KBs. However, other approaches will suffer from that too.

3   Some Ontological Elements

By default, the shared KB of WebKB-2 includes an ontology derived from the noun-related part of WordNet and various top-level ontologies [5]. A large general ontology like this is necessary to ease and normalize the cooperative construction of knowledge repositories but is still insufficient: an initial ontology on the domain of the repository is also necessary. As a proof of concept for our tools to support a cooperatively-built knowledge repository, we initially chose to model two related domains: (i) Conceptual Graphs (CGs), since this domain is the most well known to us, and (ii) ontology related tools, since Michael Denny's "Ontology editor survey"[2] attracted interest despite being purposefully shallow and poorly structured.

Modelling these two domains implies partially modelling related domains, and we soon had the problem of modularizing the information into several files to support readability, search, checking and systematic input[3]. These files are also expected to be updatable by users when our knowledge-oriented wiki is completed. In order to be generic, we have created six files[4]: "Fields of study", "Systems of logic", "Information Sciences", "Knowledge Management", "Conceptual Graph" and "Formal Concept Analysis". The last three files specialize the others. Each of the last four files is divided into sections, the uppermost ones being "Domains and Theories", "Tasks and Methodologies", "Structures and Languages", "Tools", "Journals, Conferences, Publishers and Mailing Lists", "Articles, Books and other Documents" and "People: Researchers, Specialists, Teams/Projects, ...". This is a work in progress: the content and number of files will increase but the sections seem stable. We now give examples of their content.

3.1   Domains and Theories

Names used for domains ("fields of study") are very often also names for tasks. Task categories are more convenient for representing knowledge than domain

---

[2] http://www.xml.com/pub/a/2004/07/14/onto.html

[3] Although the users of WebKB-2 can direcly update the KB one statement at a time, the documentation discourages them to do so because this is not a scalable way to represent a domain (as an analogy a line command interface is not a scalable way to develop a program). Instead, they are encouraged to create files mixing formal and informal statements and ask WebKB-2 to parse these files, and in the end when the modelling is complete and if the users wish to, integrate them to the shared KB.

[4] See http://www.webkb.org/kb/domain/

categories because (i) organizing them is easier and less arbitrary, and (ii) many relations (e.g. case relations) can then be used. Since for simplicity and normalization purposes a choice must be made, whenever suitable we have represented tasks instead of domains. When names are shared by domain categories and task categories (in WebKB-2, categories can share names but not identifiers), we advise the use of the task categories for indexing or representing resources.

When studying how to represent and relate document subjects/topics (e.g. technical domains), [10] concluded that representing them as types was not semantically correct but that mereo-topological relations between individuals were appropriate. Our own analysis confirmed this and we opted for (i) an interpretation of theories and fields of study as large "propositions" composed of many sub-propositions (this seems the simplest, most precise and most flexible way to represent these notions), and (ii) a particular part relation that we named ">part" (instead of "subdomain") for several reasons: to be generic, to remind that it can be used in WebKB-2 as if it was a specialization relation (e.g. the destination category needs not be already declared) and to mak clear that our replacement of WordNet hyponym relations between synsets about fields of study by ">part" relations refines WordNet without contradicting it. Our file on "Fields of study" details these choices. Our file on "Systems of logics" illustrates how for some categories the represented field of study *is* a theory (it does not *refer* to it) thus simplifying and normalizing the categorization. Below is an example (in the FT notation) of relations from WordNet category #computer_science, followed by an example about logical *domains/theories*. When introducing general categories in Information Sciences and Knowledge Management, we used the generic users "is" and "km". In WebKB-2, a generic user is a special kind of user that has no password: anyone can create or connect categories in its name but then cannot remove them.

```
#computer_science__computational_science  (^engineering science that ...^)
  >part:    #artificial_intelligence, //in WordNet, AI is ">part:" of CS
  >part:    is#software_engineering_science (is), //link created by "is"
  >part:    is#database_management_science (is),
  >part of: #engineering_science__engineering__applied_science__technology,
  part:     #information_theory,  //link from WordNet: "(wn)" is implicit
  part of:  #information_science;

km#substructural_logic (^system of propositional calculus weaker ...^)
  >part of: km#non-classical_logic__intuitionist_logic,
  >part: km#relevance_logic km#linear_logic,
  url: http://en.wikipedia.org/wiki/Intuitionistic_logic;

km#CG_domain__Conceptual_Graphs__Conceptual_Structures
  >part of: km#knowledge_management_science,
  object: km#CG_task km#CG_structure km#CG_tool km#CG_mailing_list,
  url: http://www.cs.uah.edu/~delugach/CG/  http://www.jfsowa.com/cg/;
```

To provide a core ontology that will guide the sharing, indexation or representation of techniques in Knowledge Management, hundreds of categories will need to be represented. We have only begun this work. On ontological issues too our approach departs from the one of the KA2 project [1] since a good part of its small predefined ontology was a specialization hierarchy of 37 Knowledge

Acquisition (KA) domains, the names of which could have been used for tasks, structures, methods (PSMs) and experiments. E.g., this hierarchy included:
`reuse_in_KA > ontologies  PSMs;`   `PSMs > Sysiphus-III_experiment;`

3.2   Tasks and Methodologies

In most model libraries for KA, each non-primitive task is linked to techniques that can be used for achieving it, and conversely, each technique combines the results of more primitive tasks. We tried this organization but at the level of generality of our current modelling it turned out to be inadequate: it led (i) to arbitrary choices between representing sometimes as a task (a kind of process) or a technique (a kind of process description), or (ii) to the representation of both notions and thus to introduce categories with names such as `KA_by_classification_from_people`; both cases are problematic for readability and normalization. Similarly, instead of representing methodologies directly, i.e. as another kind of process description, it seems better to represent *the tasks* advocated by a methodology (including their supertask: following the methodology). Furthermore, with tasks, many relations can then be used directly: similar relations do not have to be introduced for techniques or methodologies. Hence, we represented all these things as tasks and used multi-inheritance. This considerably simplified the ontology and the source files. Here are some extracts.

```
km#KM_task__knowledge_management_task__KM   < is#information_sciences_task,
 > km#knowledge_representation km#knowledge_extraction_and_modelling
   km#knowledge_comparison km#knowledge_retrieval_task
   km#knowledge_creation km#classification km#KB_sharing_management
   km#mapping/merging/federation_of_KBs km#knowledge_translation
   km#knowledge_validation
   {km#monotonic_reasoning  km#non_monotonic_reasoning}
   {km#consistent_inferencing km#inconsistent_inferencing}
   {km#complete_inferencing km#incomplete_inferencing}
   {km#structure-only_based_inferencing km#rule_based_inferencing}
   km#language/structure_specific_task  //e.g. km#CG_task and km#FCA_task
   km#teaching_a_KM_related_subject  km#KM_methodology_task,
 object of: km#knowledge_management_science,
 object: km#KM_structure; //between types, the default cardinality is 0..N
 //"object" has different meanings depending on the connected categories

   km#knowledge_retrieval_task  < is#IR_task,
    > {km#specialization_retrieval  km#generalization_retrieval}
      km#analogy_retrieval  km#structure_only_based_retrieval
      {km#complete_retrieval km#incomplete_retrieval}
      {km#consistent_retrieval km#inconsistent_retrieval};
```

3.3   Structures and Languages

In our top-level ontology [5], `pm#description_medium` (supertype for languages, data structures, ...) and `pm#description_content` (supertype for fields of studies, theories, document contents, softwares, ...) have for supertype `pm#description` because (i) such a general type grouping both notions is needed for the signatures of many basic relations, and (ii) classifying WordNet categories according to the two notions would have often led to arbitrary choices. However, we represented the default ontology of WebKB-2 as a part of WebKB-2 and hence allowed part relations between any kind of information. To ease knowledge entering and certain exploitations of it, we allow the use of generic relations such as part, object and support when, given the types of the connected objects, the relevant relations (e.g. `pm#subtask` or `pm#physical_part`) can automatically be found.

For similar reasons, to represent "sub-versions" of ontologies, softwares, or more generally, documents, we use types connected by subtype relations. Thus, for example, `km#WebKB-2` is a type and can be used with quantifiers.

```
km#KM_structure  < is#symbolic_structure,
 > {km#base_of_facts/beliefs  km#ontology  km#KB_category  km#KB_statement}
   km#KB  km#KA_model  km#KR_language  km#language_specific_structure;

   km#ontology__set_of_category_definitions/constraints
     > km#lexical_ontology  km#language_ontology  km#domain_ontology
       km#top_level_ontology  km#concept_ontology  km#relation_ontology
       km#multi_source_ontology__MSO,
     part: 1..* km#KB_category  1..* km#category_definition;

   km#KR_language__KRL__KR_model_or_notation
     > {km#KR_model/structure  km#KR_notation}
       km#frame_oriented_language  km#predicate_logic_oriented_language
       km#graph_oriented_language  km#KR_language_with_query_commands
       km#KR_language_with_scripting_capabilities,
     attribute: km#semantics;

km#CG_structure  < km#language_specific_structure,
 > km#CG_statement  km#CG_language  km#CG_ontology;
```

3.4 Tools

We first illustrate some specialization relations between tools then we use the FCG notation to give some details on WebKB-2 and Ontolingua. (The FT notation does not yet permit to enter such details. As in FT, relation names in FCG may be used instead of relations identifiers when there is no ambiguity).

```
km#CG_related_tool  < km#language/structure_specific_tool,
 > km#CG-based_KBMS  km#CG_graphical_editor  km#NL_parser_with_CG_output;

   km#CG-based_KBMS < km#KBMS,
     > {km#CGWorld  km#PROLOG\+CG  km#CoGITaNT  km#Notio  km#WebKB};

       km#WebKB  > {km#WebKB-1  km#WebKB-2},  url: http://www.webkb.org;

km#input_language (*x,*y) = [*x, may be support of: (a km#parsing,
                            input: (a statement, formalism: *y))];
[any pm#WebKB-2,   //", part:": has for part, "a ": existential quantifier
  part:(a is#user_interface, part:{a is#API, a is#HTML_based_interface,
                              a is#CGI-accessible_command_interface,
                              no is#graph_visualization_interface}),
  part: {a is#FastDB, a km#default_MSO_of_WebKB-2},
  input_language: a km#FCG,   output_language: {a km#FCG, a km#RDF},
  support of: a is#regular_expression_based_search,
  support of: a km#specialization_structural_retrieval,
  support of: a km#generalization_structural_retrieval,
  support of: (a km#specialization_structural_retrieval,
              kind: {km#complete_inferencing, km#consistent_inferencing},
              input: (a km#query, expressivity: km#PCEF_logic),
              object: (several km#statement, expressivity: km#PCEF_logic)
           )];        //"PCEF": positive conjunctive existential formula
[any km#Ontolingua,
  part: {a is#HTML_based_interface, no is#graph_visualization_interface,
         no DBMS, a km#ontolingua_library},    input_language: a km#KIF,
  output_language:{a km#KIF, no km#RDF}, support of: a is#lexical_search];
```

3.5 Articles, Books and Other Documents

This example shows how a simple document indexation using Dublin Core relations (we have done this for all the articles of ICCS 2002). Representing ideas from the article would be more valuable. For examples of representations of conferences, publishers, mailing lists, researchers and research teams, please access http://www.webkb.org/kb/domain/.

```
[an #article, dc#Coverage: km#knowledge_representation,
  pm#title: "What Is a Knowledge Representation?",
  dc#Creator: "Randall Davis, Howard E. Shrobe and Peter Szolovits",
  pm#object of: (a #publishing, pm#time: 1993,
                    pm#place: (the #object_section "14:1 p17-33",
                                      pm#part of: is#AI_Magazine)),
  pm#url: http://medg.lcs.mit.edu/ftp/psz/k-rep.html];
```

4   Example of Comparison of Two Ontology-Related Tools

Fact Guru (which is a frame-based system) permits the comparison of two objects by generating a table with the object identifiers as column headers, the identifiers of all their attributes as row headers, and for each cell either a mark to signal that the attribute does not exist for this object or a description of the destination object. The common generalizations of the two objects (possibly one of them) is also given. However, this strategy is insufficient for comparing tools. Even for people, creating detailed tool comparison tables is often a presentation challenge and involves their knowledge of which features are difficult or important and which are not. A solution could be to propose predefined tables for easing the entering of tool features and then compare them. However, this is restricting. Instead or in complement, we think that a mechanism to generate good comparison tables is necessary and can be found. The following query and generated table illustrates an approach that we propose. The idea is that a specialization hierarchy of features is generated according to (i) the uppermost relations and destination types specified in the query, and (ii) only descriptions used in at least one of the tools and their common generalizations are shown. To that end, some FCG-like descriptions of types can be generated. In the cells, '+' means "yes" (the tool has the feature), '-' means "no", and '.' means that the information has not been represented. We invite the reader to compare the content of this table with the representations given above; then, its meaning and the possibility to generate it automatically should hopefully be clear. A maximum depth of automatic exploration may be given; past this depth, the manual exploration of certain branches (like the opening or closing of sub-folders) should permit the user to give the comparison table a presentation better suited to his/her interest. Any number of tools could be compared, not just two.

```
> compare pm#WebKB-2 km#Ontolingua on
    (support of: a is#IR_task, output_language: a KR_notation), maxdepth 5

                                         WebKB-2  Ontolingua
support of:
is#IR_task                                  +         +
  is#lexical_search                         +         +
    is#regular_expression_based_search      +         .
  km#knowledge_retrieval_task               +         .
    km#specialization_structural_retrieval  +         .
       (kind: {km#complete_inferencing, km#consistent_inferencing},
        input: (a km#query, expressivity: km#PCEF_logic),
        object: (several statement, expressivity: km#PCEF_logic))
                                            +         .
    km#generalization_structural_retrieval  +         .

output_language:
km#KR_notation                              +         +
  (expressivity: km#FOL)                    +         +
    km#FCG                                  +         .
    km#KIF                                  .         +
  km#XML-based notation                     +         .
    km#RDF                                  +         -
```

In the general case, the above approach where the descriptions are put in the rows and organized in a hierarchy is likely to be more readable, scalable and easier to specify via a command than when the descriptions are put in the cells, e.g. as in Fact Guru. However, this may be envisaged as a complement for simple cases, e.g. to display {FCG, KIF} instead of '+' for the output_language relation. In addition to generalization relations, "part" relations could also be used, at least the >part relation. E.g., if Cogitant was a third entry in the above table, since it has a complete and consistent structure-based and rule-based mechanism to retrieve the specializations of a simple CG in a base of simple CGs and rules using simple CGs, we would expect the entry ending by km#PCEF_logic to be specialized by an entry ending by km#PCEF_with_rules_logic.

5 Conclusion

Knowledge repositories, as we have presented them, have many of the advantages of the "Knowledge Web" and "Digital Aristotle" but seem much more achievable. To that end, we have proposed some techniques and ontological elements, and we are: (i) implementing a knowledge oriented wiki to complement our current interfaces, (ii) experimenting on how to best support and guide semi-formal discussions, and more generally, organize technical ideas into a semantic network, (iii) implementing and refining our measures of statement/user usefulness, (iv) completing the above presented ontology to permit at least the representation of the information collected in Michael Denny's "Ontology editor survey" (we tend to think that our current ontology on knowledge management will only need to be specialized, even though we have not yet explored the categorization of the basic features of multi-user support such as concurrency control, transactions, CVS, file permissions, file importation, etc.), (v) permitting the comparison of tools as indicated above, and (vi) providing forms or tables to help tool creators represent the features of their tools.

References

1. V.R. Benjamins, D. Fensel, A. Gomez-Perez, S. Decker, M. Erdmann, E. Motta and M. Musen. Knowledge Annotation Initiative of the Knowledge Acquisition Community: (KA)2. *Proc. of KAW98*, Banff, Canada, April 1998.

2. J. Euzenat. Corporate memory through cooperative creation of knowledge bases and hyper-documents. *Proc. of 10th KAW*, (36) 1–18, Banff (CA), Nov. 1996.

3. W.D. Hillis. "Aristotle" (The Knowledge Web). *Edge Foundation, Inc.*, No 138, May 2004. *http://www.edge.org/3rd_culture/hillis04/hillis04_index.html*

4. P. Martin. Knowledge Representation, Sharing and Retrieval on the Web. *Web Intelligence* (Eds.: N. Zhong, J. Liu, Y. Yao), Springer-Verlag, Jan. 2003. *http://www.webkb.org/doc/papers/wi02/*

5. P. Martin. Correction and Extension of WordNet 1.7. *Proc. of ICCS 2003* (Dresden, Germany, July 2003), Springer Verlag, LNAI 2746, 160-173.

6. W. Schuler and J.B. Smith. Author's Argumentation Assistant (AAA): A Hypertext-Based Authoring Tool for Argumentative Texts. *Proc. of ECHT'90*, Cambridge University Press, 137–151.

7. D. Skuce and T.C. Lethbridge. CODE4: A Unified System for Managing Conceptual Knowledge. *Int. Journal of Human-Computer Studies (1995)*, 42, 413–451. Fact Guru, the commercial version of CODE4, is at www.factguru.com

8. D.A. Smith. Computerizing computer science. *Communications of the ACM (1998)*, 41(9), 21–23.

9. Vulcan Inc. Project Halo: Towards a Digital Aristotle. *www.projecthalo.com*

10. C.A. Welty and J. Jenkins. Formal Ontology for Subject. *Journal of Knowledge and Data Engineering (Sept. 1999)*, 31(2), 155-182.

What Has Happened to Ontology

Peter Øhrstrøm, Jan Andersen, and Henrik Schärfe

Aalborg University
Department of Communication
Kroghstræde 3
DK – 9220 Aalborg East
Denmark
{poe,ja,scharfe}@hum.aau.dk

Abstract Ontology as the study of being as such dates back to ancient Greek philosophy, but the term itself was coined in the early 17th century. The idea termed in this manner was further studied within academic circles of the Protestant Enlightenment. In this tradition it was generally believed that ontology is supposed to make true statements about the conceptual structure of reality. A few decades ago computer science imported and since then further elaborated the idea of 'ontology' from philosophy. Here, however, the understanding of ontology as a collection of true statements has often been played down. In the present paper we intend to discuss some significant aspects of the notion of 'ontology' in philosophy and computer science. Mainly we focus on the questions: To what extent are computer scientists and philosophers — who all claim to be working with ontology problems — in fact dealing with the same problems? To what extent may the two groups of researchers benefit from each other? It is argued that the well-known philosophical idea of ontological commitment should be generally accepted in computer science ontology.

1 Introduction

It is obvious that the term 'ontology' has become a key word within modern computer science. In particular this term has become popular in relation to the Semantic Web studies and the development of "formal ontologies" to be used in various computer systems (see [3]). It is, however, also evident that 'ontology' is a term imported from philosophy, and that the understanding of 'ontology' in computer science differs somewhat from the understanding of the term in traditional philosophy. Where philosophical ontology has been concerned with the furniture and entities of reality, i.e., with the study of "being qua being", computer scientists have been occupied with the development of formalized, semantic, and logic-based models, which can easily be implemented in computer systems. The result is that we now have two distinct branches of research dealing with 'ontology'. In the present study we intend to discuss some significant aspects of this import of 'ontology'. To what extent are computer scientists and philosophers — who all claim to be working with ontology problems — in fact dealing with

F. Dau, M.-L. Mugnier, G. Stumme (Eds.): ICCS 2005, LNAI 3596, pp. 425–438, 2005.

the same problems? To what extent may the two groups of researchers benefit from each other?

The word 'ontology' was coined in the early 17th century, apparently as an attempt to modernize the study of metaphysics. In Sect. 1 we intend to present the background of metaphysics. In Sects. 2 and 3 we intend to deal with the traditional, philosophical approach to 'ontology' taking two little known but important and historically characteristic contributions into account, the first contribution being by Jacob Lorhard (1561-1609) and the second by Jens Kraft (1720-56). In Sect. 4 we shall examine the use of ontology in some relevant parts of modern philosophy. In Sect. 5 we are going to look at the approach to ontology in modern computer science. Finally, in Sect. 6 we intend to discuss what modern philosophers and computer scientists working with ontology may possibly gain from each other.

2   The Background of Metaphysics

A detailed presentation of the history of the notions of ontology is outside the scope of this paper. However, we intend to outline some essential features of this interesting historical development. (For more detailed expositions we refer to works by José Ferrater Mora [19], Raul Corazzon [5], and Barry Smith [24]).

In short, 'ontology' may be defined as the study of being as such. As we shall see, the term ontologia, was created in the circles of German protestants sometime around 1600. However, as Dennis Bielfeldt has pointed out 'ontology is as old as philosophy itself' [4]. In fact, the construction of the word 'ontologia' may be seen as an attempt to modernize the classical debate about metaphysics. But this background of metaphysical studies of the idea of being is still essential for the understanding of modern notions of 'ontology'.

Two and a half millennia ago in his poem "On Nature" ('Περὶ φύσεως'), ontological problems were addressed by the Eleatic philosopher Parmenides [21]. He claimed that our everyday experience is incomplete and deceitful, and that the reality which comes to us through our thoughts therefore must be more reliable than the one we get to know through our senses. For thinking, Parmenides set forth two sole paths of inquiry: First, "that it is and it is impossible for it not to be", and second, "that it is not and it necessarily must not be". For Parmenides there are two categories only: "being" and "non-being".

Aristotle presented his theory of being in his *Metaphysics*. In this work he maintained that "there is a science which investigates being as being and the attributes which belong to this in virtue of its own nature" [1003a21-22]. He made it clear that this science ('ἐπιστήμη') does not restrict itself to a certain kind of being; on the contrary, it deals with everything that exists. According to Aristotle metaphysicians study "being qua being", i.e., they study those properties applied to entities by virtue of being.

Aristotle and his works *Metaphysics* and *Categories* are crucial instants in the history of formal ontology. Aristotle, through his distinction between "being potentially" (δύναμις) and "being actually" (ἐνέργεια), introduces a third category to the rigid bifurcated ontology of Parmenides. Potentiality and actuality are important explanatory terms in Aristotle's understanding of change. The idea of inactive powers or abilities, which can later become active or actual, requires an ontology that expresses ways of being as well as kinds of being, thus, the importance of his argument for δύναμις seems obvious (see [30, p. 10]).

The classical study of being as such, however, does not solely go back to Aristotelian metaphysics, but also to the ways in which Aristotle has been understood and interpreted in Antiquity and medieval scholasticism. A central character in the early period is the Greek neo-platonic thinker Porfyrios (ca. 234-305) who makes some important comments on Aristotle in his work *Isagoge*. We would also like to draw attention to the Roman philosopher Boethius (480-525) and his work *Consolation of Philosophy* as being among the sources of medieval thinkers and their work with metaphysics and conceptual formalization. Within medieval scholasticism several important contributions to the study of being come to mind. In particular, the works by Ramon Llull (1232-1336) and William of Ockham (1280-1349) should be mentioned.

In the beginning of the fourteenth century, Ramon Llull (1232-1316), in his work *Ars Magna*, developed an ingenious alternative to the hierarchic concept systems. With the aid of logical principles and graphical representation Llull attempted to chain together all kinds of knowledge in combinatorial diagrams illustrated as rotating discs. The project of Llull can be seen as an attractive attempt to construct a conceptual system that is common to all humans as well as independent of natural language, religion, and culture. The impact of Llull on Leibniz and Leibniz' further development of combinatorial thinking, along with his impact on the research on language formalization and conceptual structures, is studied by Thessa Lindof in her dissertation *From the Middle Ages to Multimedia — The Ars Magna of Ramon Llull in a New Perspective* [13].

The renewed interest in philosophical studies of the fourteenth century was dominated by the influence of Aristotle, although it followed its own path in the debate between realists on the one hand and nominalists on the other. William of Ockham, in his work *Expositio aurea*, discusses Porfyrios and develops his metaphysics as a commentary to Aristotle. Opposite to realists, Ockham assumed that those universal concepts, used by us and by science, exist as mere conventions in our minds; they are formal statements about reality rather than realities about the world itself. Ockham claimed that the objects of reality cause sense impressions, which are transformed into mental images through the active mind. The dispute concerning the status of the universals gives rise to the question as to how this debate between realists and nominalists influences the elaboration of conceptual representation of existence.

The history of ontology contains numerous other important contributors to the study of being as such; among others: Thomas Aquinas, Francisco Suarez, Johann Clauberg, Husserl and Heidegger.

3 'Ontology' in Traditional Philosophy

In his presentation of the winter curriculum of 1765-66, Kant informed the students that the course in metaphysics would deal with "ontology, the science, namely, which is concerned with the more general properties of all things". Later, during one of his lectures, Kant explained:

> Ontology is a pure doctrine of elements of all our a priori cognitions, or, it contains the summation of all our pure concepts that we can have a priori of things ... Ontology is the first part that actually belongs to metaphysics. The word itself comes from the Greek, and just means the science of beings, or property according to the sense of the words, the general doctrine of being. Ontology is the doctrine of elements of all my concepts that my understanding can have only a priori. [1, p. 307ff]

By then it is clear that the term 'ontology' as well as the study it names had become established. However, Kant's indication of the origin of the term it not very clear.

According to Barry Smith [24] the term 'ontology' (or *ontologia*) was coined in 1613, independently, by two philosophers, Rudolf Göckel (Goclenius), in his *Lexicon philosophicum* [9] and Jacob Lorhard (Lorhardus), in his *Theatrum philosophicum* [17]. This turns out to be incorrect, since the term occurred already in Jacob Lorhard's book *Ogdoas scholastica* from 1606 [16]. In fact the word "ontologia" appears on the frontispiece where it is used synonymously with "metaphysica" (see [5]).

Lorhard did not explain his use of the word "ontologia" in *Ogdoas scholastica* or his identification of this term with "metaphysica". However, his *Liber de adeptione*, which he published in 1597, appears to contain an important indication in this respect.

> Metaphysica, quae res omnes communiter considerat, quatenus sunt ὄντα, quatenus summa genera & principia, nullis sensibilibus hypothesibus subnixa. [15, p.75]
>
> Metaphysica, which considers all things in general, as far as they are existing and as far as they are of the highest genera and principles without being supported by hypotheses based on the senses. (Our translation.)

Lorhard's use of the Greek word ὄντα ('existing things') in his definition of 'metaphysica' is rather remarkable. Had the word "ontologia" at that time been part of his active vocabulary, it would certainly have been natural to use it in this context. On the other hand, his use of ὄντα in his 1597 definition of metaphysica obviously suggests that it might in fact be very straightforward for Lorhard to come up with the construction of the term "ontologia" and to use it in his 1606 book, *Ogdoas scholastica*.

According to Georg Leonhard Hartmann's careful description [10] Jacob Lorhard was born in 1561 in Münsingen in South Germany. In 1603 he became "Rektor des Gymnasiums" in the protestant city of St. Gallen. In 1607, i.e., the year after the publication of *Ogdoas scholastica*, he received a calling from Landgraf Mortiz von Hessen to become professor of theology in Marburg.

At that time Rudolph Göckel (1547-1628) was also professor in Marburg in logic, ethics, and mathematics. It seems to be a likely assumption that Lorhard and Göckel met one or several times during 1607 and that they shared some of their findings with each other. In this way the sources suggest that Göckel during 1607 may have learned about Lorhard's new term 'ontologia' not only from reading *Ogdoas scholastica* but also from personal conversations with Lorhard. For some reason, however, his stay in Marburg became very short and after less than a year he returned to his former position in St. Gallen. Lorhard died on 19 May, 1609.

Later, in 1613, Lorhard's book was printed in a second edition under the title *Theatrum philosophicum*. However, in this new edition the word "ontologia" has disappeared from the front cover but has been maintained inside the book. In 1613, however, the term is also found in Rudolph Göckel's *Lexicon philosophicum*. The word "ontologia" is only mentioned briefly in the margin on page 16 as follows: "ontologia, philosophia de ente" (i.e., "ontology, the philosophy of being"). Even though Göckel's book is intended as a lexical work, it does not include an article on "ontologia" (nor does it include articles on metaphysics or first philosophy).

Neither Lorhard nor Göckel elaborated further on the concept of "ontologia" nor did they indicate anything directly on the origin of the term. However, based on the study of Lorhard's works — in particular *Liber de adeptione* and *Ogdoas scholastica* — it appears likely that the term was found by Lorhard at some time between the writing of the two books.

Ogdoas Scholastica seems to have been intended as a textbook to be used for teaching purposes. The title indicates that the volume contains eight parts. The word 'ogdoas' in this title is in itself interesting. It is a Latin version of the Greek word 'ὀγδοάς' for eight. The word was incorporated in ecclesiastical Latin. This supports the general impression of the book as a work intended as a presentation of science and philosophy presented on the background of protestant theology. The parts of the volume may be seen as almost independent books. The topics of the eight parts are Latin and Greek grammar, logics, rhetoric, astronomy, ethics, physics and metaphysics/ontology. In the beginning of the part on metaphysics/ontology Lorhard states the following definition:

> Metaphysicae quae est 'ἐπιστήμη τοῦ νοητοῦ ἦ νοητόν' quatenus ab homine naturali rationis lumine sine ullo materiae conceptu est intelligibile. [16, Book 8, p. 1]
>
> Metaphysics is knowledge of the intelligible by which it is intelligible since it is intelligible by man with (the help of) the natural light of reasoning without conception of anything material. (Our translation.)

From a modern point of view it may seem rather surprising that metaphysics/ontology is defined as some kind of 'knowledge', since it is now commonplace to emphasize the difference between what is actually existing and what can be known. However, it would also be natural to assume that *Ogdoas Scholastica* as a schoolbook should deal with reality as it can be known by human beings. The fact that Lorhard insists that metaphysics/ontology can be obtained only by 'the natural light of reasoning' and not by the senses may be seen as an interesting parallel to Parmenides' view mentioned above according to which the ontological realities must come to us through our thoughts and not through our senses.

Ogdoas Scholastica contains a number of diagrams. In fact each of the parts in the book may be represented as a tree-like structure of concepts. The purpose of these illustrations may again be teaching. However, the use of such trees certainly also supports the ideal of logical systematisation. The first (upper) part of Lorhard's conceptual structure (his metaphysics/ontology) may be paraphrased as in Fig. 1.

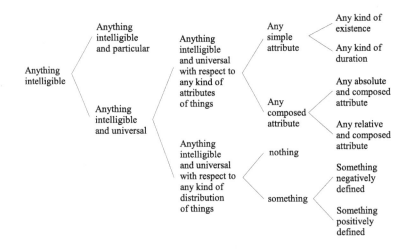

Fig. 1. Upper part of Lorhard's metaphysics/ontology

Lorhard may have been inspired by the German Calvinist and academic philosopher Clemens Timpler (1563-1624) who in his *Metaphysicae systema methodicum* (1604) [28] has suggested that "omne intelligibile" ("anything intelligibile") divides into "nihil" ("nothing") and "aliquid" ("something"), and that "aliquid" divides into "aliquid negativum" ("something negatively defined") and "aliquid positivum" ("something positively defined"), whereas "aliquid positivum" divides into "ens" ("entity") and "essential" ("the essential") and so on. According to Joseph S. Freedman [6], Timpler exemplifies the highest standards of late sixteenth and early seventeenth century European academic philosophy. Apparently, Timpler's book *could* have been one of Lorhard and Göckel's inspirations. Since Göckel wrote a preface to Timpler's book a rather close connection between Timpler and Göckel seems to have existed. However, the word "ontology", as it seems, does not occur in the book.

In the circles of protestant philosophy the study of ontology/metaphysics seems to have been linked to the interest of natural theology and to the formal and mathematical understanding of reality. One interesting example could be Lorhard's ontology of 'duration' from 1606. Lorhard suggests the conceptual structure in Fig. 2.

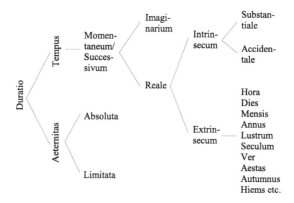

Fig. 2. Lorhard's ontology of 'Duratio' [16, Book 8, p. 5]

It should be noted that according to Lorhard both 'momentaneum' and 'successivum' may serve as genus for 'reale' and 'imaginarium'.

Unfortunately, Lorhard's books never became very famous. But Göckel has to be regarded as relatively well-known. According to Mattias Wolfes [31] he was given such flattering titles as 'the Marburg Plato', 'the Christian Aristotle', 'the teacher of Germany', and 'the light of Europe'.

After Göckel the word 'ontology' was used by Leibniz (1646-1716), but it was Christian Wolff (1679-1754) who made the word 'ontology' popular in philosophical circles. The word appears in the title of his *Philosophia prima sive ontologia* (1736) in which Wolff, displaying his systematizing skills, attempts to combine Aristotelian and scholastic ontology with the theories of Descartes and Leibniz [32]. "Ontologia" or "philosophia prima" is defined as "*scientia entis in genere, seu quantenus ens est*". Ontology, in other words, is the science of being *qua* being; it is the study of "being" understood as a "genus".

Wolff's book system is an encyclopedia of science through which he intended to demonstrate the logical structure of the world. His scientific ideals were adopted from Euclid and Descartes. Science should be built upon clearly defined concepts, on valid axioms and inferences, and logics was the formalism that should secure humans of the eighteenth century a bright future. In 1739, Wolff's pupil and successor, Alexander Gottlieb Baumgarten (1714-1762), publishes his *Metaphysica*. In *Metaphysica* [2], Baumgarten defines ontology as "scientia praedicatorum entis generaliorum", ("the science of the most general and abstract predicates of anything").

It should also be mentioned that ontologia was not the only new name suggested for the field. Johann Clauberg (1622-65) published in 1647 his *Ontosophia* (see Mora [19]). This term seems to have been attractive for some philosophers. However, the Wolffian tradition with its 'ontologia' turned out to be the winner of the name game!

4   The Ontology of Jens Kraft

In the eighteenth century Wolff was the fashionable philosopher number one in Germany as well as in the Scandinavian countries. From Wolff through Baumgarten goes a straight line to the Danish-Norwegian mathematician and academic philosopher, Jens Kraft (1720-1756). Through Kraft's *Ontologie* [12] the study of ontology was introduced to academic life in Scandinavia.

As it seems, during his student years in Copenhagen and while undertaking a study tour to Germany, Kraft had attended Wolff's lectures at Halle. Later, after having been appointed a philosophy professor at Sorø Academy, as a counterpart to Baumgarten's *Metaphysica*, Kraft published his own *Metaphysik* (1751-53) written in Danish. Imitating Baumgarten, Kraft divided his volume into the books: *Cosmologie*, *Ontologie*, *Psykologie* and *Naturlig Theologie*.

Kraft defined ontology as the "science about the most general truths, and it is using this science that I, in the following, at once intend to make a beginning of metaphysics" [§ 10]. This means that Kraft assumed that there is a truth regarding the conceptual structure of reality, and that it is in principle possible to obtain knowledge about any detail in this structure. He held that ontology will be useful background or a useful foundation for any kind of scientific activity. In the preface to *Ontologie* Kraft explained the importance of the subject in the following way:

> I hold it to be almost unuseful to mention anything regarding the usefulness of ontology, since it, in a so enlightened Seculo as ours, ought to be commonly known, that all other sciences, both theoretical and practical, have it as their foundation, and that the considerable mass of writings which from time to time come to light, and which seem partly to want to overthrow, partly to want to change and better the states and Religion, probably never would have come under the press, if their authors had had any solid grasp of the ontological truths. [12, Preface]

This means that according to Kraft, as for the whole Wolffian tradition, ontology is not just a technique, but rather a framework of a number of true statements regarding the fundamental structure of reality. He pointed out that the two most basic features of reality are the principle of contradiction and the principle of the sufficient reason.

It should be pointed out that according to Jens Kraft and the ontologists of the 18th century, the understanding of reality is also important when it comes to ethical and religious questions. They believed that dealing properly with ontology may help mankind to make a better society. According to Jens Kraft, many misunderstandings concerning social and religious improvements may in fact be avoided, if "the ontological truths" are taken properly into account.

Among other things Kraft in his *Ontologie* dealt with the notion of time. Like Lorhard he made a distinction between time and eternity. In §301 he explained that "the finite can never obtain eternity, but it can obtain an infinite time (Ævum) or a time with a beginning but without end. The infinite, by contrast, has permanence (sempiternité)". One the most interesting statements in Kraft's *Ontologie* is his description of the relation between time and existence:

> We call the order in which things follow each other 'time'; that which really exists is called 'the present', that which has existed and no longer exists is called 'the past', and that which has not yet existed, but which will, in fact, come into existence, is called 'the future time'. One part of time, as it follows upon each other, is called 'age'. [§ 298]

First of all it should be noted that Kraft, like Leibniz, does not accept any kind of 'empty time'. Time is the order of things. In consequence, there cannot be any time if there are no things. In this way time depends on the existence of the world. In addition, existence is in fact closely related to 'the present'. 'The past' is clearly explained to be that which does not exist any longer, and the future is seen as that which does not yet exist. This straightforward understanding of the relation between time and existence corresponds to the so-called A-theory, and it has later been questioned by the so-called B-theorists who would like to speak about some kind of timeless existence at an instant or over a period (see [33]). Kraft's treatment of the notion of time in his *Ontologie* is a nice illustration of the view according to which a certain ontological theory must involve some kind of ontological commitment if it is to be interesting at all.

5 'Ontology' in Modern Philosophy

As mentioned in [33] the modern emphasis on 'ontological commitment' is mainly due to Willard Van Orman Quine (1908-2000), who was one of the most influential 20th century writers in the development of modern philosophical ontology. According to Quine's view the formulation of an ontology must involve statements about actual existence or non-existence of entities discussed in the theory. This is certainly not a new position. As we have indicated above it has been essential for metaphysics/ontology since Antiquity. It is also obvious from theories of ontology such as Lorhard's and Kraft's that a number of answers to essential questions have to be incorporated in the conceptual structures to which the ontologies give rise. Peirce said something similar in (Collected Papers 5.496): "There are certain questions commonly reckoned as metaphysical, and which certainly are so, if by metaphysics we mean ontology, which as soon as pragmatism is once sincerely accepted, cannot logically resist settlement."

The study of the relation between time and existence is an important theme in modern philosophical ontology. In [33] it is argued that the four grades of tense-logical involvement suggested by A.N. Prior (1914-69) form a useful framework for discussing how various temporal notions may be related in a top-level ontology. These grades represent four different ways of relating the so-called A-notions (past, present, future) with the so-called B-notions (before, after, 'simultaneous with'). The full development of any ontology dealing with the fundamental notions of time and temporal reality presupposes the choice between these grades. But why should we choose one of the four possibilities rather than another of them? From a formal point of view all four grades are possible, i.e., they are all logically consistent. This means that each of the grades conceived as an ontological theory is logically contingent. Therefore the choice between the grades cannot be based on logical proofs only. It has to be based on something else.

This is not only the case for ontological theories dealing with ideas of time. In general, we cannot deduce a unique ontological theory from unquestionable logical or scientific principles. Ontological theories in general have to be understood as contingent. This is not to say, that no logically unrefutable argument can be given in favour of any ontological theory as compared with its competing theories. Arguments in favour of a particular theory of ontology, however, have to be based on other kinds of values than those involved in unquestionable logical or scientific principles. In many cases, the choice of a certain ontological theory among a number of consistent candidates depends on worldview and aesthetical or even ethical values.

As argued recently by Christopher P. Long [14] we may certainly question the view that ontology must precede ethics. On the contrary, the understanding of the contingency of ontology clearly gives rise to the opposite view. In Long's own words: "Ontology becomes ethical the moment it recognizes its own contingency. Responding to this recognition the ethics of ontology turns away from the quest for certainty, toward the ambiguity of individuality, seeking to do justice to that which cannot be captured by the concept." [14, p.154]

6   'Ontology' in Computer Science

In the discussions of representing knowledge in such a form that machines can process it, the word 'ontology' has undeniably held a center stage position in the last decade, and has indeed found its way into many research disciplines. The overwhelming number of conferences, papers, and books on the subject makes it indeed difficult to talk about ontology as one thing, and one thing only, in present day computer science. For an excellent overview, see Roberto Poli's: 'Framing Ontology' [22]. Likewise, the import of the term 'ontology' from the realm of philosophy into computer science is not easily accounted for, and it would be wrong to assume that we here have a case of immediate transfer. Many influences and trends in knowledge representation have contributed to this import (see [20] and [29]).

Presumably, it was John McCarthy who first introduced the term ontology in AI literature. In 1980 McCarthy used the term in his paper on circumscription in a discussion of what kinds of information should be included in our understanding of the world [18]. His use of the term may be seen as a widening of the concept, but in a non-invasive manner. In contrast, one of the most frequently cited definitions on ontology as that of 'a specification of a conceptualization', found in Gruber's paper from 1993 [7], very much changes the view on ontology and ontology research. This definition has often been debated (see for instance [8]) but clearly, it alludes to something much more subjective and changeable than the classical view allows for. In 1993 when Gruber published this paper, the term had already gained some popularity and at that time several sources were available [29]. Already in 1984, John Sowa mentions ontology in connection with knowledge engineering [25]. In this book, interestingly, the notion of ontology is closely related to the idea of possible worlds, such that a collection of possible worlds may be represented in an ontology. In some capacities, this marks a middle position, in that it is easily seen how modal propositions can be incorporated into an ontology under the classical view. And indeed, some domains cannot be formally represented without taking into account not only things that actually exists in the real world, but also things that may be imagined [23]. Possible world semantics has later played an important role in ontology research, and has been investigated, among others, by Nicola Guarino in order to formally describe ontological commitment [8]. In 2000 John Sowa suggested the following definition:

> The subject of ontology is the study of the categories of things that exist or may exist in some domain. The product of the study, called "an ontology", is a catalog of the types of things that are assumed to exist in a domain of interest, D, from the perspective of a person who uses language L for the purpose of talking about D. [26, p.492]

Viewed within knowledge engineering a theory of ontology has to be seen in relation to a certain domain of interest and it also must presuppose a certain perspective.

Even though the contemporary uses and practices in ontology research escape uniform description, certain affinities can be extracted. If we on one hand consider the philosophical discipline (sometimes referred to as capitalized 'Ontology' [8]), and on the other hand consider the growing collections of ontologies such as found on the web (e.g. at www.daml.org), we do see certain tendencies in the approaches to ontology. These tendencies are noticeable in several interdependent dimensions sustained by positions that we propose to describe as 'Ontology as philosophical discipline', and 'ontology as information practice', respectively. Some of the marks of these positions are:

Plural or Singular

In the works of Lorhard, Kraft and others, ontology is construed as one singular system of thought. In this view we may talk about the ontologies (in the plural) of Aristotle, Lorhard, Cyc, and SUMO; but we see them one at a time, as each representing something comparably stable. In other words, each of ontologies should be understood as an intended attempt to describe how reality actually is. If ontology is seen as an information practice, ontologies may refer to multiple, possibly fragmented domain descriptions relative to some selected perspectives rather than to monolithic systems. Here, ontologies are information tools, some of which are crafted for very local purposes, and they do not necessarily claim any degree of truth outside the domain or relative to other perspectives than those for which they were designed. (See [11]).

Dependency of Domain and Perspective

The philosophical approach tends to work in a top-down manner, carving up the world in segments, with the intent to provide domain-independent description. This is the view presented in Sects. 2 and 3 of this paper. In contrast, the domain-dependent approach tends to work in a bottom-up manner, centered on providing descriptions that may be populated by particular occurrences of objects of more immediate interest. The question is whether focus is given to the general question of what exists, or to the more specific question of what exists in some domain discussed relative to some selected perspectives.

Position on Product

Ontology can be viewed as closely related to the process of understanding what is (epistemology), or as an information strategy leaning towards the question of how information can be shared relative to selected communications situations. In the first view, the resulting ontology is likely to hold strong claims about the structure of the world. In the second view, top-level discussions give way to domain specific considerations, whereby the relation to top-level distinctions can be seen as problems of communication rather than as problems of existence. Here, an ontology can be viewed as an artifact declaring what is (in some particular domain).

In pure form, the Ontology as philosophical discipline is characterized by being singular, perspective- and domain-independent and oriented towards making strong claims about the world. Similarly, ontology as an information practice is characterized by fragmented pieces of knowledge, it depends on the choice of domain and perspective, and as such, makes primarily local claims, and finally, it is intended as an information strategy in which ontologies are seen as artifacts.

It is easily seen that in their pure forms we are faced with two very different strands of research, but also that this is not a dichotomy: in reality, these positions form a continuum, and specific efforts in ontology research may occur at any point between the extremes.

In [33] we have argued that any author formulating a theory on ontology dealing with temporality has to make a choice in relation to the A-/B-distinction. In general, it turns out that some kind of ontological commitment (in the philosophical sense) has to be involved in making an ontology even one made for practical purposes.

7   Relating the Positions

The introduction of conceptual ontology into computer- and communication science is an example of how knowledge from humanities can be imported and utilized in modern information technology. This emphasises the importance of an ongoing dialogue between computer science and the humanities. Some of the new findings in computer science may in fact be generated by the use of research imported from the humanities. It is obvious that mathematicians and computer scientists have developed ontology in several respects in order to make it useful for their purposes. Good examples would be the notion of multiple inheritance and the focus on classifying and structuring relations. Evidently many of these formalisms have also been used in philosophical logic. In this way a useful re-export into the humanities has taken place. The notion of ontological commitment is an important aspect of philosophical ontology. However, this aspect is sometimes ignored in computer science, since ontology in these cases has been seen as mainly an information practice. Nevertheless, it turns out by closer inspection of the ontologies used in modern computer science that they do in fact presuppose some rather specific, but hidden ontological commitments. Maybe some of the computer scientists making these ontologies are not even aware of these hidden commitments. Such a lack of awareness may turn out to be rather problematic and the analysis therefore suggests that the computer scientists involved with ontology should take the idea of ontological commitment into serious account. As argued by Erik Stubkjær [27], an ontology should not only be the result of a language study. In fact, it ought to be based on a coherent and consistent theory which deals with reality in a satisfactory manner. It is an obvious obligation on the developer of an ontology to discuss and defend his choice of theory and the ontological commitments to which it gives rise.

Acknowledgements

The authors wish to thank lektor Anders Jensen and Ulrik Petersen for valuable advice. The authors also acknowledge the stimulus and support of the 'European project on delimiting the research concept and the research activities (EURECA)' sponsored by the European Commission, DG-Research, as part of the Science and Society research programme — 6th Framework.

References

1. Ameriks, Karl and Naragon, Steve (ed.): *Lectures on Metaphysics — part III, Metaphysik Mrongovius* 1790-1791, Cambridge University Press, (1997), p. 307 and 309.
2. Baumgarten, Alexander Gottlieb: *Metaphysica.* Editio VII. Halae Magdeburgicae 1779. Georg Olms Verlag, Hildesheim, (1963).
3. Berners-Lee, T., Hendler, J., Lassila, O.: The Semantic Web, *Scientific American* May (2001), **284**(5), p34, 10p, 2 diagrams, 2c.
4. Bielfeldt, D.: 'Ontology'. in: *The Encyclopedia of Science and Religion.* Vol. 2, (2003), p. 632.
5. Corazzon, Raul: *Ontology. A resource guide for philosophers.* Updated 12/10/2004. (See http://www.formalontology.it), (2004).
6. Freedman, Joseph S.: *European Academic Philosophy in the late Sixteenth and early Seventeenth Centuries: The Life, Significance, and Philosophy of Clemens Timpler* (1563/4-1624), vol. 1-2, Georg Olms Verlag, Hildesheim, (1988).
7. Gruber, T.R.: *A Translation Approach to Portable Ontologies.* in: *Knowledge Acquisition,* **5**(2), pp. 199–220, (1993).
8. Guarino, N.: *Formal Ontology in Information Systems.* in: Guarino, N. (ed) *Proceedings of FOIS'98*, IOS Press, Amsterdam: Trento, Italy, pp. 3–15, (1998).

9. Göckel (Goclenius), Rudolf: *Lexicon philosophicum, quo tanquam clave philosophicae fores aperiuntur*. Francofurti, (1613), Reprographic reproduction, Georg Olms Verlag, Hildesheim, (1964).

10. Hartmann, Georg Leonhard: *Jacob Lorhard*. Manuscript by Georg Leonhard Hartmann (1764-1828) transcribed by Ursula Hasler, Stadtarchiv St. Gallen.

11. Hesse, Wolfgang: *Ontologie(n)*, GI Gesellschaft für Informatik e.V. — Informatik-Lexikon, (2002), available online from the GI-eV website: `http://www.gi-ev.de/informatik/lexikon/inf-lex-ontologien.shtml`

12. Kraft, Jens: *Ontologie, eller Første Deel af Metaphysik*. Sorøe Ridder-Academie, Kiøbenhavn, (1751).

13. Lindof, Thessa: *Fra middelalder til multimedier : Ramon Llull's Ars Magna i en ny belysning*. (From the Middle Ages to Multimedia. The Ars Magna of Ramon Llull in a New Perspective.) Ph.D. dissertation, Aalborg University, (1997).

14. Long, C.P.: *The Ethics of Ontology. Rethinking an Aristotelian Legacy*. State University of New York Press, (2004)

15. Lorhard, Jacob: *Liber de adeptione veri necessarii, seu apodictici*. Tubingae, (1597).

16. Lorhard, Jacob: *Ogdoas scholastica*. Sangalli, (1606).

17. Lorhard, Jacob: *Theatrum philosophicum*, Basilia, (1613).

18. McCarthy, J: *Circumscription — A Form of Nonmonotonic Reasoning*, in: Artificial Intelligence, **13**, pp. 27–39, (1980).

19. Mora, José Ferrater: *On the Early History of Ontology*. in: *Philosophy and Phenomenological research*, **24**, pp. 36–47, (1963).

20. Orbst, L. and Liu, H.: *Knowledge Representation, Ontological Engineering, and Topic Maps*. in: J. Park (ed), XML Topic Maps, Addison Wesley, pp. 103–148, (2003).

21. Parmenides: On Nature ('Περὶ φύσεως'), `http://philoctetes.free.fr/parmenidesunicode.htm` (n.d.).

22. Poli, R.: *Framing Ontology*. Available online from the Ontology resource guide for philosphers: `http://www.formalontology.it/essays/Framing.pdf`

23. Schärfe, H.: *Narrative Ontologies*, in: C.-G. Cao and Y.-F. Sui (eds), *Knowledge Economy Meets Science and Technology – KEST2004*, Tsinghua University Press, pp. 19–26, (2004).

24. Smith, Barry: *Ontology and Informations Systems*. SUNY at Buffalo, Dept. of Philosophy, `<http://www.ontology.buffalo.edu/ontology>`, (2004)

25. Sowa, John F.: *Conceptual Structures: Information Processing in Mind and Machine*, Addison–Wesley, Reading, MA, (1984).

26. Sowa, John F.: *Knowledge Representation. Logical, Philosophical, and Computational Foundations*, Brooks Cole Publishing Co., Pacific Grove, CA, (2000).

27. Stubkjær, Erik: *Integrating ontologies: Assessing the use of the Cyc ontology for cadastral applications*, in: Bjørke, J.T. and H. Tveite, (Eds): *Proceedings of ScanGIS'2001*, Agricultural University of Norway, pp. 171–84 (2001).

28. Timpler, Clemens: *Metaphysicae systema methodicum*. Steinfurt, (1604).

29. Welty, C.: *Ontology Research*. in: *AI Magazine*, **24**(3), pp. 11–12, (2003).

30. Witt, Charlotte: *Ways of Being - Potentiality and Actuality in Aristotles's Metaphysics*. Cornell University Press, London, (2003).

31. Wolfes, M.: *Biographisch-Bibliographiches Kirchenlexikon*, Band XVIII, Verlag Traugott Bautz, (2001), Spalte 514–19.

32. Wolff, Christian: *Philosophia prima sive ontologia*. Reprographic reproduction of second edition, Frankfurt/Leipzig (1736), Wissenschaftliche Buchgesellschaft, Darmstadt, (1962).

33. Øhrstrøm, P. and Schärfe, H.: *A Priorean Approach to Time Ontologies*. in: K.E. Wolff et al. (eds.): *Proceedings of ICCS 2004*, LNAI 3127, (2004), pp. 388–401.

Enhancing the Initial Requirements Capture of Multi-Agent Systems Through Conceptual Graphs

Simon Polovina and Richard Hill

Web-Based and Multi-Agent Research Group
Faculty of Arts, Computing, Engineering and Sciences, Sheffield Hallam University
·City Campus, Howard Street, Sheffield, S1 1WB, United Kingdom.
{s.polovina, r.hill}@shu.ac.uk

Abstract. A key purpose of Multi-Agent Systems (MAS) is to assist humans make better decisions given the vast and disparate information that global systems such as the Web have enabled. The resulting popularity of Agent-Oriented Software Engineering (AOSE) thus demands a methodology that facilitates the development of robust, scalable MAS implementations that recognise real-world semantics. Using an exemplar in the Community Healthcare domain and Conceptual Graphs (CG), we describe an AOSE approach that elicits the hitherto hidden requirements of a system much earlier in the software development lifecycle, whilst also incorporating model-checking to ensure robustness. The resulting output is then available for translation into Agent Oriented Unified Modelling Language (AUML) and further developed using the agent development toolkit of choice.

1. Introduction

The Multi-Agent System (MAS) paradigm is proving a popular approach for the representation of complex computer systems. Social abilities, without which agents cannot interact, interoperate and most importantly, collaborate, are fundamental to the coordination of enterprises and the systems required to support their operations. Unified Modelling Language (UML, www.uml.org) Actors, whether they are human, software or external legacy systems, appear to map readily to agents, and the notion of goal attainment also indicates a plausible link between software abstraction and the real-world [1].

The resulting popularity of Agent-Oriented Software Engineering (AOSE) has thus demanded a methodology that facilitates the development of robust, scalable MAS implementations that recognise these real-world semantics [2]. However, whilst many approaches and tools assist the generation of MAS models, most of the development has focussed upon the analysis and specification stages of the software development lifecycle. Developers typically assemble abstract models, which are iterated into a series of design models using the Unified Modelling Language (UML) or more recently, Agent-Oriented Modelling Language (AUML) [3]. An approach that utilises MAS platforms and toolkits such as JADE [4], to replicate existing systems from class model representations is also attractive to developers as relatively simple

F. Dau, M.-L. Mugnier, G. Stumme (Eds.): ICCS 2005, LNAI 3596, pp. 439-452, 2005.

mappings from classes to agents are realised, though it does require a familiarity with Java and an understanding of how real-life interaction scenarios can be decomposed and translated into program code. However, the generation of detailed program designs, and subsequent code, often results in a significant departure from the more abstract models as many important facets have not been elicited during the requirements capture phase.

AOSE methodologies such as Gaia, Prometheus, Zeus and MaSE have attempted to address these problems and provide a unifying development framework [5-8]. Except for Tropos [9] however, little work has been published that encompasses the whole cycle from initial requirements capture through to implementation of MAS.

Tropos attempts to facilitate the modelling of systems at the *knowledge* level and highlights the difficulties encountered by MAS developers, especially since notations such as UML force the conversion of knowledge concepts into program code representations [9]. As a methodology Tropos seeks to capture and specify 'soft' and 'hard' goals during an 'Early Requirements' capture stage, in order that the Belief-Desire-Intention (BDI) architectural model of agent implementation can be subsequently supported [10, 11]. Whilst model-checking is provided through the vehicle of Formal Tropos [12], this is an optional component and is not implicit within the MAS realisation process.

It is important to have a much deeper level of understanding of a system from the outset, ensuring that fundamental business concepts are captured, described and understood. Whilst conceptual modelling is often a means by which rich, flexible scenarios can be captured, there is an inherent difficulty in specifying a design later in the development lifecycle. This is compounded by the fact that flexibility often leads towards lack of discipline, or consistency, in modelling, thus there is a need for a concept-led, rigorous elicitation process, prior to MAS specification and design.

Our aim thus is to provide an extended and more rigorous means of capturing requirements for MAS at the outset, by addressing the need to scrutinise qualitative concepts that exist in the MAS environment, prior to more detailed analysis and design with existing methodologies. Extensions to the UML meta model such as AUML [3], have simplified the design and specification of agent characteristics such as interaction protocols, yet the process of gathering and specifying initial requirements is often limited to the discipline and experience of the MAS designer, using established notations such as UML's use case diagrams [13].

This paper therefore proposes and describes an improved MAS design framework that places a rigorous emphasis upon the initial requirements capture stage. Section 2 describes some of the shortcomings of requirements capture for MAS development, before explaining the proposed process in Section 3. Section 4 uses an exemplar case study in the Community Healthcare domain to explicate the process in detail, illustrating the significance of the results.

2. Capturing Requirements

Whilst AUML addresses some of the complexities of MAS model generation, the consideration of communication protocols, behaviours and allocation of tasks and roles is inherently complicated as the problem domain to be modelled is generally quite elaborate itself.

Polovina and Hill [14, 15] illustrated prior experience of modelling complex healthcare payment scenarios with the advantages of Conceptual Graphs (CG) including Peirce Logic, as explained elsewhere [16-19]. This study, conducted in comparison to a combination of the Zeus Methodology and AUML, served to illustrate that several problems remain at the requirements capture stage:

- Use case models are a convenient means of defining actors and for documenting the existing processes. Use case notation is flexible enough to capture some of the richer concepts; however there is no inherent model verification, so it is probable that some significant details will be missed from the first iteration.
- Use case analysis is a procedure that elicits process-level tasks without challenging qualitative issues.
- Whilst the process of describing and articulating use cases serves to elicit the majority of the eventual agent behaviours, generation of an ontology of terms is mostly based upon the existing processes together with the systems analyst's knowledge and experience.
- Even though actors appear to map straight to agents, the assignment of behaviours is often based on current practice, rather than from the systematic iteration from a coherent model.

We accordingly believe that the consideration of 'early requirements' enables much more capable MAS to be developed. If the above issues are thus ignored any MAS would be unduly compromised.

3.   The Process

The requirements capture process must therefore incorporate the following:
- A means of modelling the concepts in an abstract way that facilitates the consideration of qualitative issues.
- An ability to reveal more system requirements to supplement the obvious actor-to-agent mappings.
- An explicit means of model-checking before detailed analysis and design specification.
- Improved support for capturing domain terms, with less reliance upon domain experts.

We propose a process that improves the capture of requirements, in a robust and repeatable manner, whilst also eliciting an awareness of significant facets of the system much earlier during the requirements capture phase. The process is summarised in the following stages:

1. *Use Case Analysis* - Requirements are gathered initially and represented as use case models.
2. *Model Concepts* - The high level concepts are modelled as CG and used to describe the overall scenario of the problem that is being investigated.
3. *Transform with Transaction Model (TM) and Generate Ontology of Types* - The high level CG model is transformed with the Transaction Model (TM). The TM is

a pattern that enables an agent to make a much more knowledgeable decision by balancing the monetary and non-monetary costs and benefits of a candidate transaction in a single integrated, interoperable environment [14-15, 16: 110-111]. As explained later the TM thus imposes a 'balance-check' rigour upon MAS models, as well as generating an elementary hierarchy of ontological terms.

4. *Model Specific Scenarios* - Specific instances of the model are then modelled.
5. *Inference with Queries and Validate* - The model is tested by inferencing queries to elicit rules for the ontology and refine the representation.
6. *Translate to Design Specification* - The model is then transformed into a design-level specification such as AUML.

The process thus also incorporates an implicit means of validating the resulting model, both as a means to drive iterations during the modelling process, and as an overall validation prior to implementation.

4.   A Healthcare Case Study

To illustrate the process, we describe in detail the elicitation steps with respect to an ongoing exemplar in the Healthcare domain [14, 15]. Home-based community care delivery is an example of a complicated, multi-agency social care system that is plagued with inefficiencies and logistical problems. Social care systems typically comprise a large number of autonomous functions and services, each interacting and communicating with a variety of protocols. Thus the problem domain harnesses a vast number of quantitative and qualitative issues that must be captured and represented lucidly then translated into a functioning MAS so as not to limit an individual care recipient's quality of life. The first stage is to examine the use cases within the system.

1.   Use Case Analysis

Figure 1 illustrates the overall community healthcare scenario. Three actors and use cases are identified.

- *Elderly Person* - An infirm, elderly person that chooses to continue to live in their own home and request care support from the Local Authority.
- *Local Authority* - A localised representative body of the UK Government that manages and administrates the delivery of healthcare services.
- *Care Provider* - A private organisation that delivers care services into the Elderly Persons' home environment on behalf of the Local Authority.

2.   Model Concepts

Figure 2 illustrates the same scenario modelled conceptually with a CG.

3. Transform with Transaction Model (TM) and Generate Ontology of Types

The Transaction Model (TM) is a useful means of introducing model-checking to the requirements gathering process as well as a key but too often neglected component of business processes [14, 15, 16: 110-111, 20]. This capture of requirements at the outset ensures that the model-checking is not 'bolted-on' as an afterthought with all its associated consequences [20]. The models are accordingly incomplete until both sides of a transaction 'balance', and this has been shown to lucidly represent qualitative transactions such as 'quality of care received' [14, 15].

Figure 3 shows that:

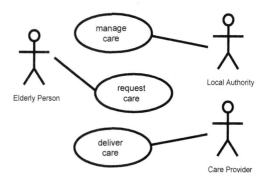

Fig. 1. Healthcare system as a use case model.

- The concepts identified within the care scenario are represented as a transaction where one or more 'economic event' and 'economic resource' are balanced against each other.

- That balance is agreed by one (or more) inside agent and one (or more) outside agent.

- Each concept is classified in terms of type. Therefore a hierarchy of types, which is an important element of an ontology, is derived.

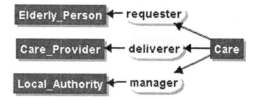

Fig. 2. Healthcare system represented as concepts.

The specialisation of the generic TM CG of Figure 3 onto the community healthcare scenario is illustrated by the CG in Figure 4. What was not clear from the outset (Figure 1) was the 'source' of the money - or which party pays the bill for the care. The UK Welfare System has three particular scenarios:

- The Local Authority pays for the care in full.
- The Elderly Person pays for the care in full.
- The Local Authority and the Elderly Person make 'part payments' that amount to 100% of the care cost.

In order to satisfy the TM the developer (or 'knowledge engineer' in the context of knowledge-based systems) would derive 'Purchase_Agent' as the supertype of 'Local_Authority' and 'Elderly_Person'. (Alternatively as indicated elsewhere this derivation could be part of an automated process, given the anticipated frequency of such a scenario, indeed as could the derivation of Figure 4 from Figure 3 given the added encoded knowledge of the healthcare domain that we describe [21].)

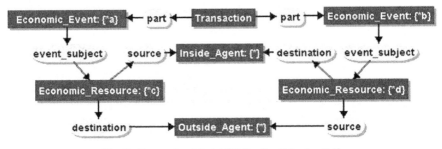

Fig. 3. Transaction Model (TM), after Polovina [14].

Fig. 4. Generic healthcare scenario after application of TM.

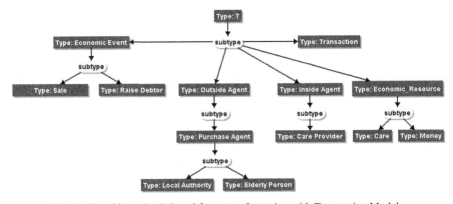

Fig. 5. Type hierarchy deduced from transformation with Transaction Model.

The elicitation of terms for the ontology is an important step during the agent realisation process. Typically developers depend upon existing processes for the most part, but the most significant contribution of this stage is the TM's explicit balance-check as noted earlier. The balance check also immediately raises the developer's awareness of the need to discover the appropriate terminology for the model. If the model does not balance then the model itself is inherently too ill-defined and the process cannot continue. Equally it is necessary to select the most appropriate terminology, or else the model will be nonsensical.

Figure 5 illustrates the type hierarchy deduced from Figure 4. (Drawn using Heaton's rationale for showing the type hierarchy in CG [22].)

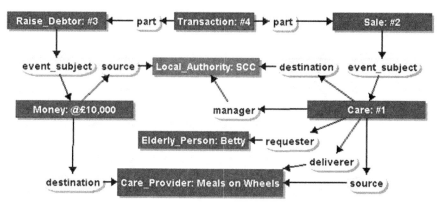

Fig. 6. Specific care scenario where the Local Authority (UK) pays.

4. Model Specific Scenarios

Once the generic model has been created, to assess its viability it is tested with some general rules. We accordingly explore the specific scenario whereby an Elderly Person has been assessed by the UK Government's Social Services and is deemed to be eligible to receive care at zero cost to the elderly person.

In this particular case (highlighted in Figure 6), we see that the 'source' of the money to pay for the care is the Local Authority, "Sheffield City Council (SCC)" who also manage the provision of the care. The care package is not delivered by the Local Authority however; this is sold to them by private organisations, hence the need for a 'Care Provider'. For our illustration the instance of the Care Provider is 'Meals on Wheels'. As the Local Authority thus incurs the cost of the care package, that is its destination. Note that each concept in this figure now has a unique reference, denoting a specific instance.

Conversely, the scenario exists where the Elderly Person is deemed to have sufficient monetary assets not to warrant a free package of care. He or she accordingly incurs its cost, and is thus its destination, rather than the Local Authority. This situation is highlighted in Figure 7, where it can also be seen that the provision of the care package is still managed by the Local Authority.

For both these scenarios (Figures 6 and 7) we also include the original use case relationships of Figure 1. These are highlighted by the lighter shaded relations 'requester', 'manager' and 'deliverer'. This ensures that these aspects of the transaction are not lost and will be recognised in subsequent development.

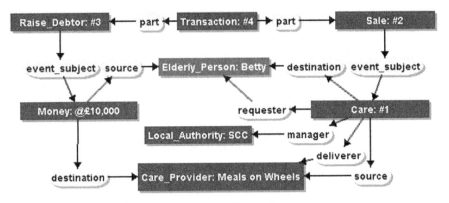

Fig. 7. Situation where Elderly Person pays.

5. Inference with Queries and Validate

We can see that from these foregoing CG that the general CG pattern shown as Figure 8 emerges. To evaluate this CG we examine the case where the Elderly Person's request has been assessed and that person's 'Assets' are deemed to be less than a particular threshold set by the Local Authority, who would therefore be the destination of the care. As such the Local Authority in turn would provide that care to the Elderly Person free of charge. Figure 9 shows this case.

Figure 10 illustrates the other situation, depicted by 'less-than-threshold' asset test being set in a negative context. Here the Elderly Person would be the care destination and costed to, as he or she is deemed to have assets that are *not* below the threshold.

So far we have explored the opposing scenarios whereby either the Local Authority or the Elderly Person settles the bill for the care in full. The model recognises that the Purchase_Agent may consist of more than one party (i.e. the Purchase_Agent would take the plural "{ * }" referent as shown in Figure 11). What about' part-payment situations, where the Elderly Person has sufficient assets to meet some of the cost?

As before, the generic model of concepts is produced, before specialising with an individual scenario. The part-payment model in Figure 12 comprises the two parties, Local_Authority and Elderly_Person, plus the Purchase_Agent derived earlier in Figure 8. However, Figure 11 does not allow joint parties to be the Purchase_Agent. Indeed we can no longer draw a co-referent link between Purchase_Agent, Local_Authority and Elderly_Person.

Therefore we must re-iterate the model further to support Figure 12. Here the Local_Authority and Elderly_Person have a split cost liability to the extent that is variable depending on an individual's circumstances from a means or assets assessment, whilst ensuring that the total liability adds up to 100%.

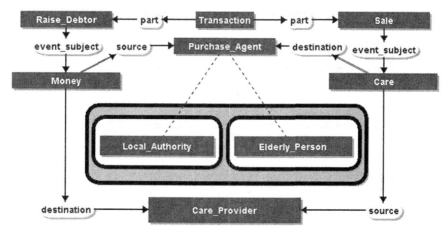

Fig. 8. Emergent general CG pattern for this TM.

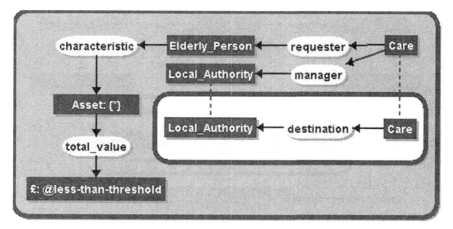

Fig. 9. Elderly Person receives care package at zero cost.

We can now see that the Elderly_Person and Local_Authority agents are no longer sub-types of the Purchase_Agent as originally illustrated, but are instead associated via 'liability' relations. Referring back to the hierarchy of types defined in Figure 5, we can now create a rule to supplant the ontology for the model. Figure 13 thus depicts an ontological component that is no longer valid, hence set in a negative context (or Peirce cut). The ontology is thus now depicted as shown by Figure 14.

Given the refinements discovered, the community care TM is updated in Figure 15 to reflect the new ontology that reflects the liability relationship. The co-referent links are now valid thus the model can now be completed, enabling all three of the payment scenarios to be accommodated.

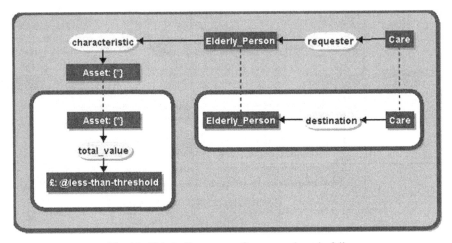

Fig. 10. Elderly Person pays for care package in full.

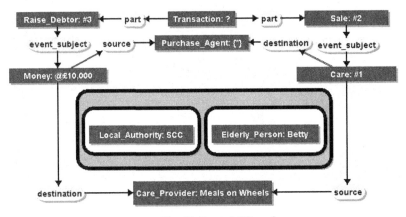

Fig. 11. Iterated CG, so far.

Fig. 12. Capturing part-payment situations.

6. Translate to Design Specification

Once the CG representations have been verified against the TM, it is then possible to perform a translation to a design specification. The 'inside' and 'outside' agents in the TM serve to provide direct mappings as follows:

- Inside Agent: Purchase_Agent, with liabilities jointly satisfied by Local_Authority ("SCC") / Elderly_Person ("Betty").
- Outside Agent: Care Provider ("Meals_on_Wheels").

Further iterations and graph joins (omitted for brevity) would illustrate the following additional agents (where LA stands for Local Authority):

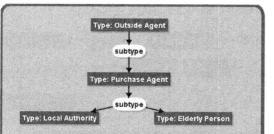

Fig. 13. Ontological component that is no longer valid.

- Care   Request   Agent:
  `Elderly_Person`
- Purchasing Agent:
  `[Local_Authority]`
  `-> (sub-agent) ->`
  `[LA_Procurement_`
  `Agent].`
- Care Assessor Agent:
  `[Local_Authority]`
  `-> (sub-agent) ->`
  `[LA_Social_Worker].`

- Finance Agent: `Local_Authority -> (sub-agent) ->`
  `[LA_Finance_Assessor].`

From these direct translations we can construct agent bodies, to which specific sub-tasks can be assigned. Each of the behaviours is informed by the relations specified within the TM. For instance, referring back to Figure 1 the key abstract definition is that: `Management of Care is Local Authority`. Accordingly, further analysis of the models results in the 'manage' role of the Local Authority Agent though its sub-agents identified above being decomposed into:

`Assess_care_needs; Confirm_financial_eligibility;`
`Procure_care_package; Manage_care_delivery.`

The process of revealing the agent behaviours is informed and contextualised by the business protocols that TM has identified and the developer needs to apply across the myriad agent protocols. For instance, the 'Procure_care_package' behaviour can accordingly be integrated from the TM to the FIPA Iterated Contract Net protocol [23], thus devolving the task of obtaining the cheapest care package available to a protocol to which a given task may be best suited.

This approach thus creates a situation whereby the method of requirements capture concentrates on what the MAS must deliver from the outset to implementation, assisting the developer in determining the extent to which the solution is influenced by the business model.

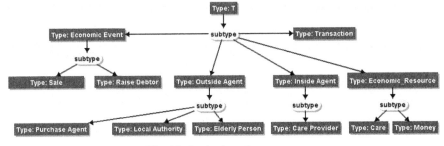

Fig. 14. Revised ontology.

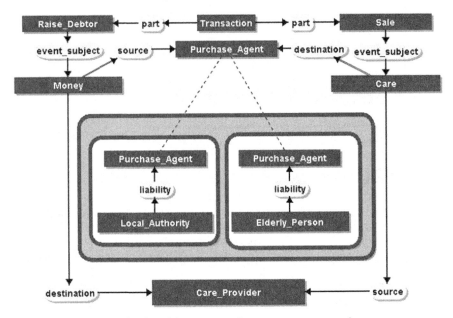

Fig. 15. Refined model to accommodate part-payment scenario.

Conclusions

We have described an approach that enhances the elicitation of MAS characteristics during initial requirements capture, whilst also developing a set of ontological terms which are relevant to the domain and its processes.

Additionally, as we demonstrate elsewhere [24] this approach does not compromise further development with AUML, rather it ensures that the qualitative issues have been captured and thoroughly considered prior to detailed system specification. We believe that this approach offers the following significant advantages:

- Abstract modelling of the concepts with CG provides a means of representing the knowledge exchange within the resulting MAS. Since the high-level, qualitative issues are addressed, developers can specify a system that includes behaviours that can accommodate transactions such as 'duty of care'.

- CG reveal subtleties during the initial requirements capture phase that are less obvious than the direct mappings from actors to agents. The community care example described clearly demonstrates the elicitation of an autonomous entity, 'Purchase_Agent' using this approach.

- After generating an initial model, specific scenarios can be inferred upon the generic model, forcing a set of balance-check rules to be tested.

- Ontological terms are derived from the TM during the process of capturing requirements. Again, the inherent balance check of the model ensures that terms are agreed upon before the model is complete. This process ensures that debates about slot names are conducted at the earliest opportunity possible, having the immediate benefit of specifying more of the system detail before further model development.

In summary we have described a framework for modelling MASs that incorporates model checking to support the development of robust systems. The use of CG and the notion of transactions enriches the requirements capture stage and serves as a precursor to existing AOSE methodologies that require a design representation such as AUML as an input. We feel that this would be a suitable primary discipline for the myriad of agent-oriented software engineering methodologies that lack the necessary detail for successful MAS requirements capture.

References

[1] M.D. Beer, I. Anderson and W. Huang, "Using agents to build a practical implementation of the INCA (intelligent community alarm) system", *in Proceedings of the fifth international conference on Autonomous agents,* 2001, pp. 106-107. http://doi.acm.org/10.1145/375735.376013

[2] N.R. Jennings, "An Agent Based Approach for Building Complex Software Systems", *Comm. of ACM,* vol. 44, pp. 35-41, 2001.

[3] B. Bauer, J.P. Muller and J. Odell, "Agent UML: A Formalism for Specifying Multiagent Software Systems," in *Agent-Oriented Software Engineering,* Lecture Notes in Computer Science, vol. 1957, P. Ciancarini and M.J. Wooldridge Eds. Springer-Verlag, 2000, pp. 91-104.

[4] F. Bellifemine, A. Poggi and G. Rimassa, "JADE: a FIPA2000 compliant agent development environment", *in AGENTS '01: Proceedings of the fifth international conference on Autonomous agents,* 2001, pp. 216-217. http://doi.acm.org/10.1145/375735.376120

[5] L. Padgham and M. Winikoff, "Prometheus: A Methodology for Developing Intelligent Agents", "citeseer.ist.psu.edu/padgham02prometheus.html"

[6] F. Zambonelli, N.R. Jennings and M. Wooldridge, "Developing multiagent systems: The Gaia methodology", *ACM Trans.Softw.Eng.Methodol.,* vol. 12, pp. 317-370, 2003.

[7] H.S. Nwana, D.T. Ndumu, L.C. Lee and J.C. Collis, "ZEUS: a toolkit and approach for building distributed multi-agent systems", *in AGENTS '99: Proceedings of the third annual conference on Autonomous Agents,* 1999, pp. 360-361. http://doi.acm.org/10.1145/301136.301234

[8] S. DeLoach, "Multiagent Systems Engineering: A Methodology and Language for Designing Agent Systems", citeseer.ist.psu.edu/deloach99multiagent.html

[9] P. Bresciani, P. Giorgini, F. Giunchiglia, J. Mylopoulos and A. Perini, "TROPOS: An Agent-Oriented Software Development Methodology", *Journal of Autonomous Agents and Multi-Agent Systems,* vol. 8, pp. 203-236, 2004.

[10] M.P. Georgeff, B. Pell, M.E. Pollack, M. Tambe and M. Wooldridge, "The Belief-Desire-Intention Model of Agency", *in ATAL '98: Proceedings of the 5th International Workshop on Intelligent Agents V, Agent Theories, Architectures, and Languages,* 1999, pp. 1-10.

[11] L.W. Harper and H.S. Delugach, "Using Conceptual Graphs to Represent Agent Semantic Constituents," in *Conceptual Structures at Work: Proc. 12th Intl. Conf. on Conceptual Structures (ICCS 2004),* Lecture Notes in Artificial Intelligence, LNAI ed., vol. 3127, K.E. Wolff, H.D. Pfeiffer and H.S. Delugach Eds. Heidelberg: Springer-Verlag, 2004, pp. 325-338.

[12] A. Fuxman, R. Kazhamiakin, M. Pistore and M. Roveri, " Formal Tropos: language and semantics (Version 1.0)". Accessed: 2005, 4th November. 2003. http://www.dit.unitn.it/~ft/papers/ftsem03.pdf

[13] L. Mattingly and H. Rao, "Writing Effective Use Cases and Introducing Collaboration Cases", *The Journal of Object-Oriented Programming,* vol. 11, pp. 77-87, 1998.

[14] S. Polovina, R. Hill, P. Crowther and M.D. Beer, "Multi-Agent Community Design in the Real, Transactional World: A Community Care Exemplar," in *Conceptual Structures at Work: Contributions to ICCS 2004 (12th International Conference on Conceptual Structures),* H. Pfeiffer, K.E. Wolff and H.S. Delugach Eds. Aachen, Germany: Shaker Verlag, 2004, pp. 69-82.

[15] R. Hill, S. Polovina and M.D. Beer, "Towards a Deployment Framework for Agent-Managed Community Healthcare Transactions", *in The Second Workshop on Agents Applied in Health Care, 23-24 Aug 2004, Proceedings of the 16th European Conference on Artificial Intelligence (ECAI 2004), Valencia, Spain, IOS Press,* 2004, pp. 13-21.

[16] J.F. Sowa, *Conceptual Structures: Information Processing in Mind and Machine,* Addison-Wesley, 1984.

[17] J.F. Sowa, *Knowledge Representation: Logical, Philosophical and Computational Foundations,* Brooks-Cole, 2000.

[18] S. Polovina and J. Heaton, "An Introduction to Conceptual Graphs", *AI Expert,* vol. 7, pp. 36-43, 1992.

[19] F. Dau, *Lecture Notes in Computer Science 2892: The Logic System of Concept Graphs with Negation: And Its Relationship to Predicate Logic,* Heidelberg: Springer-Verlag, 2003.

[20] M.N. Huhns and M.P. Singh, "Service-Oriented Computing: Key Concepts and Principles", *IEEE Internet Computing,* vol. 9, pp. 75-81, 2005.

[21] J. Lee, L.F. Lai, K.H. Hsu and Y.Y. Fanjiang, "Task-based conceptual graphs as a basis for automating software development", *International Journal of Intelligent Systems,* vol. 15, pp. 1177-1207, 2000.

[22] J.E. Heaton, *Goal Driven Theorem Proving Using Conceptual Graphs and Peirce Logic,* PhD Thesis, Loughborough University, 1994.

[23] FOUNDATION FOR INTELLIGENT PHYSICAL AGENTS, "FIPA Iterated Contract Net Interaction Protocol Specification". Accessed: 2005, 11/21. 2000. http://www.fipa.org/specs/fipa00030/PC00030D.html

[24] R. Hill, S. Polovina and M.D. Beer, "From Concepts to Agents: Towards a Framework for Multi-Agent System Modelling", *Proceedings of the Fourth International Joint Conference on Autonomous Agents and Multiagent Systems (AAMAS),* ACM Press, Utrecht University, Netherlands, *in press.* http://www.aamas-conference.org/

Outline of *trikonic* |>*k: Diagrammatic Trichotomic

Gary Richmond

City University of New York
New York, USA
garyrichmond@rcn.com

Abstract. C. S. Peirce's research into a possible applied science of whatever might be trichotomically analyzed has here been developed as *trikonic* – that is, diagrammatic trichotomic. Such a science (and its proposed adjunct tool) may have significant implications for the work of testbed collaboratories and other virtual communities concerned with pragmatically directed inquiry and knowledge representation.

1. Introduction

In a fragment entitled "Trichotomic" Charles S. Peirce, the originator of the science, introduces *trichotomic* as "the art of making three-fold divisions" [EP I: 280]. This new applied science rests upon Peirce's triadic categorial distinctions, which he outlines in many places, for example, in a letter to Victoria Lady Welby:

"Firstness is the mode of being of that which is such as it is, positively and without reference to anything else.

Secondness is the mode of being of that which is such as it is, with respect to a second but regardless of any third.

Thirdness is the mode of being of that which is such as it is, in bringing a second and third into relation to each other." [CP 8.328]

Working from suggestions in the "Trichotomic" manuscript and in such related writings of this period as "A Guess at the Riddle" and "One, Two, Three," the author has developed and here introduces *trikonic*,[1] |>*k, method and proposed tool for trichotomic analysis-synthesis in an internet age, directed especially towards catalyzing the evolution of virtual communities.

Triadic divisions figure in nearly every discipline within Peirce's purview of theoretical science, prominently in his scientific and coenoscopic philosophy, and most especially within his phenomenology and logic interpreted as semeiotic (a tripartite science – theoretical grammar, critical logic, and theoretical rhetoric – all directed to the analysis of semiosis, or "sign action" in one form or another). Additionally, for a very few years (1885-1888), Peirce made an intensive study of a

---

[1] The symbol |>*k is employed when referring to the applied science as here developed; "*trikonic*" is used whenever an adjective is needed. The author introduced a few of the ideas of this paper at the ICCS04 PORT workshop [11]

F. Dau, M.-L. Mugnier, G. Stumme (Eds.): ICCS 2005, LNAI 3596, pp. 453-466, 2005.
© Springer-Verlag Berlin Heidelberg 2005

practical science allied to, but significantly different from, the theoretical ones which tended to dominate his interests. Although he returned to theoretical researches in the years following, the development of an applied science of *trichotomic* – facilitating inquiry into anything which might be so analyzed – was seen by him as essential to the growth of a scientific humanity, a change which he imagined it might help usher in. In one of the drafts of "A Guess at the Riddle," he comments that his fully developed trichotomic analysis would constitute "one of the births of time." [CP 1.354 Fn 1 p 181]

Yet, even in those communities which value Peirce's contribution to the development of a scientific philosophy, research into an applied trichotomic science remains largely untouched by contemporary knowledge representation workers, whereas from one perspective it would seem to constitute a quasi-necessary element in the realization of Peirce's pragmatism in collaboratory testbeds and other virtual projects which incorporate this pragmatic perspective. It will be argued here that there are many places where |>*k should help in facilitating certain researches, conversations, and decision-making processes of vital interest and value to a virtual community.

2. |>*k: Diagrammatic Facilitation of Trichotomic Analysis-Synthesis

Peirce held that all thinking was essentially diagrammatic and that perhaps the most important diagrams were visual ones [CP 2.778]. The principal purpose of |>*k is to facilitate diagrammatic thinking concerning trichotomic relations of importance to a given community's research interests, making such diagrams available for (virtual) group manipulation, commentary, etc. It is certain that not all things can be trikonically analyzed – yet, in consideration of the Peircean *reduction thesis*[2] – which holds that all relations of more than three elements are reducible to triadic relations, but triadic relations are not reducible to dyadic and monadic relations [CP 3.483, 1] – there is probably not much of potential importance to a professional community which could not be incorporated into such an analysis, especially since provisions are also made within |>*k for important dyadic (and other –adic) relations where these need be considered.

Regarding such pragmatic contexts as testbed collaboratories one might argue that, given the self- and hetero-critical nature of effective collaborative work and the decision-making processes needed in preparing for and accomplishing the system- and tool-centered inquiries needed for the achievement of virtual community goals, |>*k could prove to be of some considerable value in catalyzing movement towards what has been called a Pragmatic Web [4, 5], in which definitions are provided of *the context and purpose* of information that might be handled by the continuously developing Semantic Web. It is maintained here that a tool for electronic trichotomic could expand pragmatic analysis even beyond those domains which Peirce considered

---

[2] Kelly Parker [9] gives this as "all higher order polyads can be reduced to triads; conversely, all higher order polyads can be constructed from triads." This issue is intimately connected to Peirce's *valency theory*, for which see [13]

in his trichotomic, and exactly in the direction of contemporary collaboratory needs in the projected suite of electronic tools that will support the growth of such virtual communities (and in consideration of, for example, proposals of methods and tools, discovery of the extent of participant agreement in principle and practice regarding them, identification of specific areas of consensus and disagreement, establishment of criteria for making decisions on issues relating to personnel, etc.).

In short, it will be argued that there are significant places within a pragmatically focused community where a trichotomic analysis-synthesis would prove valuable. Such analysis could be facilitated by |>*k as applied science and electronic tool for analysis-synthesis concerning matters involving such three-fold divisions as those to be considered below. While the tool itself is still in the planning stage, the terminology, symbol system, diagrammatic approach, etc. have been fairly fully fleshed out. However, only an outline and overview of |>*k will be presented here.

Despite the necessary omission for now of a further explication of the purpose, design, structure and content of the proposed trikonic tool, a primary goal of |>*k is the application of the theory and diagrammatic system to virtual inquiry and collaboratory practice, so that >*k is proposed as a small, but possibly integral component of a quasi-necessary suite of principles/meta-principles, tools/meta-tools necessary for realizing the goals of collaboratories modeled on Peircean principles.

3. |>*k Analysis and Peircean Category Theory[3]

As its theory and practice inextricably interpenetrate, it is perhaps not possible to introduce |>*k as diagrammatic system without at the same time introducing fundamental categorial distinctions and relations pertaining to it. This shall, therefore, quasi-necessarily be the approach followed below: basic trichotomic relations and terminology will be introduced as the system itself is explicated. The paper introduces a few convenient notational conventions associated with |>*k beginning with the *trikon* symbol itself, diagramming trikonic relations in relation to the three Peircean categories.

|>, the trikon symbol, represents a trichotomic relationship showing the three categorial elements of any object under consideration and, in its vectorial part (to be discussed later), the six possible paths through the three categorial elements.

Three Peircean categories are at the core of trikonic analysis-synthesis:

1^{ns} = firstness
$|{>}3^{ns}$ = thirdness
2^{ns} = secondness

When there is no vectorial movement, one should read the three categorial divisions around-the-trikon as if occurring all-at-once, each necessitating each other,

---

[3] Where plain text is used and figures not provided, |> represents the trikon symbol. Similarly, where vector analysis is employed, but bent arrows are not provided, the direction of the path is given to the left of the trikon. For example, $1/3/2$ |> represents the process vector: from 1^{ns}, through 3^{ns}, arriving at 2^{ns}.

no one more fundamental than the others. They stand in a *genuine*[4] *trichotomic relationship*, where *genuine* is used as a technical term opposed to *degenerate* (in the mathematical sense of these expressions).

The trikon symbol, |>, which rather resembles the outline of a "forward" button on an electronic device (suggesting, perhaps, process or evolving structure or time's arrow, etc.), is devised to hint at the characteristic of each categorial element within Peirce's three *Universes of Experience*

In such mature work on theory of inquiry as his "The Neglected Argument," Peirce characterizes the *Universes of Experience* in this way:

Of the three Universes of Experience familiar to us all, the first comprises all mere Ideas, those airy nothings . . . their Being consist[ing] in mere capability of getting thought, not in anybody's Actually thinking them. . .The second Universe is that of the Brute Actuality of things and facts. I am confident that their Being consists in reactions against Brute forces. . . The third Universe comprises everything whose being consists in active power to establish connections between different objects, especially between objects in different Universes.

Such is everything which is essentially a Sign – not the mere body of the Sign . . . but, so to speak, the Sign's Soul, which has its Being in its power of serving as intermediary between its Object and a Mind. Such, too, is a living consciousness, and such the life, the power of growth, of a plant. Such is a living constitution – a daily newspaper, a great fortune, a social "movement." [CP 6.455]

Three Universes of Experience:

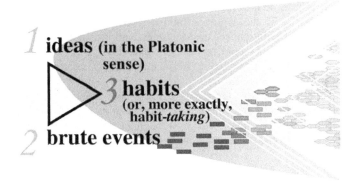

Fig. 1.

---

[4] "Every triad is either monadically degenerate, dyadically degenerate, or genuine. A monadically degenerate triad is one which results from the essence of three monads, its subjects. A dyadically degenerate triad is one which results from dyads. A genuine triad is one which cannot be resolved in any such way." [CP 1.473]

Upon such conceptions, the categories are placed around the *trikon* as follows:

1^{ns}, ideas ("airy nothings," mere possibility of being actualized, so, "floating" at the top)
|>3^{ns}, habits (tending to bring 1^{ns} and 2^{ns} into relationship *in futuro*, so "to the right")
2^{ns}, events (brute actions and reactions, existential and earthbound, so, "sinking" to the bottom of the diagram)

These, in turn, yield Peirce's three *Universal Categories*:

possibility
|>necessity
actuality

From an even more abstract perspective, one can also see the categories as representing first, *something*, second, *other*, third, *that which brings something and another into relationship*:

what is in itself
|>bringing into relation (mediate)
correlate to another

There are naturally lively associations for each of the categories, of which this selection, found in the "Trichotomic" manuscript (EP I: 280), is representative: "First is the beginning, that which is fresh, original, spontaneous, free. Second is that which is determined, terminated, ended, correlative, object, necessitated, reacting. Third is the medium, becoming, developing, bringing about."

In addition to the universal categories, there are equivalent *existential categories:*

feeling
|>*thought:*
action-reaction

The colon after "thought" above signals that that element will itself now be trikonically analyzed.

So, within *thought,* one can identify three *logical modalities*. These are more or less equivalent to the logical quantifiers:

may be == vague, $\exists$
|> will necessarily be if. . . == general, $\forall$
actually is == specific, $\exists$!

There are indeed many other important *authentic trichotomic relations* operative in virtually all the theoretical sciences and, no doubt, even beyond these in the vast trichotomic semeiotic universe Peirce conceives.[5]

4. Triadic Semeiotic[6]

Yet, as important as these singular trikonic relations are, an even more significant set of relations, at least from the standpoint of what might be called an *evolutionary pragmatism*, is made up of the various groups and complexes of trikons – "trikons of trikons," so to speak, such as those which figure prominently in the analysis of *semeiotic grammar*. Here trichotomic analysis begins to take on a richness and suggestiveness which hint at the evolutionary movement possible within a *semeiotic reality* which Peirce explicates in his metaphysical writings (see, for example, RLT). Radically different from such dyadic semiotics as, for example, Saussure's semiology and much of the twentieth century semiotics deriving from it, Peirce's semeiotic posits three essential elements within an integral whole, the *semeiotic triad* abstracting *sign, object, and interpretant*,[7] showing relationships holding between the sign and the object as it is represented for an interpretant sign (not necessarily an actually existing interpreting person: an interpretant could, for example, be a computer, or a future interpreter, etc.). This trichotomy exemplifies the "life of the sign," its vitality and semeiotic movement. The |> always represents a *genuine semeiotic relationship:*

sign
|> interpretant
object

From the standpoint of Peirce's categorial perspective, this suggests that the sign itself is a mere possibility (e.g., "cat" could be "chat" or "gato" or "gatto"), representing a brute actuality "(feline-being-in-the-world")", to some possible or actual interpretant, (e.g., a computer program generating an ontology of concepts related to "feline being"). In section 5 below, we will see how the semeiotic triadic "always-already" implies semiosis, or "sign action" as soon as vectorial movement is considered. But, for now, analyzing the sign/object/interpretant more abstractly and, as it were, "statically," the three sign elements above yield a nonadic trikonic group,

---

[5] ". . .It seems a strange thing, when one comes to ponder over it, that a sign should leave its interpreter to supply a part of its meaning; but the explanation of the phenomenon lies in the fact that the entire universe – not merely the universe of existents, but all that wider universe, embracing the universe of existents as a part, the universe which we are all accustomed to refer to as "the truth" – that all this universe is perfused with signs, if it is not composed exclusively of signs. Let us note this in passing as having a bearing upon the question of pragmaticism." [CP 5.448]

[6] Peirce's idiosyncratic spelling is employed to distinguish it from other, typically dyadic forms of semiotic.

[7] The interpretant a.k.a. the interpretant sign is the sign into which the interpreted sign is interpreted.

the 9-adic diagram of possible relations of these three in consideration of certain categorial constraints – three tripartite "parameters" for every sign.

Combining these to form embodied signs types, where each sign has an actual (triadic) relationship to the object, the sign in itself, and the interpretant, and in consideration of what Liszka has termed the "qualification rule,"[8] Peirce constructs the well known 10-adic Classification of Signs. In other words, the "parameters," as they are referred to here, are not completely independent of one another, and the combinatorial upshot is not 27 kinds of sign but ten. Here Peirce analyzes ten classes of embodied signs in what he characterizes as the order of *involution*.[9] It is a matter of some debate, but this researcher would suggest that a complete |>*k analysis strongly supports the notion that the 9-adic diagram presents only the *types of relationships* possible for *yet to be embodied sign classes*. In a word, the nine sign "parametric" choices do not themselves represent embodied signs, whereas the ten classes do; i.e., not the nine, but Peirce's ten classes of sign, are the exhaustive set of sign classes fully determinate in terms of this level of classification.

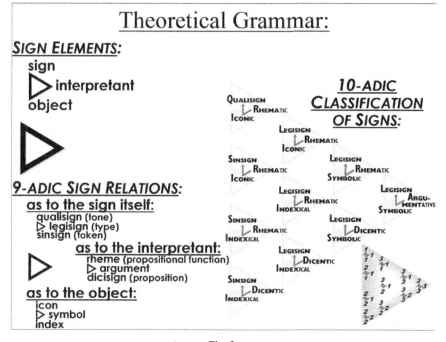

Fig. 2.

---

[8] "According to Peirce, qualisigns, icons, and rhemes are phenomenologically typed as firsts (of thirds) or possible, while legisigns, symbols, and arguments are phenomenologically typed as thirds (of thirds). Consequently . . . the total possible permutation[s] of twenty-seven [possible] classes is reduced to ten." [8]

[9] In |>*k involution is a primary expression of the *vector of analysis* – from 3^{ns} through 2^{ns} to 1^{ns}, and in the case of each of the ten sign classes, commencing at the interpretant, moving through the object, arriving at the sign itself – so that, for example, the sign type at the top of the 10-adic diagram is termed by Peirce a rhematic iconic qualisign, say, a particular red hue.

These ten classes themselves naturally form three "trikons of trikons" with an additional single central |> placed within, further illustrating the deep trichotomic nature of semeiotic as Peirce conceived it.[10]

5. |>*k Process: Vector Analysis-Synthesis

At the heart of |>*k or, better, as its methodological *advance guard,* are six vectors.[11]

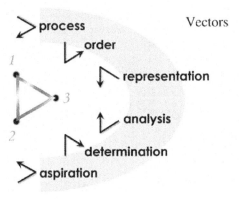

Fig. 3.

In a passage that points both toward *trichotomic vector analysis* AND the "reduction thesis," Peirce writes.

> . . . .Now the triad . . . has not for its principal element merely a certain unanalyzable quality sui generis. It makes [to be sure] a certain feeling in us. [But] the formal rule governing the triad is that it remains equally true for all six permutations of A, B, C; and further, if D is in the same relation at once to A and B and to A and C, it is in the same relation to B and C; etc. [CP 1.471]

In |>*k these six vectors (or, directions of movement through the trikon) factor *significantly,* all of them being authentic permutations of logical – and in some cases, temporal – paths of the relations holding between the 1^{ns}, 2^{ns}, 3^{ns} of the object under consideration. For instance, Parmentier [10] has noted that a *vector of determination* (whereas the object determines a sign for an interpretant) and a *vector of representation* (whereas the interpretant creates a sign to represent an object) move in opposite directions.

---

[10] A second 10-adic classification, involving specifically the interpretant element of semiosis, permits the generation of sixty-six not wholly independent sign types.

[11] Richard Parmentier uses the term "vector" in the sense of a *path* or *direction of movement* through the three categories. Parmentier, however, only identifies two vectors.

...a sign...
2/1/3|> ...for an interpretant.
The object determines...

...creates a sign (for example, a diagram)...
3/1/2 |> An interpreter (say, a scientist)...
...of a complex object.

The former, the vector of determination, figures prominently in many of Peirce's semeiotic analyses. Indeed, one finds descriptions of all six vectors ("orders") scattered through Peirce's voluminous writings, including analyses of pairs of vectors. For example, in a discussion of the categories in "The Logic of Mathematics" paper [CP1.417-520] Peirce analyzes Hegelian dialectic as following a *vector of order*, whereas thesis (1^{ns}) is followed by antithesis (2^{ns}) thus leading to synthesis (3^{ns}). He opposes to this (Hegelian) "order of evolution" an inverse "order of involution or analysis" (*vector of analysis*) by which, indeed, the three *universal categories* are in a sense themselves derived, Peirce first positing 3^{ns} which *involves* 2^{ns} which in turn *involves* 1^{ns}. Peirce held that Hegel's order, as logically significant a it undoubtedly is, yet gains its generative power and effective meaning only when seen in the light of the categories. So, despite the heuristic value of Hegel's dialectical insight, Peirce's own analysis of evolution *as such* follows a different order (*vector of process*) whereas chance sporting (1^{ns}) leads to patterns of habit-formation (3^{ns}) culminating in, say, some actual structural change in an organism (2^{ns}). The last of the six paths (*vector of aspiration*) represents in a sense the unique character of specifically human evolution, that is, the evolution of consciousness which, from a Peircean perspective, is dependent upon critical self-control.

While it is expected that the six expressions used here to name the vectors will for the most part hold, they are nevertheless offered as mere temporary expedients until the development of a trikonic tool will allow for such semeiotic inquiry as might bring about agreement concerning optimal terminology in vectorial analysis. There would indeed seem to be considerable creative potential for a semeiotic which could integrate the generative power of interpenetrating and transactional vectorial movement.

The thematization and exploration of the six vectors (especially as they interpenetrate in hierarchies of constraint, dependence/independence, etc.) aims at bringing new light, and ultimately a more systematic treatment, to some of the difficult issues which arise especially in semeiotic analysis. In short, the vector issue invites treatments involving graphically logical aspects of dependence and constraint, correlation, as well as the "living" reflection of the categories by the semeiotic triad. The proposed trikonic tool is meant to facilitate such treatment in forms suitable and editable for purposes of ongoing dialogue and comparison among many participants in any given collaborative inquiry (or activity).

While it is certainly possible that an individual analyst with nothing more than pen and paper might benefit from using Peirce's trichotomic theory and the diagrammatic system here outlined, yet the thrust of |>*k is towards empowering virtual communities employing it. It is anticipated that the approach will eventually contribute to an increase in shared understanding by the members of a given virtual community in any number of matters of interest and importance to it. Agreement in any matter so considered would in some sense represent the synthesis of the trikonic analysis in which not only various trikonic elements and relations (and complexes of trikons embedded in yet other trikonic relations, and strings of such relations, etc.) are identified, but especially as the results of these analyses are directed to the concerns of virtual communities in their attempts to reach their goals collaboratively. Of course, not all six vectors need be considered in any given analysis (or synthesis), although it is likely that several or all are operative in some ways and to some extent in most complex semioses.

Consider, for example, the forms of inference as given in the syllogism. Associating the syllogism's *rule, case, result* with the three categories results in the following: *rule* is naturally connected to 3^{ns} as expressing lawfulness, *case* (as *existential* case) with 2^{ns}, and *result* with the idea, character, feeling, image, possibility, etc. which is being considered, so 1^{ns}:

Rule/case/result:
result (these beans are white)
|>rule (all the beans from this bag are white)
case (these beans are from this bag)

The three types of inference:

abduction *(representation vector):*
this handful of beans that I find on the table are white;
3/1/2 |>All the beans from this particular bag are white,
this handful of beans are **possibly** from this bag.

deduction *(analytical vector):*
these beans will **necessarily** be white.
3/2/1|> All of the beans from this bag are white,
these beans are drawn from this bag;

induction *(determination vector):*
all these beans are white;
2/1/3 |> all the beans in the bag are **probably** white.
These beans are drawn from this bag,

In a further step these three inference patterns taken together can be seen to have their own vectorial relationship when considered as elements of an inquiry (for example, an experiment conducted on scientific principles) where, in Peircean inquiry at least, the following three steps occur: (1) an abduction – in experimental science, *retroduction* to an **hypothesis** – gives rise to (2) the **deduction** of what would necessarily follow if the hypothesis were correct, and upon which an experiment may be constructed, followed by (3) the actual **inductive testing** of the hypothesis. (Such inquiry expresses, nearly to perfection, the *process vector*).

abduction (the case is *possible*)
1/3/2|> deduction (the result is *necessary*)
induction (the rule is probable)

The point here (and by way of example) is that *trikonic vector analysis* in consideration of such complex semeiotical processes as inquiry—with its characteristic sequence of three stages following the *vector of process* (with each of these three stages being a unique form of inference having its own characteristic vectorial path)—might contribute to the advance of both theoretical and applied logic of inquiry.

6. Applications to Testbeds and Other Pragmatically Informed Virtual Communities

The principal purpose of |>*k vector analysis is to explicate this sort of vectorial movement both theoretically and as it might appear in any actual semiosis, most notably those activities directed towards the pragmatic realization of complex virtual communicative projects. Trikonic analysis – especially in conjunction with such a consensus-building instrument as Aldo de Moor's GRASS tool [3] – could lead to a kind of trikonic synthesis catalyzing the further evolution of virtual community development. How all this is to be accomplished can hardly be considered here. However, one can say with de Moor that it ought to be a "user-driven" model, participants together legitimatizing the specifications concerning the meaning of the information they consider together, the collaborative processes to be employed within their own communities, and so forth. [2]

Collaboratories built on pragmatic principles are virtual communities where users ought to play an active role, especially as many significant decisions need be made concerning, for example: the role of each member, the nature of and timing of each member's work in relation to the other's and to the project(s), the selection of tools to be employed for use individually and together, as well as individual and group reflection upon the process as a whole (including the structure and goals of the collaboratory itself). It is maintained here that all of these aspects of collaboratoriality would be facilitated and catalyzed by employing a trikonic approach to testbed inquiry, practice, decision making, etc. Wherever Peircean pragmatic logic, principles and methods factor emphatically within a virtual project, |>*k promises to be of significant intellectual and social benefit. In such collaboratories, pragmatic guidance would seem to be invaluable in support of virtual community development, for it is at the pragmatic (and *not* the semantic) level that such conversations, inquiries, and decisions are made. |>*k, facilitating such pragmatic guidance, could have a significant role to play in whole-system development.

Possible applications of |>*k to such virtual community development may be suggested by considering two examples of vectorial analysis as they relate to collaboratory activity. For example, the *process vector* might be applied strategically to testbed operations employing pragmatic approaches to inquiry and experimentation (see section 5). In this case (presented here somewhat abstractly), the retroduction of hypotheses of interest and concern to a collaboratory, such as a proposed tool being considered for use, 1^{ns}, leads to the explication of the implications for testing as this relates to the suite of tools already being used, as well as to personnel to be involved, financial considerations, etc. 3^{ns}, followed, perhaps, by the decision to test, and the actual testing (under agreed upon conditions), 2^{ns}.

At a later stage, and in further consideration of virtual community development, a given virtual collaborative community, 2^{ns}, having selected and tested a certain complete suite of tools, etc., 3^{ns}, finds that it begins to more and more fully achieve its collaborative goals, 1^{ns}.

These unit, abstract analyses can, however, only hint at the rich possibilities for employing |>*k in actual pragmatic inquiries within virtual communities.

7. Summary, Conclusion and Further Prospects

Recapitulating some of the basic themes and goals of |>*k:

Theoretically, it is meant to explicate:

- basic trichotomies.
- trikons composed of trikons
- trikons in series
- trikonic vector analysis

Pragmatically, and following upon the theoretical analysis, especially vector analysis, it is meant to support:

- collaborative inquiry
- testbeds and other collaborative projects
- consensual evolution of consciousness

As previously noted, although not everything under the sun can be trikonically analyzed, yet significant dyadic (and other –adic) relationships are also dealt with in a chapter of |>*k not considered in this brief introduction. |>*k may be capable of providing rather complex and complete analyses in areas of vital concern to virtual communities.

Peirce is considered by some to be the most original and creative genius that America has produced. His contributions – to phenomenology, to logic as semeiotic, and to a rather large number of other sciences – have been seminal in conceptual structures research.[6] Peirce's contributions can be seen to include especially his pragmatism, semeiotic, critical and graphical logic, most notably existential graphs as transformed into conceptual graphs by John Sowa.

In recent work with Arum Majumdar on analogical reasoning [12], Sowa has again demonstrated how Peirce's logical ideas are a veritable prerequisite for progress towards NLP and in establishing the conditions necessary for the further progress towards a Semantic Web.

Similarly, Mary Keeler has insisted that normative science and, especially, pragmatism *as method* be tested in collaboratories, paradigmatically in a project appropriately involving Peirce's own work, the projected digitizing of his manuscripts as the PORT project.

Both Keeler and Sowa have stressed how quintessential the theory and practice of Peirce's pragmatism is to the possible success of virtual community projects as are represented by such communities as ours. Both these researchers have allied virtual community development directly to the evolutionary, and ultimately, co-evolutionary pragmatism of Peirce. It is anticipated that Peirce's category theory – upon which the |>*k project is based – will begin to demonstrate its value to research, and to the other creative work of virtual communities, especially that relating to knowledge representation.

It is certain that in order to be effectively employed within virtual communities |>*k requires additional principles, notably methodological meta-principles, as well as advanced tools for realizing these principles in the interest of consensus building. Seen from a pragmatic perspective, collaboratories will need participants interested in such inquiry, tool building, tool and principle testing, virtual community building, etc.

And while |>*k might contribute to bringing about diagrammatically necessary and sufficient conditions for trikonic analysis-synthesis of any topic or issue amenable to such analysis, and although trikonic theory/practice may perhaps help generate themes and topics for significant discussion within the community, |>*k itself does not directly bring about agreement concerning the subject of any given analysis. This will require a companion tool permitting the individual-communal observation of changing patterns of the extent of agreement (also significant disagreement), but especially highlighting deep consensus when it does occur, in particular inquiries and experiments. Additionally, the results of the research will need to be documented in group reports summarizing the experimental findings (in some cases including the experimental process) for the purpose of developing action plans and the like.

Promising research from diverse disciplines has begun to converge on the problematic of collaboratory development. For example, building on the work of Peter Skagestad--who first distinguished Artificial Intelligence (AI) from Intelligence Augmentation (IA), showing that Peirce provided a theoretical basis for the latter much as Alan Turing had for the former--Joseph Ransdell [14] has argued that new applications of IA should perhaps be informed in their design by Peirce's notion that all thought is dialogical. |>*k means to contribute to and to catalyze exactly that manner of dialogue.

Such meta-theoretical principles as the assurance that the participants in a virtual collaboratory will be involved in specifications relating to all matters appropriate to the role(s) of each participant within the collaboratory (determining the principles underlying the community, its processes, methods, structure, tools considered and chosen together and ensemble, and so forth). One promising application of such meta-theoretical principles to group report writing, the GRASS project [3], might be suitably adapted for charting progress towards agreement in trikonic analyses. In conjunction with such meta-principles and tools |>*k would hope to play its small part in the furthering of Intelligence Augmentation.

[Note: The author is indebted to Ben Udell for creating the graphics illustrating this paper, for his help in editing the text, and for his "parametric" language in the analysis of the 9-adic sign division.]

References

[CP] *Collected Papers of Charles Sanders Peirce*, 8 vols. Edited by Charles Hartshorne, Paul Weiss, and Arthur Burks (Harvard University Press, Cambridge, Massachusetts, 1931-1958).

[EP] *The Essential Peirce*, 2 vols. Edited by Nathan Houser, Christian Kloesel, and the Peirce Edition Project (Indiana University Press, Bloomington, Indiana, 1992, 1998).

[RLT] *Reasoning and the Logic of Things: the Cambridge Conferences Lectures of 1898.* Edited by Kenneth Laine Ketner (Harvard University Press, Cambridge, Massachusetts, 1992).

1. R. Burch (1991) *A Peircean Reduction Thesis*. Texas Tech University Press.

2. A. de Moor (1999) Empowering Communities: A Method for the Legitimate User-Driven Specification of Network Information Systems. Ph.D. thesis, Tilburg University, the Netherlands.

3. A. de Moor. GRASS (Group Report Authoring Support System): Arena for Societal Discourse. http://grass-arena.net/

4. A. de Moor (2002). Making Doug's Dream Come True: Collaboratories in Context. In *Proc. of the PORT's Pragmatic Web Workshop, Borovets, Bulgaria, July 15.* 12

5. A. de Moor, M. Keeler, and G. Richmond (2002). Towards a Pragmatic Web. In *Proc. of the 10th International Conference on Conceptual Structures (ICCS 2002), Borovets, Bulgaria,* Lecture Notes in Artificial Intelligence, No. 2393, Springer-Verlag, Berlin.

6. M. Keeler. Hegel in a Strange Costume: Reconsidering Normative Science for Conceptual Structures Research. In *Proc. of the 11th International Conference on Conceptual Structures (ICCS 2003), Dresden, Germany* Lecture Notes in Artificial Intelligence, No. 2746, Springer-Verlag, Berlin.

7. M. Keeler. Using Brandom's Framework to Do Peirce's Normative Science: Pragmatism as the Game of Harmonizing Assertions? In *Proc. of the 12th International Conference on Conceptual Structures (ICCS 2004), Huntsville, Alabama,* Lecture Notes in Artificial Intelligence, No. 3127, Springer-Verlag, Berlin.

8. J. J. Liszka, *A General Introduction to the Semeiotic of Charles Sanders Peirce.* Bloomington and Indianapolis. Indiana University Press. 1996.

9. K.A. Parker, *The Continuity of Peirce's Thought.* Nashville & London: Vanderbilt University Press, 1998.

10. R. J. Parmentier,. "Signs' Place in Medias Res: Peirce's Concept of Semiotic Mediation." Semiotic Mediation: Sociocultural and Psychological Perspectives. Ed. Mertz, Elizabeth & Parmentier. 1985.

11. G. Richmond (with B. Udell), *trikonic,* slide show in ppt format of presentation at PORT Workshop, ICCS 2004 Huntsville, Alabama.
http://members.door.net/arisbe/menu/library/aboutcsp/richmond/trikonicb.ppt

12. J. Sowa and A. K. Majumdar, "Analogical Reasoning" In *Proc. of the 11th International Conference on Conceptual Structures (ICCS 2003), Dresden, Germany* Lecture Notes in Artificial Intelligence, No. 22746, Springer-Verlag, Berlin.

13. *A Thief of Peirce: The Letters of Kenneth Laine Ketner and Walker Percy.* Ed. by Patrick Samway, S.J. University Press of Mississippi, 1995.

14. J.Ransdell (2002) The Relevance Of Peircean Semiotic To Computational Intelligence Augmentation.
http://members.door.net/arisbe/menu/library/aboutcsp/ransdell/ia.htm

Author Index

Lecture Notes in Artificial Intelligence (LNAI)